Systems Biology and Bioinformatics

Systems Biology and Bioinformatics

Editor: Alexis White

Callisto Reference

www.callistoreference.com

Callisto Reference,
118-35 Queens Blvd., Suite 400,
Forest Hills, NY 11375, USA

Visit us on the World Wide Web at:
www.callistoreference.com

ISBN: 978-1-64116-064-3 (Hardback)

Cataloging-in-Publication Data

Systems biology and bioinformatics / edited by Alexis White.
p. cm.
Includes bibliographical references and index.
ISBN 978-1-64116-064-3
1. Systems biology. 2. Bioinformatics. 3. Biological systems. 4. Computational biology.
I. White, Alexis.
QH324.2 .S97 2019
570--dc23

Table of Contents

Preface

Systems biology is the modeling of biological systems by integrating the principles of computer science and mathematics. Bioinformatics and systems biology are interrelated fields, which are concerned with the construction of biological software and methods to compute biological data. Such systems of computation are called biological computers that use biological databases for use in multiple fields such as bioengineering, biotechnology, etc. This book aims to present the fundamental concepts and theories central to the fields of systems biology and bioinformatics in comprehensive detail. The objective of this book is to give a general view of different areas of these fields and their applications. It presents researches and studies performed by experts across the globe. For someone with an interest and eye for detail, this book covers the most significant topics in the fields of systems biology and bioinformatics.

The researches compiled throughout the book are authentic and of high quality, combining several disciplines and from very diverse regions from around the world. Drawing on the contributions of many researchers from diverse countries, the book's objective is to provide the readers with the latest achievements in the area of research. This book will surely be a source of knowledge to all interested and researching the field.

In the end, I would like to express my deep sense of gratitude to all the authors for meeting the set deadlines in completing and submitting their research chapters. I would also like to thank the publisher for the support offered to us throughout the course of the book. Finally, I extend my sincere thanks to my family for being a constant source of inspiration and encouragement.

Editor

1

Understanding key features of bacterial restriction-modification systems through quantitative modeling

Andjela Rodic[1,2†], Bojana Blagojevic[3†], Evgeny Zdobnov[4], Magdalena Djordjevic[3] and Marko Djordjevic[1*]

Abstract

Background: Restriction-modification (R-M) systems are rudimentary bacterial immune systems. The main components include restriction enzyme (R), which cuts specific unmethylated DNA sequences, and the methyltransferase (M), which protects the same DNA sequences. The expression of R-M system components is considered to be tightly regulated, to ensure successful establishment in a naïve bacterial host. R-M systems are organized in different architectures (convergent or divergent) and are characterized by different features, i.e. binding cooperativities, dissociation constants of dimerization, translation rates, which ensure this tight regulation. It has been proposed that R-M systems should exhibit certain dynamical properties during the system establishment, such as: *i*) a delayed expression of R with respect to M, *ii*) fast transition of R from "OFF" to "ON" state, *iii*) increased stability of the toxic molecule (R) steady-state levels. It is however unclear how different R-M system features and architectures ensure these dynamical properties, particularly since it is hard to address this question experimentally.

Results: To understand design of different R-M systems, we computationally analyze two R-M systems, representative of the subset controlled by small regulators called 'C proteins', and differing in having convergent or divergent promoter architecture. We show that, in the convergent system, abolishing any of the characteristic system features adversely affects the dynamical properties outlined above. Moreover, an extreme binding cooperativity, accompanied by a very high dissociation constant of dimerization, observed in the convergent system, but absent from other R-M systems, can be explained in terms of the same properties. Furthermore, we develop the first theoretical model for dynamics of a divergent R-M system, which does not share any of the convergent system features, but has overlapping promoters. We show that *i*) the system dynamics exhibits the same three dynamical properties, *ii*) introducing any of the convergent system features to the divergent system actually diminishes these properties.

Conclusions: Our results suggest that different R-M architectures and features may be understood in terms of constraints imposed by few simple dynamical properties of the system, providing a unifying framework for understanding these seemingly diverse systems. We also provided predictions for the perturbed R-M systems dynamics, which may in future be tested through increasingly available experimental techniques, such as re-engineering R-M systems and single-cell experiments.

Keywords: Restriction-modification, Transcription regulation, Bacterial immune systems, Biophysical modeling, Gene expression dynamics

* Correspondence: dmarko@bio.bg.ac.rs
†Equal contributors
[1]Institute of Physiology and Biochemistry, Faculty of Biology, University of Belgrade, Studentski trg 16, 11000 Belgrade, Serbia

Background

Restriction-modification systems are rudimentary bacterial immune systems, whose main components are the restriction enzyme (R), and the methyltransferase (M). We here consider Type II restriction-modification (R-M) systems [1], where R cuts the same DNA sequences that are protected by M. Consequently, R and M act, respectively, as a toxic molecule and its antidote, and analogies of R-M and toxin-antitoxin systems are often made [2]. R-M present rudimentary "bacterial immune systems", as they protect the host bacterial cell against infection by foreign DNA, such as viruses (bacteriophages) [3–6]. The protection mechanism is straightforward, as the foreign DNA entering bacterial cell is unmethylated, and is consequently cut (destroyed) by R. On the other hand, the host DNA is methylated due to presence of M, and is therefore not cut by R, which prevents autoimmunity. In fact, many bacteriophages are under pressure from R-M systems with whom they have common hosts [7, 8], and have developed different mechanisms to avoid restriction [9–11]. Consequently, expression of the toxic molecule and its antidote provides an effective protection of the bacterial cell against foreign DNA infection [12].

R-M systems are often mobile [2, 12, 13], spreading from one bacterial host to the other, so that a bacterial host, which initially did not contain the R-M system (a naïve host), can acquire it through horizontal transfer. Expression of R and M was directly observed in single cells only very recently, for the Esp1396I system [14], and it is still unclear how different R-M system features affect this expression. It is however assumed that R-M expression has to be tightly regulated during its establishment in a naïve host [15]. For example, as the naïve host genome is initially unmethylated, R must be, and where tested actually is, expressed after a delay with respect to M, so that the host's genomic DNA can be protected before the appearance of R [14, 16, 17]. To ensure such tight regulation, a significant subset of R-M systems contains a third gene, which expresses the control protein (C) [5, 6, 18–23]. C is a transcription factor, which regulates expression of genes in R-M system, including its own expression. In fact, C is typically co-transcribed with R from a common promoter (CR promoter), while M is transcribed from a separate promoter (M promoter) [5, 6, 24].

With respect to the organization of the transcription units, two different architectures are exhibited, which correspond to the convergent (Fig. 1a), and the divergent (Fig. 1b) orientation of CR and M promoters [5, 6, 14, 20, 21, 23, 25, 26]. Despite R-M systems being known for few decades now, with numerous biotechnological uses of restriction enzymes, control of expression of these systems has been insufficiently studied. Two relatively well studied examples are AhdI (a representative of the convergent architecture) [6], and EcoRV (a divergent architecture representative) [5]. For both systems, the core promoters (binding sites of RNA polymerase), and the binding sites of C protein, are experimentally mapped. In addition, for AhdI system, the transcription activity of CR promoter was measured as a function of C protein amount. We previously showed that a thermodynamic model of CR promoter regulation provides a good agreement with this measurement [6]. We also recently showed [14] that a similar thermodynamic model, coupled with a dynamical model of transcript and protein synthesis, can reasonably explain the dynamics of the enzyme synthesis measured by single-cell experiments in another convergent R-M system (Esp1396I). This strongly suggests that quantitative modeling presented here can realistically explain R-M system transcription control. Additionally, thermodynamical modeling of transcription regulation was successfully applied to a number of different biological problems [27–30], while dynamical modeling was applied to explain both more and less complex gene circuits including control of other convergent R-M systems [31–33].

As we detail below on the example of AhdI (convergent system), and EcoRV (divergent system), it is experimentally firmly established that R-M systems exhibit both different architectures, and different features that characterize their gene expression regulation [1, 15]. On the other hand, the regulation should yield the same three dynamical properties, so that the host genome is protected, while the system is efficiently established. In particular, as discussed above, there would have to be a significant expression of M before R is expressed, to ensure that the host genome is protected. Furthermore, once the host genome is protected, the system should likely turn to "ON" state as rapidly as possible, so that the host genome becomes "immune" to the virus infections – this would then require that after an initial delay, R is rapidly generated. Finally, we also previously proposed that, once the toxic molecule (R) reaches a steady-state, its fluctuations should be low – otherwise a high fluctuation in the toxic molecule (R) may not be matched by the antidote (M), which could destroy the host genome [34].

It is however unclear how the diverse system features and architectures, relate with the constraints on the dynamical response of the system stated above. Experimentally, one could, in principle, address this issue by mutating the relevant features (or introducing them in the system where they do not exist), and then measuring how the resulting system dynamics is perturbed. This would however be very hard, as the system would have to be extensively experimentally mutated and/or redesigned, and the resulting protein dynamics measured in-vivo during the system establishment. In that respect, note that the in-vivo dynamics of R and M expression were directly observed for only two Type II systems – in PvuII via nearly simultaneous introduction into a culture using bacteriophage M13 [17], and in Esp1396I, via

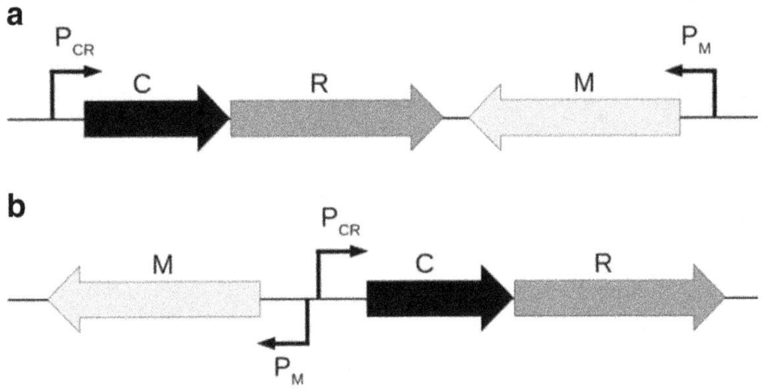

Fig. 1 Typical gene arrangement and promoter orientation in convergent and divergent R-M systems. **a** Convergent systems, a representative of which is AhdI, where other studied systems encoding C protein include Esp1396I, Kpn2I, Csp231I, PvuII [14, 23, 47–49]. Note that C and R genes are transcribed together from P_{CR} promoter. Transcription of M is exhibited from the separate P_M promoter. **b** Divergent systems, a representative of which is EcoRV, where BamHI is another studied divergent system that encods C protein [20]. C and R genes are also co-transcribed, but now share a common promoter region with M gene. In EcoRV the two divergent promoters (P_{CR} and P_M) have overlapping RNA polymerase binding sites

transformation followed by single cell analysis [14]. Even in these cases, the measurements are done only on the wild-type (wt) system, i.e. perturbations were not introduced in the system.

Therefore, the main purpose of this paper is to investigate the relationship between different system features/architectures, and the dynamical properties which the system is expected to exhibit during its establishment. In particular, it is our hypothesis that the diverse features exhibited in R-M systems may largely be explained in terms of the three dynamical properties discussed above. To start testing this hypothesis, we will here biophysically model the control of AhdI and EcoRV, and assess the resulting dynamics when the characteristic system features are either perturbed (in AhdI case) or (artificially) introduced (in EcoRV case) in the system. This is analogous to a classical approach in molecular biology, where the system is analyzed by mutating its main features, or introducing new features in the system where they do not exist, and consequently observing what effect these perturbations have on the presumed system function. The difference is that we here analyze the system *computationally* instead of *experimentally*, where we build on the fact that we previously showed that the modeling approach that we employ here can reasonably explain the available equilibrium measurements [6], and the available single cell experiments [14]. Therefore, the ability of the modeling to explain the measured wild-type data in R-M systems provides a reasonable confidence that our predictions for the perturbed system will also be realistic. Moreover, with the advancement of sophisticated experimental approaches, such as single cell experiments, or possibility to reengineer the system, there comes a prospect of directly experimentally testing these predictions in the future.

Specifically, we will here start by reviewing the relevant experimental information for AhdI and EcoRV systems (the structure of their promoter regions and their regulatory features), which will provide a bases for our theoretical modeling. We will then quantify the general principles discussed above, i.e. introduce what we here call the dynamical property observables, which will allow us quantifying the delay between R and M, how fast the system makes the transition from OFF to ON state, and the stability of R steady-state levels. We will then investigate if abolishing the characteristic features of AhdI also diminishes these observables, i.e. negatively affects the dynamical properties discussed above. Furthermore, we will also study if these dynamical properties also apply to the system (EcoRV) where AhdI features are absent, but a new feature is present (the overlapping promoters). We will then ask what happens if the AhdI features are (computationally) introduced in wild-type EcoRV system, where they originally do not exist. That is, we will investigate if introducing these features leads to (at least) some of the three dynamic property observables being diminished – therefore explaining why they are absent from EcoRV. Overall, we will here systematically investigate how perturbing (or introducing new) features in two characteristic R-M systems affects the resulting system dynamics.

Methods

In the first subsection, we provide in detail the experimentally available information on AhdI (the convergent system) and EcoRV (the divergent system), on which we base our quantitative modeling. The main properties of the model, including the observables through which we assess the system dynamical properties, are provided in the second subsection. We note that the model itself is provided in details in Additional files 1 and 2, where all the parameters (including their experimental/theoretical support) are listed.

Experimentally determined configurations of AhdI and EcoRV

For AhdI, the positions of different promoter elements (C protein and RNAP binding sites) were experimentally mapped for both CR and M promoters [6] (see Fig. 1a). In addition, the binding affinities and the transcription activities for both the wild type and mutant systems (where C protein binding sites were mutated) were measured [6]. These measured values, together with the standard literature values for the kinetic parameters (the translation and the degradation rates), were used to parameterize the model, as provided in detail in Additional file 1.

As indicated in Fig. 2a, C binds to CR promoter, regulating both its own transcription and the transcription of R [6, 19]. C binds to promoter DNA as a dimer, where binding to the distal binding site (configuration K_3), when C is present at relatively low concentration, leads to transcription activation, as C dimer bound to this position recruits RNAP binding to the promoter (configuration K_5). On the other hand, when C is present at high concentration, C dimer bound to the distal binding site recruits another C dimer to the proximal binding site (the tetramer configuration, K_4), thus repressing the transcription, as RNAP cannot bind to the promoter. Note that the configuration in which C dimer binds only to the proximal binding site (equivalent to K_3) is not shown, as the binding affinity to the proximal binding site is much lower compared to the distal binding site, making this configuration much less probable. As for M gene, its transcription is controlled by a negative feedback loop, i.e. M methylates specific sites in its own core promoter thereby repressing the transcription (Fig. 2b).

There are three features which characterize control of AhdI expression [6]. First, there is a very high cooperativity in binding of the C protein dimers to the distal and the proximal positions in CR promoter, so that C dimer bound only to the distal site (K_3 configuration) exists only very transiently in the wild-type (wt) AhdI system. That is, in the absence of RNAP, a C dimer bound to the distal position immediately recruits another C dimer to the proximal binding site. Second, the C dissociation constant of dimerization for AhdI is very high, so that almost all C protein in the solution is in the form of monomers. Finally, C protein is translated from a leaderless transcript (i.e. a transcript which does not contain a ribosome binding site), which was in *E. coli* shown to be associated with lower translation initiation rate [35, 36].

For EcoRV, CR and M promoters are divergently oriented, as schematically shown in Fig. 1b. Consequently, the promoter elements are located in the intergenic region that separates CR and M genes, and these elements are also experimentally mapped [5]. Some of the binding affinities were also measured [5], while the others were eliminated by rescaling the equations (see Additional file 2) – note that we can rescale the equations, as we are interested only in the relative protein amounts. The kinetic parameters (the

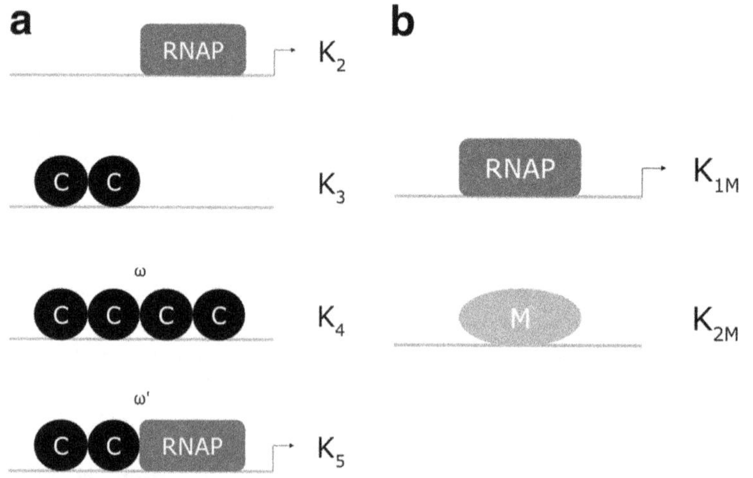

Fig. 2 AhdI R-M system promoter regions. The arrangement of the promoter elements for AhdI CR and M promoters is based on the experimental information provided in [6]. The regions which are schematically shown correspond to (**a**) P_{CR} promoter. Circles indicate C monomers, the rectangles indicates RNAP, while the arrows indicate transcriptionally active configurations. K_2 – K_5 denote the dissociation constants (see Additional file 1) corresponding to different promoter configurations (K_1 denotes the dissociation constant of dimerization), where ω and ω' denote, respectively, the binding cooperativity between the two C dimers bound to DNA, and between C dimer bound to the distal binding site and RNAP. C binds to the promoter as either dimer (K_3) or tetramer (K_4). The bound dimer recruits RNAP to the promoter (K_5). On the other hand, the tetramer configuration corresponds to the repression, as it prevents RNAP binding to the promoter. **b** Transcription is repressed by DNA methylation due to M binding [6], i.e. M methylates specific sites in M promoter that overlap RNAP binding site – for simplicity this is in the figure represented as M being bound to the promoter DNA

translation and the degradation rates), correspond to the standard literature values, and are taken to be the same as for AhdI (with the exception of C translation rate, see below).

In contrast to AhdI, the main feature of EcoRV is the partially overlapping CR and M core promoters, as schematically shown in Fig. 3. Consequently, RNAP cannot simultaneously bind to and initiate transcription from both P_M and P_{CR}. Moreover, the characteristic features of AhdI are not found in EcoRV [5]. In particular, while the transcription control of the CR promoter by C protein is similar as in AhdI, the main difference is that the large cooperativity between the C dimers at the distal and the proximal binding site is now absent, in fact it was found in EcoRV that the two dimers bind to DNA with no cooperativity [5]. Furthermore, the transcription from P_M is not directly influenced by C protein binding, i.e. C binding does not directly affect RNAP binding to P_M. However, the influence of C on P_M transcription is indirect, as the regulation by C of RNAP binding to P_{CR}, also affects when RNAP can bind to P_M. Consequently, while in AhdI transcription of CR and M was independent from each other, in EcoRV we have a more complex system where their transcription is strongly coupled. Similar regulation through overlapping CR and M core promoters is also found in CfrBI R-M system [26, 37]. Finally, C transcript is not leaderless in EcoRV, so the feature which was associated with lower translation initiation rate in *E. coli*, and which is present in AhdI, is now absent from EcoRV.

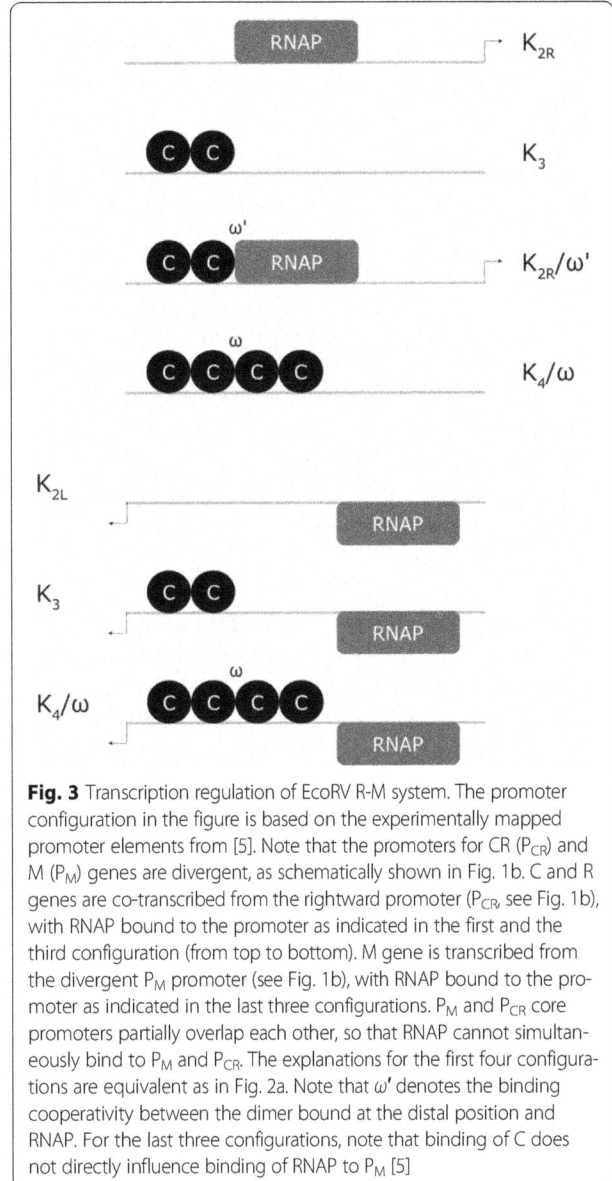

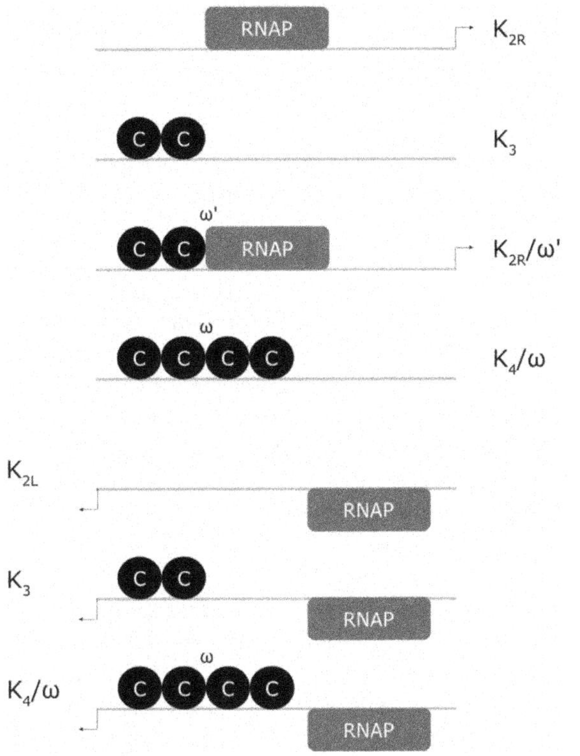

Fig. 3 Transcription regulation of EcoRV R-M system. The promoter configuration in the figure is based on the experimentally mapped promoter elements from [5]. Note that the promoters for CR (P_{CR}) and M (P_M) genes are divergent, as schematically shown in Fig. 1b. C and R genes are co-transcribed from the rightward promoter (P_{CR}, see Fig. 1b), with RNAP bound to the promoter as indicated in the first and the third configuration (from top to bottom). M gene is transcribed from the divergent P_M promoter (see Fig. 1b), with RNAP bound to the promoter as indicated in the last three configurations. P_M and P_{CR} core promoters partially overlap each other, so that RNAP cannot simultaneously bind to P_M and P_{CR}. The explanations for the first four configurations are equivalent as in Fig. 2a. Note that ω' denotes the binding cooperativity between the dimer bound at the distal position and RNAP. For the last three configurations, note that binding of C does not directly influence binding of RNAP to P_M [5]

Modeling AhdI and EcoRV dynamics

We model R and M synthesis upon introducing AhdI and EcoRV in naïve bacterial hosts. The models are based on the experimental knowledge of AhdI and EcoRV transcription regulation, which is summarized in Figs. 2 and 3, respectively. The models are provided in detail in Additional files 1 and 2, and are briefly based on:

(i) A thermodynamic model, which takes into account the activation and the repression of CR promoter by C, and the repression of M gene by its own product (which was experimentally shown in [6]). The model assumes that the promoter transcription activity is proportional to the equilibrium binding probability of RNAP to promoter, which is a general assumption initially proposed by the classical Shea-Ackers approach [38].

(ii) Equations that predict how the transcription activity of CR and M promoters depends on C-protein concentration, which further allows modeling the dynamics of transcript and protein expression. That is, the modeled transcription activities provide the main input for a kinetic model, which calculates R, C and M transcript and protein synthesis. Also, note that R-M systems are characterized by very high expression of R and M proteins [14] so that on the order of thousands of molecules are present in the cell. Consequently, the system is expected to be well in the limit where deterministic modeling can be used to realistically describe the system.

We previously showed that such modeling can well explain the wild-type measurements for AhdI [6] - in particular the measured dependence of the transcription activity on C protein concentration – as well as the most recent measurements in single-cell experiments allowing directly observing the dynamics of R and M synthesis

[14]. Our aim here is to computationally analyze how systematically abolishing individual system features affects the system's dynamics, focusing on the following properties:

i. the time delay between R and M accumulation,
ii. the transition speed of the system from "OFF" to "ON",
iii. the stability of R steady-state levels.

For this, we will introduce observables (which we call the dynamical property observables) that can quantify these properties. To reasonably define them, it is useful to visualize the predicted system dynamics, and the stability of R steady-state levels in wild-type AhdI system, which is shown in Fig. 4 and calculated from Eqs. (1.12), (1.22) and (1.24)–(1.27) (see Additional file 1).

The first dynamical property observable (delay)

From Fig. 4a, we see that the system features lead to a significant delay in the expression of R compared to M, in accordance with the first dynamical property. To quantify how the delay changes upon perturbing these features, we introduce the first dynamical property observable, which corresponds to the ratios of the shaded areas in the perturbed system and in wt AhdI, at an initial interval post-system entry.

The second dynamical property observable (OFF to ON transition speed)

Furthermore, in Fig. 4a, we see that R expression curve has a sigmoidal shape. Consequently, the maximal slope of this curve (indicated in the figure) provides a reasonable measure of transition velocity from "OFF" (low R value) to "ON" (high R value) state. Therefore, as the second dynamical property observable, we introduce the maximal slope of this curve. The changes of this slope will allow assessing how the transition velocity – which determines the time window between the host genome being methylated, and the cell being protected against viruses – will be affected when the system features are perturbed.

The third dynamical property observable (R steady-state level stability)

Finally, the third dynamical property relates with fluctuations of the toxic R molecule, which we propose should be small in the steady-state [34]. The fluctuations are directly related with the stability of the steady-state, so that smaller fluctuations imply larger steady-state stability, which we introduce as the third dynamical property observable.

Different (in-silico) perturbations of the wild-type system – i.e. gradually abolishing the existing or introducing new features – will be introduced in either the thermodynamic model, or in the kinetic equations (see Additional files 1 and 2).

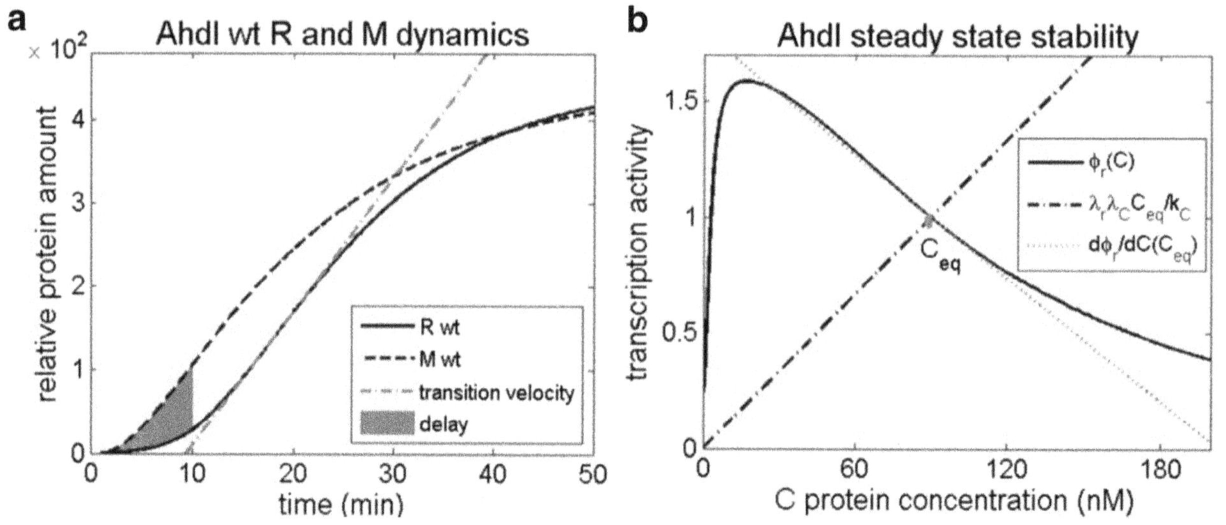

Fig. 4 a Dynamics of R and M expression. R and M expression upon the system entry in a naïve bacterial host (0 min corresponds to the system entry). The shaded area corresponds to the difference of the surface areas below M (dashed curve) and R (solid curve) expression curves for the first 10 min post-system entry; the area presents a measure of the delay between M and R expression. The dash-dot line corresponds to the maximal slope of the sigmoidal R expression curve, measuring the transition velocity from OFF to ON state. **b** Steady-state and its stability. The steady-state (indicated by C_{eq}) is obtained as an intersection of the transcription activity (the solid black line), and the dash-dot line whose slope is determined by the transcript decay and the protein translation rate (Eq. (1.33)). The stability of the steady-state is related with the difference of the dash-dot line slope, and the slope of the transcription activity (the dotted line in the figure) at the point of their intersection C_{eq} (Eq. (1.34))

Results and discussion

We will start by gradually abolishing the three characteristic AhdI features introduced above, and assess how this will affect the dynamical property observables. We will next model the dynamics of EcoRV establishment in a naïve bacterial host, to see if the proposed dynamical properties also apply to a system with different architecture and transcription regulation features. This will provide, to our knowledge, the first quantitative model of a divergent R-M system control, and an opportunity to assess dynamics of R and M expression, which was up to now not experimentally observed for the divergent systems. Finally, we will in-silico introduce to EcoRV the regulation features that exist in AhdI, but are not found in EcoRV, to investigate how this effects the dynamical property observables, and why these features are not present in EcoRV.

Perturbing AhdI system features

The three characteristic AhdI features are the high C subunit dissociation constant of dimerization, the large cooperativity between C dimers bound at the distal and the proximal position, and the low C transcript translation initiation rate. It was previously discussed that these features serve to limit the amount of the synthesized toxic molecule (R) [6]. However, it is not clear that this amount per-se should be limited, as a too small steady-state amount of R may compromise the immune response – i.e. it can lead to the virus genome being protected by M before it can be destroyed by R [39]. As we discussed above, it would be very hard to experimentally investigate the effect of these AhdI features on the system dynamics, this can be readily predicted from the model that we formulated above.

Decreasing the dissociation constant of dimerization

The dissociation constant of dimerization K_1 is very high for AhdI, leading to almost all C subunits being present as monomers in solution [6, 40] – e.g. for another convergent R-M system (Esp1396I), the measured dissociation constant of dimerization was found to be significantly (four times) lower [41]. We start by gradually decreasing this high dissociation constant of dimerization, in the range that corresponds to the wild-type (all monomers in the solution) to the opposite limit of lower K_1, in which only dimers are present in the solution. In Fig. 5a, we see that this perturbation has a significant effect on R synthesis dynamics – note that the M dynamics curve, which is also indicated in the figure for reference, is not affected by perturbing the three characteristic AhdI features. One can observe the three main effects from Fig. 5a: The decrease of the delay between R and M expression, the slower transition from OFF to ON state, and the decrease in the steady-state level of R. The first two effects are further quantified in Fig. 5b and c, as discussed below.

In Fig. 5b, we see that decreasing K_1 leads to a significant, more than twofold, decrease in the relative delay between R and M expression. This perturbation can then significantly impact the ability of the system to protect the host genome from being cut during R-M establishment, with the necessary lag also depending on the specific activity of the M protein and the propensity for R to nick hemimethylated sites. Furthermore, in Fig. 5c we see that decreasing K_1 also leads to a significantly slower transition from OFF to ON state, so that the maximal slope is decreased for almost two-fold. Therefore, decreasing the wt dissociation constant of dimerization also significantly impacts the time window in which the host will be protected from foreign DNA infection. However, perturbing K1 has no significant effect on the steady-state stability of R levels (Fig. 5d). Overall, decreasing the high dissociation constant of dimerization characteristic for wt AhdI, has a significant adverse effect on two of the three proposed design principles.

Increasing C protein translation rate

In AhdI C transcript is leaderless [6], which was in *E. coli* [35, 36] shown to be associated with a significantly smaller translation initiation rate – consequently in [6] a five times smaller C transcript translation rate k_C, compared to R and M was assumed. We now test the effect of perturbing this system feature, i.e. increasing k_C towards those of R and M transcripts, which is shown in Fig. 6. We see that the main effect of this perturbation is on decreasing the steady-state level of R and the delay between R and M expression (for ~ 40%), as shown in Fig. 6a-b. Intuitively, this can be understood that by a more efficient C transcript translation, C accumulates faster, facilitating the formation of the activating and the repressing complexes on the CR promoter, so that R is expressed with a smaller delay, and reaches the lower steady-state level. On the other hand, the effect on the other two design-observables, i.e. on the transition velocity and the stability of R steady-state levels, is rather small (Fig. 6c-d). Consequently, increasing the low C transcript translation rate adversely affects one of the dynamical property observables, i.e. the delayed expression of R with respect to M, which is considered crucial for the protection of the host genome.

Decreasing cooperativity in the dimer binding

A rather drastic feature of AhdI is a very large cooperativity ω in binding of the two dimers to the distal and the proximal position in the promoter [6], which is either not present (EcoRV) [5], or significantly smaller (Esp1396I) [41], in other R-M systems. We therefore investigate how gradually abolishing this high cooperativity affects the

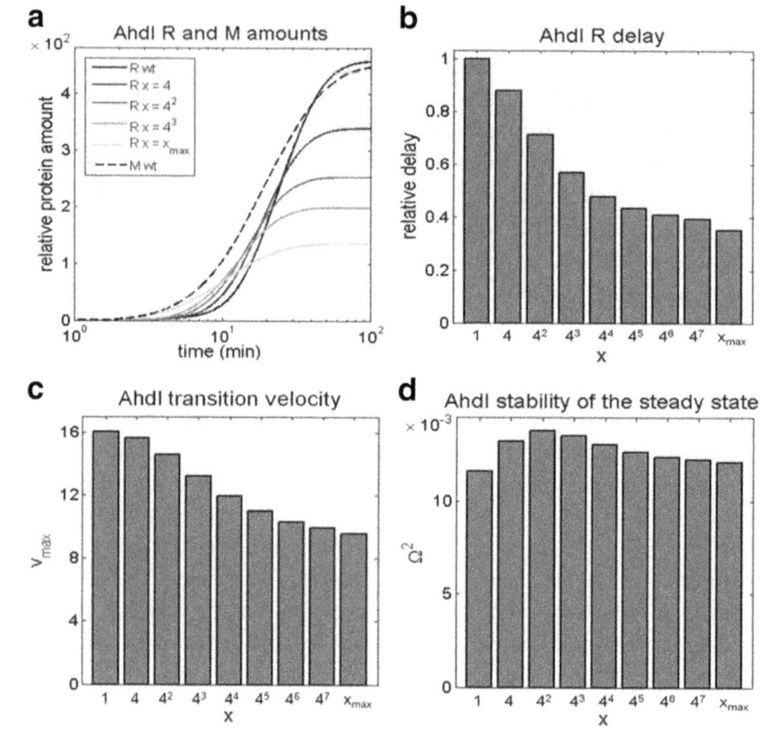

Fig. 5 Decreasing AhdI dissociation constant of dimerization. K_1 is decreased from the high value corresponding to mostly monomers in the solution, to the low value corresponding to mostly dimers in the solution, and the effect is assessed on **a** The dynamics of the protein synthesis. The black line corresponds to all monomers in the solution (wt), while the light gray line corresponds to all dimers in the solution. The curves in-between (in different shades of gray) correspond to the gradually decreasing values of K_1. The relative protein amounts for a wt system (on the vertical axis) are derived from in-vitro transcription activity measurements in [6]. x indicated in the legend corresponds to the relative decrease of K_1 (e.g. x = 4 is a four-fold decrease). **b** The first dynamical property observable, corresponding to the relative delay of R with respect to M expression. The delay is normalized with respect to the wild type (corresponding to one). **c** The second dynamical property observable, corresponding to the transition velocity from "OFF" to "ON" state, represented by the maximal slope of the R expression curve. **d** The third dynamical property observable, corresponding to the stability of R steady-state levels (see Methods)

system dynamics and the design observables. In Fig. 7a, we see that abolishing ω affects only the late dynamics of R, so that the first two dynamical properties are not affected (and not shown in Fig. 7). On the other hand, we see that the steady-state amount of R significantly increases as the cooperativity ω decreases. This can be intuitively understood by the fact that perturbing the cooperativity affects only the efficiency of forming the repressor tetramer complex. As the probability of forming this complex is proportional to C^4 (see Additional file 1), it becomes significant only in the later period, when a large enough amount of C is synthesized. Furthermore, in accordance with the perturbation affecting the late dynamics, from Fig. 7b, we see that decreasing the cooperativity significantly impacts the stability of R steady-state levels, leading to its 50% decrease.

Importantly, the first two AhdI features (the large dissociation constant of dimerization, and the small C translation initiation rate) have an opposite effect on the steady-state amount of R, as compared to the large cooperativity in C dimer binding. That is, while we showed that the first two features significantly increase the steady-state R amount, the third feature (the large cooperativity) significantly decreases it. On the other hand, all three features generally have the same effect on the three dynamical properties that we consider, i.e. abolishing these features either decreases the values of the dynamical property observables (making the corresponding dynamical property less optimal), or do not significantly affect them. This can then explain the extremely large binding cooperativity that was experimentally observed, as on the one side it allows controlling the steady-state amount of the toxic protein due to the opposite effect from the other two features, while at the same time working together with the first two features to ensure more optimal dynamical properties. In particular, note that both the large dissociation constant of dimerization and the large binding cooperativity significantly increase the stability of R steady-state levels, while having a significant - but opposite – effects on the steady-state R amounts.

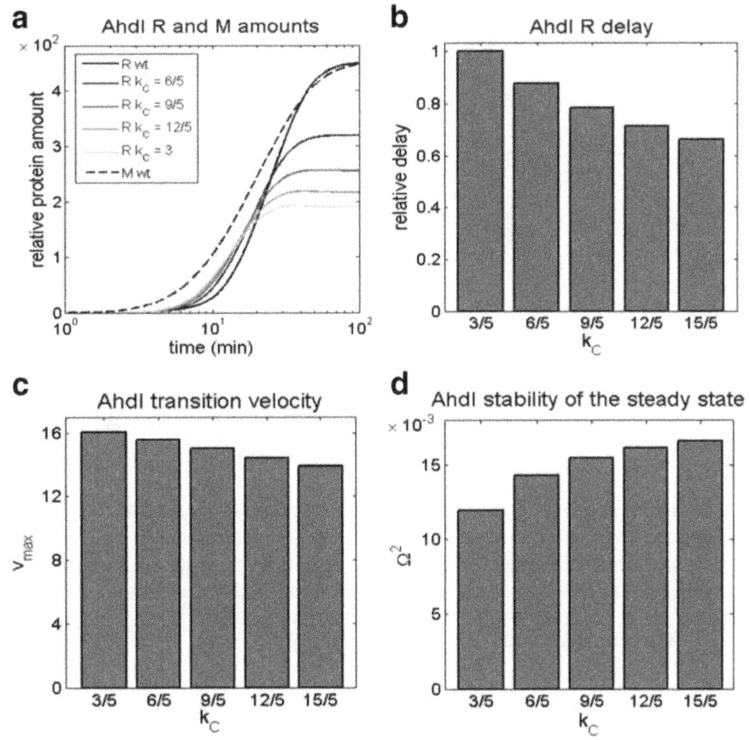

Fig. 6 Increasing C transcript translation rate: k_C is increased from the lower value (3/5 1/min) as taken in [6] to the value which equals those for R and M transcripts (3 1/min). The effect of this decrease is assessed for: **a** The dynamics of the protein synthesis, with the black curve corresponding to the lowest (wt) k_C, and the light gray curve corresponding to the highest k_C (which equals those of R and M transcripts). The curves in different shades of gray correspond to the gradually increasing k_C values. **b** The relative delay (normalized with respect to wt) of R with respect to M expression. **c** The maximal slope of the R expression curve, reflecting the transition velocity from "OFF" to "ON" state. **d** The stability of R steady-state levels, is shown on the vertical axis

EcoRV wild-type dynamics

EcoRV is an example of R-M system with a divergent organization of CR and M transcription units. Overlapping CR and M promoters is the most distinctive feature of this system (presenting its main difference with respect to AhdI), which is, together with C protein binding, responsible for control of EcoRV transcription. That is, high occupancy of M promoter by RNAP, prevents RNAP binding to CR promoter, leading to lower CR transcription activity, and vice versa. In modeling the gene expression regulation, we consider that CR promoter transcription is controlled by C, while C binding has little to none direct effect on M promoter transcription activity, as shown in [5]. In distinction to AhdI [6], which shows an extremely high cooperativity in C dimer binding, no coperativity was found in EcoRV [5]. We also assume that C dissociation constant of dimerization is significantly lower than the relevant range of C concentration, so that the majority of C molecules in solution exist as dimers. Note that in another R-M system (Esp1396I), which has a much lower cooperativity in C dimer binding compared to AhdI, a significantly lower dissociation constant of dimerization is also observed [41]. Finally, in distinction to AhdI, C transcript in EcoRV is not leaderless, so for EcoRV we assume that C has the same translation initiation rate as R and M.

Consequently, EcoRV does not have the three features that control transcription in AhdI, but has instead another characteristic feature, i.e. the overlapping CR and M promoters. We therefore ask if EcoRV, with different architecture and the regulation features, can also meet the three dynamical properties that we consider. To that end, we modeled the synthesis of R and M during the system establishment in wild-type EcoRV, under the assumptions stated above, and following the scheme of the transcription configurations shown in Fig. 3. The model is provided in detail in Additional file 2, and is based on the same thermodynamics assumptions as the one for AhdI dynamics. To our knowledge, this presents the first model of expression dynamics for a divergent R-M system, which has a more complex regulation due to overlapping nature of their promoters. This model moreover presents the first opportunity to assess the dynamics of R and M synthesis for a divergent R-M system, as, to our knowledge, either their regulation or their expression dynamics was not previously measured.

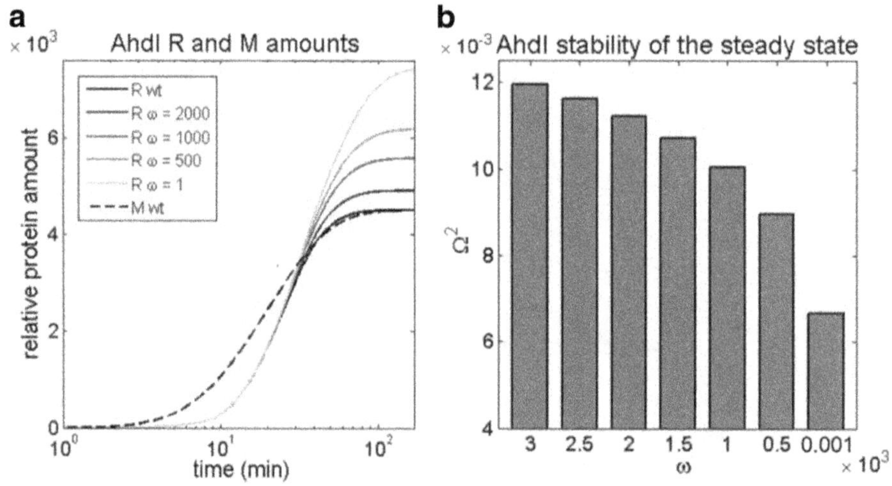

Fig. 7 Decreasing cooperativity in C dimer binding to CR promoter. The cooperativity in binding ω is gradually abolished from the very high value corresponding to wt AhdI [6] to ω corresponding to the absence of the binding cooperativity. We predict the effect of this decrease on: **a** The dynamics of R protein synthesis, where the black line corresponds to the high ω, the light gray to no cooperativity, and the values of cooperativity in-between are shown in different shades of gray. **b** The stability of R steady-state levels, corresponding to different ω values shown in **a**

The predictions for R and M accumulation in wild-type EcoRV are shown by the full black curve (for R) and by the black dashed curve (for M), in Fig. 8 below. From the figure we see that, regardless of lacking the characteristic AhdI regulatory features, the synthesis of R and M is well in accordance with the three dynamical properties. Namely, by comparing Fig. 4 (the dynamics of AhdI) with the EcoRV dynamics, we see that: *i)* the time delay for EcoRV is even larger compared to AhdI, *ii)* there is a clear switch-like behavior of R expression in EcoRV, i.e. the speed of transition from "OFF" to "ON" state is comparable to the one in AhdI, iii) the system reaches the steady-state level ($\Omega^2 > 0$), where the reached stabilities of R steady-state levels are comparable (compare Fig. 5d with Fig. 8c). Therefore, we see that the design principles which we showed are inherent to AhdI R-M system, are retained in EcoRV R-M system, despite the apparent distinction in gene expression regulation.

Introducing AhdI control features to EcoRV

Next, there is a question of why the characteristic AhdI features are absent from EcoRV. That is, could we get even more optimal design-observables if AhdI control features are introduced in wild-type EcoRV? Therefore, we next use our model, to individually introduce each of the three control features of AhdI, on the top of the existing wt EcoRV regulation (i.e. the overlapping promoters). Specifically, in the wild-type EcoRV, we will perturb: *i)* the dissociation constant of dimerization towards the high values characteristic for AhdI, *ii)* cooperativity in C dimer binding to the promoter, also towards the high values observed in AhdI, *iii)* C protein translation rate k_C, towards the low values characteristic for leaderless AhdI C transcripts.

Introducing the high dissociation constant of dimerization to EcoRV

We first perturb the wt EcoRV system by increasing the rescaled equilibrium dissociation constant of dimerization $\overline{K}_1$ (see Fig.8 and Additional file 2), which corresponds to a gradual transition from the solution containing mostly C dimers to the solution containing mostly C monomers. Note that the dynamics of both R and M expression is now affected by the perturbation, in distinction to AhdI where only R expression is changed. This is because CR and M promoters overlap in EcoRV, so that changing transcription from one promoter, necessarily impacts transcription from the other.

We observe that this perturbation does not significantly affect the early accumulation of R and M (during the first ~10 min), but that the dynamics at later times is significantly affected (see Fig. 8a). In particular, we see that increasing the dissociation constant of dimerization leads to a significantly slower switch from "OFF" to "ON" state, so that the transition velocity decreases as much as four times (Fig. 8b). Furthermore, in Fig. 8c, we see that increasing $\overline{K}_1$ also significantly decreases the stability of R steady-state levels Ω^2, which drops almost three times. Consequently, introducing the high dissociation constant of dimerization to EcoRV, which is characteristic for AhdI, has a significant adverse effect on two of the three dynamical properties.

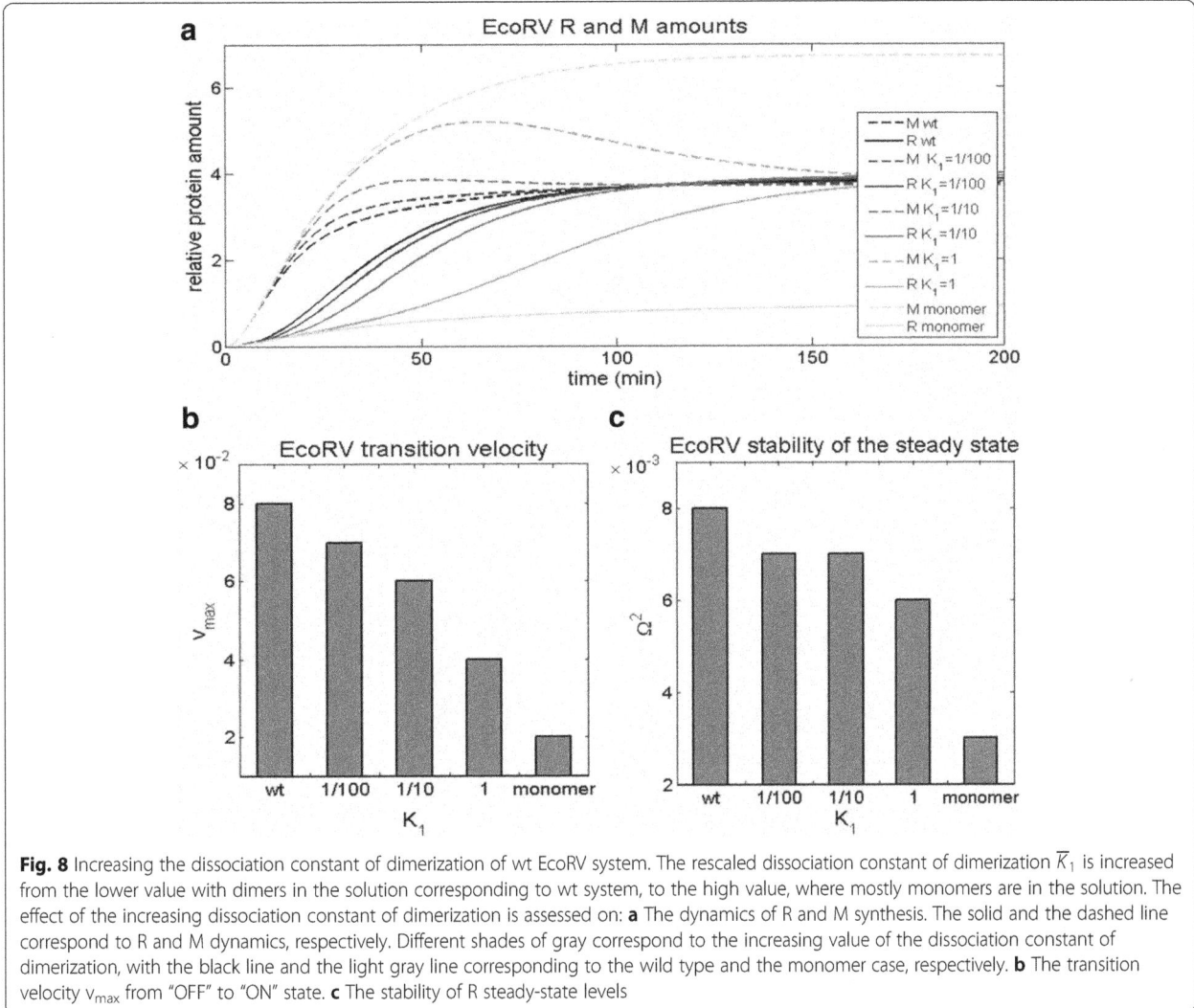

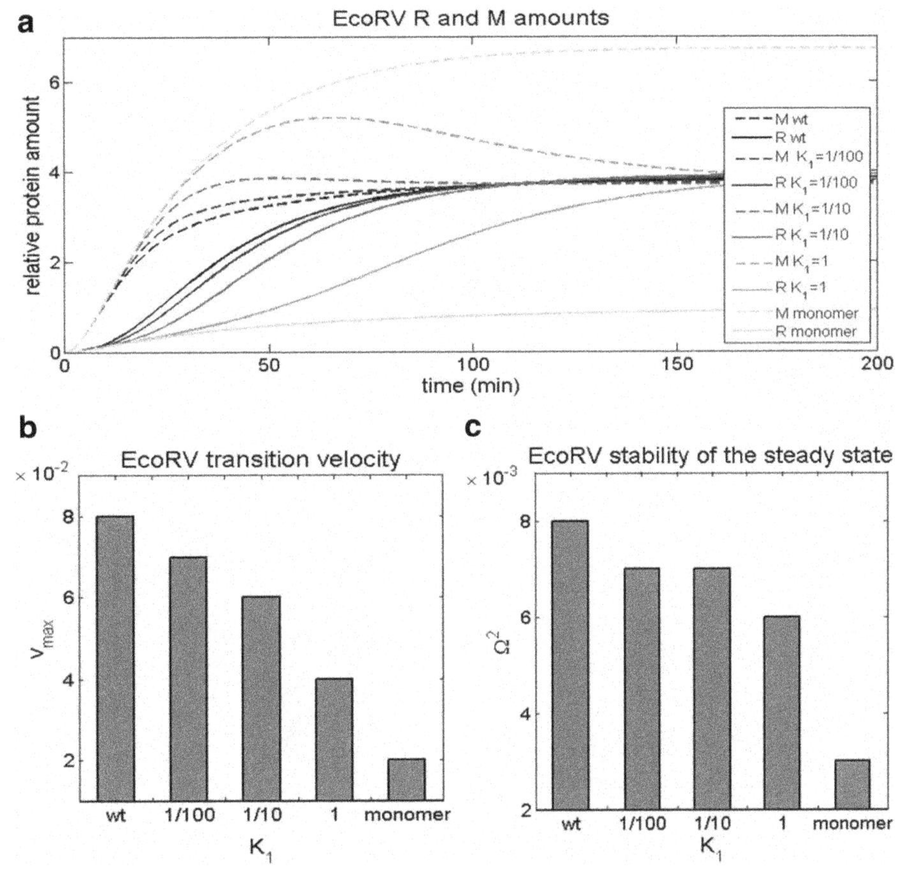

Fig. 8 Increasing the dissociation constant of dimerization of wt EcoRV system. The rescaled dissociation constant of dimerization $\overline{K}_1$ is increased from the lower value with dimers in the solution corresponding to wt system, to the high value, where mostly monomers are in the solution. The effect of the increasing dissociation constant of dimerization is assessed on: **a** The dynamics of R and M synthesis. The solid and the dashed line correspond to R and M dynamics, respectively. Different shades of gray correspond to the increasing value of the dissociation constant of dimerization, with the black line and the light gray line corresponding to the wild type and the monomer case, respectively. **b** The transition velocity v_{max} from "OFF" to "ON" state. **c** The stability of R steady-state levels

Introducing the high C dimer binding cooperativity

We next modify wt EcoRV by increasing the cooperativity ω of C dimer binding to the proximal and the distal binding site, while keeping the other wt EcoRV features unchanged. Note that the experimental measurements in wt EcoRV show an absence of C dimer binding cooperativity ($\omega = 1$) [5], as opposed to the extremely large binding cooperativity that is observed in AhdI [6]. In Fig. 9, we see that increasing ω has the following effects: *i)* the time delay remains nearly the same (Fig. 9a), *ii)* the transition velocity decreases (Fig. 9b), where we see that increasing ω for a relatively moderate factor (2^4), leads to a significant (somewhat less than twofold) decrease of v_{max}, *iii)* stability of R steady-state levels slightly increases. Consequently, we see that perturbing wt EcoRV cooperativity towards the higher values characteristic for AhdI, has a significant adverse effect on one of the dynamical properties (the transition velocity), while not significantly affecting the other two.

Decreasing C translation rate in EcoRV

Finally, we perturb wt EcoRV by decreasing C transcript translation rate k_C, towards the value characteristic for AhdI. Note that C transcript is leaderless in AhdI [6], which is not the case for EcoRV [5], so that we assume the same translation rate for all three transcripts (C, R and M) in EcoRV, while k_C is taken as five times lower in AhdI according to [6]. In Fig. 10a we observe that decreasing k_C does not impact the initial R and M accumulation (during the first ~10 min). On the other hand, at later times the perturbation significantly decreases both the transition velocity that decreases two times (see Fig. 10b), and the stability of R steady-state levels that decreases somewhat less than twofold (see Fig. 10c). Consequently, we see that again two of the three dynamical properties are significantly adversely affected by introducing a control feature from AhdI.

Overall, introducing AhdI characteristic features to EcoRV has a significant adverse effect on at least one of

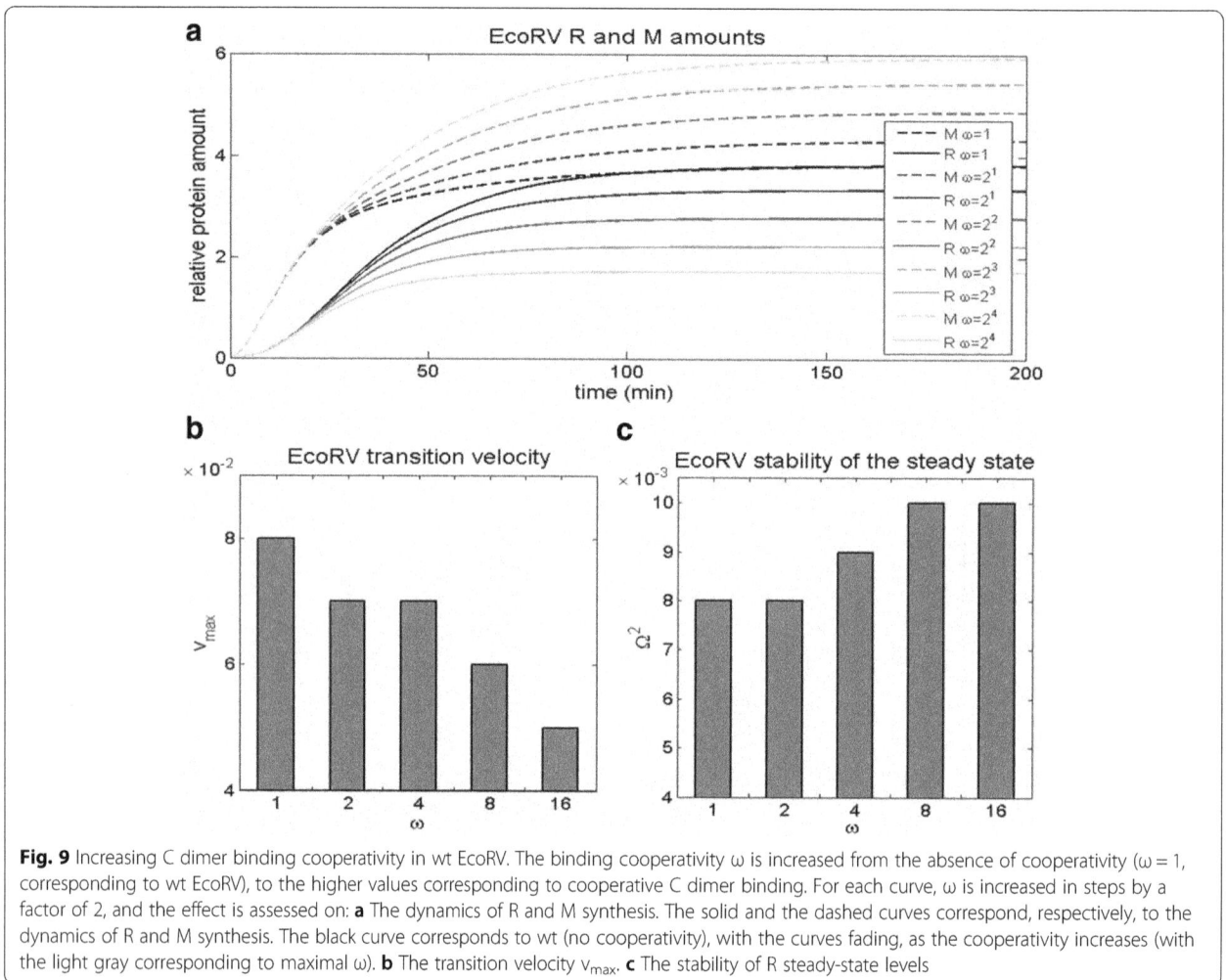

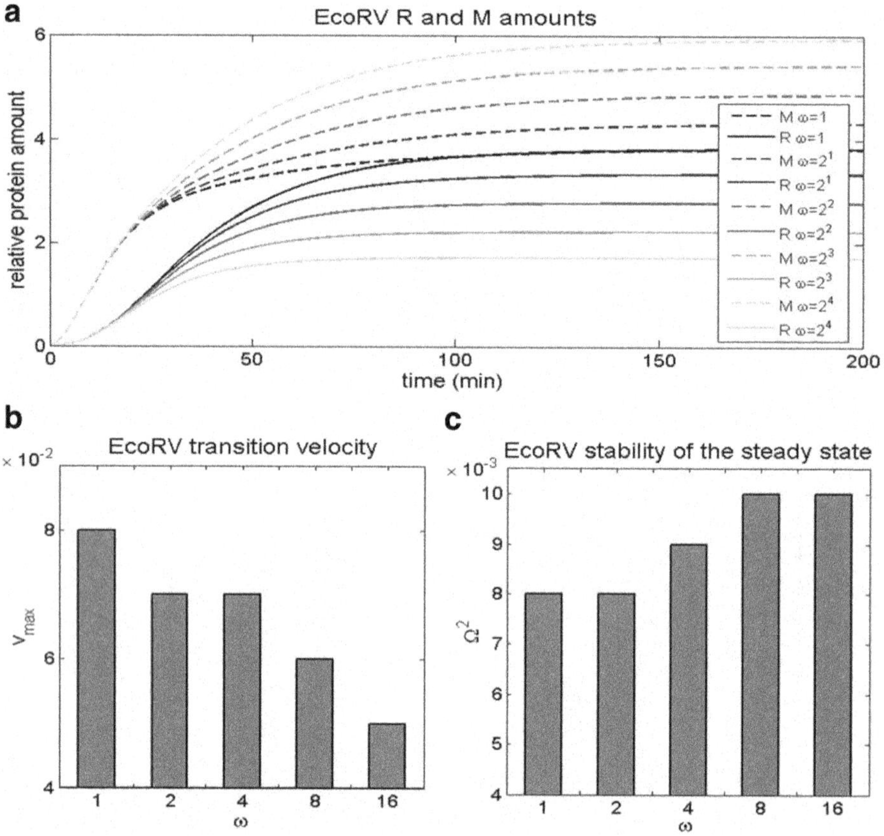

Fig. 9 Increasing C dimer binding cooperativity in wt EcoRV. The binding cooperativity ω is increased from the absence of cooperativity (ω = 1, corresponding to wt EcoRV), to the higher values corresponding to cooperative C dimer binding. For each curve, ω is increased in steps by a factor of 2, and the effect is assessed on: **a** The dynamics of R and M synthesis. The solid and the dashed curves correspond, respectively, to the dynamics of R and M synthesis. The black curve corresponds to wt (no cooperativity), with the curves fading, as the cooperativity increases (with the light gray corresponding to maximal ω). **b** The transition velocity v_{max}. **c** The stability of R steady-state levels

the dynamical properties, which may explain why those features are not found in EcoRV. Additionally, perturbing EcoRV wt parameters towards the AhdI values (Figs. 8a, 9a and 10a) changes M to R ratio in the same direction for each introduced feature (consistently increasing the ratio). This is in distinction to AhdI, where the high cooperativity of C dimer binding has an opposite effect on this ratio, compared to the other two features. Consequently, we argue that another reason for why the characteristic AhdI features are not observed in EcoRV, is because they do not allow balancing the amounts of R and M in the host cell.

Conclusion

R-M systems are characterized by different architectures and control features. We here test a hypothesis that these diverse features can be explained by constraints imposed by few dynamical properties. We started from a relatively well studied AhdI system, and computationally abolished three of its characteristic control features, showing that this has a clear adverse effect on the three dynamical properties. We then modeled a system with different architecture (EcoRV), and showed that its expression dynamics also satisfies the same properties. The EcoRV model has significance in its own right, as the expression dynamics of the divergent R-M systems was, to our knowledge, not studied before, either theoretically or experimentally. Finally, we computationally introduced to EcoRV the control features that exist in AhdI, and showed that this diminishes at least some of the proposed dynamical properties, consistent with the fact that these features do not appear in wt EcoRV. Moreover, increasing the binding cooperativity has the same effect on M to R ratio in EcoRV as increasing the dissociation constant of dimerization, or lowering the translation rate, which prevents balancing M to R ratio upon introducing these perturbations – this then provides another argument for why AhdI control features are absent from wt EcoRV.

Furthermore, dynamical properties proposed here can provide an explanation for a surprisingly large value of the cooperativity in C protein binding, accompanied by

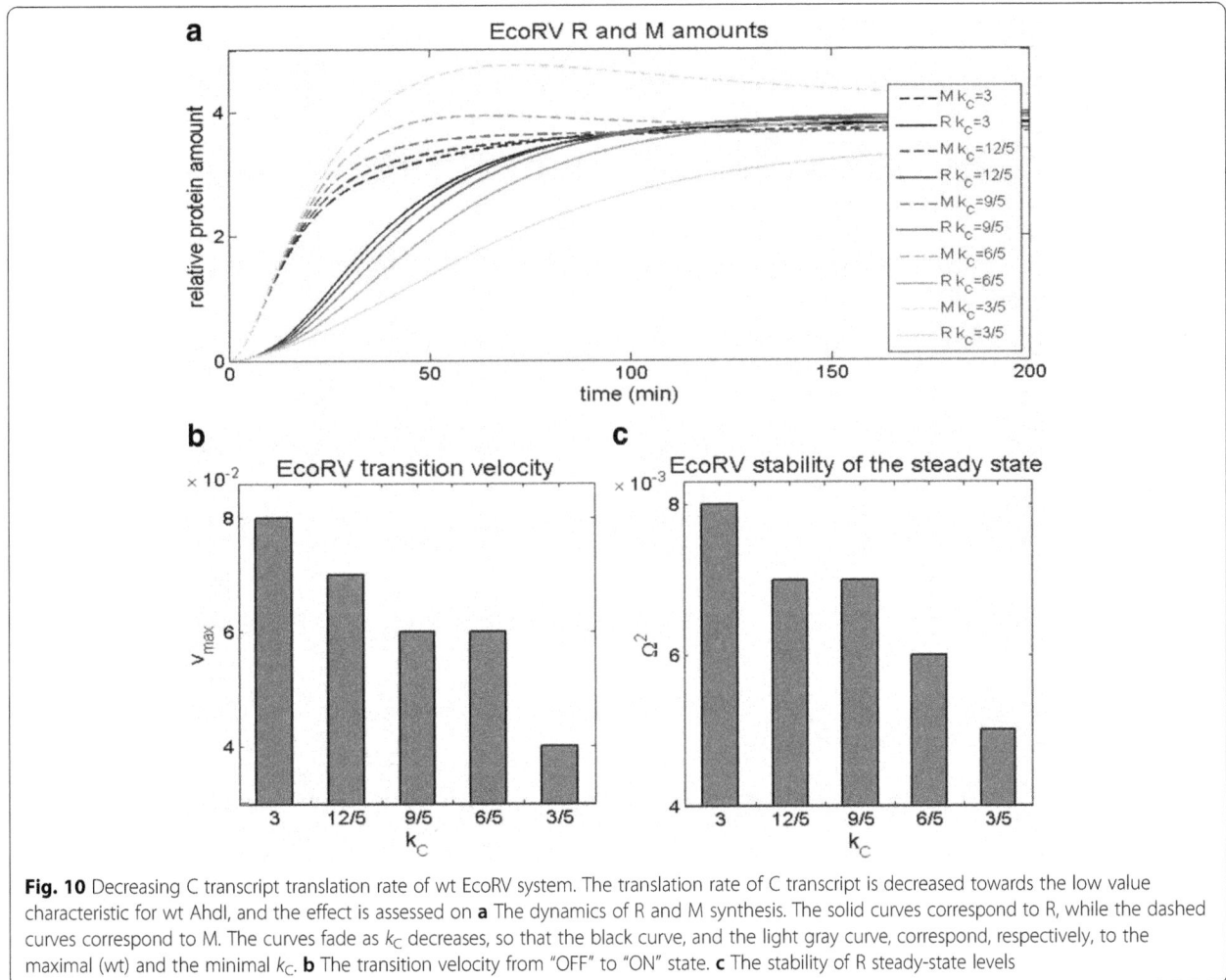

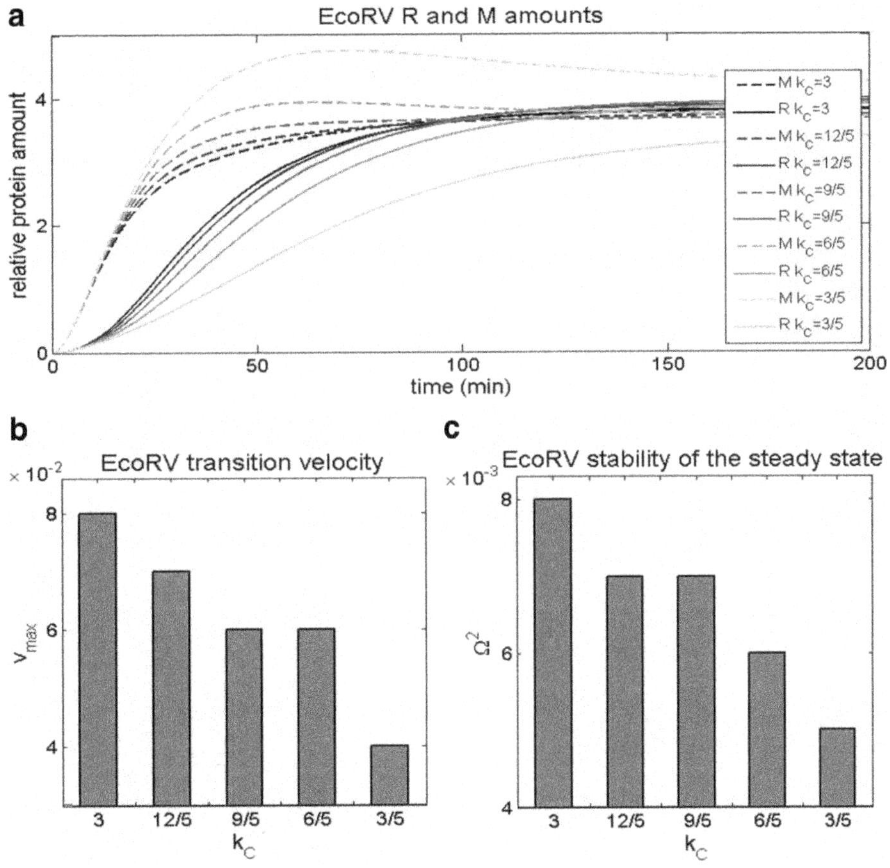

Fig. 10 Decreasing C transcript translation rate of wt EcoRV system. The translation rate of C transcript is decreased towards the low value characteristic for wt AhdI, and the effect is assessed on **a** The dynamics of R and M synthesis. The solid curves correspond to R, while the dashed curves correspond to M. The curves fade as k_C decreases, so that the black curve, and the light gray curve, correspond, respectively, to the maximal (wt) and the minimal k_C. **b** The transition velocity from "OFF" to "ON" state. **c** The stability of R steady-state levels

the large dissociation constant of dimerization that was observed in wt AhdI. We here showed that these two features have an opposite effect on the steady-state levels of the toxic molecule (R), allowing balancing the steady-state R amount, while at the same time leading to more optimal dynamical properties. In support of this proposal, a similar convergent system with lower binding cooperativity (Esp1396I) was also found to have a lower value of the dissociation constant of dimerization. As a prediction, it will be interesting to test if, in other R-M systems, the value of the dissociation constant of dimerization and the binding cooperativity are also related in this way.

Overall, this work provides an example that the system properties that may appear "random" or even surprising (such as the extremely large binding cooperativity) may be explained by constraints imposed by few general principles (in this case the system dynamical properties). Additionally, some of these system properties may serve other functions, e.g. the leaderless C transcripts might be related with a need for preferential translation under specific physiological conditions [42]. Analyzing other R-M systems can further test relation of the system features with the simple dynamical properties, where the main obstacle is that their transcription regulation is generally not well studied. In particular, investigating up to now poorly understood linear R-M systems, which have different architecture compared to the convergent and the divergent systems studied here, and which do not encode C proteins – but may exhibit control by antisense RNAs or at the level of translation initiation efficiency - may be particularly useful [43, 44]. As a further outlook, it will be interesting investigating if properties of other bacterial immune systems, such as recently discovered CRISPR/Cas systems [45], can also be explained by similar dynamical properties [34]. With that respect note that CRISPR/Cas is more advanced, i.e. adaptive bacterial immune system, which retains a memory of the past infections incorporated as spacers in the CRISPR array [46].

Also, in this work we follow a standard approach in molecular biology, where features of the system are perturbed/mutated (which is here done in-silico), and the effect of these perturbations on the presumed system function is assessed. In addition to such "single mutations", a computational equivalent of "double" or "triple" mutations can be exhibited, where more than one system feature would be simultaneously perturbed. This would address the question if perturbations in one feature, can be rescued by also perturbing the other feature(s), which is related to the system robustness. While this question is out of the scope of this work, it also provides an interesting outlook for future research.

Finally, the recent advancement of experimental techniques, such as single-cell experiments, allows directly observing the protein dynamics during the system establishment. While in principle arduous, it would be interesting to experimentally observe how the relevant dynamics is perturbed when some of the key system features are abolished. This would then directly put to test some of the prediction from the computational modelling, which we provided here.

Abbreviations

C: Control protein; M: Methyltransferase; R: Restriction enzyme; R-M: Restriction-modification; RNAP: RNA polymerase; wt: Wild-type

Acknowledgements

We thank K. Severinov for useful discussions.

Declaration

This article has been published as part of BMC Systems Biology Vol 11 Suppl 1, 2017: Selected articles from BGRS\SB-2016: systems biology. The full contents of the supplement are available online at http://bmcsystbiol.biomedcentral.com/articles/supplements/volume-11-supplement-1.

Funding

This work (including the publication costs) was funded by the Swiss National Science foundation under SCOPES project number IZ73Z0_152297, by Marie Curie International Reintegration Grant within the 7th European community Framework Programme (PIRG08-GA-2010-276996) and by the Ministry of Education and Science of the Republic of Serbia under project number ON173052.

Authors' contributions

MJD, MRD and EZ conceived the work. AR and BB performed the analysis. All the authors interpreted the results. MJD, AR and BB wrote the paper, with the help of MRD and EZ. All authors read and approved the final manuscript.

Competing interests

The authors declare that they have no competing interests.

Author details

[1]Institute of Physiology and Biochemistry, Faculty of Biology, University of Belgrade, Studentski trg 16, 11000 Belgrade, Serbia. [2]Multidisciplinary PhD program in Biophysics, University of Belgrade, Belgrade, Serbia. [3]Institute of Physics Belgrade, University of Belgrade, Belgrade, Serbia. [4]Department of Genetic Medicine and Development, University of Geneva and Swiss Institute of Bioinformatics, Geneva, Switzerland.

References

1. Pingoud A, Wilson GG, Wende W. Type II restriction endonucleases—a historical perspective and more. Nucleic Acids Res. 2014;42(12):7489–527.
2. Mruk I, Kobayashi I. To be or not to be: regulation of restriction–modification systems and other toxin–antitoxin systems. Nucleic Acids Res. 2014;42(1):70-86.
3. Gingeras TR, Brooks JE. Cloned restriction/modification system from Pseudomonas aeruginosa. Proc Natl Acad Sci. 1983;80(2):402–6.
4. Kiss A, Posfai G, Keller CC, Venetianer P, Roberts RJ. Nudeotide sequence of the BsuRI restriction-modification system. Nucleic Acids Res. 1985;13(18):6403–21.
5. Semenova E, Minakhin L, Bogdanova E, Nagornykh M, Vasilov A, Heyduk T, Solonin A, Zakharova M, Severinov K. Transcription regulation of the EcoRV restriction-modification system. Nucleic Acids Res. 2005;33(21):6942–51.
6. Bogdanova E, Djordjevic M, Papapanagiotou I, Heyduk T, Kneale G, Severinov K. Transcription regulation of the type II restriction-modification system AhdI. Nucleic Acids Res. 2008;36(5):1429–42.
7. Krüger D, Bickle TA. Bacteriophage survival: multiple mechanisms for avoiding the deoxyribonucleic acid restriction systems of their hosts. Microbiol Rev. 1983;47(3):345.
8. Tock MR, Dryden DT. The biology of restriction and anti-restriction. Curr Opin Microbiol. 2005;8(4):466–72.
9. Korona R, Korona B, Levin BR. Sensitivity of naturally occurring coliphages to type I and type II restriction and modification. Microbiology. 1993;139(6):1283–90.
10. Makino O, Saito H, Ando T. Bacillus subtilis-phage φ1 overcomes host-controlled restriction by producing BamNx inhibitor protein. Mol Gen Genet MGG. 1980;179(3):463–8.
11. Takahashi I, Marmur J. Replacement of thymidylic acid by deoxyuridylic acid in the deoxyribonucleic acid of a transducing phage for Bacillus subtilis. 1963.
12. Kobayashi I. Behavior of restriction-modification systems as selfish mobile elements and their impact on genome evolution. Nucleic Acids Res. 2001; 29(18):3742–56.
13. Jeltsch A, Pingoud A. Horizontal gene transfer contributes to the wide distribution and evolution of type II restriction-modification systems. J Mol Evol. 1996;42(2):91–6.
14. Morozova N, Sabantsev A, Bogdanova E, Fedorova Y, Maikova A, Vedyaykin A, Rodic A, Djordjevic M, Khodorkovskii M, Severinov K. Temporal dynamics of methyltransferase and restriction endonuclease accumulation in individual cells after introducing a restriction-modification system. Nucleic Acids Res. 2016;44(2):790–800.
15. Nagornykh M, Bogdanova E, Protsenko A, Solonin A, Zakharova M, Severinov K. Regulation of gene expression in a type II restriction-modification system. Russ J Genet. 2008;44(5):523–32.
16. Ichige A, Kobayashi I. Stability of EcoRI restriction-modification enzymes in vivo differentiates the EcoRI restriction-modification system from other postsegregational cell killing systems. J Bacteriol. 2005;187(19):6612–21.
17. Mruk I, Blumenthal RM. Real-time kinetics of restriction–modification gene expression after entry into a new host cell. Nucleic Acids Res. 2008;36(8):2581–93.
18. Ball NJ, McGeehan J, Streeter S, Thresh S-J, Kneale G. The structural basis of differential DNA sequence recognition by restriction–modification controller proteins. Nucleic acids research 2012;40(20):10532–42.
19. McGeehan JE, Papapanagiotou I, Streeter SD, Kneale GG. Cooperative binding of the C.AhdI controller protein to the C/R promoter and its role in endonuclease gene expression. J Mol Biol. 2006;358(2):523–31.
20. Sohail A, Ives CL, Brooks JE. Purification and characterization of C · BamHI, a regulator of the BamHI restriction-modification system. Gene. 1995;157(1):227–8.
21. Sorokin V, Severinov K, Gelfand MS. Systematic prediction of control proteins and their DNA binding sites. Nucleic Acids Res. 2009;37(2):441–51.

22. Ives CL, Nathan PD, Brooks JE. Regulation of the BamHI restriction-modification system by a small intergenic open reading frame, bamHIC, in both Escherichia coli and Bacillus subtilis. J Bacteriol. 1992;174(22): 7194–201.
23. Tao T, Bourne J, Blumenthal R. A family of regulatory genes associated with type II restriction-modification systems. J Bacteriol. 1991;173(4):1367–75.
24. Česnavičienė E, Mitkaitė G, Stankevičius K, Janulaitis A, Lubys A. Esp1396I restriction–modification system: structural organization and mode of regulation. Nucleic Acids Res. 2003;31(2):743–9.
25. Karyagina A, Shilov I, Tashlitskii V, Khodoun M, Vasil'ev S, Lau PC, Nikolskaya I. Specific binding of SsoII DNA methyltransferase to its promoter region provides the regulation of SsoII restriction-modification gene expression. Nucleic Acids Res. 1997;25(11):2114–20.
26. Zakharova M, Minakhin L, Solonin A, Severinov K. Regulation of RNA polymerase promoter selectivity by covalent modification of DNA. J Mol Biol. 2004;335(1):103–11.
27. Vilar JM, Saiz L. Systems biophysics of gene expression. Biophys J. 2013; 104(12):2574–85.
28. Tolkunov D, Morozov AV. Genomic studies and computational predictions of nucleosome positions and formation energies. Adv Protein Chem Struct Biol. 2010;79:1–57.
29. Gertz J, Siggia ED, Cohen BA. Analysis of combinatorial cis-regulation in synthetic and genomic promoters. Nature. 2009;457(7226):215–8.
30. Seo Y-J, Chen S, Nilsen-Hamilton M, Levine HA. A mathematical analysis of multiple-target SELEX. Bull Math Biol. 2010;72(7):1623–65.
31. Aguda B, Friedman A. Models of cellular regulation. New York: Oxford University Press; 2008.
32. Alon U. An introduction to systems biology: design principles of biological circuits. London: CRC press; 2006.
33. Williams K, Savageau MA, Blumenthal RM. A bistable hysteretic switch in an activator–repressor regulated restriction–modification system. Nucleic Acids Res. 2013;41(12);6045–57.
34. Djordjevic M. Modeling bacterial immune systems: strategies for expression of toxic - but useful - molecules. Biosystems. 2013;112(2):139–44.
35. O'Donnell SM, Janssen GR. The initiation codon affects ribosome binding and translational efficiency in Escherichia coli of cI mRNA with or without the 5′ untranslated leader. J Bacteriol. 2001;183(4):1277–83.
36. Shell SS, Wang J, Lapierre P, Mir M, Chase MR, Pyle MM, Gawande R, Ahmad R, Sarracino DA, Ioerger TR. Leaderless transcripts and small proteins are common features of the mycobacterial translational landscape. PLoS Genet. 2015;11(11):e1005641.
37. Beletskaya IV, Zakharova MV, Shlyapnikov MG, Semenova LM, Solonin AS. DNA methylation at the CfrBI site is involved in expression control in the CfrBI restriction–modification system. Nucleic Acids Res. 2000;28(19):3817–22.
38. Shea MA, Ackers GK. The OR control system of bacteriophage lambda. A physical-chemical model for gene regulation. J Mol Biol. 1985;181(2): 211–30.
39. Enikeeva FN, Severinov KV, Gelfand MS. Restriction–modification systems and bacteriophage invasion: Who wins? J Theor Biol. 2010;266(4):550–9.
40. Streeter SD, Papapanagiotou I, McGeehan JE, Kneale GG. DNA footprinting and biophysical characterization of the controller protein C.AhdI suggests the basis of a genetic switch. Nucleic Acids Res. 2004;32(21):6445–53.
41. Bogdanova E, Zakharova M, Streeter S, Taylor J, Heyduk T, Kneale G, Severinov K. Transcription regulation of restriction-modification system Esp1396I. Nucleic Acids Res. 2009;37(10):3354–66.
42. Vesper O, Amitai S, Belitsky M, Byrgazov K, Kaberdina AC, Engelberg-Kulka H, Moll I. Selective translation of leaderless mRNAs by specialized ribosomes generated by MazF in Escherichia coli. Cell. 2011;147(1):147–57.
43. Mruk I, Liu Y, Ge L, Kobayashi I. Antisense RNA associated with biological regulation of a restriction–modification system. Nucleic Acids Res. 2011;39(13): 5622–32.
44. Nagornykh M, Zakharova M, Protsenko A, Bogdanova E, Solonin AS, Severinov K. Regulation of gene expression in restriction-modification system Eco29kI. Nucleic Acids Res. 2011;39(11):4653–63.
45. Barrangou R, Marraffini LA. CRISPR-Cas systems: prokaryotes upgrade to adaptive immunity. Mol Cell. 2014;54(2):234–44.
46. Al-Attar S, Westra ER, van der Oost J, Brouns SJ. Clustered regularly interspaced short palindromic repeats (CRISPRs): the hallmark of an ingenious antiviral defense mechanism in prokaryotes. Biol Chem. 2011; 392(4):277–89.
47. Rezulak M, Borsuk I, Mruk I: Natural C-independent expression of restriction endonuclease in a C protein-associated restriction-modification system. Nucleic Acids Res. 2016;44(6):2646–60.
48. Lubys A, Jurenaite S, Janulaitis A. Structural organization and regulation of the plasmid-borne type II restriction-modification system Kpn2I from Klebsiella pneumoniae RFL2. Nucleic Acids Res. 1999;27(21):4228–34.
49. Knowle D, Lintner RE, Touma YM, Blumenthal RM. Nature of the promoter activated by C. PvuII, an unusual regulatory protein conserved among restriction-modification systems. J Bacteriol. 2005;187(2):488–97.

2

Delineating functional principles of the bow tie structure of a kinase-phosphatase network in the budding yeast

Diala Abd-Rabbo[1,2] and Stephen W. Michnick[1,2*]

Abstract

Background: Kinases and phosphatases (KP) form complex self-regulating networks essential for cellular signal processing. In spite of having a wealth of data about interactions among KPs and their substrates, we have very limited models of the structures of the directed networks they form and consequently our ability to formulate hypotheses about how their structure determines the flow of information in these networks is restricted.

Results: We assembled and studied the largest bona fide kinase-phosphatase network (KP-Net) known to date for the yeast *Saccharomyces cerevisiae*. Application of the vertex sort (VS) algorithm on the KP-Net allowed us to elucidate its hierarchical structure in which nodes are sorted into top, core and bottom layers, forming a bow tie structure with a strongly connected core layer. Surprisingly, phosphatases tend to sort into the top layer, implying they are less regulated by phosphorylation than kinases. Superposition of the widest range of KP biological properties over the KP-Net hierarchy shows that core layer KPs: (i), receive the largest number of inputs; (ii), form bottlenecks implicated in multiple pathways and in decision-making; (iii), and are among the most regulated KPs both temporally and spatially. Moreover, top layer KPs are more abundant and less noisy than those in the bottom layer. Finally, we showed that the VS algorithm depends on node degrees without biasing the biological results of the sorted network. The VS algorithm is available as an R package (https://cran.r-project.org/web/packages/VertexSort/index.html).

Conclusions: The KP-Net model we propose possesses a bow tie hierarchical structure in which the top layer appears to ensure highest fidelity and the core layer appears to mediate signal integration and cell state-dependent signal interpretation. Our model of the yeast KP-Net provides both functional insight into its organization as we understand today and a framework for future investigation of information processing in yeast and eukaryotes in general.

Keywords: Kinase-phosphatase signalling network, Network hierarchical structure, Topological properties, Biological properties, Vertex Sort algorithm, Functional principles of cell behaviour, *Saccharomyces cerevisiae*

Background

To maintain normal homeostasis, living cells continuously accommodate changes to their internal and external environment via signalling pathways. Protein KPs play an essential regulatory role in signalling pathways through phosphorylation and dephosphorylation interactions (PDI) that cause profound effects on substrates, affecting their turnover, localization and interactions with other proteins [1].

Numerous efforts have been made to reconstruct the budding yeast KP-Net from various types of interactions [2–7]. Despite these efforts, KP-Nets assembled so far are not fully mature to represent genuine networks in which a KP acts directly on its substrate for the following reasons. First, dephosphorylation interactions are under-represented in KP-Nets, because on one hand, dephosphorylation interactions are poorly annotated in public databases (Additional file 1: Table S1) and on the other hand, phosphatases have been modestly studied in comparison to kinases. Second, kinase networks that were assembled from in vitro phosphorylation interactions do

* Correspondence: stephen.michnick@umontreal.ca
[1]Département de Biochimie et Médecine Moléculaire, Université de Montréal, C.P. 6128, Succursale centre-ville, Montréal, Québec H3C 3J7, Canada
[2]Centre Robert-Cedergren, Bio-Informatique et Génomique, Université de Montréal, C.P. 6128, Succursale centre-ville, Montréal, Québec H3C 3J7, Canada

not include phosphatases and contain a considerable number of false positives due to non-specific phosphohorylation of proteins by kinases in vitro [5–7]. Finally, KP-Nets that were assembled from protein-protein interactions and from genetic interactions, and KP-Nets that were built by knocking out a KP lack two crucial properties: causality and directionality [2–4]. These crucial properties characterize the command-execution aspect of regulatory networks. Causality determines which KP directly acts on which substrate, whereas directionality indicates the direction of the interaction between the two interactors, which is required when substrates are themselves KPs. Interestingly, KP-Nets assembled from high quality PDIs are not characterized by the previously mentioned drawbacks and hence describe better genuine KP-Nets. Despite the large number of KP-Net studies, to our knowledge, no investigations in the budding yeast included in vivo interactions characterized by both causality and directionality [2–4]. KP-Net studies that did include interactions characterized by both causality and directionality were not performed in vivo and did not include phosphatases [5–7] (Additional file 1: Table S2). Hence, constructing a bona fide KP-Net remains an essential goal for analysis of signalling networks.

There have been a number of efforts to determine rules governing the organization and function of biological regulatory networks. For instance, a number of studies invoke command-execution organization characterizing directed networks to elucidate their hierarchical structure using network decomposition methods on various regulatory networks [5, 6, 8–13]. Decomposition methods classify network nodes into different layers to elucidate information flow in network hierarchies. The majority of these efforts were aimed at transcription networks, but rarely at other regulatory networks, including KP networks. In addition, network layers in these studies were characterized by topological and rarely by biological properties of their nodes; that is, KP-Nets are rarely characterized according to the features of the gene products that represent nodes such as stability, abundance and noise in mRNA and protein gene products (Additional file 1: Table S2). However, biological properties are the ones that profoundly affect the regulatory state of any biological network.

Despite the wealth of available evidence, deciphering the complexity of KP-Nets to gain insights into their functional principles is still challenging. Here, we overcame two basic gaps in knowledge in previous studies: first, we constructed the largest bona fide KP-Net for the yeast *Saccharomyces cerevisiae*. Second, we elucidated the KP-Net hierarchical structure using the VS algorithm and unprecedentedly, we integrated the widest range of KP biological properties within this hierarchy in order to describe the functional principles of the KP-Net with our current knowledge. We found that the KP-Net has a bow tie hierarchy formed of three layers (top, core and bottom) and that the different biological properties of KPs are unevenly distributed among KP-Net layers. This uneven distribution reveals general biological properties of KPs in each layer from which we could postulate the behaviours and information processing functions of each layer in the KP-Net hierarchy. We suggest that high protein abundances and low protein noise in KP-Net top layer could result in signal fidelity, whereas enrichment for decision-making and bottleneck proteins in the core layer may underlie signal integration. Finally, we showed that node degrees affect the way the VS algorithm sorts nodes within a network but we also showed that our results and conclusions are not biased by node degrees. We developed an R package called the VertexSort to facilitate VS algorithm application to other networks (https://cran.r-project.org/web/packages/VertexSort/index.html).

Results

The kinase-phosphatase network (KP-Net)

The kinase interaction database (KID) provides the most detailed and specialized annotation of kinase-protein interactions; its annotation is based on 31 experimental categories including genetic, biochemical, physical and phenotypic experimental evidence [14]. However, phosphatase-protein interactions are not included in and many kinase-protein interactions are missing or partially annotated in this database. Hence, we collected these interactions from different sources, then, curated, annotated and scored the collected interactions according to the KID database pipeline with minor adjustments to annotate phosphatase-protein interactions (Fig. 1a and Additional file 1: Supplementary Methods) [2, 15–20]. The KID pipeline associates a confidence score to each interaction based on the extent to which the different experimental methods that validate an interaction contribute to identifying a true positive Kinase-protein interaction. To ensure that the interactions assembled in the KP-Net represent PDIs rather than simply Kinase-protein or phosphatase-protein interactions, we selected interactions having a confidence score ≥ 4.52 (corresponding to a $P \leq 5 \times 10^{-2}$) and those validated by at least one biochemical experiment showing the occurrence of a PDI (in vitro kinase assay, in vivo or in vitro phosphosite mapping, mobility shift of phosphoproteins on gel or substrate trapping by a dead phosphatase catalytic domain). The assembled KP-Net contains 1,087 directed interactions (918 and 169 PDIs, respectively) implicating 616 proteins [101 kinases and 31 phosphatases, covering ~77% of these enzymes and 484 proteins, most of which are KP substrates that are not KPs, (Fig. 1a and Additional file 2)]. Similar to other biological networks, the KP-Net possesses a scale-free structure ($P(K) \sim K^{-2.58}$ with a goodness-of-fit test $P =$

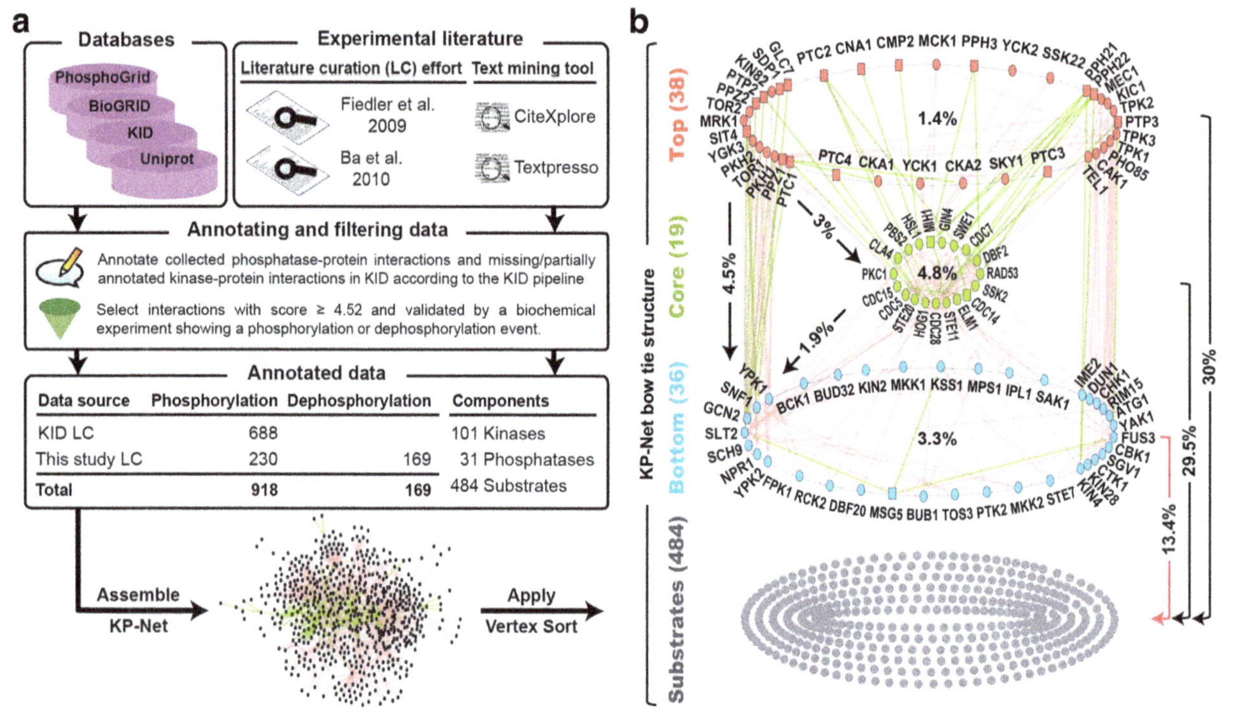

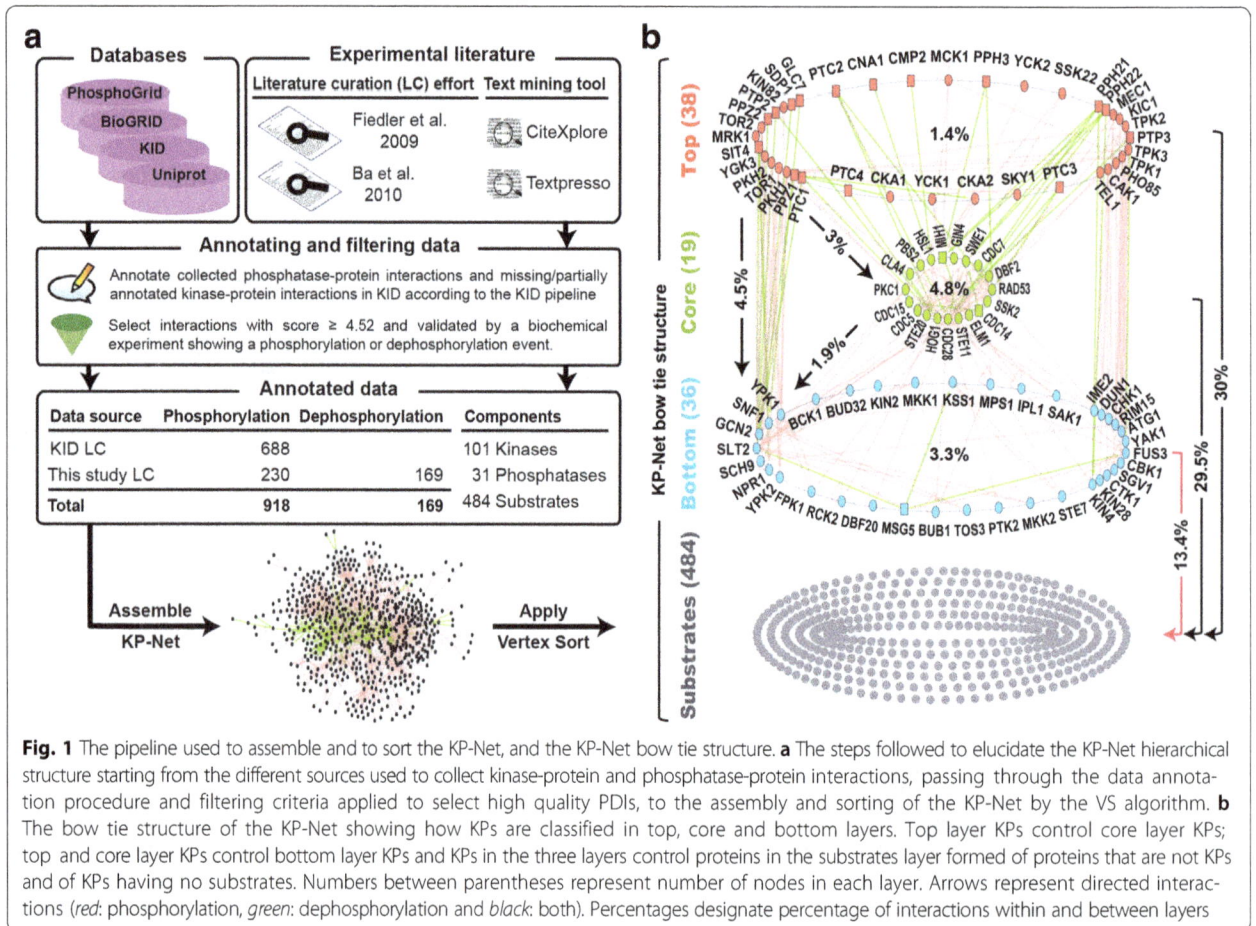

Fig. 1 The pipeline used to assemble and to sort the KP-Net, and the KP-Net bow tie structure. **a** The steps followed to elucidate the KP-Net hierarchical structure starting from the different sources used to collect kinase-protein and phosphatase-protein interactions, passing through the data annotation procedure and filtering criteria applied to select high quality PDIs, to the assembly and sorting of the KP-Net by the VS algorithm. **b** The bow tie structure of the KP-Net showing how KPs are classified in top, core and bottom layers. Top layer KPs control core layer KPs; top and core layer KPs control bottom layer KPs and KPs in the three layers control proteins in the substrates layer formed of proteins that are not KPs and of KPs having no substrates. Numbers between parentheses represent number of nodes in each layer. Arrows represent directed interactions (*red*: phosphorylation, *green*: dephosphorylation and *black*: both). Percentages designate percentage of interactions within and between layers

1.3×10^{-2}) in which most KPs regulate few proteins and few KP hubs regulate a large number of proteins (Additional file 1: Supplementary Methods and Figure S1).

The KP-Net possesses a "corporate" hierarchical structure in the form of a bow tie with a strongly connected core layer

We assessed the amount of the hierarchical structure of the KP-Net by calculating its global reaching centrality (GRC), which represents a normalized average of the proportions of nodes accessible from each node in the network [21]. The closer the GRC is to 1, the more hierarchical the network is. The KP-Net has a moderate GRC of 0.61, suggesting that the KP-Net represents a hierarchical structure that could be placed between two extremes: (i) an autocratic structure comparable to a complete tree and (ii) a democratic structure in which collaborative regulation dominates and no hierarchy exists [5]. Bhardwaj et al. observed a similar moderate hierarchy in a co-phosphorylation network and described it as a corporate hierarchy [5]. Obviously, the KP-Net does not represent a complete tree, as it is enriched for many logic motifs that do not occur in trees: feed-forward loops (a structure in which a node regulates another node and together they regulate a third one), two node feedback loops (two nodes that regulate each other), and bi-fans (a structure in which two nodes regulate two other nodes) ($P < 10^{-3}$, Methods). Moreover, the KP-Net does not represent democracies and encapsulates a hierarchical structure, as its GRC is significantly higher than that of Erdős–Rényi random networks (non-hierarchical networks) having the same number of nodes and edges as the KP-Net ($P < 10^{-4}$, Methods). Interestingly, the GRC of the KP-Net is significantly smaller than that of random networks generated by degree preserving randomization (DPR, Methods). This result is not surprising, as the degree distribution of a network is essential to determine its organizational structure, meaning networks having same degree distributions will have similar organizational structures. Thus the GRC of the KP-Net was expected to be comparable to that of DPR networks, but it was found to be significantly smaller than the GRC of DPR networks, probably indicating enrichment for feedback loops that generally exist in KP-Nets.

Subsequently, we applied the VS algorithm to the KP-Net to elucidate the network hierarchical structure and the signal flow within the elucidated hierarchy. The VS algorithm is among the best network decomposition algorithms available. It was conceived and applied by Jothi et al. to the transcription regulatory network of the budding yeast *Saccharomyces cerevisiae* to elucidate the network hierarchical structure [5, 6, 8–11, 22]. The VS algorithm sorts nodes into different levels so that nodes in upper levels control those in lower levels [8]. It first transforms a cyclic graph to an acyclic one by collapsing each strongly connected component (SCC, a sub-graph where each node pair is related by two paths of opposite directions) into a super node and then it applies the leaf removal algorithm to the resulting graph and to its transpose. This generates global solutions in which a node could span a range of levels, reflecting the huge amount of missing data in and the dynamic nature of biological networks.

Application of the VS algorithm to the KP-Net revealed a hierarchical structure in which KPs are sorted into 9 levels that we subsequently grouped into three non-overlapping layers: top, core and bottom (Additional file 1: Figure S2a). As in Jothi et al., we first identified KPs of the largest SCC and classified them as belonging to the core layer (19 KPs); we then classified KPs that regulate core layer KPs to the top layer (38 KPs) and those that are regulated by core layer KPs to the bottom layer (36 KPs) (Fig. 1b) [8]. Thirty-eight nodes, of which 33 KPs and five proteins that are not KPs, were excluded from further analysis, because the former are not connected to any KP and the latter are substrates of the excluded KPs (Additional file 1: Figure S2b). The three layers of the KP-Net generated a bow tie structure in which the core layer has relatively fewer nodes than top and bottom layers (Fig. 1b). It is important to note that the bow tie shape of the KP-Net represents an intrinsic property of this network and it is not the result of the application of the VS algorithm. More specifically it is not not the result of choosing the core layer as the SCC of the KP-Net. This is because by applying the VS algorithm in the same way, the hierarchical structure of the regulatory network elucidated by Jothi et al. do not have a bow tie shape (top, core and bottom layers contain 25, 64 and 59 nodes, respectively) [8].

Interestingly, KP-Net top, core and bottom layers regulate 235, 276 and 148 proteins, respectively, corresponding to 38, 45 and 24% of the KP-Net nodes, respectively. Although the core layer is ~2 times smaller in size than top and bottom layers, it regulates a number of substrates that is 1.2 and 1.9 times larger than that regulated by top and bottom layers, respectively, implying an essential role of the core layer in the KP-Net.

The three layers of the KP-Net have dissimilar biological roles and subcellular localizations

To unravel biological roles of the KP-Net layers, we performed a Gene Ontology (GO) enrichment/depletion analysis for KPs in each of these layers (Additional file 1: Supplementary Methods). We found that the KP-Net top layer is enriched mostly for signal regulation and transduction; interestingly, the core layer is enriched for signalling also, for metabolic processes, but mostly for cell cycle, organization processes related to cell cycle and decision-making (Additional file 1: Table S3), confirming the essential role of the core layer in the KP-Net; and the bottom layer is enriched for few GO terms, suggesting that it has a less specialized and more diverse biological roles (Fig. 2a). These results are in line with the findings of Bhardwaj et al. [5].

On another level, the top layer is depleted for, whereas the core layer is enriched for KPs located in the bud neck (Fig. 2b), a result that has been already observed by Cheng et al. [6]. We further found that the bottom layer is enriched for KPs located in the mating projection tip (Fig. 2b). The latter observations suggest that top layer KPs might remain in the mother cell to regulate signalling, while core layer KPs may be polarized towards the daughter cell to contribute to mitosis, and bottom layer KPs might reside in the cell projection to contribute in mating.

Strikingly, dephosphorylation is enriched in the top layer and depleted in the bottom layer of the KP-Net (Fig. 2a), suggesting that phosphatases are over-represented in signalling pathway upstream and depleted in downstream arms of signalling pathways. The latter results are consistent with dynamic phosphoproteomic studies showing that at least 50% of early responses to cell perturbations are dephosphorylation of phosphosites [23].

Phosphatases are less regulated by phosphorylation than kinases

Our findings confirmed our proposition that the top layer is enriched whereas the bottom layer is depleted for phosphatases (Additional file 1: Figure S3a, $P = 2.2 \times 10^{-5}$ and $P = 4.1 \times 10^{-4}$ respectively; hypergeometric test (HT)). In addition, we observed that 81% of the top layer phosphatases have a zero in-degree. Using high quality phosphoproteomic data annotated in the PhosphoGRID database, we also found that the number of phosphosites identified in phosphatase protein sequences is smaller than that identified in kinases (Additional file 1: Figure S3b, $P = 2.3 \times 10^{-3}$; randomization test (RT), Methods). These results suggest that phosphatases are less regulated by phosphorylation than kinases are. Our suggestion is also supported by the great variety of regulatory subunits controlling phosphatases [24] and by the large number of cellular mechanisms, other than phosphorylation, reported to regulate phosphatases, including phosphorylation of the regulatory subunits of phosphatases [25–30].

KP-Net upper levels are the least regulated and KP-Net lower levels are the least to regulate other KPs

Top layer KP in-degrees are on average smaller than KP in-degrees in core and bottom layers (Fig. 3a, $P < 10^{-4}$; RT,

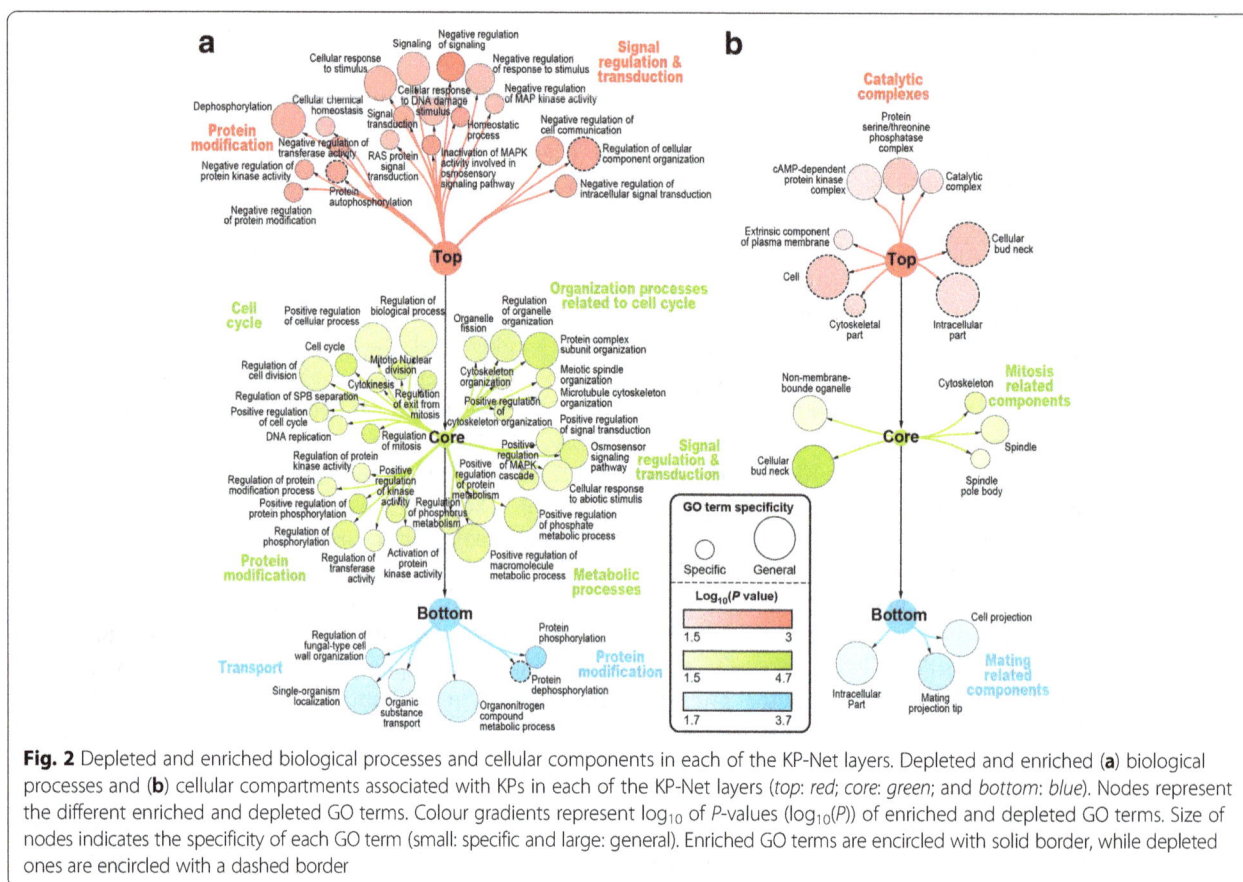

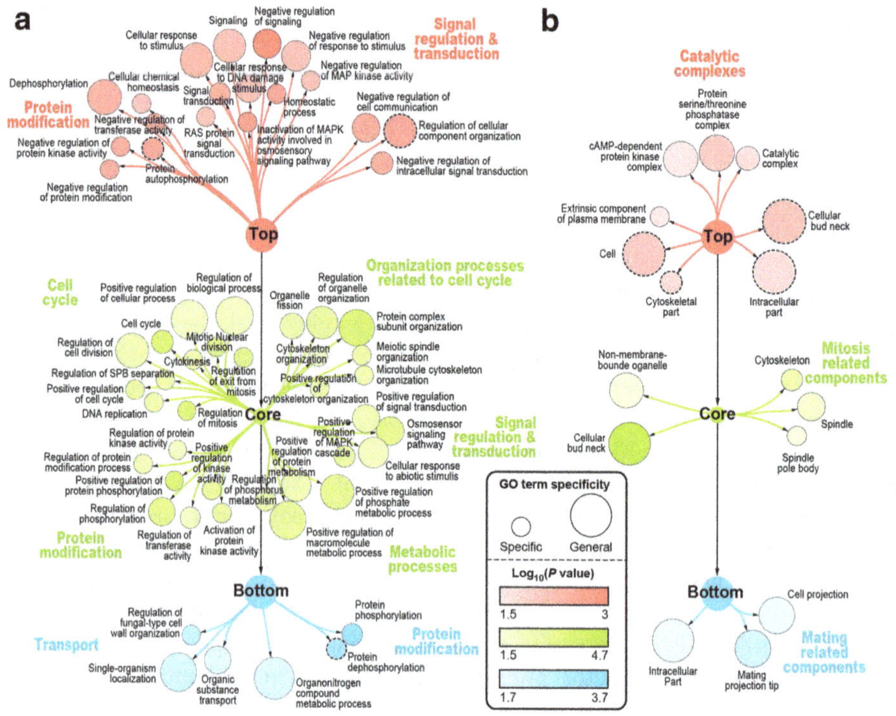

Fig. 2 Depleted and enriched biological processes and cellular components in each of the KP-Net layers. Depleted and enriched (**a**) biological processes and (**b**) cellular compartments associated with KPs in each of the KP-Net layers (*top*: *red*; *core*: *green*; and *bottom*: *blue*). Nodes represent the different enriched and depleted GO terms. Colour gradients represent $\log_{10}$ of *P*-values ($\log_{10}(P)$) of enriched and depleted GO terms. Size of nodes indicates the specificity of each GO term (small: specific and large: general). Enriched GO terms are encircled with solid border, while depleted ones are encircled with a dashed border

Methods). This observation is a direct result of the VS algorithm application ($P = 10^{-3}$; degree non-preserving randomization (DNPR), Methods) to a network, but it agrees with organizational principles found in hierarchical systems in which members of upper levels are the least regulated (e.g. pyramid networks). In contrast, the out-degree of the bottom layer is significantly smaller than that of top and core layers (Fig. 3b, $P = 3 \times 10^{-3}$; RT, Methods). This finding is independent of the VS algorithm application ($P = 0.7$; DNPR, Methods) on a network and has been previously observed in the hierarchical structure of a yeast transcriptional regulatory network elucidated by a decomposition algorithm (Breadth-First Search) different than the VS algorithm [9]. Finally, the observed features related to node in- and out-degrees were implemented in two network decomposition algorithms, other than the VS algorithm, to classify nodes in top and bottom layers, respectively [5, 12].

The KP-Net core layer is enriched for essential genes, bottlenecks, and pathway-shared components

To better grasp our knowledge of signal flow in the KP-Net, we analysed the distribution of hubs, bottlenecks, pathway-shared components (KPs involved in at least two pathways) and essential genes in the three layers of the KP-Net. Hubs and bottlenecks are defined as the 20% of KPs in the KP-Net that have, respectively, the highest degree and the highest betweenness (fraction of shortest paths between all pairs of nodes that pass through a single node; this measure captures how much signalling passes through a node). The hubs are equally distributed among the three layers, reflecting the prevalence of parallel regulation as a principle emerging from the three layers of the KP-Net (Fig. 3c). Interestingly, the core layer is enriched for bottlenecks, pathway-shared components and essential genes (Fig. 3d–f, $P = 4.3 \times 10^{-5}$, $P = 1.4 \times 10^{-2}$ and $P = 3.8 \times 10^{-2}$, respectively; HT), suggesting that most of the signal integration and cross-talk between pathways occur in the core layer.

Molecular switches are enriched in KPs in core and bottom layers

Molecular switches represent phosphosites within or adjacent to linear binding motifs (LBM) which mediate "on demand" controls switching proteins between different functional states (on-off, specificity, cumulative and sequential switches) [31]. Given their fundamental role in controlling signalling networks, we investigated the

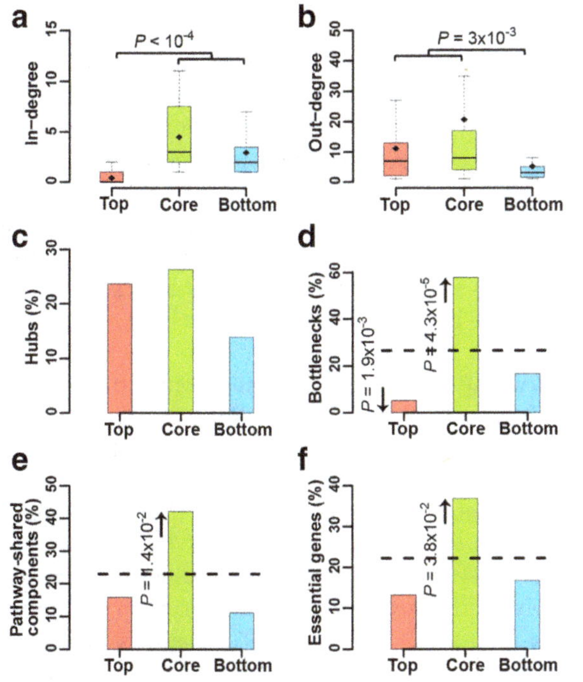

Fig. 3 Topological and biological properties of KPs in the different layers of the KP-Net. Distribution of (**a**) in-degree and (**b**) out-degree of KPs in each layer of the KP-Net and percentage of KPs representing (**c**) hubs, (**d**) bottlenecks, (**e**) shared components between pathways (KPs involved in at least two pathways), and (**f**) essential genes in each layer of the KP-Net. The *broken line* in bar plots represents the expected mean of the corresponding percentage in each layer. *Black diamonds* in box plots designate the average of the corresponding property of KPs. Outliers were omitted from box plots to simplify data representation. *P*-values were calculated by comparing property means of two layers and the enrichment/depletion of a property within a layer using the RT (Methods) and HT, respectively. For description of the used datasets see Supplementary Materials in Additional file 1

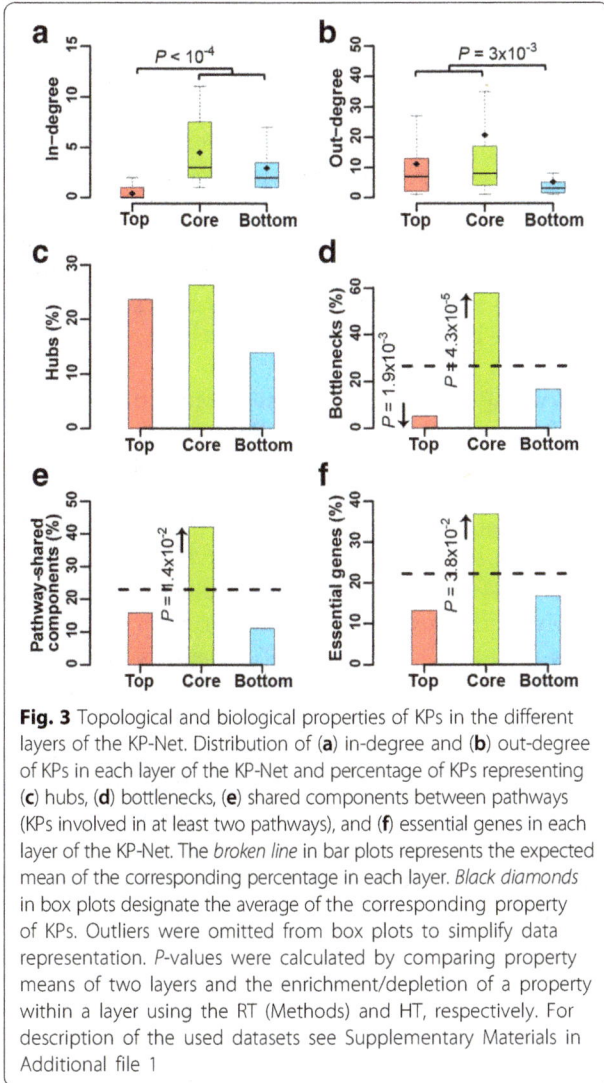

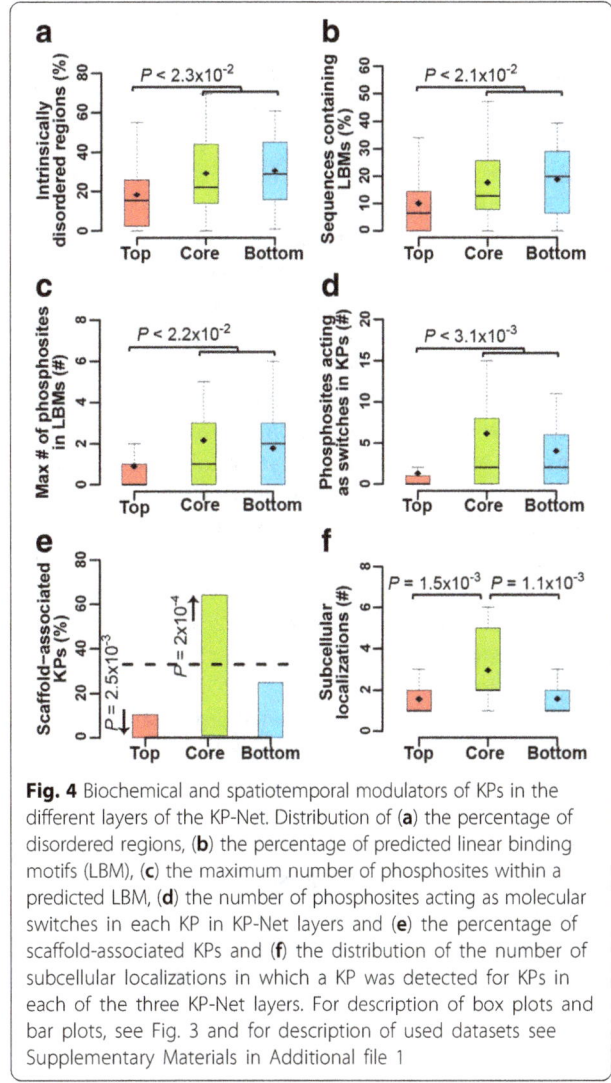

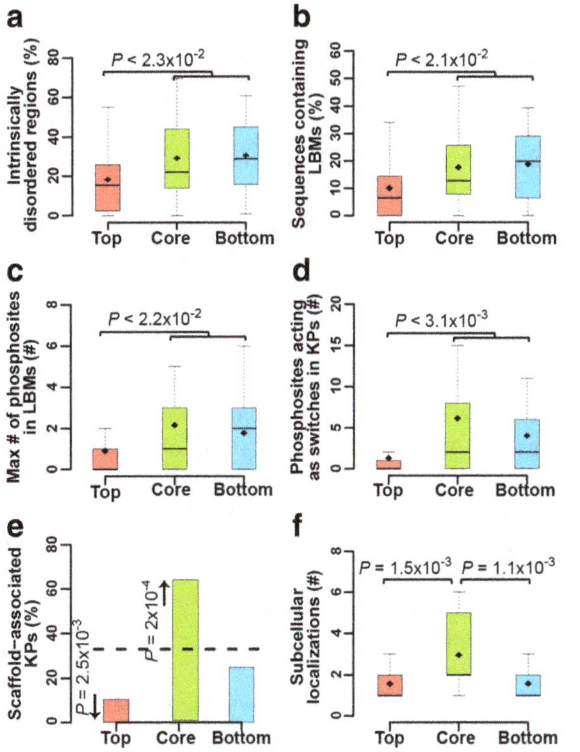

Fig. 4 Biochemical and spatiotemporal modulators of KPs in the different layers of the KP-Net. Distribution of (**a**) the percentage of disordered regions, (**b**) the percentage of predicted linear binding motifs (LBM), (**c**) the maximum number of phosphosites within a predicted LBM, (**d**) the number of phosphosites acting as molecular switches in each KP in KP-Net layers and (**e**) the percentage of scaffold-associated KPs and (**f**) the distribution of the number of subcellular localizations in which a KP was detected for KPs in each of the three KP-Net layers. For description of box plots and bar plots, see Fig. 3 and for description of used datasets see Supplementary Materials in Additional file 1

distribution of KP molecular switches in the KP-Net hierarchy. We predicted protein disordered regions in KP protein sequences and LBMs within these predicted disordered regions using the IUPred and ANCHOR algorithms, respectively (Additional file 1: Supplementary Methods) [32, 33]. We then overlaid bona fide in vivo phosphosites from the PhosphoGRID database on top of KP protein sequences (Additional file 1: Supplementary Materials). We found that percentage of predicted disordered regions in KP proteins in core and bottom layers are on average higher compared to the top layer (Fig. 4a, $P < 2.3 \times 10^{-2}$; RT, Methods). The same trend is observed for: (i), the percentage of sequences predicted to contain LBMs (Fig. 4b, $P < 2.1 \times 10^{-2}$; RT, Methods); (ii), the number of phosphosites in KP sequences generally (Additional file 1: Figure S3c, $P < 6.1 \times 10^{-4}$; RT, Methods) and (iii), in the predicted LBMs particularly (Fig. 4c, $P < 2.2 \times 10^{-2}$; RT, Methods); and (iv), the number of potential molecular switches in each KP (Fig. 4d, $P < 3.1 \times 10^{-3}$; RT, Methods). Interestingly, our findings suggest that phosphorylation of KPs in lower layers could form molecular switches important for KP temporal regulation. Two out of many examples confirming our suggestions are: (1) the specificity switch in Hsl1 (core layer kinase and morphogenesis checkpoint regulator) leading to a G2 arrest essential for cell survival upon osmotic shock and (2) the on-switch in Swe1 (core layer kinase) maintaining Cdc28 in an inhibited form essential for entry of cells into mitosis [34, 35].

Core layer KPs employ scaffolding to prevent unwanted pathway crosstalk

It is well established that redirecting information flow within signalling networks is accomplished through interactions of KP with scaffold proteins and is required for the insulation of interconnected pathways [36]. Interestingly, the KP-Net core layer is enriched for pathway-shared components (Fig. 3e) and for LBMs (Fig. 4b), suggesting that core layer KPs that are shared between pathways associate with scaffold proteins through LBMs. Indeed, although core and bottom layers are enriched for potential LBMs, only the core layer is enriched for scaffold-associated KPs (Fig. 4e, $P = 2 \times 10^{-4}$; HT). This indicates that scaffolding is extensively employed at the core layer where most pathway crosstalk occurs (Fig. 3d–e), in order to prevent inappropriate cellular responses resulting from the activation of undesired pathways. For instance, the mitogen extracellular signal-regulated kinase kinase Ste11, a core layer kinase, is involved in three pathways: high osmolarity, filamentous growth and pheromone pathway. Association of Pbs2 (a MAPK kinase and a scaffold protein implicated in the HOG signalling pathway) and Ste5 (a pheromone-responsive MAPK scaffold protein) with Ste11 reorients signal flow by activating the HOG signalling pathway and the mating pathway, respectively; whereas, unavailability of both Pbs2 and Ste5 favours filamentous growth [37].

Core layer KPs undergo more spatial organization changes than top and bottom layer KPs

Controlling spatial distribution of KPs plays an essential role in tuning KP activity and specificity towards their substrates [38, 39]. By superposing microscopic subcellular localization data of proteins in single cells under different stress conditions [40] on top of the KP-Net hierarchy, we observed that KPs in the core layer dynamically redistribute among more subcellular compartments than KPs in top and bottom layers (Fig. 4f, $P < 1.6 \times 10^{-3}$; RT, Methods). This indicates that core layer KPs might be subject to a more stringent control than top and bottom layer KPs to tightly restrict their localization. Hog1 is a relevant example of a core layer kinase that is translocated from the cytoplasm to the nucleus to trigger a wide transcriptional response on exposure to a high osmolarity stimulus [41]. Another typical example of tight localization control is Cdc14, a core layer phosphatase essential for mitotic exit, which after its sequestration in the nucleolus, is released to the nucleus and the cytoplasm where it associates with the spindle pole body during early anaphase [42].

Top layer KP proteins are more abundant and less noisy than bottom layer KPs of the KP-Net

Since KPs turnover determines their availability and thus their activity, we overlaid various information of KP turnover taken from the literature (Additional file 1: Supplementary Materials) on top of the KP-Net hierarchy [43–49]. While transcripts coding for core layer KPs are synthesized at a higher rate than top and bottom layers (Fig. 5a, $P < 3.9 \times 10^{-3}$; RT, Methods), mRNA of top layer KPs have longer half-lives than core and bottom layers (Fig. 5b, $P < 4.6 \times 10^{-3}$; RT, Methods). However, mRNA abundance has a similar trend to mRNA half-life, implying that mRNA degradation (the process that determines half-lives) is more important than synthesis rate in determining mRNA abundance (Fig. 5c, $P < 1.8 \times 10^{-2}$; RT, Methods). Similarly, mRNA of top layer KPs are translated at higher rates than core and bottom layers (Fig. 5d, $P < 4.8 \times 10^{-2}$; RT, Methods). However, half-lives of KP proteins are statistically comparable among the three layers of the KP-Net (Fig. 5e; RT,

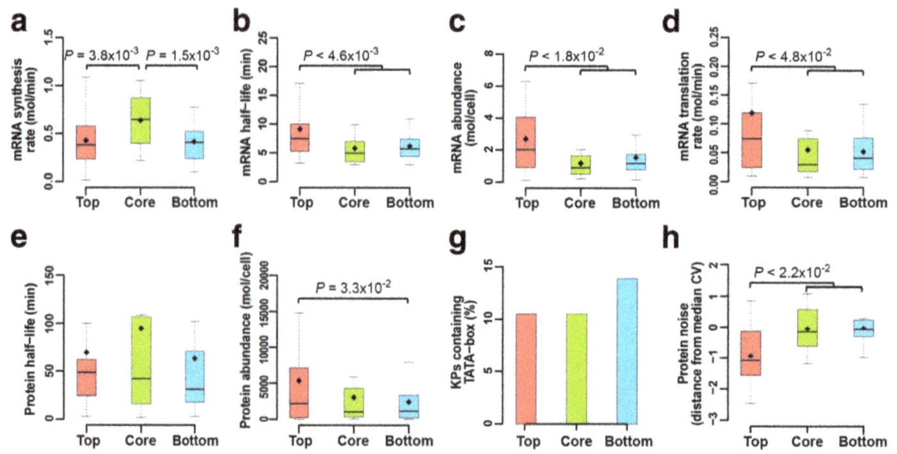

Fig. 5 mRNA and protein turnover related properties of KPs in the different layers of the KP-Net. Distribution of (**a**) mRNA synthesis rate, (**b**) mRNA half-life, (**c**) mRNA abundance, (**d**) mRNA translation rate, (**e**) protein half-life, (**f**) protein abundance, (**g**) percentage of noisy mRNA KPs and (**h**) distribution of noise in KP protein abundance of KPs in the different layers of the KP-Net. A KP is considered to be noisy at the transcriptomic level, if the promoter region of its gene was predicted to contain a TATA-box consensus sequence. Protein noise was defined as the distance of coefficient of variation (CV) of protein abundance from a running median of protein abundance CV. For description of box plots and bar plots, see Fig. 3 and for description of used datasets see Supplementary Materials in Additional file 1

Methods), suggesting that proteins abundance should have the same trend as the translation rate of mRNA molecules. This is partially true, since top layer KP proteins are more abundant than the bottom layer (Fig. 5f, $P = 3.3 \times 10^{-2}$; RT, Methods), but not more abundant than the core layer. This discrepancy might be due to the fact that KP proteins in the core layer tend to have longer half-lives (mean values are reported; 95 min, Fig. 5e) than the top layer (69 min, Fig. 5e). On another level, percentages of noisy KP genes at the mRNA level are comparable among the three KP-Net layers (Fig. 5g; HT). Moreover, top layer KP proteins are less noisy than core and bottom layers in starving *S. cerevisiae* cells (Fig. 5h, $P < 2.2 \times 10^{-2}$; RT, Methods). Interestingly although, we observed significant relative differences in each of protein abundance and noise between KP-Net layers, notably proteins were abundant (Fig. 5f, top 5,336 molecules/cell, core 3,041 molecules/cell and bottom 2,436 molecules/cell) and not noisy (Fig. 5h, top -0.94 a.u., core -0.05 a.u. and bottom 0.12 a.u.) in the three layers of the KP-Net. Taken together, these results suggest that higher protein abundance coupled with lower protein noise in the three layers and in particular in the top layer, might confer high signalling fidelity to the KP-Net.

The VS algorithm depends on node degree to classify network nodes in three layers

As the findings of this study mainly result from the application of the VS algorithm, we asked whether the VS algorithm depends on a specific node property to sort nodes into three layers and whether these findings reflect the biology underlying the KP-Net. To address these questions, we generated five sets of 1,000 random networks produced using five randomization methods: degree preserving randomization (DPR), similar degree preserving randomization (SDPR), in-degree preserving randomization (IDPR), out-degree preserving randomization (ODPR), and degree non-preserving randomization (DNPR) (Methods). We then applied the VS algorithm on these random networks and plotted means of KP properties in each layer of the KP-Net (black diamonds, Fig. 6), means of KP properties in each layer of random networks (points joined by coloured lines, Fig. 6) and the 95% confidence interval of random network means (coloured vertical segments, Fig. 6).

Strikingly, we observed that the distribution of all properties, except in-degrees, hubs and bottlenecks, of the three layers form a straight horizontal line for DNPR networks (Fig. 6, black line), showing that the VS algorithm produces a particular global signature (they peak

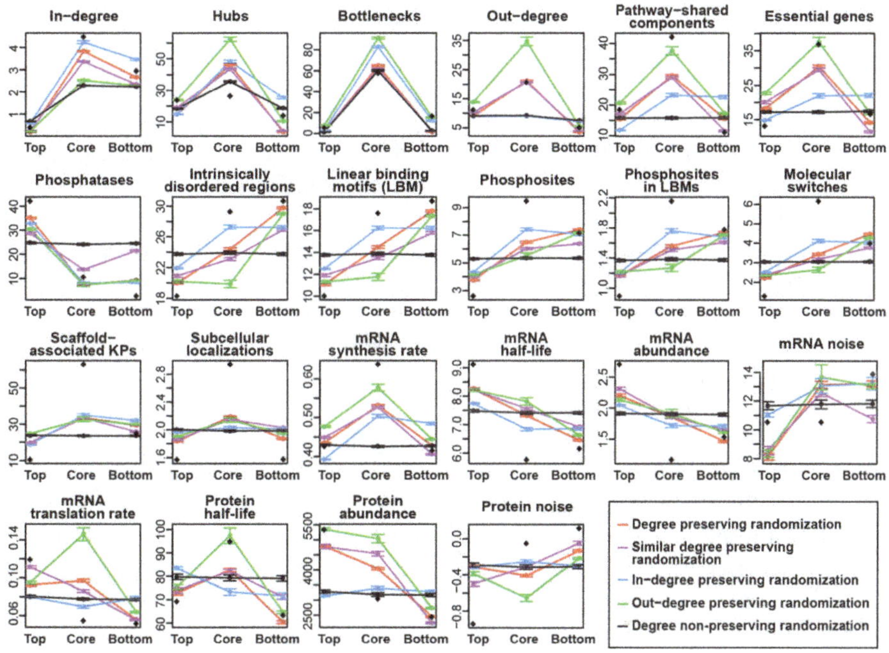

Fig. 6 The VS algorithm depends on node degrees to sort network nodes into three layers. The mean and its 95% confidence interval of the studied properties of KPs in the three layers of each of the five sets of the 1,000 random networks generated by: degree preserving randomization (DPR, *red line*), similar degree preserving randomization (SDPR, *pink line*), in-degrees preserving randomization (IDRP, *blue line*), out-degree preserving randomization (ODRP, green line) and degree non-preserving randomization (DNPR, *black line*). The black diamonds represent the mean of studied properties of KPs in the three layers of the KP-Net

at the core layer) in completely random networks for only these three properties that are all related to node degrees. Interestingly, the distribution of all properties in the DPR and SDPR networks (red and pink lines, Fig. 6) are the closest to each other when node degrees are similar to each other (DPR and SDPR cluster together in Additional file 1: Figure S4). Taken together, our observations suggest that the VS algorithm depends on node degree to sort network nodes in the different layers. Moreover, on clustering the five sets of randomized networks using the Euclidean distance between the different properties of their KPs, we found that ODPR networks are closer to DPR networks than IDPR networks (Additional file 1: Figure S4), suggesting that the VS algorithm depends on node out-degrees more than node in-degrees. However, the VS algorithm obviously depends also on node in-degrees, as any node with a zero in-degree will be automatically placed in the top layer. Therefore, the VS algorithm depends on both nodes in- and out-degrees. Nevertheless, although the VS algorithm depends on node degrees to classify network nodes into different layers, three observations suggest that KP biological properties are not associated with KP degrees and that they are not the result of a bias in the VS algorithm: (i) all biological properties showed a straight line distribution in completely random networks (Fig. 6, black line); (ii), most of the means of KP biological properties in KP-Net layers (black diamonds, Fig. 6) are outside of the 95% confidence interval of the means of the corresponding properties in random network layers; and (iii), most of the KP biological properties (12 out of 18) are neither associated with their in- nor with their out-degrees (Additional file 1: Supplementary methods).

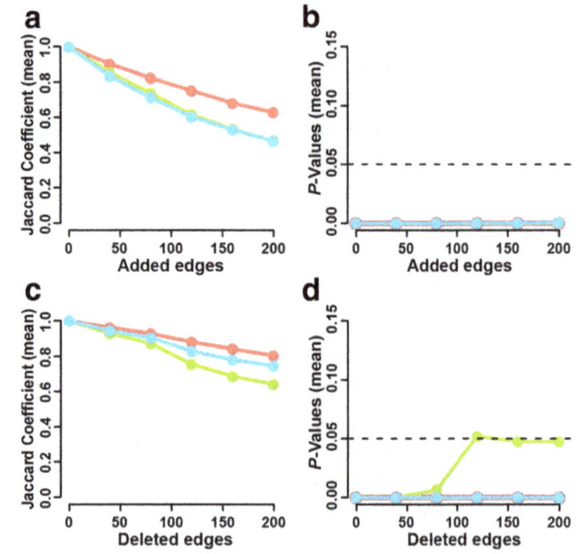

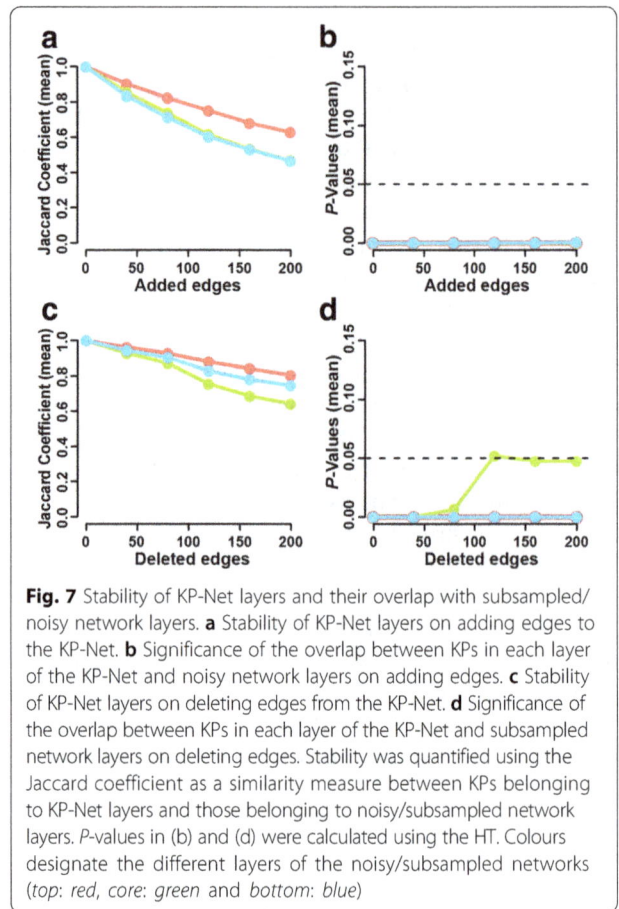

Fig. 7 Stability of KP-Net layers and their overlap with subsampled/noisy network layers. **a** Stability of KP-Net layers on adding edges to the KP-Net. **b** Significance of the overlap between KPs in each layer of the KP-Net and noisy network layers on adding edges. **c** Stability of KP-Net layers on deleting edges from the KP-Net. **d** Significance of the overlap between KPs in each layer of the KP-Net and subsampled network layers on deleting edges. Stability was quantified using the Jaccard coefficient as a similarity measure between KPs belonging to KP-Net layers and those belonging to noisy/subsampled network layers. *P*-values in (b) and (d) were calculated using the HT. Colours designate the different layers of the noisy/subsampled networks (*top*: *red*, *core*: *green* and *bottom*: *blue*)

Robustness of results and incompleteness of data

It did not escape our attention that the KP-Net that was assembled in this study represents a small snapshot of the whole phosphorylation network of the budding yeast. Therefore, we assessed the robustness of our results to missing interactions by generating noisy networks (adding edges to the KP-Net) and the robustness of our results to false positives by generating subsampled networks (deleting edges from the KP-Net) (Methods). We then assessed the stability of KP-Net layers using the Jaccard coefficient as a measure of similarity between KP-Net layers and noisy/subsampled network layers (Methods) [50]. Also, we assessed the significance of the overlap between KP-Net layers and noisy/subsampled network layers using the HT (Methods) [50]. We observed that the KP-Net is more robust to removing than to adding edges (Fig. 7a and c). Moreover, the more edges are added to and removed from the KP-Net, the more the three layers become unstable (Fig. 7a and c). However, in spite of this instability, all layers in noisy/subsampled networks significantly overlap with the KP-Net layers (Fig. 7b and d), showing that our findings are sufficiently robust to describe the KP-Net with our current knowledge. Finally, properties characterizing the KP-Net were retained to different degrees in the noisy networks (Additional file 1: Supplementary Methods and Figure S9), confirming that the characteristics of the KP-Net elucidated in this study represent the best of our knowledge to date about KP-Nets.

Using the KP-Net as a gold standard to predict kinases acting on substrates in the HOG pathway

Presently, one of the most active areas of research consists of linking each KP to its substrates. As an example, we attempted to predict the kinases that could phosphorylate substrates characterized by a change in their level of phosphorylation in cells exposed to osmotic shock. We used the KP-Net as a gold standard; we overlaid on top of it phosphorylation consensus motifs curated from the literature and proteins that undergo time-dependent phosphorylation or dephosphorylation following osmotic shock from Kanshin & Bergeron-Sandoval et al. [23]. We identified 57 interactions linking 19 kinases to 25 potential substrates (Methods and Additional file 3). The overlap between the predicted kinases in our study and the kinases that underwent changes in phosphorylation in Kanshin & Bergeron-Sandoval et al. was significant ($P = 3.8 \times 10^{-2}$; HT). This result suggests, first, that a significant number of the 19 kinases that we predicted to act on 25 potential substrates do undergo time-dependent changes in phosphorylation that may reflect their activation or deactivation in response to osmotic shock; second, that the interactions forming the KP-Net that was assembled in this study are of high confidence; and finally, that this same KP-Net could be used as a benchmark with other phosphoproteomic data to identify kinases and perhaps phosphatases that act on a set of substrates.

Discussion

In this study, we assembled the largest bona fide KP-Net known to date for the yeast *Saccharomyces cerevisiae.* We found, first, that the KP-Net has a moderate hierarchical structure made of three layers (top, core and bottom) in the form of a bow tie structure having a strongly connected core layer. Second, phosphatases are for the first time shown to be less directly regulated by kinases than are kinases by each other. Third, the observed high abundance and low noise of KP proteins in the three layers of the KP-Net, but notably in the top layer, may reflect an adaptation by which maximal sensitivity to signals at the earliest steps of signalling is assured. Finally, the tight temporal and spatial regulation that we observed for the core layer of the KP-Net could be explained by both the high load of signals received by this layer and its enrichment for KPs implicated in cell cycle and decision-making.

Recently, Cheng et al. overlaid many of the biological properties studied here on top of a kinase network assembled from in vitro phosphorylation interactions in the budding yeast (Additional file 1: Table S2) [7]. In contrast to our findings, most of the examined biological properties by Cheng et al. were statistically comparable among the three layers (gene essentiality, abundance, half-life and noise on mRNA and protein levels). It is important to note that properties of each layer depend on the identity and the properties of the proteins belonging to each layer. Difference between findings of Cheng et al. and those of this study might be due to the following reasons: (i), the lack of phosphatases in the network analysed by Cheng et al.; (ii), the high number of false positives that normally exist in any data generated in vitro, which could affect sorting of nodes in the different layers and thus directly affect layer properties; (iii), the application of a decomposition method differing from the VS algorithm, (iv), or a combination of all these reasons. Interestingly though, protein noise results of Cheng et al. concord partially with our findings as proteins in the top layer were less noisy than those in the bottom layer.

A limitation of the KP-Net generated in this study is that it cannot be used to predict novel PDIs or pathways. Note, however, that this was not among the objectives of this study. The KP-Net can serve as a gold standard in future investigations of signalling networks to suggest a set of KP candidates that might act on substrates under a given condition, as we showed in predicting kinases that act on substrates following osmotic stress. Another limitation is that although the choice of the largest SCC to represent the core layer was subjective and inspired by previous application of the VS algorithm to a transcription regulatory network, we can justify the validity of our choice by the concordance of our observations with those in the literature [8]. In the literature, a core layer of a bow tie structure is usually associated with critical decisions determining the system outputs [51]. This concords with our findings showing that 79% of the core layer KPs are implicated in cell cycle and decision-making processes, to note that the VS algorithm does not necessarily generate a bow tie structure as in reference [8] (Figure 2a; Additional file 1: Table S3). Finally, the assembled KP-Net represents a small snapshot of the real-world KP-Net affecting 60% of the proteome. Advances in high throughput technologies should eventually complete the KP-Net by unravelling missing PDIs. As with any network reconstruction exercise, there is the risk that a different sorting of KPs within the KP-Net hierarchical structure could lead to different interpretations of the KP-Net. However, when we randomly added edges to the KP-Net in order to create "noisy networks", we observed that the layers of the noisy KP-Nets became less stable by adding more edges; but at the same time, they overlap significantly with KP-Net layers (Fig. 7a and b). These results show that the properties of the KP-Net layers are robust to describe how the KP-Net functions with the best of our current knowledge, which represents the principal objective of this study.

Despite the limitations mentioned above, the functional principles of the KP-Net that are proposed in this study are consistent with other observations. Interestingly, bow tie structures are frequently associated with robustness against removal of some of their components and to external perturbations [51–54]. Robustness of the KP-Net bow tie structure could be ensured by the following factors. First, the degeneracy (overlapping functions) of many KPs in the top layer [e.g. PKAs, Tel1-Mec1 and calcineurins, (Fig. 1b)] guaranties that failure of a KP to activate a given pathway is buffered by another KP having partially redundant functions [51]. Notably, the degeneracy observed in top layer KPs concords well with the low number (13%) of KPs encoded by essential genes belonging to this layer. Second, the core layer possesses the required features for generating coordinated responses: (i), it receives and integrates various inputs (high node in-degrees and enrichment for bottlenecks, pathway-shared components, and scaffold-associated KPs (Figs. 3a, d, e and 4e); (ii), it occupies a central position in the hierarchy (Fig. 1b); (iii), it is involved in critical tasks (cell cycle and decision-making) (Fig. 2a and Additional file 1: Table S3); and most importantly, (iv) it is highly regulated at different levels in time and space. Without such a tightly regulated layer, coordinated responses would necessitate ample individual controls and any misregulation of the latter controls would easily impair cellular survival [53]. All these characteristics contribute in delineating functional principles of the KP-Net as known to date.

Conclusions

In this study, we built a KP-Net assembled from high quality PDIs in the budding yeast, determined its hierarchical structure and integrated the widest range of KP biological properties with elucidated hierarchical structure. This allowed us to formulate hypotheses about the functions of the KP-Net layers. As mentioned previously, the KP-Net assembled in this study represents a snapshot of the KP-Net that exists in the budding yeast. Advances in large-scale screens, in particular those exploring substrates of KPs will enhance coverage of the assembled KP-Net. Also, with the enhancement of high throughput technologies, integration of other type of biological properties, such as methylation, ubiquitination, and temporal PDIs, with the KP-Net might become possible, which could reveal new functional principles of the KP-Net. A better perception of how the KP-Net functions could also open new opportunities to understand the actions of KP inhibitors on normal and pathological processes such as cancers.

Methods

Over-representation of various logic motifs in the KP-Net

One thousand random networks were generated by degree preserving randomization (DPR, Methods). Each of the random networks was sorted by the VS algorithm and the number of its feed-forward loops, feedback loops, and bi-fan logic motifs was assessed. The *P*-value is the fraction of times the number of each logic motif in random networks is as large as that in the KP-Net.

Network randomization

In this study, we randomized the KP-Net using five types of network randomizations:

Degree preserving randomization (DPR): we randomly selected two edges of the KP-Net and exchanged their ends. We then removed multiple edges having the same direction between two nodes by switching each of them with randomly selected edges. The rewiring procedure was repeated 10,000 times to each random network.

Similar degree preserving randomization (SDPR): we used the matching algorithm (Methods) to generate random graphs having similar degree distributions to that of the KP-Net [55–57]. We then switched network edges using the first randomization method (DPR) to make sure that the generated random networks differ from each other.

In-degree preserving randomization (IDPR): interactions were represented as a table made of two columns: "from" and "to". We recreated the "from" column by randomly selecting KPs with replacement. We then switched network edges using the first randomization method (DPR).

Out-degree preserving randomization (ODPR): we recreated the "to" column by randomly selecting KPs with replacement. We then switched network edges using the first randomization method (DPR).

Degree non-preserving randomization (DNPR): we created a random network from scratch by connecting two nodes that were randomly selected with replacement.

The matching algorithm

In order to generate networks having a degree distribution that is similar to that of the KP-Net, we defined the degree distribution of the random network by randomly selecting three groups of KPs: the first group had the same in- and out-degrees as the KP-Net and the second and third groups had the same in- and out-degrees as the KP-Net, but incremented and decremented by 1, respectively. Second, we connected the network using a variant of the matching algorithm [58]. Briefly, each vertex of the random network was assigned a number of in- and out-stubs equal to its in- and out-degrees. In- and out-stubs were selected in pairs and joined up to make the network edges. In each step, the selection of in- and out-stubs was weighted by the square of the current in- and out-stubs that were not yet connected but should be. This procedure produced random networks that have very similar in- and out-degrees distributions to those of the KP-Net.

Testing whether the KP-Net GRC is bigger than Erdős–Rényi network GRCs

We generated 10,000 Erdős–Rényi random networks having the same number of nodes and edges as the KP-Net and calculated their GRCs [21]. The *P*-value of this test is the proportion of random network GRCs that are as large as the KP-Net GRC.

Comparing means of node properties in two layers using RT

Let L1 and L2 be the size of two layers of the KP-Net to be compared; S the set containing the nodes of these layers; S1 the set of L1 nodes randomly sampled without replacement from S; S2 the set of the remaining L2 nodes in S after sampling. The difference between the means of the node properties in S1 and S2 were calculated. These steps were repeated 10,000 times. The *P-value* is equal to the proportion of times the difference between the means of the sampled sets are as big/small as the difference between the means of the two compared layers.

Generating subsampled/noisy networks and assessing their layers stability and their overlap with KP-Net layers

We generated ten sets of 100 subsampled and noisy networks from the KP-Net. The five sets of subsampled/noisy networks were produced by randomly removing/adding 40, 80, 120, 160 and 200 edges to the KP-Net, respectively. These steps were repeated 100 times, so that each set contains 100 subsampled/noisy networks. We then applied the VS algorithm to each subsampled/noisy network to identify their three layers. Layer stability of the generated networks was assessed using the Jaccard coefficient as a similarity "cluster wise" measure between the original and the subsampled/noisy layers [50]. Overlap between original and subsampled/noisy layers were assessed using the hypergeometric test (HT).

Predicting kinases

First, we identified substrates in the KP-Net that contain a phosphorylated residue modulated by time after osmotic shock defined as dynamic phosphosite by Kanshin and Bergeron-Sandoval et al. [23]. We also identified the consensus phosphorylation motifs of each kinase in the KP-Net when possible from the literature (Additional file 4). We then connected each substrate containing a dynamic phosphosite to all kinases having a consensus motif matching the substrate phosphosite by edges to form kinase-substrate interactions. Using the KP-Net as a gold standard network, we retained kinase-substrate interactions that occur in the KP-Net.

Additional files

Additional file 1: Supplementary materials, supplementary methods, supplementary figures and supplementary tables.

Additional file 2: KP-Net phosphorylation and dephosphorylation interactions: Phosphorylation and dephosphorylation interactions (PDI) that were included in the KP-Net and the experimental methods that validated them with the Pubmed references of the articles in which PDIs were reported. The KID database pipeline was used to score and annotate these interactions (Additional file 1: Supplementary Methods).

Additional file 3: Predicted kinases implicated in the HOG pathway: The kinase-substrate interactions that were predicted to be implicated in osmotic shock in this study.

Additional file 4: Kinase consensus phosphorylation sites: The consensus phosphorylation sites of kinases and their evidence in the literature.

Abbreviations

CV: Coefficient of variation; DNPR: Degree non-preserving randomization; DPR: Degree preserving randomization; GO: Gene ontology; GRC: Global reaching centrality; HT: Hypergeometric test; IDPR: In-degree preserving randomization; KID: Kinase interaction database; KP: Kinase and phosphatase; KP-Net: Kinase-phosphatase network; LBM: Linear binding motif; ODPR: Out-degree preserving randomization; PDI: Phosphorylation and dephosphorylation interaction; RT: Randomization test; SCC: Strongly connected component; SDPR: Similar degree preserving randomization; VS: Vertex sort

Acknowledgements

The authors would like to thank Raja Jothi for helpful and instructive exchange, Abdelalli Kelil and Emmanuel Levy for helpful comments and stimulating discussions.

Funding

This work was supported by the Canadian Institutes of Health Research [MOP-GMX-152556 and MOP-GMX-231013]. Funding for open access charge: [Canadian Institutes of Health Research/MOP-GMX-152556]. The CIHR played no role in the design of the study and collection, analysis, and interpretation of data and in writing the manuscript.

Authors' contributions

DAR and SM conceived the study. DAR performed the research and the bioinformatics analyses and wrote the manuscript. SM corrected the manuscript. Both authors read and approved the final manuscript.

Competing interests

The authors declare that they have no competing interests.

References

1. Novak B, Kapuy O, Domingo-Sananes MR, Tyson JJ. Regulated protein kinases and phosphatases in cell cycle decisions. Curr Opin Cell Biol. 2010; 22(6):801–8. doi:10.1016/j.ceb.2010.07.001.
2. Fiedler D, Braberg H, Mehta M, Chechik G, Cagney G, Mukherjee P, et al. Functional organization of the S. cerevisiae phosphorylation network. Cell. 2009;136(5):952–63. doi:10.1016/j.cell.2008.12.039.
3. Breitkreutz A, Choi H, Sharom JR, Boucher L, Neduva V, Larsen B, et al. A global protein kinase and phosphatase interaction network in yeast. Science. 2010;328(5981):1043–6. doi:10.1126/science.1176495.
4. Bodenmiller B, Wanka S, Kraft C, Urban J, Campbell D, Pedrioli PG, et al. Phosphoproteomic analysis reveals interconnected system-wide responses to perturbations of kinases and phosphatases in yeast. Sci Signal. 2010; 3(153):rs4. doi:10.1126/scisignal.2001182.

5. Bhardwaj N, Yan KK, Gerstein MB. Analysis of diverse regulatory networks in a hierarchical context shows consistent tendencies for collaboration in the middle levels. Proc Natl Acad Sci U S A. 2010;107(15):6841–6. doi:10.1073/pnas.0910867107.
6. Cheng C, Andrews E, Yan KK, Ung M, Wang D, Gerstein M. An approach for determining and measuring network hierarchy applied to comparing the phosphorylome and the regulome. Genome Biol. 2015;16:63. doi:10.1186/s13059-015-0624-2.
7. Ptacek J, Devgan G, Michaud G, Zhu H, Zhu X, Fasolo J, et al. Global analysis of protein phosphorylation in yeast. Nature. 2005;438(7068):679–84. doi:10.1038/nature04187.
8. Jothi R, Balaji S, Wuster A, Grochow JA, Gsponer J, Przytycka TM, et al. Genomic analysis reveals a tight link between transcription factor dynamics and regulatory network architecture. Mol Syst Biol. 2009;5:294. doi:10.1038/msb.2009.52.
9. Yu H, Gerstein M. Genomic analysis of the hierarchical structure of regulatory networks. Proc Natl Acad Sci U S A. 2006;103(40):14724–31. doi:10.1073/pnas.0508637103.
10. Gulsoy G, Bandhyopadhyay N, Kahveci T. HIDEN: Hierarchical decomposition of regulatory networks. BMC Bioinf. 2012;13:250. doi:10.1186/1471-2105-13-250.
11. Ma HW, Buer J, Zeng AP. Hierarchical structure and modules in the Escherichia coli transcriptional regulatory network revealed by a new top-down approach. BMC Bioinf. 2004;5:199. doi:10.1186/1471-2105-5-199.
12. Gerstein MB, Kundaje A, Hariharan M, Landt SG, Yan KK, Cheng C, et al. Architecture of the human regulatory network derived from ENCODE data. Nature. 2012;489(7414):91–100. doi:10.1038/nature11245.
13. Kim D, Kim MS, Cho KH. The core regulation module of stress-responsive regulatory networks in yeast. Nucleic Acids Res. 2012;40(18):8793–802. doi:10.1093/nar/gks649.
14. Sharifpoor S, Nguyen Ba AN, Young JY, van Dyk D, Friesen H, Douglas AC, et al. A quantitative literature-curated gold standard for kinase-substrate pairs. Genome Biol. 2011;12(4):R39. doi:10.1186/gb-2011-12-4-r39.
15. Stark C, Breitkreutz BJ, Reguly T, Boucher L, Breitkreutz A, Tyers M. BioGRID: a general repository for interaction datasets. Nucleic Acids Res. 2006; 34(Database issue):D535–9. doi:10.1093/nar/gkj109.
16. Stark C, Su TC, Breitkreutz A, Lourenco P, Dahabieh M, Breitkreutz BJ, et al. PhosphoGRID: a database of experimentally verified in vivo protein phosphorylation sites from the budding yeast Saccharomyces cerevisiae. Database (Oxford). 2010;2010:bap026. doi:10.1093/database/bap026.
17. Magrane M, Consortium U. UniProt Knowledgebase: a hub of integrated protein data. Database (Oxford). 2011;2011:bar009. doi:10.1093/database/bar009.
18. Ba AN, Moses AM. Evolution of characterized phosphorylation sites in budding yeast. Mol Biol Evol. 2010;27(9):2027–37. doi:10.1093/molbev/msq090.
19. Muller HM, Kenny EE, Sternberg PW. Textpresso: an ontology-based information retrieval and extraction system for biological literature. PLoS Biol. 2004;2(11):e309. doi:10.1371/journal.pbio.0020309.
20. Institute EB. CiteXplore. http://www.ebi.ac.uk/web_guidelines/html/mitigation/frontier_test_02.html. Accessed 21 Dec 2013.
21. Mones E, Vicsek L, Vicsek T. Hierarchy measure for complex networks. PLoS ONE. 2012;7(3):e33799. doi:10.1371/journal.pone.0033799.
22. Hartsperger ML, Strache R, Stumpflen V. HiNO: an approach for inferring hierarchical organization from regulatory networks. PLoS ONE. 2010;5(11):e13698. doi:10.1371/journal.pone.0013698.
23. Kanshin E, Bergeron-Sandoval LP, Isik SS, Thibault P, Michnick SW. A cell-signaling network temporally resolves specific versus promiscuous phosphorylation. Cell Rep. 2015;10(7):1202–14. doi:10.1016/j.celrep.2015.01.052.
24. Shi Y. Serine/threonine phosphatases: mechanism through structure. Cell. 2009;139(3):468–84. doi:10.1016/j.cell.2009.10.006.
25. Trockenbacher A, Suckow V, Foerster J, Winter J, Krauss S, Ropers HH, et al. MID1, mutated in Opitz syndrome, encodes an ubiquitin ligase that targets phosphatase 2A for degradation. Nat Genet. 2001;29(3):287–94. doi:10.1038/ng762.
26. Mitchell DA, Sprague Jr GF. The phosphotyrosyl phosphatase activator, Ncs1p (Rrd1p), functions with Cla4p to regulate the G(2)/M transition in Saccharomyces cerevisiae. Mol Cell Biol. 2001;21(2):488–500. doi:10.1128/MCB.21.2.488-500.2001.
27. Trinkle-Mulcahy L, Andersen J, Lam YW, Moorhead G, Mann M, Lamond AI. Repo-Man recruits PP1 gamma to chromatin and is essential for cell viability. J Cell Biol. 2006;172(5):679–92. doi:10.1083/jcb.200508154.
28. Maeda T, Tsai AY, Saito H. Mutations in a protein tyrosine phosphatase gene (PTP2) and a protein serine/threonine phosphatase gene (PTC1) cause a synthetic growth defect in Saccharomyces cerevisiae. Mol Cell Biol. 1993;13(9):5408–17.
29. Ahn JH, McAvoy T, Rakhilin SV, Nishi A, Greengard P, Nairn AC. Protein kinase A activates protein phosphatase 2A by phosphorylation of the B56delta subunit. Proc Natl Acad Sci U S A. 2007;104(8):2979–84. doi:10.1073/pnas.0611532104.
30. Janssens V, Longin S, Goris J. PP2A holoenzyme assembly: in cauda venenum (the sting is in the tail). Trends Biochem Sci. 2008;33(3):113–21. doi:10.1016/j.tibs.2007.12.004.
31. Van Roey K, Gibson TJ, Davey NE. Motif switches: decision-making in cell regulation. Curr Opin Struct Biol. 2012;22(3):378–85. doi:10.1016/j.sbi.2012.03.004.
32. Dosztanyi Z, Csizmok V, Tompa P, Simon I. The pairwise energy content estimated from amino acid composition discriminates between folded and intrinsically unstructured proteins. J Mol Biol. 2005;347(4):827–39. doi:10.1016/j.jmb.2005.01.071.
33. Meszaros B, Simon I, Dosztanyi Z. Prediction of protein binding regions in disordered proteins. PLoS Comput Biol. 2009;5(5):e1000376. doi:10.1371/journal.pcbi.1000376.
34. Clotet J, Escote X, Adrover MA, Yaakov G, Gari E, Aldea M, et al. Phosphorylation of Hsl1 by Hog1 leads to a G2 arrest essential for cell survival at high osmolarity. EMBO J. 2006;25(11):2338–46. doi:10.1038/sj.emboj.7601095.
35. Harvey SL, Charlet A, Haas W, Gygi SP, Kellogg DR. Cdk1-dependent regulation of the mitotic inhibitor Wee1. Cell. 2005;122(3):407–20. doi:10.1016/j.cell.2005.05.029.
36. Pawson T, Scott JD. Signaling through scaffold, anchoring, and adaptor proteins. Science. 1997;278(5346):2075–80.
37. Schwartz MA, Madhani HD. Principles of MAP kinase signaling specificity in Saccharomyces cerevisiae. Annu Rev Genet. 2004;38:725–48. doi:10.1146/annurev.genet.39.073003.112634.
38. Ubersax JA, Ferrell Jr JE. Mechanisms of specificity in protein phosphorylation. Nat Rev Mol Cell Biol. 2007;8(7):530–41. doi:10.1038/nrm2203.
39. Mattison CP, Ota IM. Two protein tyrosine phosphatases, Ptp2 and Ptp3, modulate the subcellular localization of the Hog1 MAP kinase in yeast. Genes Dev. 2000;14(10):1229–35.
40. Chong YT, Koh JL, Friesen H, Duffy K, Cox MJ, Moses A, et al. Yeast Proteome Dynamics from Single Cell Imaging and Automated Analysis. Cell. 2015;161(6):1413–24. doi:10.1016/j.cell.2015.04.051.
41. Muzzey D, Gomez-Uribe CA, Mettetal JT, van Oudenaarden A. A systems-level analysis of perfect adaptation in yeast osmoregulation. Cell. 2009; 138(1):160–71. doi:10.1016/j.cell.2009.04.047.
42. Bloom J, Cristea IM, Procko AL, Lubkov V, Chait BT, Snyder M, et al. Global analysis of Cdc14 phosphatase reveals diverse roles in mitotic processes. J Biol Chem. 2011;286(7):5434–45. doi:10.1074/jbc.M110.205054.
43. Arava Y, Wang Y, Storey JD, Liu CL, Brown PO, Herschlag D. Genome-wide analysis of mRNA translation profiles in Saccharomyces cerevisiae. Proc Natl Acad Sci U S A. 2003;100(7):3889–94. doi:10.1073/pnas.0635171100.
44. Wang M, Weiss M, Simonovic M, Haertinger G, Schrimpf SP, Hengartner MO, et al. PaxDb, a database of protein abundance averages across all three domains of life. Mol Cell Proteomics. 2012;11(8):492–500. doi:10.1074/mcp.O111.014704.
45. Belle A, Tanay A, Bitincka L, Shamir R, O'Shea EK. Quantification of protein half-lives in the budding yeast proteome. Proc Natl Acad Sci U S A. 2006; 103(35):13004–9. doi:10.1073/pnas.0605420103.
46. Newman JR, Ghaemmaghami S, Ihmels J, Breslow DK, Noble M, DeRisi JL, et al. Single-cell proteomic analysis of S. cerevisiae reveals the architecture of biological noise. Nature. 2006;441(7095):840–6. doi:10.1038/nature04785.
47. Eser P, Demel C, Maier KC, Schwalb B, Pirkl N, Martin DE, et al. Periodic mRNA synthesis and degradation co-operate during cell cycle gene expression. Mol Syst Biol. 2014;10:717. doi:10.1002/msb.134886.
48. Basehoar AD, Zanton SJ, Pugh BF. Identification and distinct regulation of yeast TATA box-containing genes. Cell. 2004;116(5):699–709.
49. Miura F, Kawaguchi N, Yoshida M, Uematsu C, Kito K, Sakaki Y, et al. Absolute quantification of the budding yeast transcriptome by means of competitive PCR between genomic and complementary DNAs. BMC Genomics. 2008;9:574. doi:10.1186/1471-2164-9-574.
50. Henning C. Cluster-wise assessment of cluster stability. Comput Stat Data An. 2007;52(1):258–71. doi:10.1016/j.csda.2006.11.025.
51. Tieri P, Grignolio A, Zaikin A, Mishto M, Remondini D, Castellani GC, et al. Network, degeneracy and bow tie. Integrating paradigms and architectures

to grasp the complexity of the immune system. Theor Biol Med Model. 2010;7:32. doi:10.1186/1742-4682-7-32.
52. Whitacre JM. Biological robustness: paradigms, mechanisms, and systems principles. Front Genet. 2012;3:67. doi:10.3389/fgene.2012.00067.
53. Kitano H. Biological robustness. Nat Rev Genet. 2004;5(11):826–37. doi:10.1038/nrg1471.
54. Ma HW, Zeng AP. The connectivity structure, giant strong component and centrality of metabolic networks. Bioinformatics. 2003;19(11):1423–30.
55. Molloy M, Reed B. A critical point for random graphs with a given degree sequence. Random Struct Algoritm. 1995;6:161–79.
56. Newman MEJ, Strogatz SH, Watts DJ. Random graphs with arbitrary degree distribution and their applications. Phys Rev. 2001;6:026118.
57. Milo R, Shen-Orr S, Itzkovitz S, Kashtan N, Chklovskii D, Alon U. Network motifs: simple building blocks of complex networks. Science. 2002; 298(5594):824–7. doi:10.1126/science.298.5594.824.
58. Milo R, Kashtan N, Itzkovitz S, Newman MEJ, Alon U. On the uniform generation of random graphs with prescribed degree sequences. arXiv: cond-mat/0312028v2 [cond-matstat-mech]. 2004.

3

Markov State Models of gene regulatory networks

Brian K. Chu[1], Margaret J. Tse[1], Royce R. Sato[1] and Elizabeth L. Read[1,2*]

Abstract

Background: Gene regulatory networks with dynamics characterized by multiple stable states underlie cell fate-decisions. Quantitative models that can link molecular-level knowledge of gene regulation to a global understanding of network dynamics have the potential to guide cell-reprogramming strategies. Networks are often modeled by the stochastic Chemical Master Equation, but methods for systematic identification of key properties of the global dynamics are currently lacking.

Results: The method identifies the number, phenotypes, and lifetimes of long-lived states for a set of common gene regulatory network models. Application of transition path theory to the constructed Markov State Model decomposes global dynamics into a set of dominant transition paths and associated relative probabilities for stochastic state-switching.

Conclusions: In this proof-of-concept study, we found that the Markov State Model provides a general framework for analyzing and visualizing stochastic multistability and state-transitions in gene networks. Our results suggest that this framework—adopted from the field of atomistic Molecular Dynamics—can be a useful tool for quantitative Systems Biology at the network scale.

Keywords: Multistable systems, Stochastic processes, Gene regulatory networks, Markov State Models, Cluster analysis

Background

Gene regulatory networks (GRNs) often have dynamics characterized by multiple attractor states. This multistability is thought to underlie cell fate-decisions. According to this view, each attractor state accessible to a gene network corresponds to a particular pattern of gene expression, i.e., a cell phenotype. Bistable network motifs with two possible outcomes have been linked to binary cell fate-decisions, including the lysis/lysogeny decision of bacteriophage lambda [1], the maturation of frog oocytes [2] and a cascade of branch-point decisions in mammalian cell development (reviewed in [3]). Multistable networks with three or more attractors have been proposed to govern diverse cell fate-decisions in tumorigenesis [4], stem cell differentiation and reprogramming [5–7], and helper T cell differentiation [8]. More generally, the concept of a rugged, high-dimensional epigenetic landscape connecting every possible cell type has emerged [9–11]. Quantitative models that can link molecular-level knowledge of gene regulation to a global understanding of network behavior have the potential to guide rational cell-reprogramming strategies. As such, there has been growing interest in the development of theory and computational methods to analyze global dynamics of multistable gene regulatory networks.

Gene expression is inherently stochastic [1, 12–14], and fluctuations in expression levels can measurably impact cell phenotypes and behavior. Numerous examples of stochastic phenotype transitions have been discovered, which diversify otherwise identical cell-populations. This spontaneous state-switching has been found to promote survival of microorganisms or cancer cells in fluctuating environments [15–17], prime cells to follow alternate developmental fates in higher eukaryotes [18, 19], and generate sustained heterogeneity (mosaicism) in a homeostatic mammalian cell population [20]. These findings have motivated theoretical studies of stochastic state-switching in

* Correspondence: elread@uci.edu
[1]Department of Chemical Engineering and Materials Science, University of California Irvine, Irvine, CA, USA
[2]Department of Molecular Biology and Biochemistry, University of California Irvine, Irvine, CA, USA

gene networks, which have shed light on network parameters and topologies that promote the stability (or instability) of a given network state [20]. Characterizing the global stability of states accessible to a network is akin to quantification of the "potential energy" landscape of a network. Particularly, with the advent of stem-cell reprogramming techniques, there has been renewed interest in a quantitative reinterpretation of Waddington's classic epigenetic landscape [21], in terms of underlying regulatory mechanisms [10, 22].

A number of mathematical frameworks exist for modeling and analysis of stochastic gene regulatory network (GRN) dynamics (reviewed in [23, 24]), including probabilistic Boolean Networks, Stochastic Differential Equations, and stochastic biochemical reaction networks (i.e., Chemical Master Equations). Of these, the Chemical Master Equation (CME) approach is the most complete, in that it treats all biomolecules in the system as discrete entities, fully accounts for stochasticity due to molecular-level fluctuations, and propagates dynamics according to chemical rate laws. The CME is analytically intractable for GRNs except in some simplified model systems [25–29], but trajectories can be simulated by Monte Carlo methods such as the Stochastic Simulation Algorithm (SSA) [30]. Alternatively, methods for reducing the dimensionality of the CME, enabling numerical approximation of network behavior by matrix methods, have been developed [31–35].

Analysis of multistability and global dynamics of discrete, stochastic GRN models remains challenging. In this study, we define multistability in stochastic systems as the existence of multiple peaks in the stationary probability distribution. In such systems, the GRN dynamics can be considered somewhat analogous to that of a particle in a multi-well potential [3]. (Peaks in the probability distribution—or alternatively, basins in the potential—may or may not correspond to stable fixed points of a corresponding ODE model, as discussed in more detail further on.) Stochastic multistability is often assessed by plotting multi-peaked steady-state probability distributions (obtained either from long stochastic simulations [5, 36, 37] or from approximate CME solutions [35, 38, 39]), projected onto one or two user-specified system coordinates. However, even small networks generally have more than two dimensions along which dynamics may be projected, meaning that inspection of steady-state distributions for a given projection may underestimate multistability in a network. For example, the state-space of a GRN may comprise different activity-states of promoters and regulatory sites on DNA, the copy-number of mRNA transcripts and encoded proteins, and the activity- or multimer-states of multiple regulatory molecules or proteins. Furthermore, while steady-state distributions give a global view of system behavior, they do not directly yield dynamic information of interest, such as the lifetimes of attractor states.

In this paper, we present an approach for analyzing multistable dynamics in stochastic GRNs based on a spectral clustering method widely applied in Molecular Dynamics [40, 41]. The output of the approach is a Markov State Model (MSM)—a coarse-grained model of system dynamics, in which a large number of system states (i.e., "microstates") is clustered into a small number of metastable (that is, relatively long-lived) "macrostates", together with the conditional probabilities for transitioning from one macrostate to another on a given timescale. The MSM approach identifies clusters based on separation of timescales, i.e., systems with multistability exhibit relatively fast transitions among microstates within basins and relatively slow inter-basin transitions. By neglecting fast transitions, the size of the system is vastly reduced. Based on its utility for visualization and analysis of Molecular Dynamics, the potential application of the MSM framework to diverse dynamical systems, including biochemical networks, has been discussed [42].

Biochemical reaction networks present an unexplored opportunity for the MSM approach. Herein, we applied the method to small GRN motifs and analyzed their global dynamics using two frameworks: the quasipotential landscape (based on the log-transformed stationary probability distribution), and the MSM. The MSM approach distilled network dynamics down to the essential stationary and dynamic properties, including the number and identities of stable phenotypes encoded by the network, the global probability of the network to adopt a given phenotype, and the likelihoods of all possible stochastic phenotype transitions. The method revealed the existence of network states and processes not readily apparent from inspection of quasipotential landscapes. Our results demonstrate how MSMs can yield insight into regulation of cell phenotype stability and reprogramming. Furthermore, our results suggest that, by delivering systematic coarse-graining of high-dimensional (i.e., many-species) dynamics, MSMs could find more general applications in Systems Biology, such as in signal-transduction, evolution, and population dynamics. In our implementation, the MSM framework is applied to the CME, thus mapping all enumerated molecular states onto long-lived system macrostates. We anticipate that the method could in future studies be used to analyze more complex systems where enumeration of the CME is intractable, if implemented in combination with stochastic simulation or other model reduction approaches.

Methods

Gene regulatory network motifs

We studied two common GRN motifs that are thought to control cell fate-decisions. The full lists of reactions and associated rate parameters for each network are given in the Additional file 1. Both motifs consist of two mutually-inhibiting genes, denoted by A and B. In the Exclusive Toggle Switch (ETS) motif, each gene encodes a transcription factor protein; the protein forms homodimers, which are capable of binding to the promoter of the competing gene, thereby repressing its expression. One DNA-promoter region controls the expression of both genes; when a repressor is bound, it excludes the possibility of binding by the repressor encoded by the competing gene. Therefore, the promoter can exist in three possible binding configurations, P_{00}, P_{10}, and P_{01}, denoting the unbound, a_2-bound, or b_2-bound states, respectively. Production of new protein molecules (including all processes involved in transcription, translation, and protein synthesis) occurs at a constant rate, which depends on the state of the promoter. When the gene is repressed, the encoded protein is produced at a low rate, denoted g_0. When the gene is not repressed, protein is produced at a high rate, g_1. For example, when the promoter state is P_{10} the a protein is produced at rate g_1, and the b protein is produced at g_0. When the promoter is unbound, neither gene is repressed, causing both proteins to be produced at rate g_1.

In the Mutual Inhibition/Self-Activation (MISA) motif, each homodimeric transcription factor also activates its own expression, in addition to repressing the other gene. The A and B genes are controlled by separate promoters, and each promoter can be bound by repressor and activator simultaneously. Therefore, the A-promoter can exist in four possible states, A_{00}, A_{10}, A_{01} and A_{11}, denoting unbound, a_2-activator bound, b_2-repressor bound, and both transcription factors bound, respectively (and similarly for the B-promoter). Proteins are produced at rate g_1 only when the activator is bound and the repressor is unbound. For example, the A_{10} promoter state allows a protein to be produced at g_1. The other three A promoter states result in a protein being produced at rate g_0. Similarly, the rate of b protein production depends only on the binding configuration of the B-promoter. In both the ETS and MISA networks, protein dimerization is assumed to occur simultaneously with binding to DNA. All rate parameters are given in Additional file 1: Tables S1 and S2.

Chemical master equation

The stochastic dynamics are modeled by the discrete, Markovian Chemical Master Equation, which gives the time-evolution of the probability to observe the system in a given state over time. In vector–matrix form, the CME can be written

$$\frac{d\mathbf{p}(\mathbf{x},t)}{dt} = \mathbf{K}\mathbf{p}(\mathbf{x},t)$$

where $\mathbf{p}(\mathbf{x}, t)$ is the probability over the system state-space at time t, and $\mathbf{K}$ is the reaction rate-matrix. The off-diagonal elements K_{ij} give the time-independent rate of transitioning from state $\mathbf{x}_i$ to $\mathbf{x}_j$, and the diagonal elements are given by $K_{ii} = -\sum_{j\neq i} K_{ji}$. We assume a well-mixed system of reacting species, and the state of the system is fully specified by $\mathbf{x} \in \mathbb{N}^S$, a state-vector containing the positive-integer values of all S molecular species/configurations. We hereon denote these state-vectors as "microstates" of the system. In the ETS network, $\mathbf{x} = [n_A, n_B, P_{ab}]$, where n_A is the copy-number of a molecules (protein monomers expressed by gene A, and similar for B), and P_{ab} indexes the promoter binding-configuration. In the MISA network, $\mathbf{x} = [n_A, n_B, A_{ab}, B_{ba}]$, which lists the protein copy numbers and promoter configuration-states associated with both genes.

The reaction rate matrix $\mathbf{K} \in \mathbb{R}^{N\times N}$ is built from the stochastic reaction propensities (Additional file 1: Eq. 1), for some choice of enumeration over the state-space with N reachable microstates. In general, if a system of S molecular species has a maximum copy number per species of $n_{\max}$, then $N \sim n_{\max}^{S}$. To enumerate the system state-space, we neglect microstates with protein copy-numbers larger than a threshold value, which exceeds the maximum steady-state gene expression rate, g_1/k, (where g_1 is the maximum production rate of protein and k is the degradation rate), as these states are rarely reached. This truncation of the state-space introduces a small approximation error, which we calculate using the Finite State Projection method [31] (Additional file 1: Figure S1).

Stochastic simulations

Stochastic simulations were performed according to the SSA method, implemented by the software package StochKit2 [43].

Quasipotential landscape

The steady-state probability $\boldsymbol{\pi}(\mathbf{x})$ over N microstates is obtained from $\mathbf{K}$ as the normalized eigenvector corresponding to the zero-eigenvalue, satisfying $\mathbf{K}\boldsymbol{\pi}(\mathbf{x}) = 0$ [44]. Quasipotential landscapes were obtained from $\boldsymbol{\pi}(\mathbf{x})$ using a Boltzmann definition, $U(\mathbf{x}) = -\ln(\boldsymbol{\pi}(\mathbf{x}))$ [22]. All matrix calculations were performed with MATLAB [45].

Markov State Models: mathematical background

The last 15 years have seen continual progress in development of theory, algorithms, and software implementing the MSM framework. We briefly summarize the theoretical background here; the reader is referred to other works (e.g., [41, 46–49]) for more details.

The MSM is a highly coarse-grained projection of system dynamics over N microstates onto a reduced space of selected size C (generally, $C \ll N$). The C states in the projected dynamics are constructed by clustering together microstates that experience relatively fast transitions among them. The C clusters, also called "almost invariant aggregates" [48], are hereon denoted "macrostates".

The MSM approach makes use of Robust Perron Cluster Analysis (PCCA+), a spectral clustering algorithm that takes as input a row-stochastic transition matrix, $\mathbf{T}(\tau)$ which gives the conditional probability for the system to transition between each pair of microstates within a given lagtime τ. The lagtime determines the time-resolution of the model, as expressed by the transition matrix. Off-diagonal elements T_{ij} give the probability of finding the system in microstate j at time $t + \tau$, given that it was in microstate i at time t. Diagonal elements T_{ii} give the conditional probability of again finding the system in microstate i at time $t + \tau$, and thus rows sum to 1. $\mathbf{T}(\tau)$ is directly obtained from the reaction rate matrix by [50]:

$$\mathbf{T}(\tau) = \exp\left(\tau \mathbf{K}^{\mathrm{T}}\right),$$

(where exp denotes the matrix exponential). The evolution of the probability over discrete intervals of τ is given by the Chapman-Kolmogorov equation,

$$\boldsymbol{p}^T(\mathbf{x}, t + k\tau) = \boldsymbol{p}^T(\mathbf{x}, t)\mathbf{T}^k(\tau).$$

For an ergodic system (i.e., any state in the system can be reached from any other state in finite time), $\mathbf{T}(\tau)$ will have one largest eigenvalue, the Perron root, $\lambda_1 = 1$. The stationary probability is then given by the normalized left-eigenvector corresponding to the Perron eigenvalue,

$$\boldsymbol{\pi}^{\mathrm{T}}(\mathbf{x})\mathbf{T}(\tau) = \boldsymbol{\pi}^{\mathrm{T}}(\mathbf{x}).$$

If the system exhibits multistability, then the dynamics can be approximately separated into fast and slow processes, with fast transitions occurring between microstates belonging to the same metastable macrostate, and slow transitions carrying the system from one macrostate to another. Then $\mathbf{T}(\tau)$ is nearly decomposable, and will exhibit an almost block-diagonal structure (for an appropriate ordering of microstates) with C nearly uncoupled blocks. In this case, the eigenvalue spectrum of $\mathbf{T}(\tau)$ shows a cluster of C eigenvalues near $\lambda_1 = 1$, denoting C slow processes (including the stationary process), and for $i > C$, $\lambda_i \ll \lambda_C$, corresponding to rapidly decaying processes. The system timescales can be computed from the eigenvalue spectrum according to $t_i = -\tau/\ln|\lambda_i(\tau)|$.

The PCCA+ algorithm obtains fuzzy membership vectors $\boldsymbol{\chi} = [\chi_1, \chi_2, \ldots, \chi_C] \in \mathbb{R}^{N \times C}$, which assigns microstates $i \in \{1, \ldots, N\}$ to macrostates $j \in \{1, \ldots, C\}$ according to grades (i.e., probabilities) of membership, $\chi_j(i) \in [0, 1]$. The membership vectors satisfy the linear transformation:

$$\boldsymbol{\chi} = \boldsymbol{\psi}\mathbf{B}$$

Where $\boldsymbol{\psi} = [\psi_1, \ldots, \psi_C]$ is the $N \times C$ matrix constructed from the C dominant right-eigenvectors of $\mathbf{T}(\tau)$, and $\mathbf{B}$ is a non-singular matrix that transforms the dominant eigenvectors into membership vectors. The coarse-grained $C \times C$ transition matrix $\tilde{\mathbf{T}}(\tau) \in \mathbb{R}^{C \times C}$ (i.e., the Markov State Model) is then obtained as the projection of $\mathbf{T}(\tau)$ onto the C sets by:

$$\tilde{\mathbf{T}}(\tau) = \tilde{\boldsymbol{D}}^{-1}\boldsymbol{\chi}^T\boldsymbol{D}\mathbf{T}(\tau)\boldsymbol{\chi}$$

where $\boldsymbol{D}$ is the diagonal matrix obtained from the stationary probability vector, $\boldsymbol{D} = \mathrm{diag}(\pi_1, \ldots, \pi_N)$. The coarse-grained probability $\tilde{\boldsymbol{\pi}}(x)$ is obtained by $\tilde{\boldsymbol{\pi}}(\mathbf{x}) = \boldsymbol{\chi}^{\mathrm{T}}\boldsymbol{\pi}(\mathbf{x})$, and $\tilde{\boldsymbol{D}} = \mathrm{diag}(\tilde{\pi}_1, \ldots, \tilde{\pi}_C)$. The elements of the linear transformation matrix $\mathbf{B}$ are obtained by an optimization procedure, with "metastability" of the resultant coarse-grained projection as the objective function to be maximized. The trace of the coarse-grained transition matrix, $trace[T]$ has been taken to be the measure of metastability, because it expresses the probabilities for the system to remain in metastable states over the lagtime (i.e., maximizing the sum over the diagonal elements). The original PCCA method [48] used the sign structure of the eigenvectors to identify almost invariant aggregates (instead of this optimization procedure), and more recent work has identified an alternative objective function [49]. The results of this paper were generated using the PCCA+ implementation of MSMBuilder2 [51].

Construction of Markov State Models and pathway decomposition

The PCCA+ algorithm generates a fuzzy discretization. We convert fuzzy values into a so-called "crisp" partitioning of N states into C clusters, which entirely partitions the space with no overlap, by assigning $\chi_j^{crisp}(i) \in \{0, 1\}$. That is, $\chi_j^{crisp}(i) = 1$ if the jth element of the row vector $\chi(i)$ is maximal, and 0 otherwise. Transition probabilities are estimated over the C coarse-grained sets by summing over the fluxes, or equivalently:

$$\tilde{\mathbf{T}}(\tau) = \tilde{\boldsymbol{D}}^{-1}\boldsymbol{\chi}^T\boldsymbol{D}\mathbf{T}(\tau)\boldsymbol{\chi},$$

where $\tilde{\mathbf{T}}(\tau) \in \mathbb{R}^{C \times C}$ is the coarse-grained Markov State Model and $\boldsymbol{D}$ is the diagonal matrix obtained from the stationary probability vector, $D = \text{diag}(\pi_1, \ldots, \pi_N)$. The coarse-grained probability $\tilde{\boldsymbol{\pi}}(\mathbf{x})$ is obtained by $\tilde{\boldsymbol{\pi}}(\mathbf{x}) = \boldsymbol{\chi}^T \boldsymbol{\pi}(\mathbf{x})$, and $\tilde{D} = \text{diag}(\tilde{\pi}_1, \ldots, \tilde{\pi}_C)$.

The Markov State Model is visualized using the PyEmma 2 plotting module [46], where the magnitude of the transition probabilities and steady state probabilities are represented by the thickness of the arrows and size of the circles, respectively.

Upon construction of the Markov State Model, transition-path theory [52–54] was applied in order to compute an ensemble of transition paths connecting two states of interest, along with their relative probabilities. This was achieved by applying a pathway decomposition algorithm adapted from Noe, et al. in a study of protein folding pathways [54] (details in Additional file 1). A summary of the workflow used in generating the results of this paper is included in the Additional file 1: Supplement S5.

Results

Eigenvalues and Eigenvectors of the stochastic transition matrix reveal slow dynamics in gene networks

In order to explore the utility of the MSM approach for analyzing global dynamics of gene networks, we studied common motifs that control lineage decisions. The MISA network motif (Fig. 1a, Additional file 1: Supplement S1, and Methods) has been the subject of previous theoretical studies and is thought to appear in a wide variety of binary fate-decisions [5, 55, 56]. In the network model, the *A/B* gene pair represents known antagonistic pairs such as Oct4/Cdx2, PU.1/Gata1, and GATA3/T-bet, which control lineage decisions in embryonic stem cells, common myeloid progenitors, and naïve T-helper cells, respectively [9, 57, 58]. In general, a particular cell lineage will be associated with a phenotype in which one of the genes is expressed at a high level, and the other is expressed at a low (repressed) level. The MISA network as an ODE model has been reported to have

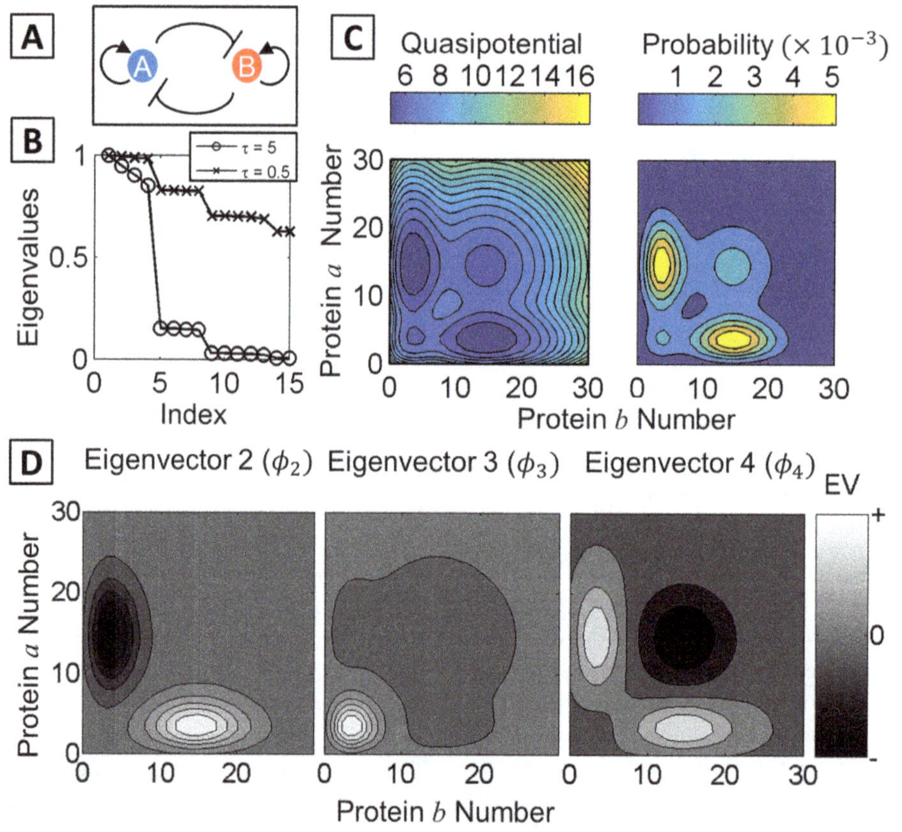

Fig. 1 Eigenvalue and eigenvector analysis of the Mutual Inhibition/Self Activation (MISA) network. **a** Schematic of the MISA network motif. **b** The fifteen largest eigenvalues of the stochastic transition matrix $\mathbf{T}(\tau)$, indexed in descending order, for $\tau = 5$ (*circles*) and $\tau = 0.5$ (*crosses*) (time units of inverse protein degradation rate, k^{-1}). Gaps indicate separation between processes occurring on different timescales. Network parameter values are listed in Additional file 1: Table S1. **c** The quasipotential landscape (*left*) and probability landscape (*right*) for the MISA motif, projected onto the A vs. B protein copy number subspace, showing four visible basins Landscapes were obtained from $\boldsymbol{\phi}_1$, the eigenvector associated with the largest eigenvalue of $\mathbf{T}(\tau)$. **d** *Left* to *right*: second, third, and fourth eigenvectors ($\boldsymbol{\phi}_2$, $\boldsymbol{\phi}_3$, $\boldsymbol{\phi}_4$) of $\mathbf{T}(\tau)$. The sign structure reveals the nature of the slowest dynamical processes (see text)

up to four stable fixed-points corresponding to the A/B gene pair expression combinations Lo/Lo, Lo/Hi, Hi/Lo, and Hi/Hi. We computed the probability and quasipotential landscape of the MISA network. For a symmetric system with sufficiently balanced rates of activator and repressor binding and unbinding from DNA, four peaks (or basins) can be distinguished in the steady state probability (quasipotential) landscape, plotted as a function of protein a copy number vs. protein b copy number (Fig. 1a, b). Quasipotentials computed from $\boldsymbol{\pi}(\mathbf{x})$, the Perron eigenvector of the transition matrix (see Methods) and from a long stochastic simulation showed agreement (Additional file 1: Figure S2).

The Markov State Model framework has been applied in studies of protein folding, where dynamics occurs over rugged energetic landscapes characterized by multiple long-lived states (reviewed in [40, 41]). Therefore, we reasoned that the approach could be useful for studying global dynamics of multistable GRNs. The method identifies the slowest system processes based on the dominant eigenvalues and eigenvectors of the stochastic transition matrix, $\mathbf{T}(\tau)$, which gives the probability of the system to transition from every possible initial state to every possible destination state within lagtime τ (with τ having units of k^{-1} and k being the rate of protein degradation). Inspection of the eigenvalue spectrum of $\mathbf{T}(\tau = 5)$ for the MISA network in Fig. 1b reveals four eigenvalues near 1 followed by a gap, indicating four system processes that are slow on this timescale. Decreasing τ to 0.5 reveals a step-structure in the eigenvalue spectrum, suggesting a hierarchy of system timescales. The timescales are related to the eigenvalues according to $t_i = -\tau/\ln|\lambda_i(\tau)|$. The Perron eigenvalue $\lambda_1 = 1$ is associated with the stationary (infinite time) process, and the lifetimes t_2 through t_5 are computed to be {95.6, 49.4, 30.8, 2.6} (in units of k^{-1}). Thus, the first gap in the eigenvalue spectrum arises from a more than ten-fold separation in timescales between t_4 and t_5. The original PCCA method [48] used the sign structure of the eigenvectors to assign cluster memberships. Plotting the left-eigenvectors corresponding to the four dominant eigenvalues in the MISA network is instructive: the stationary landscape is obtained from the first left-eigenvector ($\phi_1 = \boldsymbol{\pi}(\mathbf{x})$), which is positive over all microstates, while the opposite-sign regions in ϕ_2, ϕ_3, ϕ_4 reveal the nature of the slow processes (Fig. 1d). An eigenvector with regions of opposite sign corresponds to an exchange between those two regions (in both directions, since eigenvectors are sign-interchangeable). For example, the slowest process corresponds to exchange between the $a > b$ and $b > a$ regions of state-space, i.e., switching between B-gene dominant and A-gene-dominant expression states. Eigenvectors ϕ_3 and ϕ_4 show that somewhat faster timescales are associated with exchange in and out of the Lo/Lo and Hi/Hi basins.

The Markov State Model approach identifies multistability in GRNs

Reduced models of the MISA network

The MSM framework utilizes a clustering algorithm known as PCCA+ (see Methods and Additional file 1) to assign every microstate in the system to a macrostate (i.e., a cluster of microstates) based on the slow system processes identified by the eigenvectors and eigenvalues of $\mathbf{T}(\tau)$. Applying the PCCA+ algorithm to the MISA network for the parameter set of Fig. 1 resulted in a mapping from $N = 15,376$ $(31 \times 31 \times 4 \times 4)$ microstates onto $C = 4$ macrostates. The N microstates were first enumerated by accounting for all possible system configurations with $0 \le a \le 30$ and $0 \le b \le 30$. This enumeration assumes a negligible probability for the system to ever exceed 30 copies of either protein, which introduces a small approximation error of $1E-5$ (details in Additional file 1: Figure S1). Because the promoters of each gene can take four possible configurations—that is, two binding sites (for the repressor and activator) that can be either bound or unbound—a total of 16 gene configuration states are possible, giving $N = 15,376$ enumerated microstates. For this parameter set, the highest probability densities within the four macrostates obtained correspond closely to the visible peaks (basins) in the probability (quasipotential) landscape. This can be seen by the ellipsoids in Fig. 2a, which show the highest probability-density regions of each macrostate (according to the stationary probability), projected onto the protein subspace. The average expression levels of proteins in each macrostate indicate the four distinct cell phenotypes (Lo/Lo, Lo/Hi, Hi/Lo, Hi/Hi). The complete microstate-to-macrostate mapping is detailed in Additional file 1: Figure S3 and Table S3. In this parameter regime, since the protein binding and unbinding rates are slow relative to protein production and degradation, the promoter configurations determine the macrostate assignment exactly. That is, the algorithm partitions microstates according to the promoter configuration, rather than the protein copy number. Each of the four macrostates contains microstates from four distinct promoter configurations out of the possible sixteen, along with microstates with all possible protein copy number (a/b) combinations. A representative gene promoter configuration for each macrostate (i.e., the configuration contributing the most probability density to each macrostate) is shown schematically (Fig. 2b).

Parameter-dependence of landscapes and MSMs

To determine whether the MSM approach can robustly identify gene network macrostates, we applied it over a

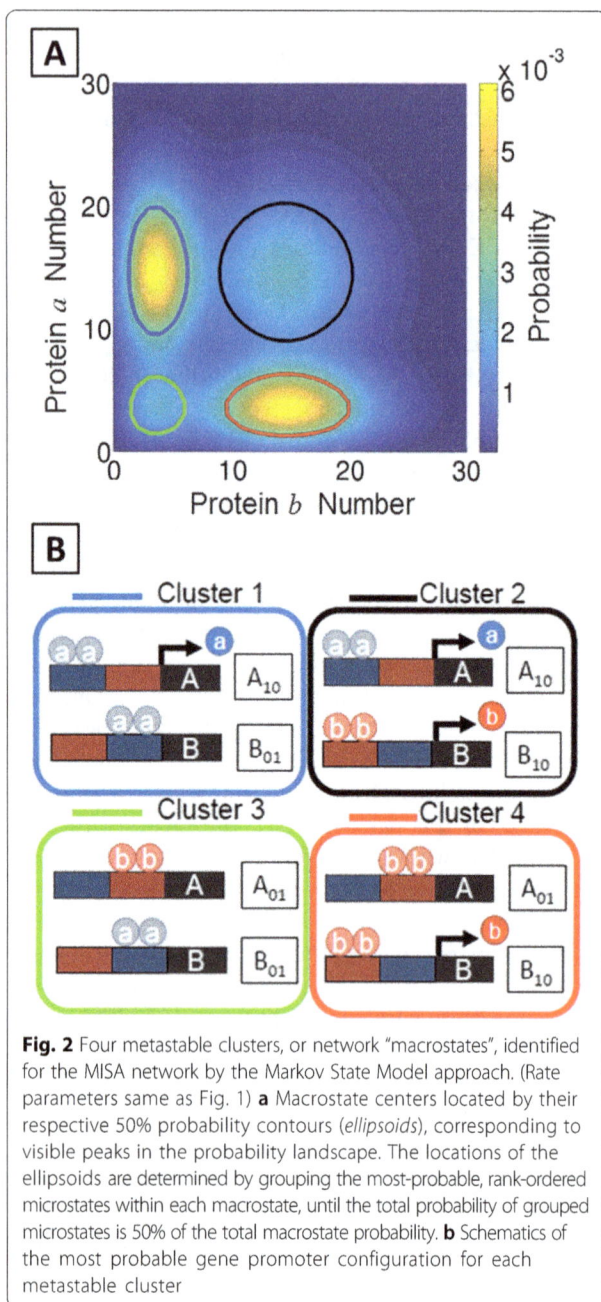

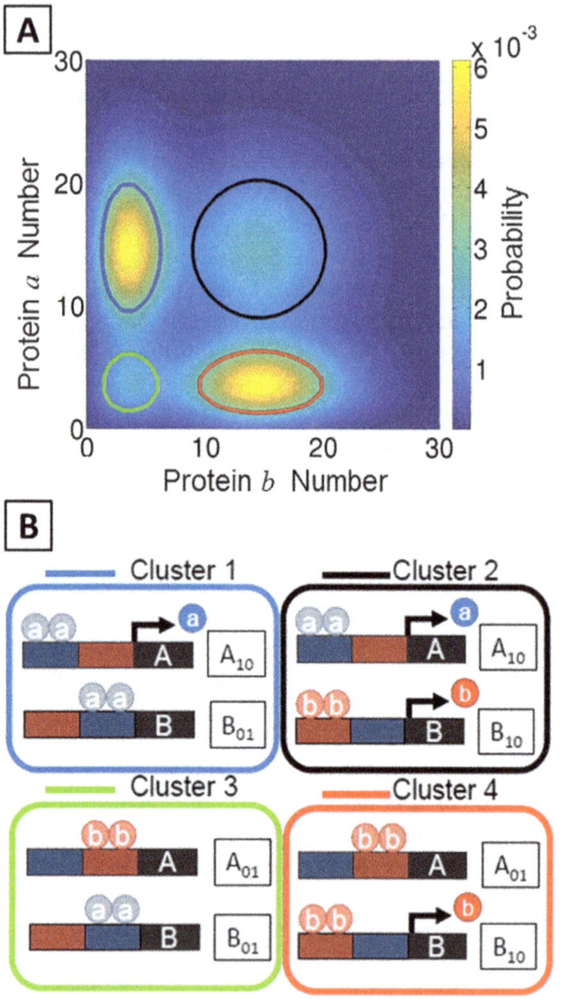

Fig. 2 Four metastable clusters, or network "macrostates", identified for the MISA network by the Markov State Model approach. (Rate parameters same as Fig. 1) **a** Macrostate centers located by their respective 50% probability contours (*ellipsoids*), corresponding to visible peaks in the probability landscape. The locations of the ellipsoids are determined by grouping the most-probable, rank-ordered microstates within each macrostate, until the total probability of grouped microstates is 50% of the total macrostate probability. **b** Schematics of the most probable gene promoter configuration for each metastable cluster

range of network parameters by varying the repressor unbinding rate f_r (all parameters defined in Additional file 1: Table S1). Increasing f_r relative to other network parameters modulates the quasipotential landscape by increasing the probability of the Hi/Hi phenotype, in which both genes express at a high level simultaneously (Fig. 3b). This occurs as a result of weakened repressive interactions, since the lifetimes of repressor occupancy on promoters are shortened when f_r is increased. The eigenvalue spectra show a corresponding shift: when $f_r = 1E-3$, four dominant eigenvalues are present. When f_r is increased to $f_r = 1$, the largest visible gap in the eigenvalue spectrum shifts to occur after the first eigenvalue ($\lambda = 1$), indicating loss of multistability on the timescale of τ (here, $\tau = 5$) (Fig. 3a). Correspondingly, for this parameter set, the landscape shows only a single visible Hi/Hi basin.

The PCCA+ algorithm seeks C long-lived macrostates, where C is user-specified. We constructed Markov State Models for the MISA network over varying f_r, specifying four macrostates. The MSMs are shown graphically in Fig. 3d. The sizes of the circles are proportional to the relative steady-state probability of the macrostate, and the thickness of the directed edges are proportional to the relative transition probability within τ. In agreement with the landscapes, the MSMs over this parameter regime show increasing probability of the Hi/Hi state, as a result of an increasing ratio of transition probability "into" versus "out of" the Hi/Hi state. The locations of the clusters in the state-space (according to 50% (of the total) stationary probability contours) do not change appreciably. The choice of lagtime τ sets the timescale on which metastability is defined in the system. However, in practice, the PCCA+ seeks an assignment of C clusters regardless of whether C metastable states exist in the system on the τ timescale, and the resulting aggregated macrostates are generally invariant to τ. Thus, for $f_r = 1$, the algorithm locates four macrostates, although the (low-probability) Hi/Lo, Lo/Lo, and Lo/Hi macrostates are likely to experience transitions away, into the Hi/Hi macrostate, within τ. These low-probability states appear in the landscape as shoulders on the outskirts of the Hi/Hi basin. Overall, Fig. 3 demonstrates that, for this parameter regime, the quasipotential landscape and the MSM yield similar information on the global system dynamics in terms of the number and locations of long-lived states, and their relative probabilities as a function of the unbinding rate parameter f_r. The MSM further provides quantitative information on the probabilities (and thus timescales) of transitioning between each pair of macrostates.

MSM identifies purely stochastic multistability

Multistability in gene networks is often analyzed within an ordinary differential equation (ODE) framework, by graphical analysis of isoclines and phase portraits, or by linear stability analysis [4, 8]. ODE models of gene networks treat molecular copy numbers (i.e., proteins, mRNAs) as continuous variables and apply a quasi-steady-state approximation to neglect explicit binding/unbinding of proteins to DNA. This approximation is valid in the so-called "adiabatic" limit, where binding and unbinding of regulatory proteins to DNA is fast, relative to protein production and degradation. Previous studies have shown that such ODE models can give rise to landscape structures that are qualitatively different

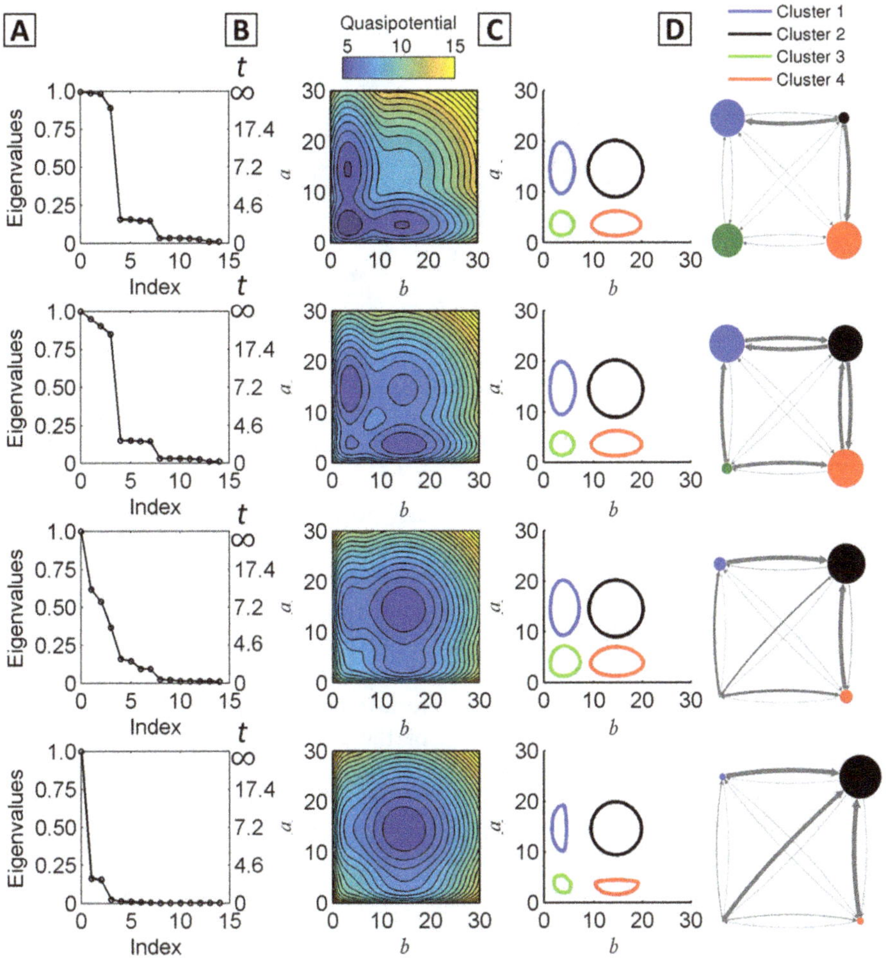

Fig. 3 Dependence of the MISA network eigenvalues, landscape, and MSM on the repressor unbinding parameter f_r. Top to Bottom: increasing $fr = \{1E-3, 1E-2, 1E-1, 1\}$ in units of protein degradation rate, k^{-1} (complete parameter list in Additional file 1: Table S1). **a** The eigenvalue spectrum of $\mathbf{T}(\tau)$ for $\tau = 5$, and associated timescales. **b** The quasipotential landscape. **c** The Markov State Model with four macrostates, visualized by the 50% probability contour for each metastable state. **d** The state transition graph. Nodes and edges denote macrostates and transition probabilities, respectively. The size of each node is proportional to the steady-state probability, and edge thickness is proportional to the probability of transition within $\tau = 5$

from those of their corresponding discrete, stochastic networks. For example, multistability in an ODE model of the genetic toggle switch requires cooperativity—i.e., multimers of proteins must act as regulators of gene expression [59]. However, it was found that monomer repressors are sufficient to give bistability in a stochastic biochemical model [55, 60]. We compared the dynamics of the monomer ETS network (shown schematically in Fig. 4a) as determined by analysis of the ODEs, along with the corresponding stochastic quasipotential landscape and the MSM. In a small-number regime, the ODEs predict monostability (Fig. 4c), while the stochastic landscape shows tristability—that is, three basins corresponding to the Hi/Lo, Hi/Hi, and Lo/Hi expressing phenotypes (Fig. 4d) (The dominant eigenvectors are shown in Additional file 1: Figure S4). This type of discrepancy has been shown to occur in systems with small number effects, i.e., extinction at the boundaries [55] or slow transitions between expression states [29].

The MSM approach identifies three metastable macrostates for the monomer ETS in this parameter regime, as seen in the eigenvalue spectrum, which shows a gap after the third index. The reduced Markov State Model constructed for this network thus reduces the system from $N = 7,803$ ($51 \times 51 \times 3$) microstates to $C = 3$ macrostates (Fig. 4b), corresponding to the same Hi/Lo, Hi/Hi, and Lo/Hi metastable phenotypes seen in the quasipotential landscape. Figure 4 demonstrates that the MSM approach can accurately identify purely stochastic multistability in systems where continuous models predict only a single stable fixed-point steady state. Similar results were found for a self-regulating, single-gene network (Additional file 1: Figure S5 and Table S4). This network, which has been solved analytically, gives rise to

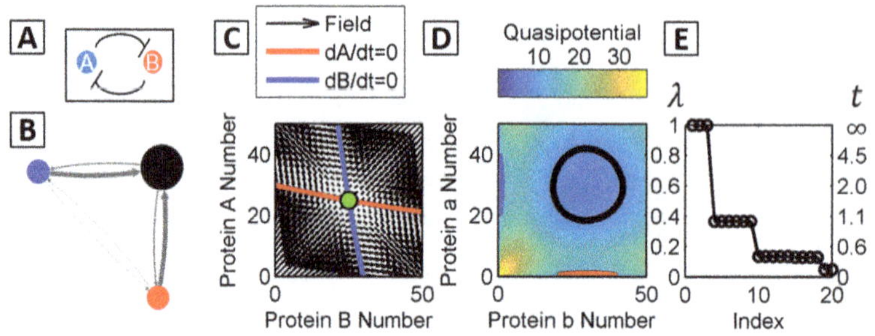

Fig. 4 Comparison of ODE and MSM analysis of the monomer Exclusive Toggle Switch (ETS) network. **a** Schematic of the ETS network motif. **b** The Markov State Model identifies three macrostates corresponding to the Hi/Lo, Hi/Hi, and Lo/Hi phenotypes. Parameter values are listed in Additional file 1: Table S2. **c** The nulllclines and vector field of the deterministic ODEs show a single fixed point steady-state, with both genes expressing at the maximum rate (Hi/Hi phenotype). **b**, **d**, **e** The corresponding landscape and MSM show tristability: **d** The quasipotential landscape shows three visible basins corresponding to the Hi/Lo, Hi/Hi, and Lo/Hi phenotypes. Macrostate centers located by their respective 50% probability contours (ellipsoids), as in Fig. 2. **e** The 20 dominant eigenvalues reveal timescale separation, including a gap after λ_3

a bimodal or monomodal stationary distribution depending on the protein binding/unbinding rates [28, 29, 61].

Analyzing global gene network dynamics with the Markov State Model

MSM provides good approximation to relaxation dynamics from a given initial configuration

Figures 1, 2, 3 and 4 demonstrate the utility of the MSM approach for analyzing stationary properties of networks—that is, for identifying the number and locations of multiple long-lived states. Additionally, the MSM can be used to make dynamic predictions about transitions among macrostates. Dynamics for either the "full" transition matrix (with all system states enumerated up to a maximum protein copy number) or reduced transition matrix (i.e., the MSM) is propagated according to the Chapman-Kolmogorov equation (see Methods and Additional file 1). We sought to determine the accuracy of the dynamic predictions obtained from the MSM. Applying the methods proposed by Prinz, et al. ([47]) (details in Additional file 1), we compared the dynamics propagated by the fully enumerated transition matrix $\mathbf{T}(\tau)$, which is then projected onto the coarse-grained macrostates, to the dynamics of the coarse-grained system propagated by $\tilde{\mathbf{T}}(\tau)$ (i.e., the MSM). We thus computed the error in dynamics of relaxation out of a given initial system configuration. The system relaxation from a given initial microstate can also be computed by running a large number of brute force SSA simulations. Relaxation dynamics for the full, brute-force, and reduced MSM methods, applied to the MISA with $f_r = 1E-2$, all show good agreement (Fig. 5a, b, and c). The error computed between the reduced MSM vs. full dynamics (i.e., $\tilde{\mathbf{T}}(\tau)$ vs $\mathbf{T}(\tau)$), is maximally $7.8E-3$, varies over short times, and decreases continuously after time $t = 140$. Alternatively, the error of the MSM can be quantified by comparing the autocorrelation functions of the MSM and brute force simulation [50, 62]. In Additional file 1: Figure S6, we show that the derived autocorrelation functions of the MSM and brute force, and the relaxation constants τ_r, which describes the amount of time to reach equilibrium, are close in value ($\tau_r = 1E3$, for the MSM, and $\tau_r = 1.1E3$ for the brute force). Overall, these results demonstrate that the most accurate predictions of the coarse-grained MSM can be obtained on long

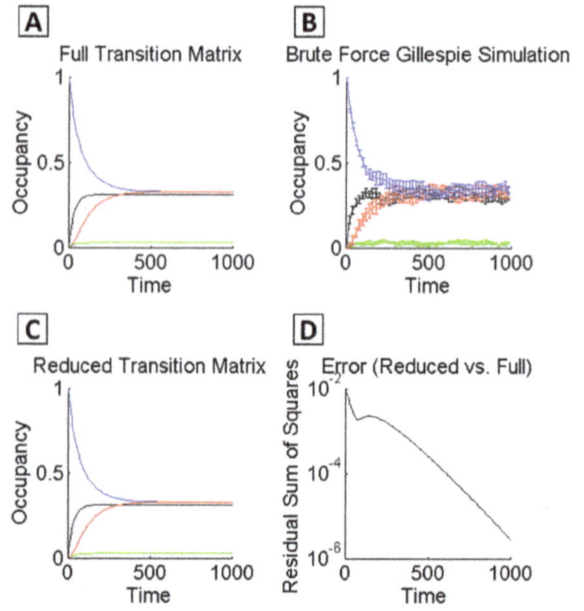

Fig. 5 MSM approximation error for the MISA motif. Relaxation of the system from a particular initial configuration (see text), as obtained from **a** the full transition matrix, **b** brute force SSA simulation, and **c** the reduced transition matrix obtained from the MSM. Color-coding is according to the macrostates, as in Figs. 1, 2 and 3: *blue*, *black*, *red*, *green* correspond to A/B expression phenotypes Hi/Lo, Hi/Hi, Lo/Hi, and Lo/Lo, respectively. **d** Calculated approximation error as a function of time, comparing the reduced MSM to the full CME dynamics. Network parameter values are same as Figs. 1 and 2

timescales, but dynamic approximations with reasonable accuracy can also be obtained for short timescales.

Parameter-dependence of MSM error

The accuracy of the MSM dynamic predictions depends on whether inter-macrostate transitions can be treated as memory-less hops. Previous theoretical studies of gene network dynamics found that the height of the barrier separating phenotypic states, and the state-switching time associated with overcoming the barrier, depends on the rate parameters governing DNA-binding by the protein regulators [5, 6, 55, 63]. We reasoned that a larger time-scale separation between intra- and inter-basin transitions (corresponding to a larger barrier height separating basins) should result in higher accuracy of the MSM approximation. Thus, we hypothesized that the accuracy of the MSM dynamic predictions should depend on the DNA-binding and unbinding rate parameters. We demonstrated this using the dimeric ETS motif, by computing the error of the MSM approximation for a range of repressor unbinding rates f. We varied the binding kinetics without changing the overall relative strength of repression, by varying f together with the repressor binding rate h, to maintain a constant binding equilibrium $\left(X_{eq} = \frac{f}{h} = 100\right)$. By varying f and h in this way over eight orders of magnitude, we found that the barrier height and timescale of the slowest system process (t_2) had a non-monotonic dependence on the binding/unbinding parameters. Thus, the fastest inter-phenotype switching was observed in the regime with intermediate binding kinetics, in agreement with previous work [5]. The system also exhibits a shift from three visible basins in the quasi-potential landscape in the small f regime to two basins in the large f regime. We performed clustering by selecting $C = 2$ (dashed lines, Fig. 6) and $C = 3$ clusters (solid lines, Fig. 6), and computed the total error over all choices of system initialization, as well as the error associated with relaxation from a particular system microstate. In general, we find that the 3-state MSM approximation is more accurate than the 2-state partitioning. The 3-state MSM dynamic predictions are highly accurate when the DNA-binding/unbinding kinetics is slow. As such, in this regime the Markovian assumption of memory-less transitions between the three phenotypic states is most accurate. As hypothesized, the accuracy of the MSM approximation is lowest (highest error) when the lifetime t_2 is shortest (intermediate regime, $f = 1$), and the error decreases modestly with further increase in f (i.e., increase in t_2).

Decomposition of state-transition pathways in gene networks using the MSM framework

Quantitative models of gene network dynamics can shed light on transition paths connecting phenotypic states. The MSM approach coupled with transition path theory [52, 53, 64] enables decomposition of all major pathways linking initial and final macrostates of interest. This type of pathway decomposition has previously shed light on mechanisms of protein folding [54]. We demonstrate this pathway decomposition on the MISA network, by computing the transition paths linking the polarized A-dominant (Hi/Lo) and B-dominant (Lo/Hi) phenotypes. Multiple alternative pathways linking these phenotypes are possible: for the 4-state coarse-graining, the system can alternatively transit through the Hi/Hi or Lo/Lo phenotypes when undergoing a stochastic state-transition from one polarized phenotype to the other. Not all possible paths are enumerated since only transitions with net positive fluxes are considered (see Additional file 1: Equation S18). The hierarchy of pathway probabilities for successful transitions depends on the kinetic rate parameters (Fig. 7a). It could be tempting to intuit pathway intermediates based on visible basins in the quasipotential landscape. However, we found that the steady-state probability of an intermediate macrostate (i.e., the Hi/Hi or Lo/Lo states) does not accurately predict if it serves as a pathway intermediate for successful transitions, because parameter regimes are possible in which successful transitions are likely to transition through intermediates with high potential/low probability (Fig. 7c). This occurs because the relative probability of transiting through one intermediate macrostate versus another is based on the balance of probabilities for entering and exiting the intermediate: intermediate states that can be easily reached—but not easily exited—as a result of stochastic fluctuations can act as "trap" states. Therefore, it is shown that the pathway probability cannot be inferred from the steady state probability of the intermediates alone.

MSMs can be constructed with different resolutions of coarse-graining

The eigenvalue spectrum of the MISA network shows a step-structure, with nearly constant eigenvalue clusters separated by gaps. These multiple spectral gaps suggest a hierarchy of dynamical processes on separate timescales. A convenient feature of the MSM framework is that it can build coarse-grained models with different levels of resolution by PCCA+, in order to explore such hierarchical processes. We applied the MSM framework to a MISA network with very slow rates of DNA-binding and unbinding ($f_r = 1E-4$, $h_r = 1E-6$), comparing the macrostates obtained from selecting $C = 4$ versus $C = 16$ clusters. For $\mathbf{T}(\tau = 1)$, a prominent gap occurs in the eigenvalue spectrum between λ_{16} and λ_{17}, corresponding to an almost 30-fold separation of timescales between $t_{16} = 27.8$ and $t_{17} = 0.99$ (Fig. 8a). Applying PCCA+ with $C = 16$ clusters uncovered a 16-macrostate network with four highly-interconnected subnetworks consisting of

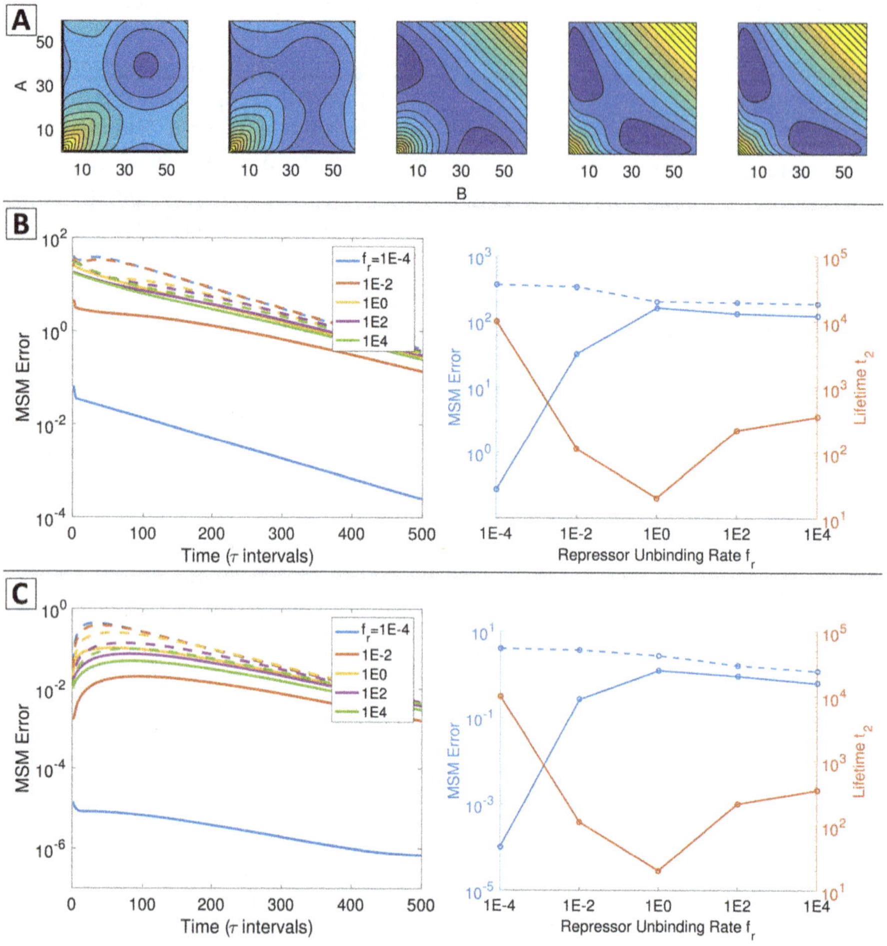

Fig. 6 The MSM approximation accuracy for the ETS motif depends on rate parameters and number of macrostates in the reduced model. **a** Quasipotential landscape for the exclusive dimeric repressor toggle switch, with increasing DNA-binding rates (left to right: $fr = \{1E-4, 1E-2, 1E0, 1E2, 1E4\}$, all parameter values listed in Additional file 1: Table S2), demonstrating the dependence of basin number and barrier height on network parameters. **b** Global error of the MSM approximation. *Left*: Global error as a function of time (in intervals of τ) for different f_r and numbers of macrostates. *Solid lines*: global error of the 3-state MSM. Dashed lines: global error of the 2-state MSM. *Right*: Total global error over kτ, $k = 0$ to 500, for a 3-state (*solid blue*) or 2-state (*dashed blue*) MSM. *Solid orange line*: the longest system lifetime t_2. **c** Error of the MSM approximation when the system is initialized in a particular microstate. *Left*: Error as a function of time (in intervals of τ) for different adiabaticities and different numbers of macrostates. *Solid lines*: error of the 3-state MSM. *Dashed lines*: error of the 2-state MSM. *Right*: Total error from a particular microstate over kτ where $k = 0$ to 500, for a 3-state (*solid blue*) or 2-state (*dashed blue*) MSM. *Orange line*: the longest system lifetime t_2

four states each (Fig. 8c). The identities of the sixteen macrostates showed an exact correspondence to the sixteen possible *A/B* promoter binding configurations. This correspondence reflects the fact that, in the slow binding/unbinding, so-called non-adiabatic regime [65], the slow network dynamics are completely determined by unbinding and binding events that take the system from one promoter configuration macrostate to another, while all fluctuations in protein copy number occur on much faster timescales.

Each subnetwork in the MSM constructed with $C = 16$ corresponds to a single macrostate in the MSM constructed with $C = 4$. Thus, in the $C = 4$ MSM, four different promoter configurations are lumped together in a single macrostate, and dynamics of transitions among them is neglected. Counterintuitively, the locations of the $C = 4$ macrostates do not correspond directly to the four basins visible in the quasipotential landscape (Fig. 8b, d). Instead, the clusters combine distinct phenotypes—e.g., the red macrostate combines the *A/B* Lo/Lo and Lo/Hi phenotypes, because it includes the promoter configurations A_{01} B_{10} and A_{11} B_{10} (corresponding to Lo/Hi expression) and A_{01} B_{00} and A_{11} B_{00} (corresponding to Lo/Lo expression) (Fig. 8b, Additional file 1: Table S5 and Figure S7). This result demonstrates that the barriers visible in the quasipotential landscape do not reflect the slowest timescales in the system. This occurs because of the loss of information inherent to visualizing global dynamics via the quasipotential landscape, which often projects

A

f_r	Pathway Probability (%)				Stationary Probability (%)			
	blue→red	blue→green→red	blue→black→red	blue→black→green→red	blue	black	green	red
5E-4	1.0	61.8	36.9	2.3E-01	27.4	1.3	44.0	27.4
1E-3	1.5	44.7	53.5	3.3E-01	34.5	3.2	27.8	34.5
5E-3	2.3E-01	13.8	83.4	5.0E-01	37.9	17.9	6.3	37.9
1E-2	2.4E-01	7.3	89.7	5.2E-01	33.0	31.1	2.9	33.0
1.5E-2	2.5E-01	4.9	92.1	5.2E-01	28.8	40.7	1.7	28.8

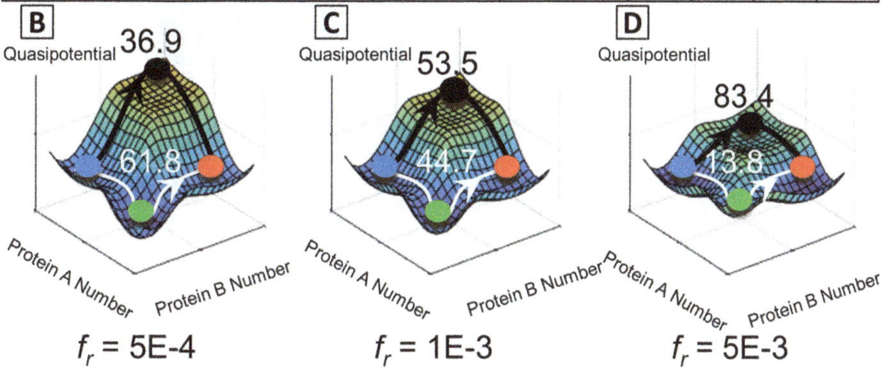

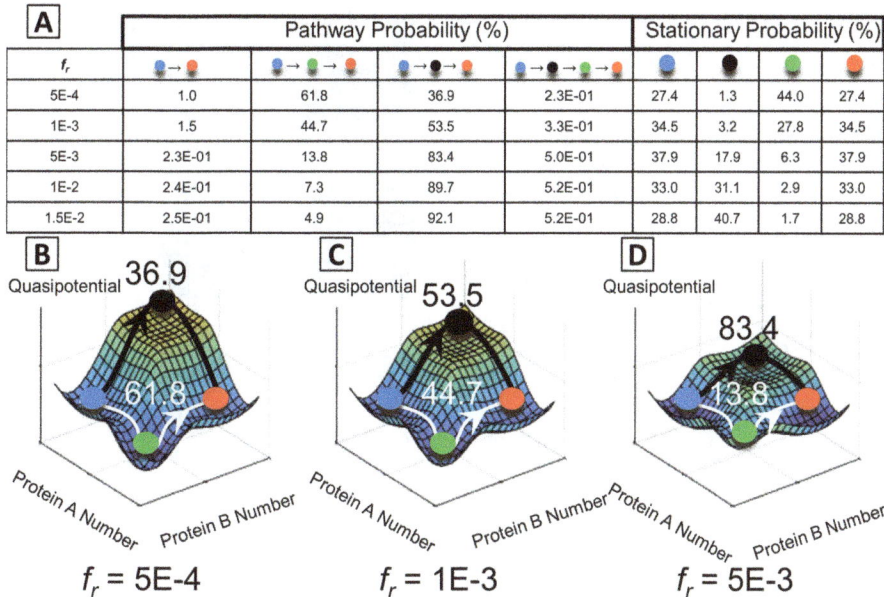

Fig. 7 Dependence of stochastic transition paths on the repressor unbinding rate parameter f_r in the MISA network (parameter values listed in Additional file 1: Table S1). **a** Table of all possible transition paths starting from the Hi/Lo (blue) and ending in the Lo/Hi (red) macrostate (color coding is same as Figs. 1, 2, 3 and 5). Relative probabilities of traversing a given path are shown, along with the stationary probabilities of the system to be found in a given macrostate. **b-d** Dominant transition paths superimposed on the 3D quasipotential surfaces for $fr = \{5E-4, 1E-3, 5E-3\}$, demonstrating how dominant paths can traverse high-potential areas of the landscape. For example, when $f_r = 1E-3$, (panel **c**), successful transitions most likely go through the Hi/Hi state (3.2% populated at steady state), though this requires a large barrier crossing. Pathway percentages are superimposed on the landscapes

dynamics onto two system coordinates. In this case, projecting onto the protein *a* and protein *b* copy numbers loses information about the sixteen promoter configurations, obscuring the fact that barrier-crossing transitions can occur faster than some within-basin transitions. Plotting a time trajectory of brute force SSA simulations for this network supports the findings from the MSM: the dynamics shows frequent transitions within subnetworks, and less-frequent transitions between subnetworks, indicating the same hierarchy of system dynamics as was revealed by the 4- and 16-state MSMs (Fig. 8e).

Transition path decomposition reveals nonequilibrium dynamics

Mapping the most probable paths forward and backward between macrostate "1" (promoter configuration: $A_{01}B_{00}$) and macrostate "11" (promoter configuration: $A_{00}B_{01}$) revealed that a number of alternative transition paths are accessible to the network, and the paths typically transit between three and five intermediate macrostates. The decomposition shows three paths with significant (i.e., >15%) probability and 12 distinct paths with >1% probability (for both forward and backward transitions, Additional file 1: Tables S3-S4). The pathway decomposition also reveals a great deal of irreversibility in the forward and reverse transition paths, which is a hallmark of nonequilibrium dynamical systems. For example, the most probable forward and reverse paths both transit three intermediates, but have only one intermediate (macrostate 5) in common (Fig. 8c and Additional file 1: Tables S6-S7). Thus, the complete process of transitioning away from macrostate 1, through macrostate 11, and returning to 1 maps a dynamic cycle.

Discussion

Our application of the MSM method to representative GRN motifs yielded dynamic insights with potential biological significance. Decomposition of transition pathways revealed that stochastic state-transitions between phenotypic states can occur via multiple alternative routes. Preference of the network to transition with higher likelihood through one particular pathway depended on the stability of intermediate macrostates, in a manner not directly intuitive from the steady-state probability landscape. The existence of "spurious attractors", or metastable intermediates that act as trap states to hinder stem cell reprogramming, has been discussed previously [11] as a general explanation for the existence of partially reprogrammed cells. By analogy, MSMs constructed in protein folding studies predict an ensemble of folding pathways, as well as the existence of misfolded trap states that reduce folding speed [54]. Our results

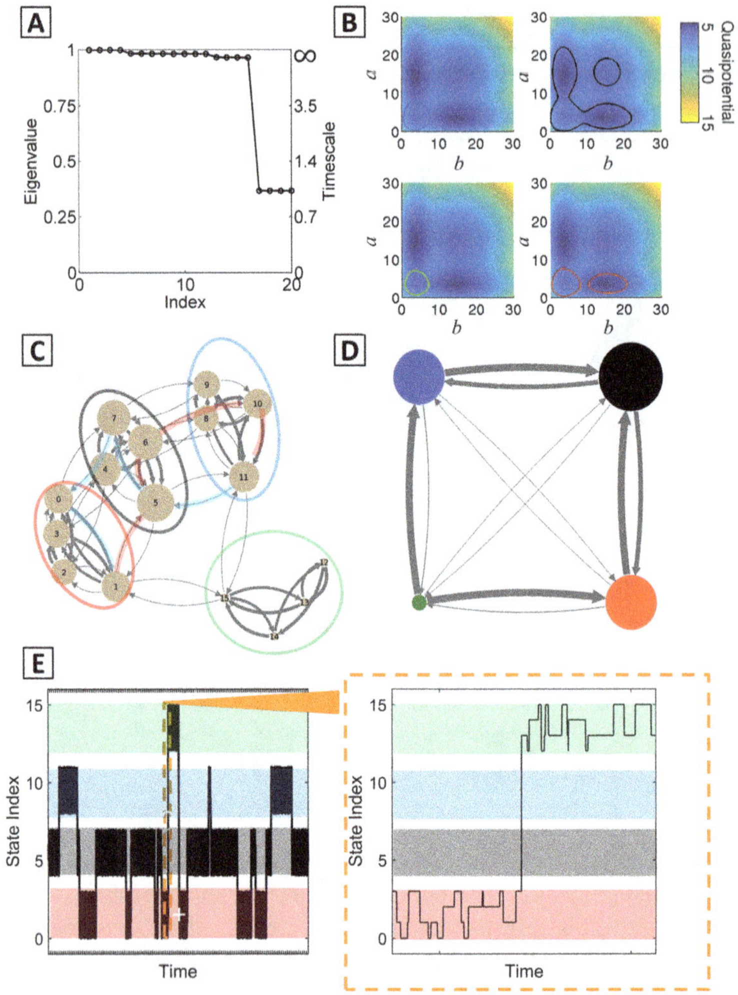

Fig. 8 Hierarchical dynamics revealed by MSM analysis of the MISA network in the slow DNA-binding/unbinding parameter regime. All network parameters listed in Additional file 1: Table S1. **a** Eigenvalue spectrum of **T**(τ), $\tau = 1$, showing 16 dominant eigenvalues. **b** 4-macrostate MSM: 70% probability contours superimposed onto the quasipotential surface. In this parameter regime, separate attractors in the landscape are kinetically linked in the same subnetwork (see text). **c** 16-macrostate MSM showing 4 highly connected subnetworks (*colored ovals*). Each macrostate corresponds to a particular promoter binding-configuration (see numbering scheme in Additional file 1: Table S5). A pair of representative transition paths through the network are highlighted. *Red path*: most probable forward transition path from macrostate 1 to macrostate 11. *Blue path*: most probable reverse path from 11 to 1. **d** State transition graph for the 4-macrostate MSM. **e** Brute force SSA simulation of the MISA network over time. Trajectory is plotted according to the 16-macrostate (promoter configuration) indexing as in panel C and Additional file 1: Table S5. Colored panels reflect the four subnetworks/C = 4 macrostates. *Orange inset*: zoomed in trajectory segment, showing a switching event between the *red* and *green* subnetworks

suggest that multiple partially reprogrammed cell types could be accessible from a single initial cell state. Successful phenotype-transitions can occur predominantly through high-potential (unstable)—and thus difficult to observe experimentally—intermediate cell types. In future applications to specific gene GRNs, the MSM approach could predict a complex map of cell-reprogramming pathways, and thus potentially suggest combinations of targets towards improved safety and efficiency of reprogramming protocols. In synthetic biology applications, the method could be potentially used to optimize biochemical parameters in the design of synthetic gene circuits. For example, it may be desirable to realize synthetic switches with a very crisp on/off macrostate partitioning (i.e., lacking spurious intermediate states) to give a highly digital response.

Our study revealed that the two-gene MISA network can exhibit complex dynamic phenomena, involving a large number of metastable macrostates (up to 16), cycles and hierarchical dynamics, which can be conveniently visualized using the MSM. The quasipotential landscape has been used recently as a means of visualizing global dynamics and assessing locations and relative stabilities of phenotypic states of interest, in a manner that is quantitative (deriving strictly from underlying gene regulatory interactions), rather than qualitative or metaphorical (as was the case for the original Waddington epigenetic landscape) [21]. However, our study highlights the potential difficulty of interpreting global network dynamics based solely on the steady-state landscape, which is often projected onto one or two degrees of freedom. We found that phenotypically identical cell states—that is, network states marked by identical patterns of protein expression, inhabiting the same position in the projected landscape—can be separated by kinetic barriers, experiencing slow inter-conversion due to slow timescales for update to the epigenetic state (or promoter binding occupancy). Conversely, phenotypically distinct states marked by different levels of protein expression can be kinetically linked, experiencing relatively rapid inter-conversion. This type of stochastic inter-conversion is thought to occur in embryonic stem cells—for example, fluctuations in expression of the Nanog gene have been proposed to play a role in maintaining pluripotency [66, 67]. The hierarchical dynamics revealed by our study supports the idea that the phenotype of a cell could be more appropriately defined by dynamic patterns of regulator or marker expression levels [67], rather than on single-timepoint levels alone. This was seen in the 16-state MSM for the MISA network, where a given expression pattern (e.g., the Lo/Lo peak) comprised multiple macrostates from separate dynamic subnetworks.

Complex, high-dimensional dynamical systems call for systematic methods of coarse-graining (or dimensionality reduction), for analysis of mechanisms and extraction of information that can be compared with experimental results. In the field of Molecular Dynamics, the complexity of, e.g., macromolecular conformational changes—involving thousands of atomic degrees of freedom and multiple dynamic intermediates—has driven the development of automated methods for prediction and analysis of essential system dynamics from simulations [68, 69]. In that field, coarse-graining has been achieved based on a variety of so-called geometric (structural) or, alternatively, kinetic clustering methods [70, 71]. Noe, et al. [71], discussed that geometric (or structure-based) coarse-graining methods can fail to produce an accurate description of system dynamics when structurally similar molecular conformations are separated by large energy barriers or, conversely, when dissimilar structures are connected by fast transitions, as they found in a study of polypeptide folding dynamics. In such cases, kinetic (i.e., separation-of-timescale-based) coarse-graining methods such as the MSM approach are more appropriate. Our application of the MSMs to GRNs demonstrates how similar complex dynamic phenomena can manifest at the "network"-scale.

The challenge of solving the CME due to the curse-of-dimensionality is well known. The MSM approach is related to other projection-based model reduction methods that aim to reduce the computational burden of solving the CME directly by projecting the rate (or transition) matrix onto a smaller subspace or aggregated state-space with fewer degrees of freedom. Such approaches include the Finite State Projection algorithm [31], and methods based on Krylov subspaces [33, 72, 73], sparse-gridding [74], and separation-of-timescales [34, 74, 75] (related timescale-separation-based reduction methods have also been developed to analyze complex ODE models of biochemical networks, e.g., [76, 77]). The MSM is distinct from other timescale-based model reductions in that, rather than partitioning the system into categories of slow versus fast reactions [78] or species [34], or basing categories on physical intuition [75], it systematically groups microstates in such a way that maximizes metastability of aggregated states [40]. The practical benefit of this approach is its capacity to describe a system compactly in terms of long-lived, perhaps experimentally observable, states. Another important distinction between the MSM approach and other CME model reduction methods is that its primary end-goal is *not* to solve the CME per se. Rather, the emphasis in studies employing MSMs has generally been on gaining mechanistic, physical, or experimentally-relevant insights to complex system dynamics [79–81]. As such, the approach does not optimally balance the tradeoff between computational expense versus quantitative accuracy of the solution, as other methods have done explicitly [82]. Instead, the method can be considered to balance the tradeoff between accuracy and "human-interpretability", where decreasing the number of macrostates preserved in the MSM coarse-graining tends to favor the latter over the former.

A potential drawback of the workflow presented in this paper is that it requires an enumeration of the system state-space in order to construct the biochemical rate matrix **K**. Networks of increased complexity or molecular copy numbers will lead to prohibitively large matrix sizes. Here, we restricted our study to model systems with a relatively small number of reachable microstates (i.e., $\sim 10^4$ microstates permitted tractable computations on desktop computers with MATLAB [45]). However, it is important to point out that in typical applications of the MSM framework in Molecular Dynamics, the computational complexity of the coarse-graining procedure is largely decoupled from the full dimensionality of the system state-space, because it is often applied as part of a suite of tools for post-processing atomistic simulation data. An advantage of the MSM approach is its use of the stochastic transition matrix $\mathbf{T}(\tau)$ (rather than $\mathbf{K}$), which can be estimated from simulations by sampling transition counts between designated regions of state-space in trajectories of length τ [47]. Systems of increased complexity/dimensionality are generally more accessible to simulations, because the size of the state-space is automatically restricted to those states visited within finite-length simulations. Furthermore, in macromolecular systems with high-dimensional configuration spaces, clustering algorithms have been applied in order to obtain a tractable partitioning of state-space, prior to application of the MSM coarse-graining [47]. Typically, a large number of sampled configurations (10^4-10^7) is lumped into a more tractable number of 'microstates' (10^2-10^4), and the MSM framework subsequently identifies ~ tens of metastable macrostates. A recent study of G-protein-coupled receptor activation showcased the high complexity of systems that can be analyzed by MSMs: 250,000 sampled molecular structures were projected to coarse-grained MSMs with either 3000 or 10 states [83]. Based on these previous studies in Molecular Dynamics, we anticipate that the MSM framework will likewise prove useful in analysis of highly complex biochemical networks, particularly when coupled with stochastic simulations and thus bypassing the need for enumerating the CME. In ongoing work (Tse, et al., in preparation), we find that the MSM approach interfaces well with SSA simulations of biochemical network dynamics, combined with enhanced sampling techniques [84–86]. We anticipate that the approach could also potentially interface with other numerical approximation techniques that have been developed in recent years for reduction of the CME.

A potential challenge for the application of the PCCA + -based spectral clustering method to biochemical networks is that, as open systems, biochemical networks generally do not obey detailed balance. This means that the stochastic transition matrices do not have the property of irreversibility, which was originally taken to be a requirement for application of the PCCA algorithm [48]. However, later work by Roblitz et al. [49] found that the PCCA+ method also delivers an optimal clustering for irreversible systems. In this study, we found that the PCCA+ method could determine appropriate clusters in GRNs, and could furthermore uncover nonequilibrium cycles, as seen in the irreversibility (distinct forward and backward) of transition paths in the 16-state system. Newer methods of MSM building, which are specifically designed to treat nonequilibrium dynamical systems, have appeared recently [87]. It may prove fruitful to explore these alternative methods in order to identify the most appropriate, general MSM framework for application to various biochemical networks. On a separate note, another possible area for future study could be the relationship between the MSM framework, specifically its estimation of switching times in multistable networks, to the results from other theoretical approaches to GRNs, such as Large Deviation Theory [88] or Wentzel-Kramers-Brillouin theory [89].

Conclusions

In this work, we present a method for analyzing multistability and global state-switching dynamics in gene networks modeled by stochastic chemical kinetics, using the MSM framework. We found that the approach is able to: (1) identify the number and identities of long-lived phenotypic-states, or network "macrostates", (2) predict the steady-state probabilities of all macrostates along with probabilities of transitioning to other macrostates on a given timescale, and (3) decompose global dynamics into a set of dominant transition pathways and their associated relative probabilities, linking two system states of interest. Because the method is based on the discrete-space, stochastic transition matrix, it correctly identified stochastic multistability where a continuum model failed to find multiple steady states. The quantitative accuracy of the dynamics propagated by the coarse-grained MSM was highest in a parameter regime with slow DNA-binding and unbinding kinetics, indicating that in GRNs the assumption of memory-less hopping among a small number of macrostates is most valid in this regime. By projecting dynamics encompassing a large state-space onto a tractable number of macrostates, the MSMs revealed complex dynamic phenomena in GRNs, including hierarchical dynamics, nonequilibrium cycles, and alternative possible routes for phenotypic state-transitions. The ability to unravel these processes using the MSM framework can shed light on regulatory mechanisms that govern cell phenotype stability, and inform experimental reprogramming strategies. The MSM provides an intuitive representation of complex biological dynamics operating over multiple timescales, which in turn can provide the key to decoding biological mechanisms. Overall, our results demonstrate that the MSM framework—which has been generally applied thus far in the context of molecular dynamics via atomistic simulations—can be a useful tool for visualization and analysis of complex, multistable dynamics in gene networks, and in biochemical reaction networks more generally.

Abbreviations

CME: Chemical master equation; ETS: Exclusive toggle switch; GRN: Gene regulatory network; GRNs: Gene regulatory networks; MISA: Mutual inhibition/Self-activation; MSM: Markov state model; ODE: Ordinary Differential Equation; PCCA+: Robust Perron Cluster Analysis; SSA: Stochastic simulation algorithm

Acknowledgements
We thank Jun Allard for helpful discussions.

Funding
We acknowledge financial support from the UC Irvine Henry Samueli School of Engineering.

Authors' contributions
BC and ER designed and performed research. MT contributed to data analysis and manuscript preparation. RS contributed to data analysis. BC and ER wrote the manuscript. All authors read and approved the final manuscript.

Competing interests
The authors declare that they have no competing interests.

References

1. Arkin A, Ross J, McAdams HH. Stochastic kinetic analysis of developmental pathway bifurcation in phage lambda-infected Escherichia coli cells. Genetics. 1998;149(4):1633–48.
2. Xiong W, Ferrell JE. A positive-feedback-based bistable 'memory module' that governs a cell fate decision. Nature. 2003;426(6965):460–5.
3. Zhou JX, Huang S. Understanding gene circuits at cell-fate branch points for rational cell reprogramming. Trends Genet. 2011;27(2):55–62.
4. Lu M, Jolly MK, Gomoto R, Huang B, Onuchic J, Ben-Jacob E. Tristability in Cancer-Associated MicroRNA-TF Chimera Toggle Switch. J Phys Chem B. 2013;117(42):13164–74.
5. Feng H, Wang J. A new mechanism of stem cell differentiation through slow binding/unbinding of regulators to genes. Sci Rep. 2012;2:550.
6. Zhang B, Wolynes PG. Stem cell differentiation as a many-body problem. Proc Natl Acad Sci. 2014;111(28):10185–90.
7. Wang P, Song C, Zhang H, Wu Z, Tian X-J, Xing J. Epigenetic state network approach for describing cell phenotypic transitions. Interface Focus. 2014; 4(3):20130068.
8. Hong T, Xing J, Li L, Tyson JJ. A mathematical model for the reciprocal differentiation of T helper 17 cells and induced regulatory T cells. PLoS Comput Biol. 2011;7(7):e1002122.
9. Graf T, Enver T. Forcing cells to change lineages. Nature. 2009;462(7273): 587–94.
10. Huang S. The molecular and mathematical basis of Waddington's epigenetic landscape: a framework for post-Darwinian biology? Bioessays. 2012;34(2):149–57.
11. Lang AH, Li H, Collins JJ, Mehta P. Epigenetic landscapes explain partially reprogrammed cells and identify key reprogramming genes. PLoS Comput Biol. 2014;10(8):e1003734.
12. Elowitz MB. Stochastic gene expression in a single cell. Science. 2002; 297(5584):1183–6.
13. Ozbudak EM, Thattai M, Kurtser I, Grossman AD, van Oudenaarden A. Regulation of noise in the expression of a single gene. Nat Genet. 2002; 31(1):69–73.
14. Golding I, Paulsson J, Zawilski SM, Cox EC. Real-time kinetics of gene activity in individual bacteria. Cell. 2005;123(6):1025–36.
15. Balaban NQ. Bacterial persistence as a phenotypic switch. Science. 2004; 305(5690):1622–5.
16. Acar M, Mettetal JT, van Oudenaarden A. Stochastic switching as a survival strategy in fluctuating environments. Nat Genet. 2008;40(4):471–5.
17. Sharma SV, Lee DY, Li B, Quinlan MP, Takahashi F, Maheswaran S, McDermott U, Azizian N, Zou L, Fischbach MA, et al. A chromatin-mediated reversible drug-tolerant state in cancer cell subpopulations. Cell. 2010;141(1):69–80.
18. Chang HH, Hemberg M, Barahona M, Ingber DE, Huang S. Transcriptome-wide noise controls lineage choice in mammalian progenitor cells. Nature. 2008;453(7194):544–7.
19. Dietrich JE, Hiiragi T. Stochastic patterning in the mouse pre-implantation embryo. Development. 2007;134(23):4219–31.
20. Yuan L, Chan GC, Beeler D, Janes L, Spokes KC, Dharaneeswaran H, Mojiri A, et al. A role of stochastic phenotype switching in generating mosaic endothelial cell heterogeneity. Nat Commun. 2016;7:10160.
21. Waddington CH. The Strategy of the Genes. London: Allen & Unwin; 1957.
22. Wang J, Zhang K, Xu L, Wang E. Quantifying the Waddington landscape and biological paths for development and differentiation. Proc Natl Acad Sci. 2011;108(20):8257–62.
23. Karlebach G, Shamir R. Modelling and analysis of gene regulatory networks. Nat Rev Mol Cell Biol. 2008;9(10):770–80.
24. Kepler TB, Elston TC. Stochasticity in transcriptional regulation: origins, consequences, and mathematical representations. Biophys J. 2001;81(6): 3116–36.
25. Shahrezaei V, Swain PS. Analytical distributions for stochastic gene expression. Proc Natl Acad Sci. 2008;105(45):17256–61.
26. Mackey MC, Tyran-Kamińska M, Yvinec R. Dynamic behavior of stochastic gene expression models in the presence of bursting. SIAM J Appl Math. 2013;73(5):1830–52.
27. Jiao F, Sun Q, Tang M, Yu J, Zheng B. Distribution modes and their corresponding parameter regions in stochastic gene transcription. SIAM J Appl Math. 2015;75(6):2396–420.
28. Schultz D, Onuchic JN, Wolynes PG. Understanding stochastic simulations of the smallest genetic networks. J Chem Phys. 2007;126(24):245102.
29. Ramos AF, Innocentini GCP, Hornos JEM. Exact time-dependent solutions for a self-regulating gene. Phys Rev E. 2011;83(6):062902.
30. Gillespie DT. Exact stochastic simulation of coupled chemical reactions. J Phys Chem. 1977;81(25):2340–61.
31. Munsky B, Khammash M. The finite state projection algorithm for the solution of the chemical master equation. J Chem Phys. 2006;124(4):044104.
32. Cao Y, Liang J. Optimal enumeration of state space of finitely buffered stochastic molecular networks and exact computation of steady state landscape probability. BMC Syst Biol. 2008;2(1):30.
33. Wolf V, Goel R, Mateescu M, Henzinger TA. Solving the chemical master equation using sliding windows. BMC Syst Biol. 2010;4(1):42.
34. Pahlajani CD, Atzberger PJ, Khammash M. Stochastic reduction method for biological chemical kinetics using time-scale separation. J Theor Biol. 2011; 272(1):96–112.
35. Sidje RB, Vo HD. Solving the chemical master equation by a fast adaptive finite state projection based on the stochastic simulation algorithm. Math Biosci. 2015;269:10–6.
36. Huang S, Guo YP, May G, Enver T. Bifurcation dynamics in lineage-commitment in bipotent progenitor cells. Dev Biol. 2007;305(2):695–713.
37. Ma R, Wang J, Hou Z, Liu H. Small-number effects: a third stable state in a genetic bistable toggle switch. Phys Rev Lett. 2012;109(24):248107.
38. Cao Y, Lu H-M, Liang J. Probability landscape of heritable and robust epigenetic state of lysogeny in phage lambda. Proc Natl Acad Sci. 2010; 107(43):18445–50.
39. Munsky B, Fox Z, Neuert G. Integrating single-molecule experiments and discrete stochastic models to understand heterogeneous gene transcription dynamics. Methods. 2015;85:12–21.
40. Pande VS, Beauchamp K, Bowman GR. Everything you wanted to know about Markov State Models but were afraid to ask. Methods. 2010;52(1):99–105.
41. Chodera JD, Noé F. Markov state models of biomolecular conformational dynamics. Curr Opin Struct Biol. 2014;25:135–44.
42. Bowman GR, Huang X, Pande VS. Network models for molecular kinetics and their initial applications to human health. Cell Res. 2010;20(6):622–30.
43. Sanft KR, Wu S, Roh M, Fu J, Lim RK, Petzold LR. StochKit2: software for discrete stochastic simulation of biochemical systems with events. Bioinformatics. 2011;27(17):2457–8.
44. van Kampen NG. Stochastic processes in physics and chemistry. Amsterdam; Boston; London: Elsevier; 2007.
45. The MathWorks. MATLAB Release. Natick: Massachusetts;2015a.
46. Scherer MK, Trendelkamp-Schroer B, Paul F, Perez-Hernandez G, Hoffmann M, Plattner N, Wehmeyer C, Prinz J, Noé F. PyEMMA 2: a software package for estimation, validation, and analysis of Markov Models. J Chem Theory Comput. 2015;11(11):5525–42.
47. Prinz JH, Wu H, Sarich M, Keller B, Senne M, Held M, Chodera JD, Schütte C, Noé F. Markov models of molecular kinetics: generation and validation. J Chem Phys. 2011;134(17):174105.

48. Deuflhard P, Huisinga W, Fischer A, Schütte C. Identification of almost invariant aggregates in reversible nearly uncoupled Markov chains. Linear Algebra Its Appl. 2000;315(1–3):39–59.
49. Röblitz S, Weber M. Fuzzy spectral clustering by PCCA+: application to Markov state models and data classification. Adv Data Anal Classif. 2013;7(2):147–79.
50. Buchete NV, Hummer G. Coarse master equations for peptide folding dynamics †. J Phys Chem B. 2008;112(19):6057–69.
51. Beauchamp KA, Bowman GR, Lane TJ, Maibaum L, Haque IS, Pande VS. MSMBuilder2: modeling conformational dynamics on the picosecond to millisecond scale. J Chem Theory Comput. 2011;7(10):3412–9.
52. W. E and Vanden-Eijnden E. Towards a Theory of Transition Paths. J Stat Phys. 2006;123(3):503–523.
53. Metzner P, Schütte C, Vanden-Eijnden E. Transition path theory for Markov jump processes. Multiscale Model Simul. 2009;7(3):1192–219.
54. Noe F, Schutte C, Vanden-Eijnden E, Reich L, Weikl TR. Constructing the equilibrium ensemble of folding pathways from short off-equilibrium simulations. Proc Natl Acad Sci. 2009;106(45):19011–6.
55. Schultz D, Walczak AM, Onuchic JN, Wolynes PG. Extinction and resurrection in gene networks. Proc Natl Acad Sci. 2008;105(49):19165–70.
56. Morelli MJ, Tănase-Nicola S, Allen RJ, ten Wolde PR. Reaction coordinates for the flipping of genetic switches. Biophys J. 2008;94(9):3413–23.
57. Huang S. Reprogramming cell fates: reconciling rarity with robustness. Bioessays. 2009;31(5):546–60.
58. Huang S. Hybrid T-helper cells: stabilizing the moderate center in a polarized system. PLoS Biol. 2013;11(8):e1001632.
59. Gardmer T, Cantor C, Collins J. Construction of a genetic toggle switch in Escherichia coli. Nature. 2000;403(6767):339–42.
60. Lipshtat A, Loinger A, Balaban NQ, Biham O. Genetic toggle switch without cooperative binding. Phys Rev Lett. 2006;96(18):188101.
61. Hornos JEM, Schultz D, Innocentini GC, Wang JA, Walczak AM, Onuchic JN, Wolynes PG. Self-regulating gene: An exact solution. Phys Rev E. 2005;72(5):051907.
62. Lane TJ, Bowman GR, Beauchamp K, Voelz VA, Pande VS. Markov State Model reveals folding and functional dynamics in ultra-long MD trajectories. J Am Chem Soc. 2011;133(45):18413–9.
63. Tse MJ, Chu BK, Roy M, Read EL. DNA-binding kinetics determines the mechanism of noise-induced switching in gene networks. Biophys J. 2015; 109(8):1746–57.
64. Berezhkovskii A, Hummer G, Szabo A. Reactive flux and folding pathways in network models of coarse-grained protein dynamics. J Chem Phys. 2009; 130(20):205102.
65. Walczak AM, Onuchic JN, Wolynes PG. Absolute rate theories of epigenetic stability. Proc Natl Acad Sci. 2005;102(52):18926–31.
66. Chambers I, Silva J, Colby D, Nichols J, Nijmeijer B, Robertson M, Vrana J, Jones K, Grotewold L, Smith A. Nanog safeguards pluripotency and mediates germline development. Nature. 2007;450(7173):1230–4.
67. Kalmar T, Lim C, Hayward P, Muñoz-Descalzo S, Nichols J, Garcia-Ojalvo J, Arias AM. Regulated fluctuations in nanog expression mediate cell fate decisions in embryonic stem cells. PLoS Biol. 2009;7(7):e1000149.
68. Chodera JD, Singhal N, Pande VS, Dill KA, Swope WC. Automatic discovery of metastable states for the construction of Markov models of macromolecular conformational dynamics. J Chem Phys. 2007;126(15):155101.
69. Bowman GR, Beauchamp KA, Boxer G, Pande VS. Progress and challenges in the automated construction of Markov state models for full protein systems. J Chem Phys. 2009;131(12):124101.
70. Deuflhard P, Weber M. Robust Perron cluster analysis in conformation dynamics. Linear Algebra Its Appl. 2005;398:161–84.
71. Pérez-Hernández G, Paul F, Giorgino T, De Fabritiis G, Noé F. Identification of slow molecular order parameters for Markov model construction. J Chem Phys. 2013;139(1):015102.
72. Burrage K, Hegland M, Macnamara S, Sidje R. A Krylov-based finite state projection algorithm for solving the chemical master equation arising in the discrete modelling of biological systems, Proceedings of the Markov 150th Anniversary Conference. 2006.
73. Cao Y, Terebus A, Liang J. Accurate chemical master equation solution using multi-finite buffers. Multiscale Model Simul. 2016;14(2):923–63.
74. Hegland M, Burden C, Santoso L, MacNamara S, Booth H. A solver for the stochastic master equation applied to gene regulatory networks. J Comput Appl Math. 2007;205(2):708–24.
75. Peleš S, Munsky B, Khammash M. Reduction and solution of the chemical master equation using time scale separation and finite state projection. J Chem Phys. 2006;125(20):204104.
76. Anna L, Csikász-Nagy A, Gy Zsély I, Zádor J, Turányi T, Novák B. Time scale and dimension analysis of a budding yeast cell cycle model. BMC Bioinformatics. 2006;7:494.
77. Surovtsova I, Simus N, Lorenz T, Konig A, Sahle S, Kummer U. Accessible methods for the dynamic time-scale decomposition of biochemical systems. Bioinformatics. 2009;25(21):2816–23.
78. Haseltine EL, Rawlings JB. Approximate simulation of coupled fast and slow reactions for stochastic chemical kinetics. J Chem Phys. 2002;117(15):6959.
79. Kuroda Y, Suenaga A, Sato Y, Kosuda S, Taiji M. All-atom molecular dynamics analysis of multi-peptide systems reproduces peptide solubility in line with experimental observations. Sci Rep. 2016;6:19479.
80. Jayachandran G, Vishal V, Pande VS. Using massively parallel simulation and Markovian models to study protein folding: Examining the dynamics of the villin headpiece. J Chem Phys. 2006;124(16):164902.
81. Singhal N, Snow CD, Pande VS. Using path sampling to build better Markovian state models: predicting the folding rate and mechanism of a tryptophan zipper beta hairpin. J Chem Phys. 2004;121(1):415.
82. Tapia JJ, Faeder JR, Munsky B. Adaptive coarse-graining for transient and quasi-equilibrium analyses of stochastic gene regulation. 2012. p. 5361–6.
83. Kohlhoff KJ, Shukla D, Lawrenz M, Bowman GR, Konerding DE, Belov D, Altman RB, Pande VS. Cloud-based simulations on Google Exacycle reveal ligand modulation of GPCR activation pathways. Nat Chem. 2013;6(1):15–21.
84. Bhatt D, Bahar I. An adaptive weighted ensemble procedure for efficient computation of free energies and first passage rates. J Chem Phys. 2012; 137(10):104101.
85. Zhang BW, Jasnow D, Zuckerman DM. The 'weighted ensemble' path sampling method is statistically exact for a broad class of stochastic processes and binning procedures. J Chem Phys. 2010;132(5):054107.
86. Adelman JL, Grabe M. Simulating rare events using a weighted ensemble-based string method. J Chem Phys. 2013;138(4):044105.
87. Marcus W, Fackeldey K. G-pcca: Spectral clustering for non-reversible markov chains. ZIB Rep. 2015;15(35).
88. Lv C, Li X, Li F, Li T. Constructing the energy landscape for genetic switching system driven by intrinsic noise. PLoS One. 2014;9(2):e88167.
89. Assaf M, Roberts E, Luthey-Schulten Z. Determining the Stability of Genetic Switches: Explicitly Accounting for mRNA Noise. Phys Rev Lett. 2011;106(24):248102.

4

SEQUOIA: significance enhanced network querying through context-sensitive random walk and minimization of network conductance

Hyundoo Jeong and Byung-Jun Yoon*

Abstract

Background: Network querying algorithms provide computational means to identify conserved network modules in large-scale biological networks that are similar to known functional modules, such as pathways or molecular complexes. Two main challenges for network querying algorithms are the high computational complexity of detecting potential isomorphism between the query and the target graphs and ensuring the biological significance of the query results.

Results: In this paper, we propose SEQUOIA, a novel network querying algorithm that effectively addresses these issues by utilizing a context-sensitive random walk (CSRW) model for network comparison and minimizing the network conductance of potential matches in the target network. The CSRW model, inspired by the pair hidden Markov model (pair-HMM) that has been widely used for sequence comparison and alignment, can accurately assess the node-to-node correspondence between different graphs by accounting for node insertions and deletions. The proposed algorithm identifies high-scoring network regions based on the CSRW scores, which are subsequently extended by maximally reducing the network conductance of the identified subnetworks.

Conclusions: Performance assessment based on real PPI networks and known molecular complexes show that SEQUOIA outperforms existing methods and clearly enhances the biological significance of the query results. The source code and datasets can be downloaded from http://www.ece.tamu.edu/~bjyoon/SEQUOIA.

Background

Protein-protein interaction (PPI) plays pivotal roles in understanding biological systems. Diverse functional modules in cells, such as signaling pathways and protein complexes, involve numerous proteins and their functions are governed by the intertwined interactions among these proteins. For this reason, to better understand the functions and roles of proteins in cells, it is critically important to investigate how groups of proteins collaborate with each other to perform certain biological functions and achieve common goals, in addition to studying the functions of individual proteins. Recent advances in technologies for high throughput measurement of protein-protein interactions have enabled genome-scale studies of protein interactions, and systematic analyses of the available PPI networks may reveal new functional network modules and unveil novel functionalities of the proteins that are involved in such modules. Recent investigations of PPI networks show that functionally important network modules (e.g., molecular complexes and pathways) are often well conserved across networks of different species [1, 2]. These observations clearly point to comparative network analysis [3] as a promising solution for effectively analyzing large-scale PPI networks, detecting common

*Correspondence: bjyoon@ece.tamu.edu
Department of Electrical and Computer Engineering, Texas A&M University, College Station, TX, USA

functional modules that are embedded in the networks, and predicting the functions of proteins that comprise these modules.

Network querying is one possible way of comparatively analyzing biological networks, which can be especially useful when prior knowledge of functional modules is available for a given species. As implied in its name, network querying aims to find out whether a target network (typically, belonging to another species) contains network modules that resemble the module that is being used as the query [3]. This provides an efficient way of transferring knowledge between species, since we could use computational means to predict potential network modules in a new (or less-studied) species that may have similar functions, structures, and underlying mechanisms to well-studied modules in other species.

Several network querying algorithms have been proposed so far [4–10]. PathBLAST [4] has been designed to identify conserved signaling pathways. However, it can only handle linear pathways and its high computational complexity places a stringent restriction on the maximum length of the pathway that could be searched. QPath [5] can search for longer pathways and QNet [6] can search for linear pathways as well as trees, but both algorithms are not suitable for large queries due to their high computational cost. To overcome restrictions on the topology of the query network, several network querying algorithms have been proposed that can identify network modules with arbitrary topology [7–10]. For example, TORQUE [7] finds a connected subnetwork of matching proteins in the target network based on sequence similarity, without explicitly utilizing the topological structure of the query network in identifying conserved functional modules. NatalieQ [10] formulates the network alignment problem as an integer linear programming problem, and solves the optimization problem using Lagrangian relaxation combined with a branch-and-bound approach. RESQUE [8] adopts a semi-Markov random walk (SMRW) model to estimate the node correspondence between the query and the target networks, based on which it iteratively reduces the target network by removing irrelevant nodes. Once the target network has been sufficiently reduced, RESQUE identifies the best matching subnetwork either by the Hungarian method or by identifying the largest connected subnetwork. Another recent algorithm, called Corbi [9], measures the node correspondence between networks based on a conditional random field (CRF), after which the matching subnetwork is identified through an iterative bi-directional mapping.

Most of the aforementioned network querying methods consider both *node similarity* and *topological similarity* between the query and the target networks to detect matching subnetworks in the target network. Node similarity between nodes that belong to different networks is typically measured based on sequence similarity. Topological similarity between (sub)networks are measured in various ways to capture the molecular interaction patterns that are conserved across networks. Incorporating both types of similarities has been shown to be crucial in making biologically relevant predictions about conserved functional modules [1–3, 11]. However, one important aspect of network module detection that is often neglected in network querying is that such modules are often well separated from the rest of the network. In fact, this separability has played critical roles in "non-comparative" network analysis methods that aim to detect modules or sub-communities in a given network [12–14], since molecules in a functional module tend to be densely connected to other molecules in the same module but loosely connected to nodes that are not part of the module. Although identifying densely connected subnetwork modules is not the main objective of network querying, explicitly incorporating separability criterion into comparative network analysis methods has strong potentials to enhance the quality of the predictions [15].

In this paper, we propose a novel network querying algorithm called SEQUOIA (Significance Enhanced QUerying Of InterAction networks). The proposed algorithm is built on the following important concepts: (i) effective estimation of *node correspondence* – or overall functional similarity between nodes in different networks – by sensibly combining sequence similarity and interaction pattern similarity through a random walk model; and (ii) minimization of network conductance of potential network modules, thereby identifying matching modules in the target network that are well separated from the rest of the network. In our proposed algorithm, we first estimate the node correspondence based on a context-sensitive random walk model [16, 17], and select a seed network based on the estimated node correspondence scores. Then, the seed network is iteratively extended by adding the nodes that maximally reduce the conductance of the subnetwork. Finally, the significance enhanced querying result is achieved by keeping the nodes with acceptable extension reward scores, which are updated for every node at each extension step. Through extensive evaluations based on real biological complexes, we show that SEQUOIA can remarkably enhance the biological significance of the network querying results by estimating the node correspondence based on the CSRW model and minimizing the conductance of matching network modules.

Methods

Problem formulation and overview of the proposed method

Suppose that we have a query protein-protein interaction (PPI) network represented by a graph $\mathcal{G}_Q = (\mathcal{V}_Q, \mathcal{E}_Q)$, which has a set of nodes $\mathcal{V}_\mathcal{Q} = \{v_1, v_2, \ldots\}$ and set of edges $\mathcal{E}_\mathcal{Q} = \{e_{i,j}\}$. A protein in the query network is represented as a node $v_i \in \mathcal{V}_\mathcal{Q}$ in the graph $\mathcal{G}_\mathcal{Q}$ and the interaction between two proteins v_i and v_j is represented by an edge $e_{i,j}$, whose weight $w_{i,j}$ reflects the strength (or confidence) of the interaction. Similarly, suppose we are also given a target PPI network represented by a graph $\mathcal{G}_\mathcal{T} = (\mathcal{V}_\mathcal{T}, \mathcal{E}_\mathcal{T})$. We define the size of a network as the number of nodes in the given network, hence the size of the query network is $|\mathcal{V}_\mathcal{Q}|$ and that of the target network is $|\mathcal{V}_\mathcal{T}|$. Typically, in a network querying problem, the size of the target network is significantly larger than the query network (i.e., $|\mathcal{V}_\mathcal{Q}| \ll |\mathcal{V}_\mathcal{T}|$). We assume that a pairwise node similarity score $s\left(v_q, v_t\right)$ is available $\forall v_q \in \mathcal{V}_\mathcal{Q}$ and $\forall v_t \in \mathcal{V}_\mathcal{T}$, reflecting the molecular level similarity between the proteins in the query network and the target PPI network. In this study, we use the BLAST bit score as the pairwise node similarity score as in most network querying and alignment algorithms.

The main objective of network querying is to find the conserved subnetwork $\hat{\mathcal{G}}_\mathcal{T} = \left(\hat{\mathcal{V}}_\mathcal{T}, \hat{\mathcal{E}}_\mathcal{T}\right)$ within the target PPI network $\mathcal{G}_\mathcal{T} = (\mathcal{V}_\mathcal{T}, \mathcal{E}_\mathcal{T})$ that bears the largest overall functional similarity to the given query network $\mathcal{G}_\mathcal{Q}$. Therefore, we can formulate the network querying problem as the following optimization problem:

$$\hat{\mathcal{G}}_\mathcal{T}^* = \underset{\hat{\mathcal{G}}_\mathcal{T} \in \mathbf{G}_\mathrm{T}}{\arg\max} f\left(\hat{\mathcal{G}}_\mathcal{T}, \mathcal{G}_\mathcal{Q}\right), \tag{1}$$

where $\mathbf{G}_\mathrm{T}$ is the set of all possible subnetworks of the target PPI network, and $f\left(\mathcal{G}_x, \mathcal{G}_y\right)$ is a function that measures the overall functional similarity between two networks $\mathcal{G}_x$ and $\mathcal{G}_y$.

The network querying problem can be reformulated as a subgraph isomorphism problem, whose goal is to find a bijection between two graphs. In order to find a one-to-one mapping, deleted nodes can be modeled as dummy nodes so that an inserted node in the query network can be mapped to a dummy node in the target network, and vice versa. The subgraph isomorphism problem is known to be NP-complete [18], hence the existence of a polynomial time algorithm for solving the problem is unknown. Furthermore, it is also not straightforward to quantitatively estimate the overall functional similarity $f\left(\mathcal{G}_x, \mathcal{G}_y\right)$ between two networks $\mathcal{G}_x$ and $\mathcal{G}_y$ in such a way that is biologically meaningful. As a result, it is practically challenging to effectively formulate the optimization problem in (1) and solve the problem for large-scale networks in a computationally efficient manner [6–8]. A reasonable way to estimate this functional similarity is to define $f\left(\mathcal{G}_x, \mathcal{G}_y\right)$ by sensibly combining the node similarity and the topological similarity between the networks under comparison [3]. Given a reasonable $f\left(\mathcal{G}_x, \mathcal{G}_y\right)$, heuristic optimization schemes may have to be employed to make the optimization problem (1) computationally tractable.

In our proposed network querying algorithm SEQUOIA, we first pre-process the target network by removing non-homologous nodes and inserting pseudo-edges between nodes that are likely to share similar functionalities. Next, the query and the target networks are compared and node correspondence scores are estimated using the context-sensitive random walk (CSRW) model [16]. The resulting scores are used to select a "seed network" that consists of target nodes that have strong correspondence to query nodes. The seed network is extended by iteratively adding the nodes that maximally reduce the network conductance of the extended network, through which SEQUOIA aims to find a subnetwork that is densely connected within the subnetwork while sparsely connected to the rest of the target network. This has the effect of identifying a subnetwork in the target PPI network that closely matches the query, and at the same time, has strong potential to be a functional network module. Finally, the extended subnetwork is pruned by removing potentially irrelevant nodes that contribute little to making the network dense, which improves the functional coherence of the querying results, as will be demonstrated later.

The context-sensitive random walk (CSRW) model

Here, we briefly review the CSRW model [16] that is used for estimating the correspondence between nodes in the query and the target networks. To accurately estimate the node correspondence, it is desirable to effectively integrate the node similarity (sequence similarity between proteins) and topological similarity (similarity between interaction patterns for different proteins), as mentioned previously. However, as depicted in Fig. 1, inserted and deleted nodes in the conserved network can make effective estimation of the node correspondence

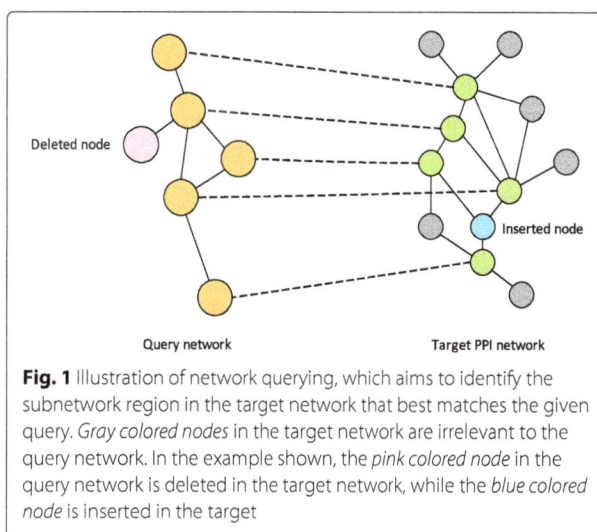

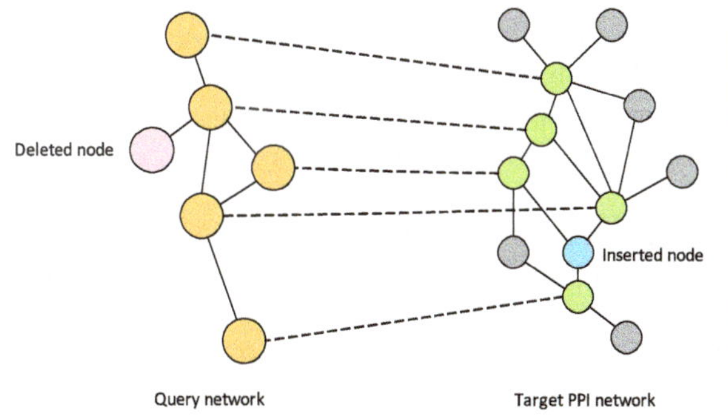

Fig. 1 Illustration of network querying, which aims to identify the subnetwork region in the target network that best matches the given query. *Gray colored nodes* in the target network are irrelevant to the query network. In the example shown, the *pink colored node* in the query network is deleted in the target network, while the *blue colored node* is inserted in the target

difficult. The CSRW model has been recently proposed to explicitly model such node insertions and deletions, while integrating the two types of similarities to compute the node correspondence scores.

At a given moment, the random walker in the CSRW model is located at a node pair (v_q, v_t), where v_q is a node in the query network and v_t is a node in the target network. At each step, the walker makes a random move to neighboring nodes, where it switches between two different types of modes of random walk – namely, *simultaneous walk* on both networks and *individual walk* on one of the networks – depending on the surrounding "context" of the walker's current location. For example, if v_q and v_t have neighboring nodes with positive node similarity, the random walker simultaneously moves on both networks to such nodes. Otherwise, it randomly selects one of the networks and randomly moves only on the selected network. Further details of the CSRW model can be found in the supplementary material (see Additional file 1: Section S1). In the long run, the random walker is designed to simultaneously visit node pairs with better correspondence (i.e., with higher node similarity and topological similarity) more frequently. Based on the design, the long-run proportion of time that the random walker simultaneously visits a given pair of nodes can be used as a probabilistic measure of the correspondence between the nodes [16]. This long-run proportion of time, or the steady-state probability of the CSRW model, can be efficiently computed in practice using the power method, as real PPI networks tend to be very sparse [8, 19]. We use the steady-state probability of the context-sensitive random walker as the node correspondence score $c\left(v_q, v_t\right), \forall v_q \in \mathcal{V}_Q$ and $\forall v_t \in \mathcal{V}_T$, and the node correspondence scores for all node pairs can be concisely written in a $|\mathcal{V}_Q| \times |\mathcal{V}_T|$ dimensional matrix **C**. The context-awareness of the CSRW model makes it robust to potential node insertions/deletions, and the model has been shown to be useful for estimating node correspondence [16]. In fact, the CSRW-based node correspondence scores have been recently applied to multiple network alignment [17], where they have been shown to clearly enhance the overall alignment accuracy.

SEQUOIA network querying algorithm

Before computing the node correspondence scores based on the CSRW model, we perform two pre-processing steps. First, we reduce the target network by removing potential non-homologous nodes. Specifically, we remove every node v_t in the target network whose node similarity $s\left(v_q, v_t\right)$ never exceeds a given threshold T_h for any of the query nodes $v_q \in \mathcal{V}_Q$. In this study, we set the threshold T_h as 0, such that a node is kept in the target network if it has at least one query node with nonzero similarity score. Removing target nodes that do not have any homologous node in the query network can significantly reduce the computation time as well as the memory requirement. Second, since removing non-homologous nodes may make the target network disconnected, we insert a pseudo-edge between nodes that are likely to share similar functionalities, motivated by the fact that proteins with direct interactions are more likely to share similar functionalities [20]. For this purpose, we assumed that any two nodes in the target network are likely to share similar functionalities and may potentially have a direct interaction if they have a common node in the query network with high node similarity. However, to refrain from inserting too many false-positive pseudo edges, we only insert a pseudo edge if the two nodes under consideration belong to different subnetworks that are disconnected from each other. Since current PPI networks are incomplete and noisy – with many false positive interactions as well as false negative interactions [21, 22] – adding pseudo-edges to the reduced target network can lead to more reliable querying results, as will be demonstrated in our simulation results. Further details of the pre-processing step can be found in the supplementary material (see Additional file 1: Section S2) with an illustrative example.

After pre-processing the target network, the CSRW model is used to estimate the correspondence between nodes in the query and the target networks. The resulting node correspondence score matrix **C** is normalized to obtain the normalized score matrix $\bar{\mathbf{C}}$ using the normalization method proposed in [19]:

$$\bar{\mathbf{C}} = \frac{1}{2}\left[\mathbf{J_L} \cdot \mathbf{C} + \mathbf{C} \cdot \mathbf{J_R}\right]. \tag{2}$$

The matrix $\bar{\mathbf{C}}$ is a $|\mathcal{V}_Q| \times |\mathcal{V}_T|$ dimensional matrix containing the normalized node correspondence scores, $\mathbf{J_L}$ is a $|\mathcal{V}_Q| \times |\mathcal{V}_Q|$ dimensional diagonal matrix with the diagonal term $\mathbf{J_L}(q,q) = 1 \Big/ \sum_{t=1}^{|\mathcal{V}_T|} c\left(v_q, v_t\right)$, and $\mathbf{J_R}$ is a $|\mathcal{V}_T| \times |\mathcal{V}_T|$ dimensional diagonal matrix with the diagonal term $\mathbf{J_R}(t,t) = 1 \Big/ \sum_{q=1}^{|\mathcal{V}_Q|} c\left(v_q, v_t\right)$. This normalization step aims to estimate the *relative* significance between corresponding nodes, which has been shown to be useful for comparing networks of different size [19]. Based on the normalized correspondence score $\bar{\mathbf{C}}$, we iteratively select N_Q seed nodes in the target network based on the following rule:

$$\arg\min_{v_t} \left[\prod_{v_q \in \mathcal{V}_Q} \left(1 - \bar{c}\left(v_q, v_t\right)\right) \right]. \tag{3}$$

The above selection rule aims to identify the nodes in the target network that have a large number of highly corresponding nodes in the query network. The score $\bar{c}(v_q, v_t)$ will be close to 1 for a highly corresponding node pair (v_q, v_t). Therefore, the product $\prod_{v_q \in \mathcal{V}_Q} (1 - \bar{c}(v_q, v_t))$ will approach 0 for a target node v_t (i.e., a potential seed node) that has a large number of query nodes $v_q \in \mathcal{V}_Q$ with a high node correspondence score $\bar{c}(v_q, v_t)$. This is based on an assumption that a target node with a larger number of relevant nodes in the query network may be more likely to be involved in similar functions as the query network compared to a node that has fewer corresponding nodes. After selecting the N_Q seeds, we find the largest connected subnetwork based on the N_Q seed nodes, which is referred to as the seed network. In this work, we set $N_Q = |\mathcal{V}_Q|$ so that the size of the seed network does not exceed the size of the query network.

Once the seed network is obtained, we iteratively extend the network by adding nodes that can make the extended network well-separated from the rest of the network. To this aim, we estimate the conductance of the subnetwork and define the extension reward score for each node as follows. First, given a network $\mathcal{G} = (\mathcal{V}_\mathcal{G}, \mathcal{E}_\mathcal{G})$, suppose that we have a Gaussian surface enclosing the subnetwork $\mathcal{H} = (\mathcal{V}_\mathcal{H}, \mathcal{E}_\mathcal{H})$ such that $\mathcal{H} \subseteq \mathcal{G}$. Then, the conductance φ of the subnetwork $\mathcal{H}$ is defined as the number of edges that pass through the surface divided by the volume of the subnetwork (i.e., the number of edges that are enclosed by the surface) [23, 24]. The conductance of the subnetwork $\mathcal{H}$ is given by

$$\phi(\mathcal{H}) = \frac{\left|\left\{e_{i,j} | i \in \mathcal{V}_\mathcal{H}, j \in \mathcal{V}_{\bar{\mathcal{H}}}\right\}\right|}{\min\left(vol(\mathcal{V}_\mathcal{H}), vol\left(\mathcal{V}_{\bar{\mathcal{H}}}\right)\right)}, \quad (4)$$

where $\bar{\mathcal{H}} = (\mathcal{V}_\mathcal{G} \backslash \mathcal{V}_\mathcal{H}, \mathcal{E}_\mathcal{G} \backslash \mathcal{E}_\mathcal{H})$, and $vol(\mathcal{V}_\mathcal{X}) = \sum_{u \in \mathcal{V}_\mathcal{X}} d(u)$, where $d(u)$ is the degree of the node u. In a network querying problem, since the conserved subnetwork is typically significantly smaller than the rest of the target PPI network, the volume of the querying result is also much smaller than the volume of the rest of the target network, i.e., $vol(\mathcal{V}_\mathcal{H}) \ll vol\left(\mathcal{V}_{\bar{\mathcal{H}}}\right)$. Hence, the conductance of the subnetwork $\mathcal{H}$ becomes

$$\phi(\mathcal{H}) = \frac{\left|\left\{e_{i,j} | i \in \mathcal{V}_\mathcal{H}, j \in \mathcal{V}_{\bar{\mathcal{H}}}\right\}\right|}{vol(\mathcal{V}_\mathcal{H})} = \frac{\left|\left\{e_{i,j} | i \in \mathcal{V}_\mathcal{H}, j \in \mathcal{V}_{\bar{\mathcal{H}}}\right\}\right|}{\left|\left\{e_{i,j} | i, j \in \mathcal{V}_\mathcal{H}\right\}\right|}. \quad (5)$$

Second, we define the extension reward score for a given node as the number of newly added neighboring nodes during the extension step. That is, in each extension step, when we add a new node, all neighboring nodes in the extended subnetwork will get an extra extension reward score of 1. Based on the extension reward score, we can measure the contribution of each node towards making the subnetwork dense. A node with a higher extension reward score interacts with a larger number of newly added nodes, playing a more significant role in making the subnetwork dense after adding the new nodes.

In each extension step, we add the node which is densely connected to the nodes within the extending network and loosely connected to the nodes out of the extending network, in order to minimize the conductance defined in (5). We repeat the extension steps until there is no more neighboring node that can reduce the current conductance by more than 5 percent or until the size of extending network exceeds twice the size of the query network, whichever occurs first. Once the extension process comes to an end, we remove all nodes whose extension reward score does not exceed a certain threshold. This is to enhance the functional coherence of the final querying result, since nodes with fewer interactions are relatively less likely to share similar functionalities with other neighbors. However, the original seed nodes are kept in the final result, even if their extension reward score is not large, since those nodes have high node correspondence to nodes in the query network. In this study, we set the threshold for node removal as 0, so that nodes that do not interact with any of the newly added nodes are removed in the final querying result. The overall procedure of the proposed SEQUOIA network querying algorithm is summarized in Algorithm 1.

Results and discussion

Datasets and experimental set-up

To assess the performance of SEQUOIA, we carried out network querying experiments based on the real PPI networks of three different species – *H. sapiens* (human), *S. cerevisiae* (yeast), and *D. melanogaster* (fly) – obtained from [25]. PPI networks in [25] were originally obtained from the STRING database [26], but interactions between proteins without experimental validation were removed. The human PPI network contains 12,575 proteins and 86,890 interactions, the fly PPI network contains 8624 proteins and 39,466 interactions, and the yeast PPI network contains 6136 proteins and 166,229 interactions.

As the query networks, we used protein complexes obtained from [7], comprised of complexes in three species: *H. sapiens*, *S. cerevisiae*, and *D. melanogaster*. Furthermore, we expanded the query set by adding the latest version of human complexes obtained from CORUM [27], and yeast complexes from SGD [28] (as of Jan. 5, 2015). Finally, as in [7, 8], we selected connected complexes of size 5∼25 and used them as our query networks (863 complexes in total). We assessed the performance of SEQUOIA based on the 863 real protein complexes, where 293 human complexes were searched against the fly PPI network, 289 human complexes were searched against the yeast PPI network, 141 yeast complexes were searched

Algorithm 1: SEQUOIA network querying algorithm

Data: Query and target network, pairwise node similarity score
Result: Best matching subnetwork in the target network for the given query
begin
  Data pre-processing: i) Removing non-homologous nodes and ii) Inserting pseudo-edges
  Compute the normalized node correspondence $\bar{\mathbf{C}}$ using Eq. (2)
  Select the seed network $\mathcal{G}_S = \{\mathcal{V}_S, \mathcal{E}_S\}$ using Eq. (3)
  while $|\mathcal{G}_S| \leq 2 \cdot N_Q$ *or* $\varphi_{current} \leq \beta \cdot \varphi_{previous}$ **do**
    Find the set of neighboring nodes $\mathcal{N}$ of the network $\mathcal{G}_S$
    Compute the conductance φ_t for the extended network $\{\mathcal{V}_S \cup v_t\}$, for each v_t, $\forall v_t \in \mathcal{N}$
    Find the node $v_{t^*} = \arg\min_t \varphi_t$
    Extend the network $\mathcal{G}_S$, i.e., $\mathcal{V}_S = \{\mathcal{V}_S \cup v_{t^*}\}$ and $\mathcal{E}_S = \left\{\mathcal{E}_S \cup e_{i,j}\right\}, \forall i \in \mathcal{V}_S, \forall j \in v_{t^*}$
    Update the current conductance $\varphi_{current} = \varphi_{t^*}$
    Update the extension reward score $r(v_t) = r(v_t) + 1, \forall v_t \in \mathcal{N}(v_{t^*})$
  end
  Remove nodes in $\mathcal{G}_S$ whose extension reward score is 0 while keeping the initial seed nodes.
end

against the human PPI network, and 140 yeast complexes were searched against the fly PPI network. Since there are only a small number of test cases for querying fly complexes against human and yeast PPI networks, we excluded those experiments in this study.

The performance of SEQUOIA was compared against several state-of-the-art algorithms, which include: RESQUE [8], Corbi [9], NatalieQ [10], HubAlign [29], and LocalAli [30]. Although HubAlign and LocalAli are global and local network alignment algorithms, respectively, we used those algorithms to identify conserved subnetworks as network querying can be viewed as a special case of pairwise network alignment. For Corbi, we used the default parameters for the gap penalty and set the option for the query type as 1, which is for general network querying. For HubAlign, we used the default parameters (i.e., $\lambda = 0.1$ and $\alpha = 0.7$). We also used the default parameter for NatalieQ. For LocalAli, we set the minimum number of extension (-minext) to 0 and the maximum number of extension (-maxext) to 25, since the size of the query networks ranged between 5 to 25. Default values were used for other parameters. Since LocalAli identifies multiple local complexes as its output, we selected the complex with the best score as the querying result of LocalAli.

Performance assessment metrics

To assess various aspects of the network querying algorithms, we defined several performance metrics. First, we used the matching score to count the number of matches for each query and target species pair [31]. Given two biological complexes Q and C, the matching score is computed based on the Jaccard index between the nodes in the two biological complexes as follows:

$$match\_score\,(Q, C) = \frac{|\mathcal{V}_Q \cap \mathcal{V}_C|}{|\mathcal{V}_Q \cup \mathcal{V}_C|}, \qquad (6)$$

where $\mathcal{V}_X$ is the set of nodes in the complex X. If the matching score is greater than the threshold, the two complexes were regarded to be a match. As in [31], we set the threshold for the matching score as 0.5. To count the number of matches, we used the known biological complexes as our gold standard reference $\mathcal{C} = \{C_1, C_2, \ldots, C_N\}$. Given the querying result Q_i, if there is at least one matching complex C_j in the gold standard reference, we counted Q_i as a match. Then, we report the total number of matches for each query and target species pair. That is, given the querying results $\mathcal{Q} = \{Q_1, Q_2, ..., Q_M\}$ for the M query complexes, we count the total number of querying results $\left|\left\{Q_i | match\_score\left(Q_i, C_j\right) \geq 0.5, \forall C_j \in \mathcal{C}, \forall Q_i \in \mathcal{Q}\right\}\right|$.

Next, we defined two different types of hits that respectively measure: 1) the accuracy of the obtained querying results and 2) the capability of detecting novel functional network modules with strong biological significance. The former counts the number of querying results whose annotation is identical to the functional annotation of the query network so that it can assess the capability of a given algorithm to identify the conserved functional modules. The latter counts the number of querying results with strong biological significance, regardless of whether or not they have the same functional annotation as the query, so that it can be used to assess the ability of the network querying algorithm to predict novel potential functional modules in the target PPI network.

To evaluate the accuracy of the querying results, we picked the most significantly enriched GO term of the query network (referred to as the significant GO term). Note that the most significantly enriched GO term denotes the GO term with the lowest false discovery rate (FDR) corrected p-value. To this aim, we performed GO enrichment tests for the query network and the querying result. If the significant GO term in the query is also enriched in the network querying result and if its FDR corrected p-value is less than a threshold, we regarded the querying result as a significant hit. However, a higher number of significant hits does not necessarily imply that the network querying algorithm yields accurate results, since the querying results may potentially include a large number of functionally irrelevant proteins (i.e., proteins whose annotation does not include the significant GO term). For this reason, in order to assess the accuracy of the querying results, we additionally defined two important performance metrics: the significant specificity (SPE) and the significant functionally coherent (FC) hit. Significant SPE is defined as the relative proportion of the proteins annotated with the significant GO term among the proteins included in the querying result. Based on this definition, an accurate querying result with fewer irrelevant proteins will have a higher significant SPE. Significant FC hits were defined as hits that satisfy the following two conditions: 1) FDR corrected p-value should be less than a certain threshold and 2) at least 50% of the proteins included in the querying result should be annotated with the significant GO term. A network querying algorithm that can yield a larger number of significant FC hits can be viewed as being more accurate and being capable of making better predictions that are biologically more significant.

Next, in order to assess the capability of detecting novel potential functional network modules, we investigated the biological significance of the querying results. To this aim, we performed the GO enrichment test only for the querying result (i.e., not for the query network) and selected the GO term with the smallest FDR corrected p-value as the most significantly enriched GO term. If the FDR corrected p-value of the most significantly enriched GO term of the querying result is less than a threshold, we regarded the querying result as a hit. A querying result with a small FDR corrected p-value can be viewed as being biologically significant, even if the most significantly enriched GO term of the querying result and that of the query network do not match. As a result, for a given network querying algorithm, we can assess its capability of detecting potential functional network modules by measuring the number of hits. Furthermore, we defined the specificity as the relative proportion of proteins (in the querying result) that are annotated with the most significantly enriched GO term among all proteins included in the querying result. As before, we defined a hit as being functionally coherent (FC) – hence called a FC hit – if the FDR corrected p-value is less than a certain threshold and if more than 50% of the proteins in the retrieved result are annotated with the most significantly enriched GO term.

We used the latest version of GO::TermFinder [32] for the GO enrichment test, and analyzed the querying results based on three different ontology aspects: 1) cellular component (CC, GO:0005575), 2) biological process (BP, GO:0008150), and 3) molecular function (MF, GO:0003674). In the following, we mainly present the assessment results based on the ontology aspect of "cellular component", and simulation results for other ontology aspects – i.e., "biological process" and "molecular function" – are included in the supplementary material (see Additional file 1: Section S4). The ontology and annotation files for the three species considered in our study have been downloaded from Gene Ontology Consortium [33, 34] (as of Feb. 9 2015). Then, we removed all GO terms without experimental evidence. That is, we only used GO terms having one of the following evidence codes: 'EXP', 'IDA', 'IPI', 'IMP', 'IGI', and 'IEP'. Additionally, due to the hierarchical structure of GO terms, certain GO terms are annotated to a large number of proteins, where such commonly appearing GO terms would not be very informative. In order to use the GO terms that are informative, we computed the information content (IC) for each GO term as recommended in [33]. IC is defined as

$$IC(g) = -\log_2 \frac{|g|}{|root(g)|}, \tag{7}$$

where $|g|$ is the total number of proteins with the GO term g, and $|root(g)|$ is the number of proteins under the root GO term of the GO term g. Note that there are three root GO terms: cellular component (CC, GO:0005575), biological process (BP, GO:0008150), and molecular function (MF, GO:0003674). In this study, we only used the GO terms whose information content is at least 2.

Comparison of the querying results to the gold standard reference sets

Figure 2 shows the number of matches for each query-target species pair. The figure shows that SEQUOIA yields the largest number of matches among all tested algorithms for all query-target pairs. When querying human

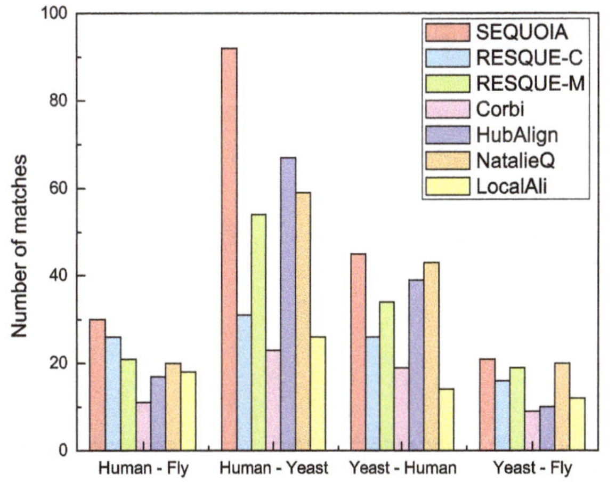

Fig. 2 Number of matches for each query and target species pair (i.e., query species – target species)

complexes against the fly and the yeast PPI networks, SEQUOIA clearly outperforms other methods. When querying yeast complexes against the human and the fly PPI networks, NatalieQ shows comparable performance to SEQUOIA, although SEQUOIA still yields a larger number of matches compared to all other methods. Overall, SEQUOIA resulted in 188 matches, which is almost 32 percent more compared to the number of matches achieved by the next best algorithm, NatalieQ.

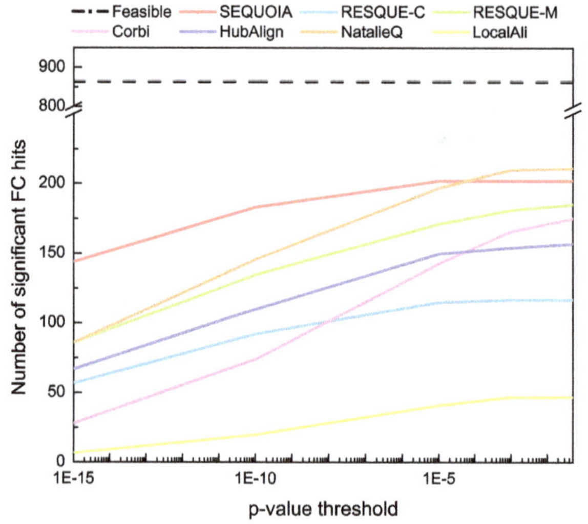

Fig. 4 Number of significant functionally coherent (FC) hits for the 863 query complexes

Assessing the accuracy of the network querying results

Figures 3 and 4 shows the number of significant hits and significant FC hits for all 863 querying results. As we can see in Fig. 3, SEQUOIA yields a larger number of significant hits compared to other algorithms. This means that SEQUOIA can more accurately identify conserved functional network modules with the significant GO term, (i.e., the most significantly enriched GO term in the query network). RESQUE family yielded similar number of significant hits at the p-value threshold of 0.05, but SEQUOIA outperformed both RESQUE-C and RESQUE-M when a smaller p-value threshold was used. Except for SEQUOIA and RESQUE-C, the number of nodes in the querying result is generally smaller than that in the query network for other tested algorithms. As a consequence, many algorithms may fail to identify inserted nodes and yield fewer significant hits.

Figure 4 shows that SEQUOIA yields a larger number of significant FC hits compared to other algorithms. This implies that SEQUOIA produces more accurate querying results that are functionally more coherent. Compared to SEQUOIA, the number of significant FC hits for Corbi decreases quickly as the p-value threshold decreases. Interestingly, although RESQUE family shows similar performance in terms of the number of significant hits, the number of significant FC hits for RESQUE-C is much smaller than RESQUE-M. This result shows that using a more sophisticated method to predict the best matching subnetwork would be needed to obtain better querying results that are functionally more coherent. In fact, RESQUE-C uses a relatively simple approach to find the best matching subnetwork, which is to find the largest connected subnetwork in the reduced target network, and this may increase the chances of including a larger number of functionally irrelevant nodes in the final querying result. SEQUOIA results in higher significant hits as well as higher significant FC hits by minimizing the network conductance of the matching subnetwork and filtering out potentially irrelevant nodes based on the extension reward score. Detailed querying results for different query and target species pairs can be found in the supplementary material (see Additional file 1: Section S4).

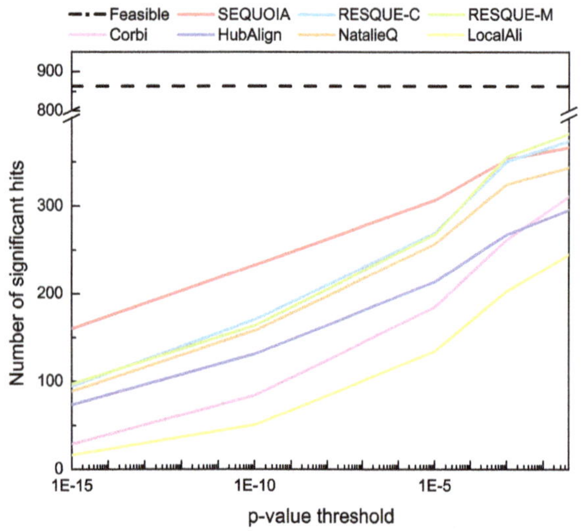

Fig. 3 Number of significant hits for the 863 query complexes

The number of identified nodes included in the querying results and the number of nodes annotated with the most significant GO term are summarized in Table 1. The table shows that NatalieQ and RESQUE-M achieve higher significant SPE compared to SEQUOIA, but it should be noted that SEQUOIA can identify a much larger number of "annotated nodes" while keeping relatively higher significant SPE compared to other algorithms. The total number of identified nodes is comparable for SEQUOIA and RESQUE-C, although SEQUOIA results in a much higher significant SPE compared to RESQUE-C. From the perspective of potential knowledge transfer from a well-studied species to a less-studied species, the ability to

Table 1 Significant SPE for the ontology aspect of "cellular component"

	Identified nodes	Annotated nodes[a]	Significant SPE
SEQUOIA	9537	2568	0.269
RESQUE-C	10,213	2115	0.207
RESQUE-M	7000	1941	0.277
Corbi	4761	1149	0.241
HubAlign	7342	1526	0.208
NatalieQ	5452	1745	0.320
LocalAli	6220	892	0.143

[a] Annotation corresponding to the most significantly enriched GO term in the query network

achieve higher significant SPE is critical, as it implies that the network querying algorithm may be able to annotate the proteins in the querying result more accurately.

Capability of detecting novel functional network modules

Figures 5 and 6 shows the number of hits and the number of FC hits for various FDR corrected *p*-value thresholds. Feasible hits in each figure correspond to the total number of query complexes, which is the maximum number of hits that can be achieved. As we can see in Fig. 5, SEQUOIA clearly outperforms other algorithms for various *p*-value thresholds. For example, at a *p*-value threshold of 1E-10, SEQUOIA yields 29% more hits than RESQUE-C, which is the next best algorithm. This results indicate that SEQUOIA has stronger potentials to identify novel protein complexes compared to other state-of-the-art algorithms.

Next, we compared the number of FC hits for different network querying algorithms. Figure 6 shows that

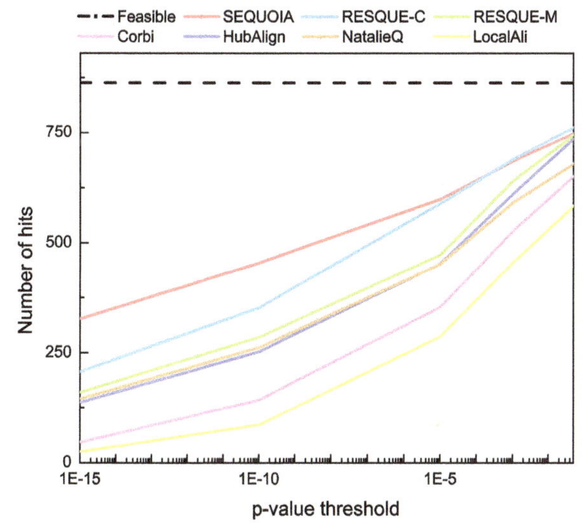

Fig. 5 Number of hits for querying 863 biological complexes

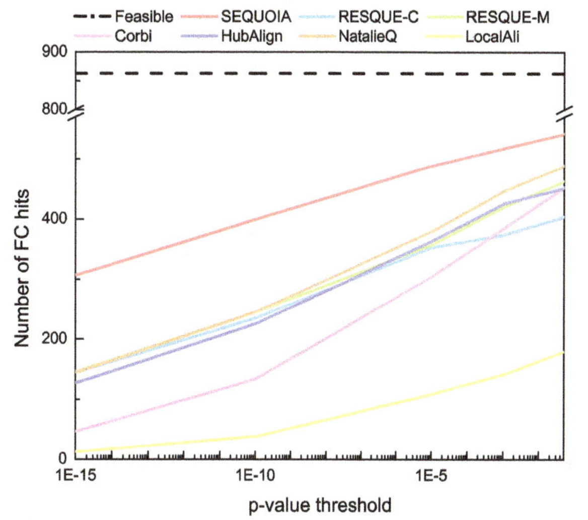

Fig. 6 Number of functionally coherent (FC) hits for querying 863 biological complexes

SEQUOIA clearly outperforms other algorithms. For example, SEQUOIA can identify 11% more FC hits than NatalieQ at a *p*-value threshold of 0.05 and almost twice as many FC hits compared to RESQUE and NatalieQ at a *p*-value threshold of 1E-15. LocalAli and NatalieQ fail to yield querying results in some test cases (i.e., these algorithms cannot identify any protein node in the target network). LocalAli and NatalieQ may not perform robustly under certain conditions (e.g., for certain query topology), which may result in a smaller number of hits. The results in Fig. 6 show that SEQUOIA's performance is more robust compared to many other algorithms, and that SEQUOIA can more effectively detect conserved network modules with high functional coherence.

Finally, we also evaluated the functional coherence of the querying results for each algorithm. To this aim, we selected the most significantly enriched GO term in the querying result obtained by each algorithm for each query, and compute the relative proportion of proteins annotated with the most significantly enriched GO term. The results are summarized in Table 2. With the exception of NatalieQ, SEQUOIA achieves the highest SPE compared to all other algorithms. Although NatalieQ results in the highest SPE, SEQUOIA can identify about 66% more annotated nodes (i.e., proteins annotated with the most significant GO term) compared to NatalieQ, while achieving a comparable SPE. This indicates that SEQUOIA can effectively identify a larger number of protein nodes that are functionally coherence than the other tested algorithms.

Computation time

Figure 7 shows the box plot for the computation time for each network querying algorithm. For RESQUE,

Table 2 SPE for the ontology aspect of "cellular component"

	Identified nodes	Annotated nodes[a]	SPE
SEQUOIA	9537	5531	0.580
RESQUE-C	10,213	5002	0.492
RESQUE-M	7000	3856	0.551
Corbi	4761	2486	0.522
HubAlign	7342	3822	0.521
NatalieQ	5452	3324	0.610
LocalAli	6220	2170	0.349

[a] Annotation corresponding to the most significantly enriched GO term in the querying result

we used the MATLAB script version 1.0 and MATLAB version 2014b. Executable binaries for NatalieQ, HubAlign, and LocalAli were obtained by compiling their source code using a C++ compiler. For Corbi, we used its R package and tested the algorithm on Windows. Except for Corbi, all other algorithms were tested on Mac OS X. All computer simulations were performed on a desktop computer equipped with a 2.4 GHz Intel i7 processor and 8 GB memory. For certain queries, NatalieQ and LocalAli may require a very long time (which is significantly longer than the average computation time), and such outliers were excluded when drawing the box plot for readability. As shown in Fig. 7, the computation time of SEQUOIA is comparable to that of the RESQUE family, but it is much faster compared to other algorithms. On average, SEQUOIA yields the querying result in less than 0.06 second, and in 98% of the test cases, the algorithm needs less than a second to find the subnetwork that best matches the query.

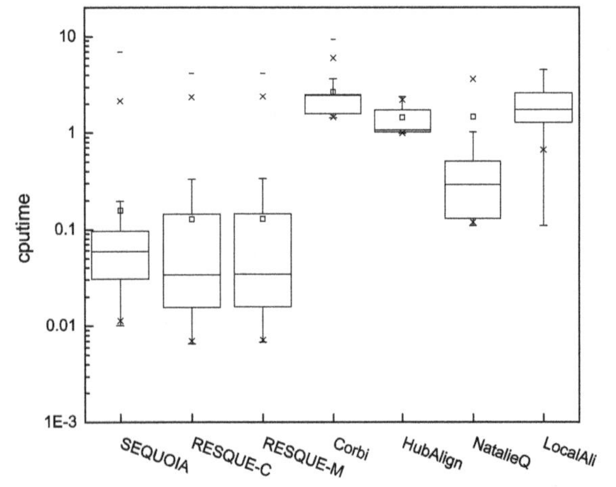

Fig. 7 *Boxplot* of the computation time for different algorithms for the 863 queries

Conclusions

In this paper, we proposed SEQUOIA, a novel network querying algorithm that can enhance the biological significance of the query results. In order to identify conserved subnetwork regions in the target network that are similar to a given query network, the algorithm compares the two networks and estimates the node correspondence scores by using the context-sensitive random walk model. Inspired by the pair hidden Markov model that has been widely used in the comparative sequence analysis, the CSRW model effectively captures the similarities between graphs by explicitly accounting for potentially inserted/deleted nodes. Based on the estimated CSRW node correspondence scores, SEQUOIA identifies high-scoring regions (referred to as the seed networks) in the target network that bear considerable similarity with the query network. The seed network is further extended by adding neighboring nodes that reduce the network conductance of the extended network by the largest amount. This extension step identifies nearby proteins that are densely connected to other nodes in the potential network module, thereby effectively recruiting proteins that are likely to share similar functions with other proteins in the module. The final query result is obtained after pruning the matching subnetwork by removing any irrelevant nodes, thereby enhancing the separability and coherence of the identified network module. As we have shown through extensive numerical simulations based on 863 real biological complexes, our network querying algorithm SEQUOIA yields accurate query results with enhanced biological significance.

Additional file

Additional file 1: Section S1. Review of the context-sensitive random walk model. This section provides detailed description of the context-sensitive random walk model. **Section S2:**Illustration of the pre-processing step. This section provides detailed description of the pre-processing step with an example. **Section S3**: Flow chart for SEQUOIA with a toy example. **Section S4**: Performance assessment for various GO ontology aspects. This section presents performance assessment results for various GO ontology aspects: cellular component, biological process, and molecular function. It also shows results for various query and target network pairs. **Section S5**: Performance improvement through post-filtering based on extension reward scores. Results in this section show the effectiveness of the pruning step based on the extension reward scores for enhancing the biological significance of the querying results.

Acknowledgements
This work was supported in part by the National Science Foundation through the NSF Award CCF-1149544.

Funding
Publication cost for this article was funded by the National Science Foundation through the NSF Award CCF-1149544.

Authors' contributions
Conceived the method: HJ, BJY. Developed the algorithm and performed the simulations: HJ. Analyzed the results and wrote the paper: HJ, BJY. Both authors read and approved the final manuscript.

Competing interests
The authors declare that they have no competing interests.

About this supplement
This article has been published as part of *BMC Systems Biology* Volume 11 Supplement 3, 2017: Selected original research articles from the Third International Workshop on Computational Network Biology: Modeling, Analysis, and Control (CNB-MAC 2016): systems biology. The full contents of the supplement are available online at http://bmcsystbiol.biomedcentral.com/articles/supplements/volume-11-supplement-3.

References
1. Sharan R, Ideker T. Modeling cellular machinery through biological network comparison. Nat Biotechnol. 2006;24(4):427–33.
2. Sharan R, Suthram S, Kelley RM, Kuhn T, McCuine S, Uetz P, Sittler T, Karp RM, Ideker T. Conserved patterns of protein interaction in multiple species. Proc Nat Acad Sci. 2005;102(6):1974–9.
3. Yoon BJ, Qian X, Sahraeian SME. Comparative analysis of biological networks: hidden markov model and markov chain-based approach. IEEE Signal Proc Mag. 2012;1(29):22–34.
4. Kelley BP, Yuan B, Lewitter F, Sharan R, Stockwell BR, Ideker T. PathBLAST: a tool for alignment of protein interaction networks. Nucleic Acids Res. 2004;32(suppl 2):83–8.
5. Shlomi T, Segal D, Ruppin E, Sharan R. QPath: a method for querying pathways in a protein-protein interaction network. BMC Bioinforma. 2006;7(1):1.
6. Dost B, Shlomi T, Gupta N, Ruppin E, Bafna V, Sharan R. QNet: a tool for querying protein interaction networks. J Comput Biol. 2008;15(7):913–25.
7. Bruckner S, Hüffner F, Karp RM, Shamir R, Sharan R. Topology-free querying of protein interaction networks. J Comput Biol. 2010;17(3): 237–52.
8. Sahraeian SME, Yoon BJ. RESQUE: network reduction using semi-markov random walk scores for efficient querying of biological networks. Bioinformatics. 2012;28(16):2129–36.
9. Huang Q, Wu LY, Zhang XS. Corbi: a new r package for biological network alignment and querying. BMC Syst Biol. 2013;7(Suppl 2):6.
10. Klau GW. A new graph-based method for pairwise global network alignment. BMC Bioinforma. 2009;10(Suppl 1):59.
11. Qian X, Sahraeian SM, Yoon BJ. Enhancing the accuracy of HMM-based conserved pathway prediction using global correspondence scores. BMC Bioinforma. 2011;12(Suppl 10):6.
12. Girvan M, Newman ME. Community structure in social and biological networks. Proc Nat Acad Sci. 2002;99(12):7821–6.
13. Newman ME. Finding community structure in networks using the eigenvectors of matrices. Phys Rev E. 2006;74(3):036104.
14. Spirin V, Mirny LA. Protein complexes and functional modules in molecular networks. Proc Nat Acad Sci. 2003;100(21):12123–8.
15. Zamani Dadaneh S, Qian X. Bayesian module identification from multiple noisy networks. EURASIP J Bioinforma Syst Biol. 2016;2016(1):5.
16. Jeong H, Yoon BJ. Effective estimation of node-to-node correspondence between different graphs. IEEE Signal Proc Lett. 2015;22(6):661–5.
17. Jeong H, Yoon BJ. Accurate multiple network alignment through context-sensitive random walk. BMC Syst Biol. 2015;9(Suppl 1):7.
18. Cook SA. The complexity of theorem-proving procedures. In: Proceedings of the Third Annual ACM Symposium on Theory of Computing. New York: ACM; 1971. p. 151–8.
19. Sahraeian SME, Yoon BJ. SMETANA: accurate and scalable algorithm for probabilistic alignment of large-scale biological networks. PLoS ONE. 2013;8(7):67995.
20. Sharan R, Ulitsky I, Shamir R. Network-based prediction of protein function. Mol Syst Biol. 2007;3(1):88.
21. Hakes L, Pinney JW, Robertson DL, Lovell SC. Protein-protein interaction networks and biology–what's the connection? Nat Biotechnol. 2008;26(1): 69–72.
22. Kuchaiev O, Rašajski M, Higham DJ, Pržulj N. Geometric de-noising of protein-protein interaction networks. PLoS Comput Biol. 2009;5(8): 1000454.
23. Kannan R, Vempala S, Vetta A. On clusterings: good, bad and spectral. J ACM (JACM). 2004;51(3):497–515.
24. Leskovec J, Lang KJ, Mahoney M. Empirical comparison of algorithms for network community detection. In: Proceedings of the 19th International Conference on World Wide Web. New York: ACM; 2010. p. 631–40.
25. Micale G, Pulvirenti A, Giugno R, Ferro A. GASOLINE: a greedy and stochastic algorithm for optimal local multiple alignment of interaction networks. PLoS ONE. 2014;9(6):98750.
26. Szklarczyk D, Franceschini A, Kuhn M, Simonovic M, Roth A, Minguez P, Doerks T, Stark M, Muller J, Bork P, et al. The STRING database in 2011: functional interaction networks of proteins, globally integrated and scored. Nucleic Acids Res. 2011;39(suppl 1):561–8.
27. Ruepp A, Brauner B, Dunger-Kaltenbach I, Frishman G, Montrone C, Stransky M, Waegele B, Schmidt T, Doudieu ON, Stümpflen V, et al. CORUM: the comprehensive resource of mammalian protein complexes. Nucleic Acids Res. 2008;36(suppl 1):646–50.
28. Cherry JM, Hong EL, Amundsen C, Balakrishnan R, Binkley G, Chan ET, Christie KR, Costanzo MC, Dwight SS, Engel SR, et al. Saccharomyces genome database: the genomics resource of budding yeast. Nucleic Acids Res. 2011;40:D700–705.
29. Hashemifar S, Xu J. HubAlign: an accurate and efficient method for global alignment of protein–protein interaction networks. Bioinformatics. 2014;30(17):438–44.
30. Hu J, Reinert K. LocalAli: an evolutionary-based local alignment approach to identify functionally conserved modules in multiple networks. Bioinformatics. 2014;31(3):363-72.
31. Liu G, Wong L, Chua HN. Complex discovery from weighted ppi networks. Bioinformatics. 2009;25(15):1891–7.
32. Boyle EI, Weng S, Gollub J, Jin H, Botstein D, Cherry JM, Sherlock G. GO::TermFinder–open source software for accessing gene ontology information and finding significantly enriched gene ontology terms associated with a list of genes. Bioinformatics. 2004;20(18):3710–5.
33. Consortium GO, et al. Gene ontology consortium: going forward. Nucleic Acids Res. 2015;43(D1):1049–56.
34. Ashburner M, Ball CA, Blake JA, Botstein D, Butler H, Cherry JM, Davis AP, Dolinski K, Dwight SS, Eppig JT, et al. Gene ontology: tool for the unification of biology. Nat Genet. 2000;25(1):25–9.

5

Dynamic genome-scale metabolic modeling of the yeast *Pichia pastoris*

Francisco Saitua, Paulina Torres, José Ricardo Pérez-Correa and Eduardo Agosin*

Abstract

Background: *Pichia pastoris* shows physiological advantages in producing recombinant proteins, compared to other commonly used cell factories. This yeast is mostly grown in dynamic cultivation systems, where the cell's environment is continuously changing and many variables influence process productivity. In this context, a model capable of explaining and predicting cell behavior for the rational design of bioprocesses is highly desirable. Currently, there are five genome-scale metabolic reconstructions of *P. pastoris* which have been used to predict extracellular cell behavior in stationary conditions.

Results: In this work, we assembled a dynamic genome-scale metabolic model for glucose-limited, aerobic cultivations of *Pichia pastoris*. Starting from an initial model structure for batch and fed-batch cultures, we performed pre/post regression diagnostics to ensure that model parameters were identifiable, significant and sensitive. Once identified, the non-relevant ones were iteratively fixed until a priori robust modeling structures were found for each type of cultivation. Next, the robustness of these reduced structures was confirmed by calibrating the model with new datasets, where no sensitivity, identifiability or significance problems appeared in their parameters. Afterwards, the model was validated for the prediction of batch and fed-batch dynamics in the studied conditions.

Lastly, the model was employed as a case study to analyze the metabolic flux distribution of a fed-batch culture and to unravel genetic and process engineering strategies to improve the production of recombinant Human Serum Albumin (HSA). Simulation of single knock-outs indicated that deviation of carbon towards cysteine and tryptophan formation improves HSA production. The deletion of methylene tetrahydrofolate dehydrogenase could increase the HSA volumetric productivity by 630%. Moreover, given specific bioprocess limitations and strain characteristics, the model suggests that implementation of a decreasing specific growth rate during the feed phase of a fed-batch culture results in a 25% increase of the volumetric productivity of the protein.

Conclusion: In this work, we formulated a dynamic genome scale metabolic model of *Pichia pastoris* that yields realistic metabolic flux distributions throughout dynamic cultivations. The model can be calibrated with experimental data to rationally propose genetic and process engineering strategies to improve the performance of a *P. pastoris* strain of interest.

Keywords: dFBA, *Pichia pastoris*, Pre/post regression diagnostics, Sensitivity, Identifiability, Significance, Genome-scale metabolic modeling, Fed-batch, MOMA, Bioprocess optimization, Reparametrization

* Correspondence: agosin@ing.puc.cl
Department of Chemical and Bioprocess Engineering, School of Engineering, Pontificia Universidad Católica de Chile, Avenida Vicuña Mackenna 4860, Santiago, Chile

Background

Recombinant protein production is a multibillion-dollar business, mainly comprised by therapeutic agents (i.e. recombinant biologic drugs) and industrial enzymes [1–3]. These compounds are commonly synthesized in *Escherichia coli, Saccharomyces cerevisiae* and Chinese Hamster Ovary cells (CHO) [1, 4–6]; however, there is strong pressure to find cost-effective alternatives to overcome technical and economic disadvantages of the aforementioned cell factories, especially in downstream processing [7].

Among the unconventional cell factories used for recombinant protein production, the methylotrophic yeast *Pichia pastoris* (syn. *Komagataella phaffii*) has received special attention thanks to its convenient physiology and easy handling [8]. There are strong promoters for this cell factory which are commercially available and that allow for the controlled expression of heterologous proteins [8]. Unlike *E. coli, P. pastoris* naturally performs post-translational modifications [6, 9], which are essential for most eukaryotic protein functionality [7, 10, 11]. In contrast to *S. cerevisiae, P. pastoris* exhibits a Crabtree-negative phenotype, showing a reduced synthesis of undesirable products, like ethanol, in glucose-limited conditions [12, 13]. It also shows a lower basal secretion of proteins when compared to other yeasts, which makes downstream processing easier [13, 14]. Finally, *P. pastoris* can be efficiently cultivated up to high cell densities using fed-batch technology [8], achieving high titers and productivities. For these desirable features, *P. pastoris* has been widely used for the expression of recombinant proteins, reaching grams per liter concentrations in several cases [9, 15–18]. Most remarkably, and as proof of its technical feasibility and adequacy, two recombinant proteins produced in this cell factory have already been approved by the FDA for medical purposes [10, 19].

Despite its growing acceptance and actual successful applications, recombinant protein production in *P. pastoris* can be undermined by several cellular processes, where protein folding and secretion are the most recurrent bottlenecks [14, 20, 21]. In addition, limitations may also be caused by the codon usage of the recombinant protein [22], promoter selection [23], carbon and oxygen availability in the culture [24, 25] and fed-batch operational parameters [26], seriously hampering protein yield, productivity and the economic feasibility of the process.

Industrially, *P. pastoris* is commonly grown in fed-batch cultures in order to maximize the titer and volumetric productivity of a desired compound, often a recombinant protein [27, 28]. This is achieved by adding a culture medium in such a way that the microorganism grows at a desired specific growth rate, which is chosen to maximize the synthesis of the target product and to limit the formation of inhibitory compounds [29]. During this and other cultivation systems, the cells adapt constantly to the changing extracellular environment and to the limited mass transfer conditions observed at high densities [30, 31]. Therefore, it is critical to understand how the cell metabolism interacts with the nutritional and environmental stresses exerted by process conditions to improve bioreactor performance [32]. This is a complex task, however, since the strain's characteristics and process variables often require significant amounts of time and money for characterization and fine-tuning [12]. Therefore, it is desirable to have a platform to integrate different levels of information from dynamic cultivations of *P. pastoris* that can be used to elaborate rational hypotheses to increase process productivity.

Systems biology offers a quantitative and comprehensive approach to address this task [33]. In particular, Genome-Scale dynamic Flux Balance Analysis (GS-dFBA) [34–36] is a modeling framework that allows the simulation of metabolism during non-stationary (batch or fed-batch) cultures. GS-dFBA models couple the dynamic mass balances of the extracellular environment of the bioreactor with comprehensive mathematical representations of cellular metabolism called Genome Scale Metabolic Models (GSMs). These structures represent the cell's entire metabolism as a set of underdetermined constrained mass-balances [30, 37, 38]. GSMs have been employed to understand cellular behavior under different environmental conditions, to map over omics data, and to define a metabolic engineering targets [39, 40]. There are currently five published GSMs of *P. pastoris* [41–45] which have been developed to help the strain optimization process with a special emphasis on recombinant protein production. Moreover, one of these models has been successfully employed to improve recombinant protein production in *P. pastoris* [46], validating these frameworks as strain engineering tools for this particular yeast.

GS-dFBA models usually contain several parameters, whose values can be obtained by regression of experimental data. These parameters are used as inputs to obtain flux distributions throughout cultivations, so their values need to be reliable. To ensure this, pre- and post-regression diagnostics have been employed to determine if a certain parameter is supported by the observed data or not [47, 48]. These analyses consist in verifying the model's capacity to explain the behavior of a system (goodness-of-fit) and the presence of the following parametric limitations: (i) low or no impact on the state variables (sensitivity), (ii) strong correlations with other parameters of the model (identifiability) and (iii) lack of statistical significance (significance). A model is considered robust if it has the capacity to explain different conditions, while containing only sensitive, identifiable and significant parameters.

Here, we present a robust dynamic genome-scale metabolic model of *P. pastoris* in glucose-limited, aerobic batch and fed-batch cultivations. To assemble the dynamic modeling framework, we started by selecting one of the available genome-scale metabolic models [43] and manually curated it to yield realistic flux distributions. Then, we included it in a set of mass balances representing the main compounds present in culture supernatant. Once assembled, the model was calibrated using experimental data from eight batch and three fed-batch cultivations. Next, we employed pre/post regression diagnostics to determine sensitivity, significance and identifiability problems in the model. In order to avoid the aforementioned statistical limitations, problematic parameters were fixed (i.e. removed from the adjustable parameter set) based on the pre/post regression diagnostics, yielding reduced and potentially robust model structures. Potentially robust model structures consisted in the original model formulation with less adjustable parameters. After evaluating these reduced models for each type of cultivation, we chose the one that presented fewer parametric limitations after being re-calibrated with the available data. These reduced models yielded no (or just a few) significance, sensitivity or identifiability problems when calibrating new data and they could predict bioreactor dynamics in conditions like the ones used for their determination. Finally, we carried out simulations to assess the potential of the model to study *P. pastoris* metabolism under industrially relevant conditions, and to select molecular and process engineering strategies to improve recombinant protein production.

Methods

Model construction

The structure of the model was based on an existing dFBA framework developed by Sanchez et al. for *S. cerevisiae* [48], which divides the fermentation time into short integration periods where a metabolic steady state could be assumed [35, 49]. The model considers the evolution of seven state variables throughout batch and fed-batch glucose-limited aerobic cultivations: culture volume as well as the concentrations of glucose, biomass, ethanol, arabitol, citrate and pyruvate. It consists of three linked blocks that are solved iteratively; (i) the kinetic block, (ii) the metabolic block and (iii) the dynamic block (Fig. 1). First, the initial conditions of the system enter into the kinetic block to determine the specific consumption and production rates of the species involved in the analysis according to kinetic expressions. These rates are included as constraints to the corresponding exchange reactions of the metabolic model. The constrained model is then passed to the metabolic block of the framework, where the flux distribution inside the cell is determined. This procedure includes the calculation of the specific growth rate, which is passed along with the other exchange rates to the dynamic block as consumption and production terms in the mass balances. Here, the concentration of the state variables is updated and then incorporated into the kinetic block for the calculation of instantaneous exchange rates. This cycle iterates throughout the cultivation yielding the culture profile and instantaneous flux distributions that can be saved for further analysis. The model is included in Additional file 1 and its latest version can be found online at https://github.com/fjsaitua/RY-dFBA/tree/master/main%20P_pastoris%20dFBA:

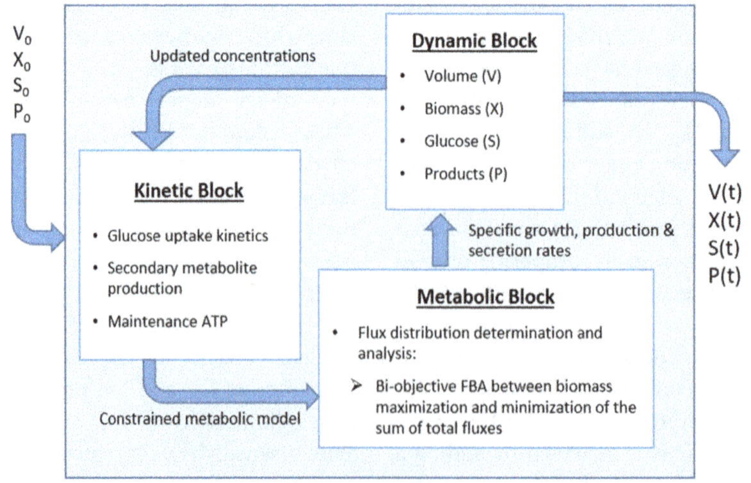

Fig. 1 Iterative structure of the model. V refers to culture volume [L], F_{IN} is the feeding policy used in fed-batch cultures, X, S and P are biomass, limiting substrate and Product concentration in [g/L] respectively

Kinetic block

The kinetic block sets the uptake and production rates for all the compounds in the model. First, the glucose uptake rate (v_G) is determined using Michaelis-Menten kinetics [50].

$$v_G = \frac{v_{G,\,Max} \cdot G}{K_G + G} \tag{1}$$

Here, G is the glucose concentration in the medium [g/L], $v_{G,Max}$ is the maximum glucose uptake rate [mmol/$g_{DCW} \cdot h$] and K_G is the uptake half activity constant of this substrate [g/L]. Once determined, $-v_G$ [mmol/$g_{DCW} \cdot h$] is included as the lower bound of the corresponding exchange reaction in the model since substrate consumption is represented with a negative flux through this reaction.

Then, the lower bounds of the exchange reactions (lb) associated with the remaining k compounds (lb_k) are fixed. We considered ethanol, pyruvate, arabitol and citrate dynamics, besides glucose consumption and biomass formation.

$$lb_k = v_{P_k} \; k = 1 \ldots 4 \tag{2}$$

These parameters are redefined during the fed-batch phase; therefore, they have two values during this type of cultivation.

Finally, the kinetic block fixes the non-growth associated maintenance ATP (m_{ATP}, a flux through the cytosolic ATP hydrolysis reaction in the model), which accounts for the energy drain caused by cellular processes not related with the generation of new cell material, such as osmoregulation, shifts in metabolic pathways, cell motility, etc. [51, 52].

Metabolic block

The metabolic block receives a constrained GSM from the kinetic block and solves an optimization problem to determine specific growth rate and the flux distribution in the cell. The GSM consists of a set of m metabolites and n reactions grouped in a Stoichiometric Matrix, S (m x n), that represents the cell's entire metabolism. If accumulation of metabolites is neglected, a mass balance can be stated according to equation (3):

$$\begin{aligned} &S \cdot v = 0 \\ &s.\,t. \\ &lb < v < ub \end{aligned} \tag{3}$$

Where v is a vector of metabolic fluxes in [mmol/$g_{DCW} \cdot h$], and lb and ub are the lower and upper bounds for each component of the flux vector.

The metabolic block solves a bi-objective Quadratic Programming (QP) problem between maximization of growth rate and minimization of the total absolute sum of fluxes [53], subjected to the constraints imposed by the stoichiometric matrix mentioned above [51]:

$$\begin{aligned} &Min \;\; \alpha \cdot \sum_{i=1}^{n} v_i^2 - (1-\alpha) \cdot \mu \\ &s.t. \\ &S \cdot v = 0 \\ &lb_i \le v_i \le ub_i \;\; i = 1 \ldots n \end{aligned} \tag{4}$$

In this formulation, α, the suboptimal growth coefficient, is an adjustable parameter from the model used to modulate the importance of the two – biologically relevant – competing objectives [48, 52, 54]. In our analysis, "optimal growth" occurs when the objective function of the cell is biomass maximization ($\alpha = 0$). However, when $\alpha > 0$, the calculated growth rate is lower than the theoretical maximum derived from biomass maximization, at the same glucose uptake rate. In this sense α is considered as a "suboptimal growth coefficient"; it is worthy to note that we do not refer to the optimality of the flux distribution vector, which is actually optimal, given the convexity of the problem in the metabolic block (Equation 4 - See Additional file 2 for details).

The minimization of total fluxes adds a quadratic term to the objective function, which has the practical benefit of eliminating Type III pathways [55] from the flux distribution, which arise from the multiplicity of solutions of a LP problem. These pathways appear as high fluxes (often taking the value of the upper bound of a particular flux) through closed cycles of reactions. This misleads pathway analysis because despite the mass balance around each participating metabolite is satisfied, the fluxes are thermodynamically infeasible [55]. The use of Quadratic Programing makes pathway analysis easier since these large cycling fluxes undermine the minimization of the total fluxes term in the objective function (Equation 4), so they will be forced to a minimum by the optimization software and the flux distribution will be "cleaned" from these unrealistic fluxes. This is especially significant in large networks because these cycles are more recurrent.

In this study, we employed a curated version of the iPP668 model developed by Chung and collaborators [43], called iFS670 (Additional files 3 and 4). In this updated version, we incorporated the arabitol biosynthesis pathway and the stoichiometric reactions for the production of three recombinant proteins (FAB fragment, Human Serum Albumin and Thaumatin). The arabitol synthesis pathway was included because it was a major compound in the culture supernatant of our experiments. Moreover, the reversibility of cytosolic reactions involving redox cofactors and mitochondrial symporters was checked according to Pereira et al. [56] in order to obtain a more realistic flux distribution through the central metabolism. This was done because the initial flux distributions obtained with the un-modified iFS670 model presented the exact same problems as the iMM904 model of *Saccharomyces cerevis*iae on Pereira's work, suggesting that the central metabolism structure of the iPP668 model was based upon the aforementioned *S. cerevisiae* model. These problems were caused by: (i) the lack of a flux through the oxidative branch of the Pentose Phosphate Pathway; (ii) the presence of a flux of a cytosolic NAPDH dependent isocitrate dehydrogenase (which was the responsible of producing cytosolic NADPH); (iii) an unrealistic flux through mitochondrial symporters; and (iv) almost no mitochondrial formation of α-ketoglutarate. These model limitations are inconsistent with previous *P. pastoris* fluxomic studies in glucose-limited aerobic conditions [24, 57, 58].

FBA problems were solved using the Constraint-Based Reconstruction and Analysis (COBRA) toolbox [59, 60], which employs the programming library libSBML [61] and the SBML toolbox [62]. Finally, we used Gurobi 6.0.2 as an optimization solver.

Dynamic block

The dynamic block consists of a set of ordinary differential equations (ODEs) that account for the volume change of the culture and the mass balances of biomass and the species considered by the model:

$$\frac{dV}{dt} = F(t) - SR \tag{5}$$

$$\frac{d(V \cdot X)}{dt} = \mu \cdot (V \cdot X) - SR \cdot X \tag{6}$$

$$\frac{d(V \cdot G)}{dt} = F(t) \cdot G_F - v_G \cdot MW_G \cdot (V \cdot X) - SR \cdot G \tag{7}$$

$$\frac{d(V \cdot P_k)}{dt} = v_{P_k} \cdot MW_{P_k} \cdot (V \cdot X) - SR \cdot P_k \tag{8}$$

Where V is volume [L], t is time [h], $F(t)$ is the feed function for the fed-batch phase in [L/h]. SR is a constant sampling rate [L/h] determined from each cultivation to emulate the remaining volume of the culture considering sampling, since this value is used for the calculation of the feeding profile during the feed phase. During the batch phase of the fed-batch cultures, we collected between 15 and 20% of the reactor volume in samples. For batch cultivations, $F(t)$ was eliminated from the mass balances. X is the biomass concentration [g/L], μ is the specific growth rate [h^{-1}] (obtained from equation 4), G is the extracellular concentration glucose [g/L], G_F is the feed's glucose concentration [g/L], P_K is the k-th extracellular product concentration in [g/L], v_{P_k} is the corresponding production rate [$mmol/g_{DCW} \cdot h$] and MW accounts for the corresponding molecular weight [g/mmol].

The set of equations was solved in Matlab 2013a (Mathworks, USA) using the solvers ode113 and ode15s for batch and fed-batch cultures respectively.

Model parameters

The lower, upper and initial values of the parameters of the model used in all the calibrations are presented Table 1. The lower and upper bounds of $v_{G,max}$, K_s, and m_{ATP} were chosen according to literature [29, 43, 63] while the rest of the bounds were selected to ensure that the algorithm had enough search space. To do this, the upper bounds of the rest of the parameters were set at higher values than the observed experimental rates, also taking into account reported values [24, 57, 58]. In addition, initial estimated parameter values were chosen to attain a feasible simulation.

Model calibration with experimental data

Strains

Four *P. pastoris* strains were employed in this study: a parental GS115 strain (Invitrogen) and three recombinant strains constructed according to the instructions of the manufacturers harboring respectively one, five and eight copies of the gene encoding for the sweet protein thaumatin. Even though the strains were transformed, thaumatin was not detected at concentrations higher than 100 μg/L in the cultivations. Therefore, due to its small contribution to the overall mass balance, thaumatin production was left out of the analysis and none of the parameters of the model were associated with it. Nevertheless, a mass balance for a recombinant protein can be easily added to the framework.

Experiments

The batch model was calibrated with aerobic glucose limited cultivations of the four strains available; each cultivation was performed twice. On the other hand, the fed-batch model was calibrated with data from three cultures of the strain with one copy the recombinant gene, under the same environmental conditions of the batch cultivations.

Cultivation conditions

Each batch or fed-batch culture started from a 2 [mL] cryotube of the corresponding strain kept at −80 °C. A pre-culture was grown overnight at 30 °C in shake flasks with 50 [mL] of the inoculum medium. After reaching 1 OD_{600}, the whole broth was added to 450 [mL] of fresh medium to reach an initial volume of 500 [mL] in 1 L bioreactors. Culture conditions were kept at 30 °C and pH = 6.0. Dissolved Oxygen was maintained above 40% saturation during all the cultivation period. Aerobiosis was achieved by a triple split-range control action,

Table 1 Parameters of the model

Symbol	Name	Units	LB	Initial value	UB
$v_{G,max}$	Maximum glucose uptake rate	$mmol/g_{DCW}h$	0	2.5	10
K_G	Half saturation constant for glucose uptake	g/L	0	10^{-4}	10^{-3}
$v_{EtOH,B}$	Ethanol minimum secretion rate (batch)	$mmol/g_{DCW}h$	0	0.5	3
$v_{Pyr,B}$	Pyruvate minimum secretion rate (batch)	$mmol/g_{DCW}h$	0	0.1	2
$v_{Arab,B}$	Arabitol minimum secretion rate (batch)	$mmol/g_{DCW}h$	0	0.2	2
$v_{Cit,B}$	Citrate minimum consumption rate (batch)	$mmol/g_{DCW}h$	0	0	2
$v_{EtOH,FB}$	Ethanol minimum consumption rate (fed-batch)	$mmol/g_{DCW}h$	0	0	2
$v_{Pyr,FB}$	Pyruvate minimum consumption rate (fed-batch)	$mmol/g_{DCW}h$	0	0	2
$v_{Arab,FB}$	Arabitol minimum consumption rate (fed-batch)	$mmol/g_{DCW}h$	0	0	2
$v_{Cit,FB}$	Citrate minimum consumption rate (fed-batch)	$mmol/g_{DCW}h$	0	0	2
a_B	Sub-optimal growth coefficient (batch)	[–]	0	0	10^{-3}
a_{FB}	Sub-optimal growth coefficient (fed-batch)	[–]	0	0	10^{-3}
m_{ATP}	Non-growth associated ATP	$mmol/g_{DCW}h$	0	2	10
T_{Fed}	Time when secondary metabolite consumption starts in fed-batch cultures	h	20	25	32

including agitation (200–800 [RPM]), air flow (0.25–1.0 [L/min]) and pure oxygen flow (0–1.0 [L/min]) [64]. pH was controlled using phosphoric acid 20% [v/v] and sodium hydroxide 20% [v/v]. The temperature was controlled with a mixture of hot and cold water, using the glass jacket of the reactors. Lastly, foam was controlled manually using silicone antifoam 10% [v/v]. Glucose starvation was detected when a sudden decrease of the CO_2 composition in the off-gas occurred, and it was confirmed each time using Benedict's reagent. For fed-batch experiments, the feed F(t) was designed to track a variable growth rate for a predefined time. This feed can be calculated from the reactor's glucose and biomass mass balances, as detailed in the literature [65]:

$$F(t) = \frac{\mu_{set(t)}}{G_F \cdot Y_{SX}} \cdot V_i X_i \cdot \exp\left(\int_{t_i}^{t} \mu_{set}(t)dt\right) \quad (9)$$

with G_F the glucose feed concentration [g/L], Y_{SX} the experimental glucose-biomass yield [g_{DCW}/g] calculated using the genome-scale model, t_i the time at which the feed started for a given cultivation [h], V_i and X_i the volume [L] and biomass [g/L] values at t_i, respectively, and $\mu_{SET}(t)$ is the time-dependent user-defined growth rate at which the fed-batch culture is grown. The latter was defined as follows:

$$\mu_{set(t)} = \left(\mu_{max} - \mu_{min}\right) \cdot e^{-Ct} + \mu_{min} \quad (10)$$

Where $\mu_{MAX} = 0.1$ [1/h], $\mu_{MIN} = 0.07$ [1/h] and $C = 0.07$ [1/h]. Therefore, $\mu_{SET}(t)$ decays exponentially from 0.1 to 0.07 [1/h], which has been found to increase (in contrast to constant growth rates in the feed phase) the final biomass concentration in fed-batch cultivations of *E. coli* and *S. cerevisiae* performed in our laboratory [66].

Culture media

The culture media employed in these studies were based on Tolner et al. [67]. **Inoculum**: Glucose 10 [g/L], $(NH_4)_2SO_4$ 1.8 [g/L], $MgSO_4 \cdot 7H2O$ 2.3 [g/L], K_2SO_4 2.9 [g/L], trace elements solution 0.8 [ml/L], histidine 0.08 [g/L], sodium hexametaphosphate 5 [g/L] and biotin 0.32 [mg/L]. **Batch cultures:** Glucose 50 [g/L], $(NH_4)_2SO_4$ 9 [g/L], $MgSO_4 \cdot 7H_2O$ 11.7 [g/L], K_2SO_4 14.7 [g/L], trace elements solution 4 [ml/L], histidine 0.4 [g/L], sodium hexametaphosphate 25.1 [g/L] and biotin 1.6 [mg/L] and sodium hydroxide NaOH 1 [g/L]. **Feeding medium:** Glucose 500 [g/L], $MgSO_4 \cdot 7H2O$ 9 [g/L], trace solution 12.5 [g/L], histidine 4 [g/L] and biotin 0.1 [g/L]. Sodium hydroxide was added to all the media until a pH of 6 was reached.

Analytical procedures

Sampling and biomass determination Samples of ~6 mL were periodically collected (every 2–3 h) from all fermentations. Biomass was measured by optical density (OD) at 600 nm using an UV-160 UV-visible spectrophotometer (Shimadzu, Japan). Biomass concentration was determined using the linear relationship: 1 $OD_{600} = 0.72$ [g/L] using the methodology from [68]. Then, samples were centrifuged at 10.000 rpm for 3 min and the supernatant stored at −80 °C for further analysis.

Extracellular metabolite concentration analyses Glucose, ethanol, arabitol, citrate and pyruvate extracellular concentrations were quantified in duplicate by High-Performance Liquid Chromatography (HPLC), as detailed in Sánchez et al. [48], with the exception of the working temperature of the Anion-Exchange Column (Bio-Rad, USA), which was lowered from 55 °C to 35 °C for better resolution.

Objective Function For model calibration, we minimized the sum of square errors between the experimental data (Additional files 5 and 6) and the simulation output by searching the parameter space, with the enhanced scatter search algorithm (eSS) [69], which has been successfully used to solve complex bioprocess optimization problems [70–72]. The objective function J used in the minimization was normalized by the maximum corresponding measured variable to give all data a similar weight:

$$J = \min_{\theta} \sum_{i=1}^{m} \sum_{j=1}^{n} \left(\frac{X_{ij}^{mod} - X_{ij}^{exp}}{\max_j \left(X_{ij}^{exp} \right)} \right)^2 \tag{11}$$

With θ representing the parameter space, m the number of measured variables, n the number of measurements per variable, X_{ij}^{mod} the dFBA output of variable i and measurement j, X_{ij}^{exp} the corresponding experimental value and $\max_j\left(X_{ij}^{exp}\right)$ the maximum value measured for variable i.

Pre/Post regression analysis

Once the initial calibration of the model was completed, statistical tests were performed in order to determine if the initial model formulation had sensitivity, identifiability or significance problems [47].

Sensitivity corresponds to the impact that model parameters have on the state variables or process output. The relative sensitivity of parameter k on the state variable i (g_{ik}) was calculated according to the following formula

$$g_{ik}(t, \theta_k) = \frac{\theta_k}{X_i(t)} \cdot \frac{dX_i(t)}{d\theta_k} \tag{12}$$

Where $X_i(t)$ is the *i*th state variable in time t and θ_k is the *k*th parameter. With all g_{ik} values, we formed a sensitivity matrix g(t) for each experimental time, in which the *k*th column denotes the sensitivity of the kth parameter on the state variables. These matrices were averaged to obtain a single normalized score of the sensitivity of parameter k on the state variable i during the cultivation. Furthermore, if the score of each variable was under 0.01 for a given parameter, this parameter was considered insensitive and a candidate to be fixed (or left out of the adjustable parameter set) in the reparametrization stage.

Identifiability refers to the possibility of unambiguously determining the parameter values by fitting a model to experimental data. If parameter identifiability is not properly assessed, misleading parameter values can be obtained after model calibration. To calculate identifiability, we determined the correlation between the columns of the sensitivity matrix using the *corrcoef* function from Matlab, which yielded a correlation coefficient matrix (C). A pair of parameters j and k was considered to be correlated (therefore not-identifiable) if the absolute value of the number at the (j, k) position in the correlation coefficient matrix was higher than 0.95 ($\left(\left|C_{jk}\right| \geq 0.95\right)$).

To determine parameter significance, we started by calculating the Fisher Information Matrix (FIM) [73]

$$FIM = \sum_{j=1}^{n} g_j^T Q_j g_j \tag{13}$$

Here, g_j is the sensitivity matrix for measurement j, n is the number of samples, and Q_j is a weighting matrix given by the inverse of the measurement error covariance matrix assuming white and uncorrelated noise. Hence, the variances for each estimated parameter were calculated as in [73, 74]

$$\sigma_k^2 = FIM_{kk}^{-1} \tag{14}$$

which was used to determine the confidence interval (CI) with 5% significance for the kth parameter as follows:

$$CI_k = \left[\hat{\theta}_k \pm 1.96 \sigma_k \right] \tag{15}$$

Here, θ_k is the estimated value of the corresponding parameter. Finally, coefficients of confidence (CC) were calculated as follows:

$$CC_k = \frac{\Delta(CI_k)}{\hat{\theta}_k} = \frac{2 \cdot 1.96 \sigma}{\hat{\theta}_k} \tag{16}$$

$\Delta(CI_k)$ is the CI's length. A parameter was not significant if the confidence interval contained zero, i. e. if the absolute value of the CC was equal or larger than 2.

Reparametrization

A reparametrization procedure called HIPPO [75] (Heuristic Iterative Procedure for Parameter Optimization, http://www.systemsbiology.cl/tools/) was applied to overcome parametric statistical limitations in the model.

First, HIPPO performed sensitivity and identifiability tests on the initial calibration results for each dataset.

Then, model parameters were fixed one by one until the non-fixed subset presented none of the statistical limitations. Finally, significance was determined for the remaining parameter set, also called the reduced model structure. If all the remaining parameters were significantly different from zero, the resulting structure is considered to be an a priori robust candidate for cross calibration with the available data.

Cross calibration of robust structure candidates derived from the reparametrization stage using the available datasets

After reparametrization of the model derived from each dataset, a potentially robust structure was generated. This structure was recalibrated with the rest of the datasets to assess its robustness. It is worthy to note that the parameters left out of the calibration were either fixed according to values reported in literature, assumed to be zero or fixed at the mean value achieved in the calibrations. This was done to avoid assuming a minimum production of compounds in batch cultivations and to ensure model convergence for parameters that had no reported values in literature (Table 2). For example, fixing feed phase consumption rates at zero does not allow consumption of batch by-products and yielded poor fed-batch fittings (data not shown).

The reduced modeling structures were evaluated according to four parameters:

Table 2 Values at which problematic parameters were fixed in the cross-calibration stage

Parameter	Fixation value	Units	Reference
$v_{G,max}$	6	$mmol/g_{DCW}h$	[63]
K_G	0.0027	g/L	[63]
$v_{EtOH,B}$	0	$mmol/g_{DCW}h$	-
$v_{Pyr,B}$	0	$mmol/g_{DCW}h$	-
$v_{Arab,B}$	0	$mmol/g_{DCW}h$	-
$v_{Cit,B}$	0	$mmol/g_{DCW}h$	-
$v_{EtOH,FB}$	1.21	$mmol/g_{DCW}h$	*
$v_{Pyr,FB}$	0.14	$mmol/g_{DCW}h$	*
$v_{Arab,FB}$	0.15	$mmol/g_{DCW}h$	*
$v_{Cit,FB}$	0.008	$mmol/g_{DCW}h$	*
a_B	0	[-]	[85]
a_{FB}	0	[-]	[85]
m_{ATP}	2.18	$mmol/g_{DCW}h$	[43]
T_{Fed}	22	h	*

Parameters marked with '-' in the reference column indicate that no a priori value was assumed for that particular parameter, which is the case for the batch minimum secretion rates. '*' means that the value of a particular parameter was fixed at the mean value achieved in the calibrations, because no information about them could be found in the literature

I. **Relative difference between calibration objective functions (J_{DIFF}):**

$$J_{DIFF} = \frac{1}{n} \cdot \sum_{i=1}^{n} \frac{J_{i,Reduced} - J_{i,Original}}{J_{i,Original}} \quad (17)$$

Where n corresponds to the number of cultures of each type, $J_{i,Original}$ is the calibration objective function (Equation 11) achieved for dataset i using the original model structure and $J_{i,Reduced}$ is the calibration objective function achieved in dataset i using a reduced, a priori robust, modeling structure.

II. **Percentage of Significance issues**; refers to the number of times a parameter is found to be non-significant out of the total of significance determinations performed for a structure. For instance, if a model structure had 6 parameters and 8 datasets were used to calibrate it, a total of 48 significance determinations were performed for that particular model.

III. **Percentage of Sensitivity issues;** refers to the number of times one of the estimated parameters shows low or no impact over state variables (average relative sensitivity ≤ 0.01) out of the total sensitivity determinations performed.

IV. **Percentage of Identifiability issues**; corresponds to the number of times a pair of parameters presents a strong correlation (≥0.95), out of the total parameter pairs of a modeling structure. If p is the number of parameters of the model and n is the number of datasets used for its calibration, the total of parameter pairs for which identifiability was determined is:

$$Total\ pairs = \frac{p \cdot (p-1)}{2} \cdot n \quad (18)$$

Finally, the modeling structure that presented the lowest J_{DIFF} and fewest statistical limitations was used as a robust structure candidate for the corresponding type of culture.

Robustness check of the chosen modeling structure

Once a candidate for a robust structure was determined for the batch and fed-batch configurations, we tested its robustness (absence of parametric problems) by calibrating it with new experimental data. For the batch model, we employed fermentation data from *P. pastoris* GS115 strain grown with 40 [g/L] of glucose as carbon source at T° = 25 °C and pH = 6. The robustness of the fed-batch model was assessed with a glucose-limited cultivation consisting of a 60 [g/L] glucose batch phase and an exponential feed using 500 [g/L] of glucose. The medium was added in the feeding phase in order to achieve an exponentially decreasing growth rate from 0.1 to 0.07 [1/h].

Model validation

Finally, the predicting capability of the model was evaluated for conditions similar to the ones used in the initial calibrations (training set).

The robust batch model was first calibrated with the two cultivations of the strain harboring one copy of the thaumatin gene, obtaining a characteristic parameter set for that strain. Then, these parameters were used to predict the course of a different batch cultivation performed in the same conditions (30 °C and pH 6).

This procedure was also applied for the fed-batch model. Here, the bioreactor dynamics was simulated using the parameters obtained in the best calibration within the training dataset (the one in which the calibration objective function was minimal compared to the rest of the calibrations) using the robust modeling structure obtained previously. This prediction was compared with experimental data of a different fed-batch cultivation.

Goodness of fit

For both the robustness check and validation datasets, the goodness of fit was determined by two scores: the mean normalized error (MNE) and the Anderson-Darling test [76]. The MNE quantifies the difference between model simulations and experimental data; the closer the difference is to zero, the better the fit. In addition, the sign of MNE shows whether the model over (+) or underestimates (–) the observed data (equation 19).

$$MNE_i = \frac{\sum_{j=1}^{n}\left(X_{ij}^{mod} - X_{ij}^{exp}\right)}{n \cdot \max_j\left(X_{ij}^{exp}\right)} \tag{19}$$

with n the number of time points measured for variable i.

The Anderson-Darling test was used to verify if the residuals between simulations and experimental data $\left(X_{ij}^{mod} - X_{ij}^{exp}\right)$ were normally distributed. If they were, the differences between them can be attributed to measurement noise and not to model inadequacy. The failure of this test by one of the model's state variables (p-value < 0.05) indicates that a different mathematical relation than the one used in the model may underlie its dynamics. Therefore, the results of this test may be used to confirm or update the kinetic expressions associated with the consumption and production of compounds.

Simulation

Analysis of the metabolic flux distribution during key stages of a dynamic cultivation

After the calibration of the fed-batch model with the dataset used for checking its robustness, we evaluated the central metabolic flux distributions at three different stages of the cultivation: exponential growth during the batch phase (~20 h), ethanol and arabitol consumption during glucose starvation phase (~27.5 h) and controlled growth during the feeding phase (~45 h).

Discovery of beneficial knock-out targets for the overproduction of recombinant Human Serum Albumin (HSA)

To show the potential applications of the model, gene targets for the overproduction of the recombinant Human Serum Albumin (HSA) were determined by simulating the growth and protein secretion of single knock-out strains of *P. pastoris* in batch cultivations. To do this, we included in the Metabolic Block a second quadratic programing problem consisting in the Minimization of Metabolic Adjustment (MOMA) algorithm [77], which states that, after a genetic perturbation, the cell will attempt to redistribute its metabolic fluxes as similar as possible to the parental strain. Mathematically, equation 4 of the metabolic block is employed in order to obtain the parental flux distribution v_0 at a given instant.

$$\begin{aligned} & Min\ \alpha \cdot \sum_{i=1}^{n} v_{0,i}^2 - (1-\alpha) \cdot \mu_0 \\ & s.t. \\ & S \cdot v_0 = 0 \\ & lb_{0,i} \le v_{0,i} \le ub_{0,i} \quad i = 1 \ldots n \end{aligned} \tag{20}$$

Then, the k reactions associated with gene j are blocked:

$$lb_{l,j} = ub_{l,j} = 0 \; l = 1 \ldots k \tag{21}$$

Finally, the MOMA algorithm was applied using the flux distribution of the parental strain v_0 to calculate the knockout distribution v_{KO} as the Euclidean distance between them, considering that the actual model has the corresponding deletion.

$$\begin{aligned} \text{MOMA}: & \\ & Min\ \left(v_0 - v_{KO,j}\right)^2 \\ & s.t. \\ & S \cdot v_{KO,j} = 0 \\ & lb_i \le v_{i,KO,j} \le ub_i \qquad i = 1 \ldots n \end{aligned} \tag{22}$$

The hypothetical parental strain was characterized using the parameters obtained above plus the growth rate dependent specific HSA productivity (q_P) of *P.*

pastoris strain SMD1168H grown on glucose, as reported by Rebnegger et al. [78], (Fig. 2). In each iteration of the model, the minimum HSA production was fixed according to this relationship, which was fitted with a third degree polynomial. Other kinetic expressions could be employed to represent the q_P vs μ relationship, depending on the strain and protein being produced [26].

We simulated one batch cultivation for each gene in the model and compared their final protein and biomass concentrations with those of the parental strain. The candidates that reached a higher HSA concentration than the parental strain were manually analyzed and some of them were proposed as candidates to improve HSA production. It is important to mention that we used a set of parameters derived in this study to characterize the growth kinetics of the HSA producing strain used in the simulations. Therefore, the predictions derived from this work should be assessed carefully and considered only as an example of the applicability of our modeling framework.

Evaluation of different feeding policies in silico to improve recombinant protein production considering specific information about the strain and process setup

Simulations were run using the parameters obtained in the calibration used for intracellular flux analysis and adding the q_P vs μ relation for HSA biosynthesis in the mass balances. The process limitations (based on our setup) were a maximum reactor volume of 1 L, and a maximum oxygen transfer rate of 10.9 [g/L · h]. If any of these limits were violated by either the feeding rate of medium or the oxygen uptake rate (extracted from the model), the integration stopped.

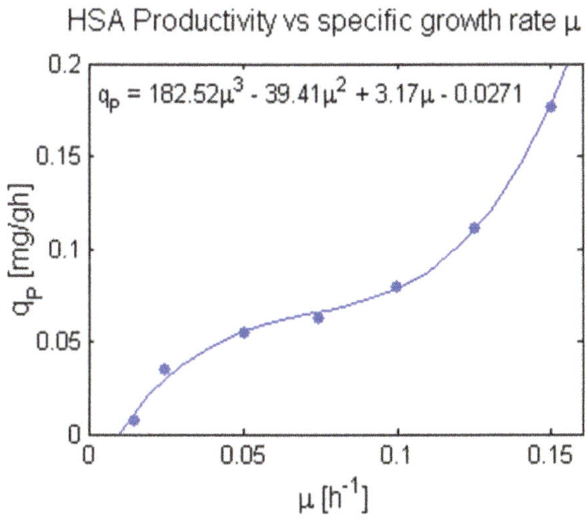

Fig. 2 Relation between Human Serum Albumin specific production rate (q_P) and growth rate (μ) in glucose limited chemostats, taken from Rebnegger et al. [78]. This relation was included to simulate the specific protein productivity for a given growth rate, allowing the assessment of the impact of different feeding profiles on process productivity

We assessed 13 exponential feeding policies. Five of them maintained a constant growth rate during the feeding phase and the rest considered a decreasing growth rate throughout the culture (Additional file 7). After the simulation, we ranked the strategies according to the volumetric productivity of recombinant HSA and chose the best one as a cultivation strategy that could potentially improve bioreactor performance.

Results and discussion

The batch and fed-batch models were developed in four steps: (i) determination of initial parametric problems, (ii) reparametrization and cross calibration, (iii) robustness evaluation and (iv) validation of predictive potential under the studied conditions.

Once the models were developed, three applications were proposed to improve recombinant |protein production using Human Serum Albumin as a case study.

Initial parametric problems

Batch model

The initial structure of the batch model comprised eight parameters (Table 3). The model was able to successfully accommodate different cellular dynamics from eight glucose-limited aerobic cultivations. In these calibrations, several statistical parametric limitations were found (Additional file 8). m_{ATP} was the parameter that presented the strongest correlation with other parameters, such as maximum specific glucose uptake rate ($v_{G,Max}$), ethanol and arabitol specific secretion rates ($v_{EtOH,B}$ *and* $v_{Arab,B}$), and with the sub-optimal growth coefficient (α_B). This might result from the fact that a change in m_{ATP} directly impacts the ATP-producing pathways in the metabolic model, affecting the biomass

Table 3 Potential Robust Structures Tested in the Cross-Calibration Stage for the batch model

Structure	Parameters included
Original	$v_{G,\ Max}$, K_G $v_{EtOH,B}$, $v_{Pyr,B}$, $v_{Arab,B}$, $v_{Cit,B}$, m_{ATP} and α_B
1	$v_{G,Max}$, $v_{EtOH,B}$, $v_{Pyr,B}$, $v_{Arab,B}$, $v_{Cit,B}$ and α_B
2	$v_{G,\ Max}$, $v_{Cit,B}$ and α_B
3	K_G, $v_{EtOH,B}$, $v_{Pyr,B}$, $v_{Arab,B}$ and $v_{Cit,B}$
4	$v_{EtOH,B}$ and $v_{Cit,B}$
5	$v_{G,Max}$, $v_{Pyr,B}$, $v_{Arab,B}$
6	$v_{G,Max}$, $v_{EtOH,B}$, $v_{Pyr,B}$, $v_{Arab,B}$, $v_{Cit,B}$
7	$v_{G,Max}$, K_G, $v_{EtOH,B}$, $v_{Pyr,B}$, $v_{Cit,B}$
8	K_G, $v_{Pyr,B}$, $v_{Arab,B}$, α_B and m_{ATP}

Each one of these structures was derived using HIPPO after model calibration using each dataset and product yields, which are also influenced by other parameters of the model. In addition, the glucose uptake saturation constant K_G was the only parameter with frequent sensitivity and significance problems, making it a potential candidate to be left out of the adjustable parameter set.

Fed batch model

Data from three aerobic, glucose-limited fed-batch cultivations was successfully calibrated with the initial model of fourteen parameters. As in the batch model, several statistical parametric limitations arose (Additional file 8). The most frequent correlation (in two out of the three calibrations) was between $v_{G,Max}$ and the $v_{EtOH,B}$ during the batch phase. Also, $v_{EtOH,B}$ and $v_{Arab,FB}$ showed 5 and 6 strong correlations with other parameters of the model, respectively.

Finally, the citrate minimum secretion rate during the fed-batch phase and the suboptimal growth during the feeding phase (α_{FB}) were the parameters that presented more sensitivity and significance limitations.

Reparametrization and cross calibration

After model calibration and the subsequent determination of the parametric problems for each dataset, the non-relevant parameters were fixed (left out of the adjustable set) using HIPPO [75] to achieve robust modeling structures.

Batch model

The reduced batch models derived from the initial calibrations (Table 3) were recalibrated with the available data (eight batch cultivations) to determine if they could reproduce *P. pastoris* behavior appropriately. The persistence of parametric problems in the reduced models was compared to the original model.

Structures 1 and 6 were the only parameter sets whose fitting capabilities were similar to the original eight parameters model (Table 4), showing the importance of including the specific uptake and production rates of the compounds considered in the model. On the contrary, m_{ATP} and K_G were left out of these structures because of the frequent identifiability and sensitivity associated problems.

Structure 6 lacks the sub-optimal growth parameter α_B, which forces the solution of a linear programming (LP) problem of specific growth rate maximization in the metabolic block. This is because this parameter was assumed to be zero if it was left out of the adjustable parameter set (Table 2), which eliminates the total flux minimization term from the objective function. This structure showed a significant increase in significance and sensitivity compared to the original model; however, identifiability was a major problem (Table 4). Probably, the multiple solutions associated with an underdetermined LP problem may hamper the possibility to unambiguously infer parameter values from the data.

Therefore, due to the recurrent identifiability issues found in Structure 6, it was preferable to apply Structure 1 to fit a different dataset to check its robustness in aerobic, glucose-limited batch cultures of *P. pastoris*.

Fed-batch model

In the fed-batch model, three potentially robust model structures were found after its calibration with three datasets (Table 5).

All the candidate structures considered the following parameters: K_G, $v_{Pyr,B}$, $v_{Cit,B}$, α_B, $v_{Pyr,FB}$ and m_{ATP}. Contrary to the batch model, K_G plays an important role in this cultivation system. This parameter, which usually lies in the micromolar range [79], can directly modulate substrate uptake under glucose-limited conditions.

Table 4 Batch cross calibration summary

Structure	N° parameters	J_{DIFF}	Significance Issues	Sensitivity Issues	Identifiability Issues
Original	**8**	**0**	**23.6**	**16.7**	**17.4**
1	6	-0.10	22.9	18.8	15.0
2	3	2.77	29.1	8	25
3	5	0.18	90.0	23	68
4	2	0.56	62.5	13	0
5	3	4.10	29.2	0	54
6	5	0.18	0	2.5	61.3
7	5	2.93	18.8	20.8	60.0
8	5	2.82	22.5	25.0	33.8

Structures that reduced de frequency of parametric problems with respect to the original model are highlighted

Table 5 Potential robust structures for a fed-batch model

Structure	Parameters included
Original	$v_{G,Max}$, K_G, $v_{EtOH,B}$, $v_{Pyr,B}$, $v_{Arab,B}$, $v_{Cit,B}$, $v_{EtOH,FB}$, $v_{Pyr,FB}$, $v_{Arab,FB}$, $v_{Cit,FB}$, a_B, a_{FB}, m_{ATP}, T_{Cons}
1	$v_{G,Max}$, K_G, $v_{Pyr,B}$, $v_{Cit,B}$, $v_{EtOH,FB}$, $v_{Pyr,FB}$, $v_{Arab,FB}$, $v_{Cit,FB}$, a_B, m_{ATP}, T_{Cons}
2	K_G, $v_{EtOH,B}$, $v_{Pyr,B}$, $v_{Arab,B}$, $v_{Cit,\ B}$, $v_{EtOH,FB}$, $v_{Pyr,FB}$, a_B, m_{ATP}
3	$v_{G,Max}$, K_G, $v_{Pyr,B}$, $v_{Arab,B}$, $v_{Cit,B}$, $v_{Pyr,FB}$ a_B, a_{FB}, m_{ATP}, T_{Cons}

Therefore, when glucose concentration is close to zero (like in the feeding phase), slight variations in the value of K_G can change glucose uptake significantly, which has a direct impact in the specific growth rate. Also, m_{ATP} appears to have a relevant role since it might act as an energy sink when the glucose from the batch phase is depleted. Here, secondary product consumption occurs with a slower or null biomass formation prior to the addition of glucose (Fig. 3 in Additional file 8). This indicates that the substrates were consumed to maintain basic cellular functions to survive, instead of being used for cell division.

The three reduced structures improved the initial fittings (lower J_{DIFF}) and reduced the frequency of fitting problems observed in the initial model of 14 parameters

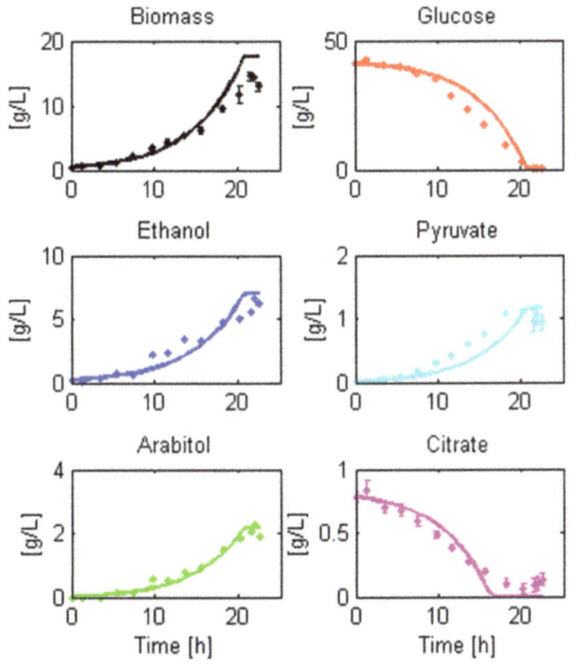

Fig. 3 Robustness check of Structure 1 as modeling framework for aerobic, glucose-limited batch cultures of *Pichia pastoris*. The figure shows the capacity of the reduced model structure to be calibrated with new data despite having fewer parameters than the original model structure (6 instead of 8 parameters). Points with whiskers represent experimental data and continuous lines correspond to the model approximation

(Table 6). Among these, Structure 3 performed the best in the cross-calibration in terms of fitting capability compared to the original model. On average, this structure improved in initial calibrations by 25%. It is worthy to note that, even though Structure 3 did not include the minimum production rate of ethanol during the batch phase, it could adequately reproduce the profiles of this compound by adjusting the objective function and the maintenance ATP. Finally, we chose to apply Structure 3 to fit new fed-batch data to check its robustness for modeling glucose-limited aerobic fed-batch cultivations of *P. pastoris*.

Robustness check

Batch model

On this new dataset, Structure 1 showed a good fit to the data and did not yield identifiability nor significance problems. However, $v_{G,Max}$ had no impact on the state variables. Therefore, after the initial calibration (data not shown), we fixed this parameter at 6 [mmol/g_{DCW}h] [63]. Figure 3 illustrates the model fit and Table 7 presents the parameter values with their 95% confidence intervals achieved in the second calibration, which also had no identifiability, significance or sensitivity limitations. This calibration also yielded mean normalized errors close to zero and normally distributed residuals for all the state variables except for glucose (Additional file 9).

Despite the sensitivity problem associated with $v_{G,Max}$ for this particular dataset, we included this parameter in the proposed robust modeling structure. This is because for some calibrations, e.g. the batch cultivations of strains harboring 8 copies of the thaumatin gene, the state variables were very sensitive to this parameter (average sensitivity > 0.7, recall that the sensitivity threshold is 0.01); hence, it should be included to achieve a close fit to the data. Therefore, if this parameter is found insensitive in future calibrations, it could be easily fixed at reported values.

We achieved a robust modeling structure for glucose-limited, aerobic batch cultivations of *Pichia pastoris*, composed of six parameters that estimate specific consumption and production rates of all the species involved in the mass balances. The modeling structure also allows us to determine the specific growth rate by solving a bi-objective optimization problem, which reduces the identifiability issues arising between parameters (comparison between candidate batch model robust structures 1 and 6).

Fed-batch model

Structure 3 shows a good fit to new experimental fed-batch data (Fig. 4) and did not yield identifiability or significance problems (Table 8 and Additional file 9). The

Table 6 Summary of the cross calibration of the fed-batch datasets

Structure	N° parameters	J_{DIFF}	Significance Issues (%)	Sensitivity Issues (%)	Identifiability Issues (%)
Original	**16**	**0**	**33**	**18.8**	**3.9**
1	11	-2%	27.2	2.6	1.8
2	9	-15%	25.1	7.4	0.9
3	10	-25%	26.7	3.7	0.9

Structures that reduced de frequency of parametric problems with respect to the original model are highlighted

profile of some of the state variables still depends on the fixed values assigned. For example, arabitol was consumed at a slower rate than the profile observed in the experiment because the parameter representing this consumption ($v_{Arab,FB}$) was fixed as the mean of the training datasets (not included in the adjustable parameter set). Thus, the model assumed a faster consumption rate than observed in the cultivations. Also, pyruvate was found at such low concentrations that the parameters associated to its production ($v_{Pyr,B}$ *and* $v_{Pyr,FB}$) were ignored in this analysis.

The chosen model structure showed a strong fitting capacity and a limited occurrence of parametric identifiability, sensitivity and significance problems. Therefore, we selected it as the most robust model structure for fed-batch cultivations of *P. pastoris*.

Model validation

Batch model

The parameters found for the strain harboring one copy of the thaumatin gene were used to predict the dynamics of a different batch cultivation using the same strain (Fig. 5). Biomass and glucose profiles were correctly predicted by the model (MNEs close to zero and p-values of the Anderson-Darling test > 0.05, see Additional file 10). Ethanol, pyruvate, citrate and arabitol dynamics also showed an overall concordance with the data, however the simulated profiles overestimated their final concentrations (see associated MNEs in Additional file 10). These differences occurred probably because in the training datasets the initial concentration of glucose was higher than the one used in the validation experiment (~60 g/L vs. ~40 g/L), which might have increased the formation of secondary products [80]. Therefore, future versions of the model may consider more elaborate kinetic expressions for the secretion of secondary products in order to accurately predict their formation in different circumstances.

Table 7 Parameter values achieved in the validation of the batch model structure

Parameter	Value	Units
$v_{G,Max}$	6	$mmol/g_{DCW} \cdot h$
$v_{EtOH,B}$	1.47 ± 0.07	$mmol/g_{DCW} \cdot h$
$v_{Pyr,B}$	0.13 ± 0.05	$mmol/g_{DCW} \cdot h$
$v_{Arab,B}$	0.14 ± 0.06	$mmol/g_{DCW} \cdot h$
$v_{Cit,B}$	0.09 ± 0.04	$mmol/g_{DCW} \cdot h$
a_B	$4.1 \pm 0.9 \cdot 10^{-4}$	[-]

Values of the parameters are presented together with their 95% confidence intervals. In this calibration, $v_{G,Max}$ was fixed at a known value to avoid sensitivity issues. Finally, the calibration yielded no parametric problems

Fed-batch model

The prediction of biomass, glucose, ethanol and arabitol concentrations during the culture agreed with experimental data, whereas pyruvate and citrate dynamics were inaccurate (Fig. 6). Specifically, the simulation

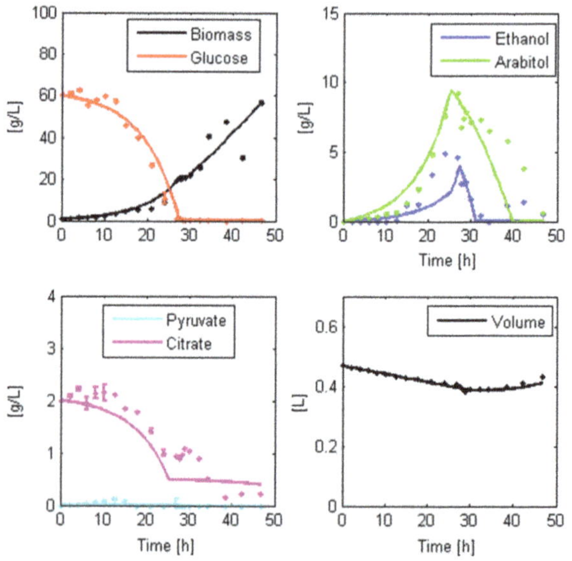

Fig. 4 Robustness check of Structure 3 as a modeling framework of aerobic glucose-limited fed-batch cultures of *Pichia pastoris*. The figure shows the capacity of the reduced model structure to be calibrated with new data, despite having fewer parameters than the original model structure (10 instead of 14 parameters). Points with whiskers represent experimental data and continuous lines correspond to the model approximation

Table 8 Parameter values achieved in the calibration to check the robustness of the fed-batch model. The confidence interval on the time where the consumption of secondary metabolites started T_{CONS}, could not be determined due to the stiffness of the solution caused by a sudden consumption of arabitol and ethanol

Parameter	Value	Units
v_{MAX}	2.09 ± 0.46	$mmol/g_{DCW} \cdot h$
K_S	$5.55 \cdot 10^{-2} \pm 0.0000004 \cdot 10^{-2}$	g/L
$v_{Pyr,B}$	0	$mmol/g_{DCW} \cdot h$
$v_{Arab,B}$	0.42 ± 0.17	$mmol/g_{DCW} \cdot h$
$v_{Cit,B}$	0.04 ± 0.00	$mmol/g_{DCW} \cdot h$
$v_{Pyr,FB}$	0	$mmol/g_{DCW} \cdot h$
α_B	$2.6 \cdot 10^{-4} \pm 0.4 \cdot 10^{-4}$	[-]
α_{FB}	$2.455 \cdot 10^{-5} \pm 0.003 \cdot 10^{-5}$	[-]
m_{ATP}	7.0 ± 1.4	$mmol/g_{DCW} \cdot h$
T_{Cons}	25.73	H

predicted that pyruvate was generated during the batch phase but experimental data did not show pyruvate production. In the experimental culture we saw that there was no generation of citrate in the feed phase, contrary to what the simulated predicted. These differences arose because in the culture from where the parameters were derived (Fed-batch culture 1, see Additional file 8), pyruvate formation occurred in the batch phase and citrate was formed during the feed phase; therefore, the model assumed that these compounds were generated in the respective phase of the culture. Nevertheless, for the major compounds found in the culture, the model had a low mean normalized error.

Potential applications of the model

Analysis of the metabolic flux distribution at different stages of a dynamic cultivation

Once we confirmed the robustness of the fed-batch model, we analyzed the redistribution of central carbon metabolic fluxes at three different stages of the cultivation (Fig. 7), i.e. exponential growth during the batch phase (~20 h, $\mu = 0.12\ h^{-1}$); co–consumption of arabitol and ethanol during the glucose starvation phase (~27.5 h, $\mu = 0.02\ h^{-1}$); and controlled exponential growth during the feeding phase (~45 h, $\mu = 0.06\ h^{-1}$) (Fig. 7).

During exponential growth in the batch phase, the carbon reaching the glucose-6-phosphate node is split between carbohydrate production (11%), glycolysis (63%) and the oxidative branch of the PPP (24%). Furthermore, the latter is the main source of cytosolic NADPH. Cytosolic ATP is formed by the activity of the ATP synthase and substrate-level phosphorylation (glycolysis and synthesis of arabitol and ethanol) (data not shown). In the iPP618 model, which is the basis of the iFS670, cytosolic NADPH was produced by a NADP dependent isocitrate dehydrogenase, and no flux appeared through the oxidative branch of the PPP. Using the proposals from Pereira et al. [56], the flux through this pathway was restored and overall agreement

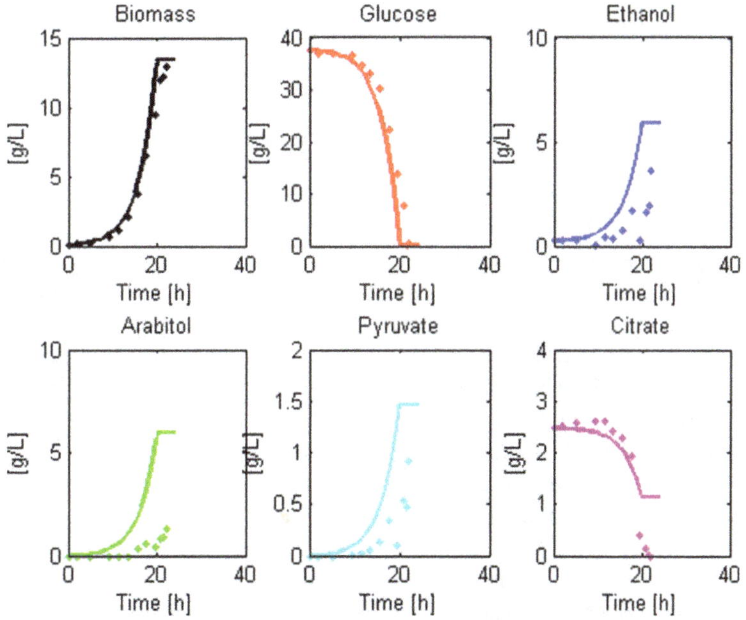

Fig. 5 Batch model preliminary validation. This figure shows how well the model predicts the course of a batch cultivation. To do this, we used the derived robust model structure to determine the characteristic parameters of a recombinant strain. Then, we simulated a batch culture (*continuous line*) and compared it with the experimental data (*filled circles*)

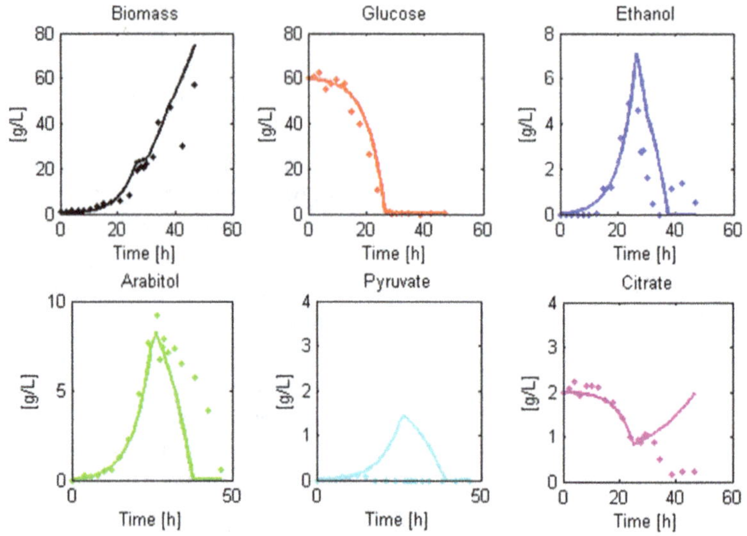

Fig. 6 Fed-batch model validation. This figure shows how well the model predicts the course of a fed-batch cultivation. To do this, we used the derived robust model structure to determine the characteristic parameters of a recombinant strain. Then, we simulated a fed-batch culture (*continuous line*) and compared it with the experimental data (*filled circles*)

in directionality to fluxomic studies performed in similar conditions was achieved (Additional file 3).

During the starvation phase, ethanol and arabitol are co-consumed with limited formation of biomass ($\mu = 0.02\ h^{-1}$). As indicated by the negative fluxes, both compounds are directed towards the TCA cycle in order to synthesize the necessary reducing equivalents to fuel oxidative phosphorylation. The ATP formed in this pathway - ~ 7 mmol/g_{DCW}·h -, is mostly employed for maintenance. Even though this m_{ATP} is high compared to other reported values for *P. pastoris* (2.2 – 5 mmol/g_{DCW}·h) [43], it is required to account for the fast consumption of both secondary metabolites under limited cellular growth. The use of a recombinant strain for model calibration, which might have higher maintenance requirements, could further explain this result.

Finally, during controlled growth at the feed phase, neither ethanol nor arabitol are produced. All the carbon is directed towards biomass formation and the energy necessary for its synthesis and maintenance. This result agrees with previous fluxomic studies carried out in aerobic, glucose-limited chemostats [57, 58], where significant carbon fluxes through the oxidative and non-oxidative branches of the PPP were found, without arabitol formation. Furthermore, the model shows significant oxaloacetate transport from the cytosol to the mitochondria, which was also observed in the cited studies. The most distinguishable feature of this phase is the high activity of the TCA cycle, which almost doubles the flux through this pathway reported under glucose limited conditions in chemostats ([24, 57, 58]). This higher activity in the TCA is probably associated with the need to cope with maintenance and growth-associated energy requirements under stressful conditions, such as high cell density, especially when no significant substrate level phosphorylation besides glycolysis occurs.

This analysis could have been performed using the genome-scale model in static conditions by deriving instantaneous exchange rates from contiguous samples and determining the flux distributions by specific growth rate maximization. Nevertheless, the inspection of flux distributions after model calibration has the advantage of considering the overall behavior of the cells during the cultivations. This provides more experimental support for the determination of parameters such as m_{ATP}, K_G, that cannot be directly estimated but that have a strong impact on the model output.

Discovery of single knock-outs to improve recombinant Human Serum Albumin production using Minimization of Metabolic Adjustment (MOMA) as the objective function to simulate mutant behavior

We performed 670 (number of genes in the model) batch simulations of single knock-out strains to discover beneficial deletions for the production of recombinant Human Serum Albumin (HSA), a 66 kDa protein with 16 disulfide bridges, that comprises about one half of the total blood serum protein [81].

The two main clusters (Fig. 8) show the relation between the final HSA and the final biomass concentration of the 130 mutations that improved HSA production (>30 mg/L at the end of the batch). The first cluster consists of strains that privilege HSA production over biomass formation; whereas the second one presents a trade-off between both.

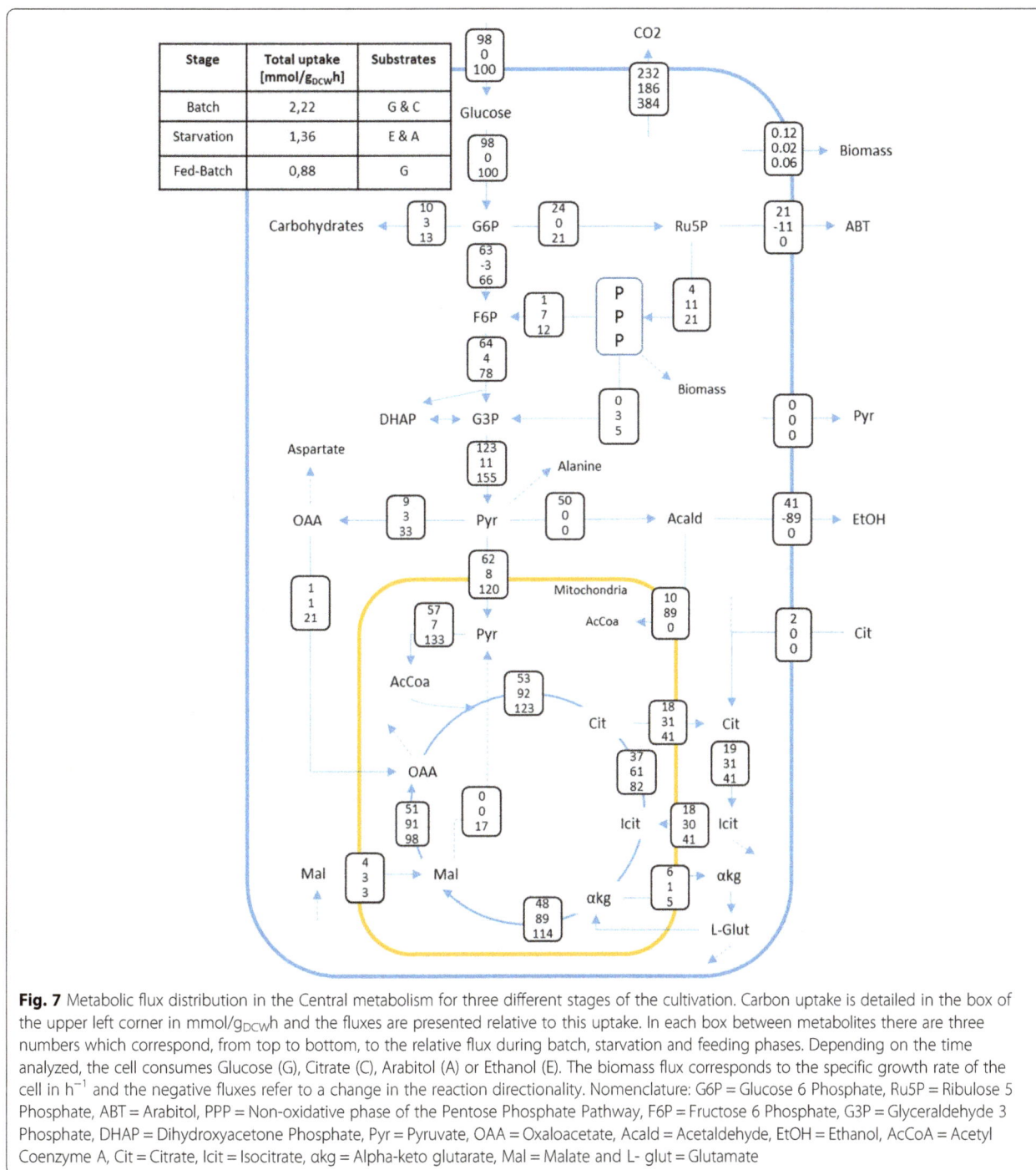

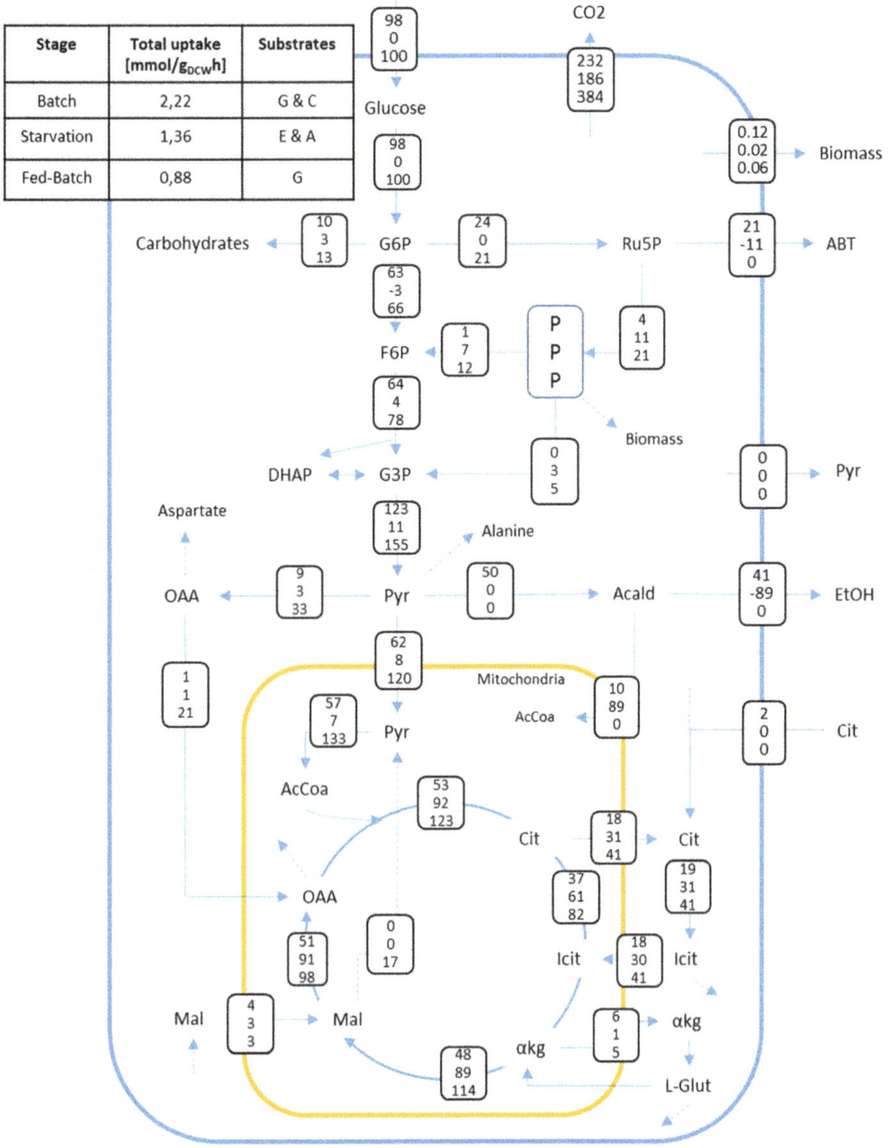

Fig. 7 Metabolic flux distribution in the Central metabolism for three different stages of the cultivation. Carbon uptake is detailed in the box of the upper left corner in mmol/g$_{DCW}$h and the fluxes are presented relative to this uptake. In each box between metabolites there are three numbers which correspond, from top to bottom, to the relative flux during batch, starvation and feeding phases. Depending on the time analyzed, the cell consumes Glucose (G), Citrate (C), Arabitol (A) or Ethanol (E). The biomass flux corresponds to the specific growth rate of the cell in h^{-1} and the negative fluxes refer to a change in the reaction directionality. Nomenclature: G6P = Glucose 6 Phosphate, Ru5P = Ribulose 5 Phosphate, ABT = Arabitol, PPP = Non-oxidative phase of the Pentose Phosphate Pathway, F6P = Fructose 6 Phosphate, G3P = Glyceraldehyde 3 Phosphate, DHAP = Dihydroxyacetone Phosphate, Pyr = Pyruvate, OAA = Oxaloacetate, Acald = Acetaldehyde, EtOH = Ethanol, AcCoA = Acetyl Coenzyme A, Cit = Citrate, Icit = Isocitrate, αkg = Alpha-keto glutarate, Mal = Malate and L- glut = Glutamate

We decided to leave Cluster I out of the analysis because of the impaired growth observed in the simulations, mainly due to the deletion of reactions associated to lipid biosynthesis. However, candidates from Cluster II (32 in total) were manually analyzed to identify the cause of HSA overproduction (Additional file 11).

A relative increase in the formation of cysteine and tryptophan was found for most of the candidates for Cluster II when compared to the parental strain, a trend that was not observed for the rest of the amino acids (Fig. 9). These energetically costly residues [82] are formed from serine. Therefore, re-routing carbon through this pathway could be beneficial to improve HSA production.

After manually analyzing the candidates, we found that one possible strategy could be the deletion of the cytosolic NAD-dependent methylene tetrahydrofolate dehydrogenase (Fig. 10). When compared to the parental strain, the knock-out results in a 6.3 fold improvement of the final

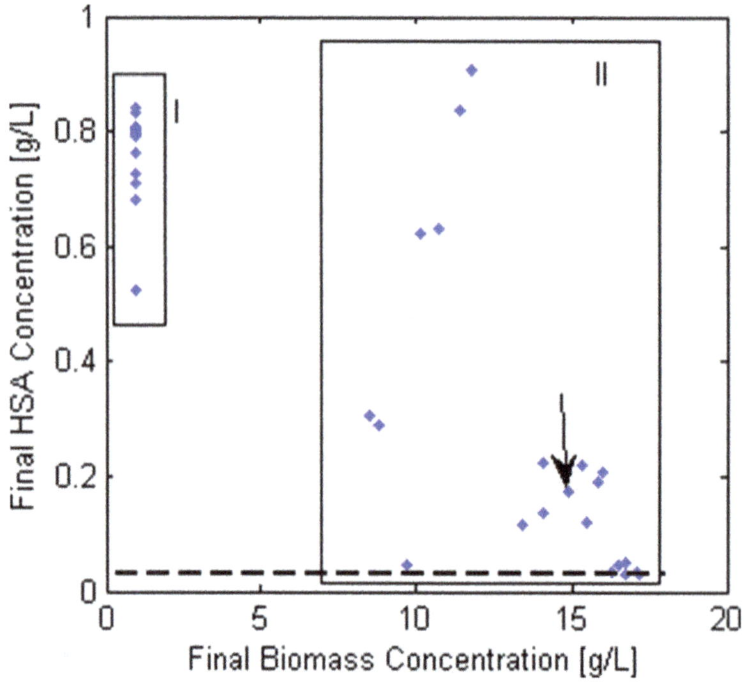

Fig. 8 Final HSA vs. final biomass concentrations of simulated batch cultivations of single knock-out-strains. Blue dots correspond to the output of strains that improved the initial final HSA concentration (30 mg/L). Candidates out of Cluster II were manually analyzed. The red circle indicates the performance of the parental strain and the black arrow points to the methylene tetrahydrofolate dehydrogenase knock-out strain

concentration of the recombinant protein with a 5.8-fold increase in protein volumetric productivity (arrow in Fig. 8). This deletion eliminates the transformation of serine to 5–10 methylene tetrahydrofolate; hence, serine can be re-routed to two cysteine reactions. This gene is non-essential in *S. cerevisiae* [83] and, to the best of our knowledge, its essentiality has not been determined in *P. pastoris.* Therefore, it constitutes an interesting knock-out candidate to improve recombinant HSA production.

Bioprocess optimization for HSA overproduction

Here, we evaluated 13 feeding strategies of a fed-batch cultivation to improve the production of recombinant HSA. After the simulations, we selected a strategy that considered a slow decrease in the growth rate from $\mu = 0.14\ h^{-1}$ to $\mu = 0.08\ h^{-1}$ during the feeding phase (Table 9). The selected policy allows a 25% increase in volumetric productivity and reaches almost the same final HSA concentration as the constant growth rate

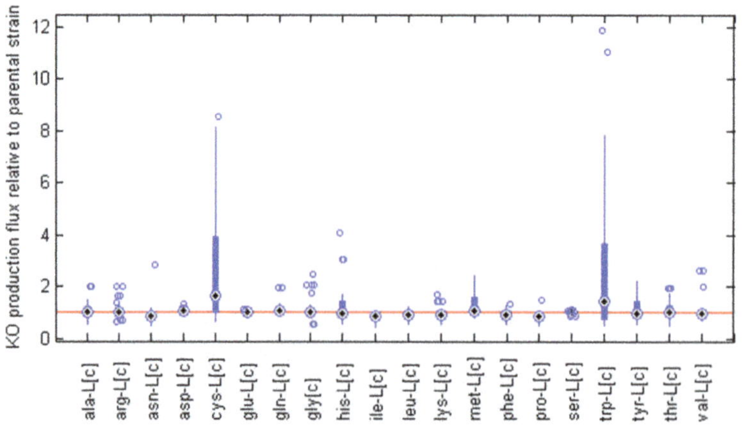

Fig. 9 Turnover of key amino acids in knock-out strains relative to the parental strain. Each box summarizes how the production of each amino acid changed in the 32 knock out strains of Cluster II relative to the production in the parental strain (*Red Line*). Black dots correspond to the sample median, the extreme of the boxes to the 25th and 75th percentiles, the whiskers extend to the most extreme data points and circles mark outliers

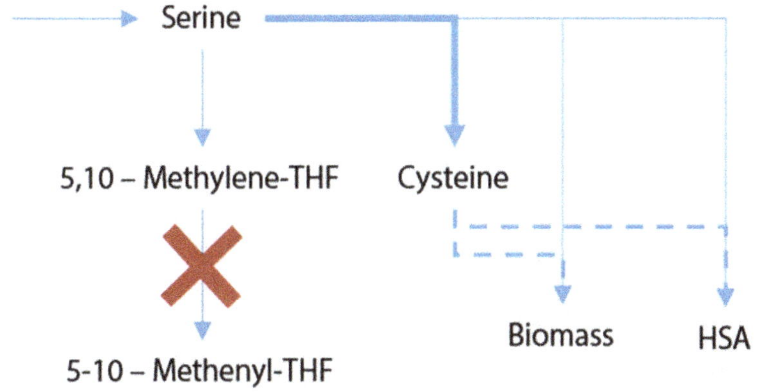

Fig. 10 Rationale behind the knockout of the Methylene tetrahydrofolate (THF) dehydrogenase. By deleting this enzyme, the flux from Serine to 5-10-Methylene THF is blocked and redirected towards cysteine formation, whose availability increases the productivity of HSA

strategy that reached the highest concentration ($\mu = 0.06\ h^{-1}$).

The improvement in process productivity by modifying substrate addition during the feed phase is less efficient than the one attained by genetic modifications. However, other process variables such as reactor volume and oxygen transfer may be modified to further improve HSA production.

Conclusions

Current GSMs of *P. pastoris* have been employed to address cellular behavior in stationary conditions. They have been successfully used for predicting production and consumption rates of different compounds and even achieving a 40% improvement of recombinant protein production by model-discovered knock-outs [42]. However, little attention has been given to the actual metabolic flux distribution that these reconstructions yield and how they evolve in a dynamic environment. Resulting flux distributions are important for two reasons: (i) they help to understand the cellular response to the different stresses to which the cell is subjected to and (ii) they can serve as input for several algorithms whose aim is to find metabolic engineering targets to improve the production of a certain compound.

In this work, we developed a robust dynamic GSM of glucose-limited aerobic cultivations of *P. pastoris*, linking and showing the impact that the model formulation process has over flux balance analysis. The assembled platform can fit several datasets with minimum significance, sensitivity and identifiability problems in its parameters. Moreover, if properly trained, it can be used to predict bioreactor dynamics. The model could also be employed to obtain realistic flux distributions throughout dynamic cultivations and to determine metabolic and process engineering strategies to improve the production of a target compound.

To broaden its applications to other relevant conditions for *P. pastoris*, the model could be calibrated with data from cultures with different carbon sources and feeding strategies, such as glycerol batch phase followed by a methanol induction phase. Also, the model could be used to study perturbations such as oxygen limitation, which is a common problem in industrial *P. pastoris* cultivations [84]. Moreover, it would be desirable to calibrate the model with data from a strain capable of producing high concentrations of a recombinant protein to understand and quantify the metabolic burden caused by this production.

Table 9 Feeding policies evaluated to improve the production of Serum Albumin in a particular bioreactor setup

Strategy	μ_{MAX}	Rate	μ_{MIN}	q_P [mg/g · h]	X_{FINAL} [g/L]	P_{FINAL} [mg/L]	Limitation
1	0.14	-	-	2.85	164.8	138	Oxygen
2	0.12	-	-	2.59	187.8	135	Oxygen
3	0.1	-	-	2.32	195.3	130	Volume
4	0.08	-	-	2.29	191.3	138	Volume
5	0.06	-	-	2.28	184.7	154	Volume
Best	0.14	0.1	0.08	2.83	197.5	150	Volume

This table shows the process indicators for the constant feeding Strategies (1–5) plus the best decreasing growth rate strategy. μ_{MAX} is the maximum growth rate in the feeding police. μ_{MIN} is the minimum growth rate in the feeding police. Rate is the rate of decreasing of set growth rate in feeding police. q_P is the protein productivity. X_{FINAL} and P_{FINAL} refer to the final concentration of biomass and serum albumin in the reactor when the simulation stops, which happened by either violating user-defined volume or Oxygen Transfer thresholds

Finally, it is expected that the incorporation of more curated metabolic reconstructions [44], gas mass balances and the knowledge derived from testing the hypotheses proposed using the model would improve its accuracy and broaden its applicability.

Additional files

Additional file 1: Dynamic genome-scale metabolic model of *Pichia pastoris*.

Additional file 2: Demonstration of the convexity of the solution space of the QP problem in the metabolic block.

Additional file 3: Construction and evaluation of the iFS670 model. This file explains the modifications made on the iPP668 model to obtain the iFS670. Also, model performance is evaluated in terms of the internal flux distribution and the capacity of the model to predict experimental chemostat data in comparison to other available models at the beginning of the study.

Additional file 4: iFS670 model for COBRA.

Additional file 5: Batch cultivation data used for model calibration, cross calibration, robustness check and validation.

Additional file 6: Fed-Batch cultivation data used for model calibration, cross calibration, robustness check and validation.

Additional file 7: Evaluation of Feeding policies. This file describes the performance of thirteen different feeding strategies on improving recombinant HSA production.

Additional file 8: Cross Calibration Summary. Shows the performance of each one of the modeling structures derived using HIPPO (See Tables 3 and 5) after their calibration using the available datasets.

Additional file 9: Absence of Parametric problems in the Robustness Check Datasets and Goodness of Fit. This file shows how the reduced (robust) modeling structures derived for the batch and fed-batch configurations presented no identifiability or sensitivity problems after being calibrated with new fermentation data.

Additional file 10: Goodness of fit analysis of the validation datasets.

Additional file 11: Knockout candidates for the overproduction of Human Serum Albumin. This file contains the details of the 32 candidates from Cluster II (Fig. 8) that could theoretically improve recombinant protein production. Expected final protein and biomass concentrations are reported.

Acknowledgements

We would like to acknowledge Alexandra Lobos for her support during the experiments and for facilitating the *P. pastoris* strains. We are also grateful to Dr. Dong-Yup Lee for facilitating the iPP668 model for COBRA, from which our version was built upon. English edition of the final manuscript by Lisa Gingles is highly appreciated.

Funding

This project was funded by CORFO (Project 11CEII-9568). F.S. was recipient of a M.Sc. scholarship from CONICYT N° 22140230 and P.T. obtained a Ph.D scholarship from the same institution, N° 21140759.

Authors' contributions

FS, PT, RP and EA conceived the experiments and simulations. FS and PT performed the experiments. FS assembled the model, performed the parametric analysis and simulations. All authors read and approved the final manuscript.

Competing interest

The authors declare that they have no competing interests.

References

1. Walsh G. Biopharmaceutical benchmarks 2014. Nat Biotechnol. 2014;32(7):992–1000.
2. BCC Research. Global markets for enzymes in industrial applications - report overview. 2014 [Online]. Available: http://www.bccresearch.com/market-research/biotechnology/enzymes-industrial-applications-bio030h.html. [Accessed: 07 Dec 2015].
3. Markets and Markets. Industrial Enzymes Market by Type (Carbohydrases, Proteases, Non-starch Polysaccharides & Others), Application (Food & Beverage, Cleaning Agents, Animal Feed & Others), Brands & by Region - Globlar Trends and Forecasts to 2020. 2015 [Online]. Available: http://www.marketsandmarkets.com/Market-Reports/industrial-enzymes-market-237327836.html. [Accessed: 07 Dec 2015].
4. Overton TW. Recombinant protein production in bacterial hosts. Drug Discov Today. 2014;19(5):590–601.
5. Maccani A, Landes N, Stadlmayr G, Maresch D, Leitner C, Maurer M, Gasser B, Ernst W, Kunert R, Mattanovich D. Pichia pastoris secretes recombinant proteins less efficiently than Chinese hamster ovary cells but allows higher space-time yields for less complex proteins. Biotechnol J. 2014;9(4):526–37.
6. Ferrer-Miralles N, Domingo-Espín J, Corchero JL, Vázquez E, Villaverde A. Microbial factories for recombinant pharmaceuticals. Microb Cell Fact. 2009;8:17.
7. Corchero JL, Gasser B, Resina D, Smith W, Parrilli E, Vázquez F, Abasolo I, Giuliani M, Jäntti J, Ferrer P, Saloheimo M, Mattanovich D, Schwartz S, Tutino ML, Villaverde A. Unconventional microbial systems for the cost-efficient production of high-quality protein therapeutics. Biotechnol Adv. 2013;31(2):140–53.
8. Daly R, Hearn MTW. Expression of heterologous proteins in Pichia pastoris: a useful experimental tool in protein engineering and production. J Mol Recognit. 2005;18(2):119–38.
9. Cereghino JL, Cregg JM. Heterologous protein expression in the methylotrophic yeast Pichia pastoris. FEMS Microbiol Rev. 2000;24(1):45–66.
10. Ciofalo V, Barton N, Kreps J, Coats I, Shanahan D. Safety evaluation of a lipase enzyme preparation, expressed in Pichia pastoris, intended for use in the degumming of edible vegetable oil. Regul Toxicol Pharmacol. 2006;45(1):1–8.
11. Masuda T, Ide N, Ohta K, Kitabatake N. High-yield secretion of the recombinant sweet-tasting protein Thaumatin I. Food Sci Technol Res. 2010;16(6):585–92.
12. Çalık P, Ata Ö, Güneş H, Massahi A, Boy E, Keskin A, Öztürk S, Zerze GH, Özdamar TH. Recombinant protein production in Pichia pastoris under glyceraldehyde-3-phosphate dehydrogenase promoter: from carbon source metabolism to bioreactor operation parameters. Biochem Eng J. 2015;95:20–36.
13. Mattanovich D, Graf A, Stadlmann J, Dragosits M, Redl A, Maurer M, Kleinheinz M, Sauer M, Altmann F, Gasser B. Genome, secretome and glucose transport highlight unique features of the protein production host Pichia pastoris. Microb Cell Fact. 2009;8:29.
14. Delic M, Valli M, Graf AB, Pfeffer M, Mattanovich D, Gasser B. The secretory pathway: exploring yeast diversity. FEMS Microbiol Rev. 2013;37(6):872–914.
15. Hasslacher M, Schall M, Hayn M, Bona R, Rumbold K, Lückl J, Griengl H, Kohlwein SD, Schwab H. High-level intracellular expression of hydroxynitrile lyase from the tropical rubber TreeHevea brasiliensisin microbial hosts. Protein Expr Purif. 1997;11(1):61–71.
16. Heyland J, Fu J, Blank LM, Schmid A. Quantitative physiology of Pichia pastoris during glucose-limited high-cell density fed-batch cultivation for recombinant protein production. Biotechnol Bioeng. 2010;107(2):357–68.
17. Čiplys E, Žitkus E, Gold LI, Daubriac J, Pavlides SC, Højrup P, Houen G, Wang W-A, Michalak M, Slibinskas R. High-level secretion of native recombinant human calreticulin in yeast. Microb Cell Fact. 2015;14(1):165.
18. Wang Y, Liang ZH, Zhang YS, Yao SY, Xu YG, Tang YH, Zhu SQ, Cui DF, Feng YM. Human insulin from a precursor overexpressed in the methylotrophic yeast Pichia pastoris and a simple procedure for purifying the expression product. Biotechnol Bioeng. 2001;73:74–9.
19. Thompson CA. FDA approves kallikrein inhibitor to treat hereditary angioedema. Am J Health Pharm. 2010;67:93.

20. Delic M, Göngrich R, Mattanovich D, Gasser B. Engineering of protein folding and secretion-strategies to overcome bottlenecks for efficient production of recombinant proteins. Antioxid Redox Signal. 2014;21(3):414–37.
21. Gasser B, Prielhofer R, Marx H, Maurer M, Nocon J, Steiger M, Puxbaum V, Sauer M, Mattanovich D. Pichia pastoris : protein production host and model organism for biomedical research. Future Microbiol. 2013;8(2):191–208.
22. Wang J-R, Li Y-Y, Liu D-N, Liu J-S, Li P, Chen L-Z, Xu S-D. Codon optimization significantly improves the expression level of α -Amylase gene from *Bacillus licheniformis* in *Pichia pastoris*. Biomed Res Int. 2015;2015:1–9.
23. Prielhofer R, Maurer M, Klein J, Wenger J, Kiziak C, Gasser B, Mattanovich D. Induction without methanol: novel regulated promoters enable high-level expression in Pichia pastoris. Microb Cell Fact. 2013;12(1):5.
24. Heyland J, Fu J, Blank LM, Schmid A. Carbon metabolism limits recombinant protein production in Pichia pastoris. Biotechnol Bioeng. 2011;108(8):1942–53.
25. Baumann K, Maurer M, Dragosits M, Cos O, Ferrer P, Mattanovich D. Hypoxic fed-batch cultivation ofPichia pastoris increases specific and volumetric productivity of recombinant proteins. Biotechnol Bioeng. 2008;100(1):177–83.
26. Maurer M, Kühleitner M, Gasser B, Mattanovich D. Versatile modeling and optimization of fed batch processes for the production of secreted heterologous proteins with Pichia pastoris. Microb Cell Fact. 2006;5:37.
27. Looser V, Bruhlmann B, Bumbak F, Stenger C, Costa M, Camattari A, Fotiadis D, Kovar K. Cultivation strategies to enhance productivity of Pichia pastoris : a review. Biotechnol Adv. 2015;33:1177–93.
28. Riesenberg D, Guthke R. High-cell-density cultivation of microorganisms. Appl Microbiol Biotechnol. 1999;51(4):422–30.
29. Villadsen J, Nielsen J, Lidén G. Bioreaction engineering principles. 3rd ed. Nueva York: Springer; 2011.
30. Vargas F, Pizarro F, Pérez-Correa JR, Agosin E. Expanding a dynamic flux balance model of yeast fermentation to genome-scale. BMC Syst Biol. 2011;5(1):75.
31. Landi C, Paciello L, De Alteriis E, Brambilla L, Parascandola P. High cell density culture with S. cerevisiae CEN.PK113-5D for IL-1β production: optimization, modeling, and physiological aspects. Bioprocess Biosyst Eng. 2015;38(2):251–61.
32. Graf A, Dragosits M, Gasser B, Mattanovich D. Yeast systems biotechnology for the production of heterologous proteins. FEMS Yeast Res. 2009;9(3):335–48.
33. Kitano H. Systems biology: a brief overview. Science. 2002;295(5560):1662–4.
34. Varma A, Palsson B. Stoichiometric flux balance models quantitatively predict growth and metabolic by-product secretion in wild-type Escherichia coli W3110. Appl Environ Microbiol. 1994;60(10):3724–31.
35. Mahadevan R, Edwards JS, Doyle FJ. Dynamic flux balance analysis of diauxic growth in Escherichia coli. Biophys J. 2002;83(3):1331–40.
36. Höffner K, Harwood SM, Barton PI. A reliable simulator for dynamic flux balance analysis. Biotechnol Bioeng. 2013;110(3):792–802.
37. Thiele I, Palsson BØ. A protocol for generating a high-quality genome-scale metabolic reconstruction. Nat Protoc. 2010;5(1):93–121.
38. Palsson BO. Systems biology: constraint-based reconstruction and analysis. Cambridge: Cambridge University Press; 2015.
39. Park JH, Lee KH, Kim TY, Lee SY. Metabolic engineering of Escherichia coli for the production of L-valine based on transcriptome analysis and in silico gene knockout simulation. Proc Natl Acad Sci U S A. 2007;104(19):7797–802.
40. Asadollahi M a, Maury J, Patil KR, Schalk M, Clark A, Nielsen J. Enhancing sesquiterpene production in Saccharomyces cerevisiae through in silico driven metabolic engineering. Metab Eng. 2009;11(6):328–34.
41. Sohn SB, Graf AB, Kim TY, Gasser B, Maurer M, Ferrer P, Mattanovich D, Lee SY. Genome-scale metabolic model of methylotrophic yeast Pichia pastoris and its use for in silico analysis of heterologous protein production. Biotechnol J. 2010;5(7):705–15.
42. Caspeta L, Shoaie S, Agren R, Nookaew I, Nielsen J. Genome-scale metabolic reconstructions of Pichia stipitis and Pichia pastoris and in silico evaluation of their potentials. BMC Syst Biol. 2012;6(1):24.
43. Chung BKS, Selvarasu S, Camattari A, Ryu J, Lee H, Ahn J, Lee H, Lee D. Genome-scale metabolic reconstruction and in silico analysis of methylotrophic yeast Pichia pastoris for strain improvement. Microb Cell Fact. 2010;9(50):2–15.
44. Tomàs-Gamisans M, Ferrer P, Albiol J. Integration and validation of the genome-scale metabolic models of Pichia pastoris: a comprehensive update of protein glycosylation pathways, lipid and energy metabolism. PLoS One. 2016;11(1):e0148031.
45. Irani ZA, Kerkhoven EJ, Shojaosadati SA, Nielsen J. Genome-scale metabolic model of Pichia pastoris with native and humanized glycosylation of recombinant proteins. Biotechnol Bioeng. 2015;113(5):961–9.
46. Nocon J, Steiger MG, Pfeffer M, Sohn SB, Kim TY, Maurer M, Rußmayer H, Pflügl S, Ask M, Haberhauer-Troyer C, Ortmayr K, Hann S, Koellensperger G, Gasser B, Lee SY, Mattanovich D. Model based engineering of Pichia pastoris central metabolism enhances recombinant protein production. Metab Eng. 2014;24:129–38.
47. Jaqaman K, Danuser G. Linking data to models: data regression. Nat Rev Mol Cell Biol. 2006;7(11):813–9.
48. Sánchez BJ, Pérez-Correa JR, Agosin E. Construction of robust dynamic genome-scale metabolic model structures of Saccharomyces cerevisiae through iterative re-parameterization. Metab Eng. 2014;25:159–73.
49. Stephanopoulos GM, Aristidou AA, Nielsen J. Metabolic engineering principles and methodologies. San Diego: Academic; 1998.
50. Postma E, Verduyn C, Scheffers W a, Van Dijken JP. Enzymic analysis of the crabtree effect in glucose-limited chemostat cultures of Saccharomyces cerevisiae. Appl Environ Microbiol. 1989;55(2):468–77.
51. Feng X, Xu Y, Chen Y, Tang YJ. Integrating flux balance analysis into kinetic models to decipher the dynamic metabolism of shewanella oneidensis MR-1. PLoS Comput Biol. 2012;8:2.
52. Schuetz R, Zamboni N, Zampieri M, Heinemann M, Sauer U. Multidimensional optimality of microbial metabolism. Science. 2012; 336(6081):601–4.
53. Holzhütter HG. The principle of flux minimization and its application to estimate stationary fluxes in metabolic networks. Eur J Biochem. 2004; 271(14):2905–22.
54. Schuetz R, Kuepfer L, Sauer U. Systematic evaluation of objective functions for predicting intracellular fluxes in Escherichia coli. Mol Syst Biol. 2007; 3(119):119.
55. Price ND, Famili I, Beard D a, Palsson BØ. Extreme pathways and Kirchhoff's second law. Biophys J. 2002;83(5):2879–82.
56. Pereira R, Nielsen J, Rocha I. Improving the flux distributions simulated with genome-scale metabolic models of Saccharomyces cerevisiae. Metab Eng Commun. 2016;3:153–63.
57. Baumann K, Carnicer M, Dragosits M, Graf AB, Stadlmann J, Jouhten P, Maaheimo H, Gasser B, Albiol J, Mattanovich D, Ferrer P. A multi-level study of recombinant Pichia pastoris in different oxygen conditions. BMC Syst Biol. 2010;4(1):141.
58. Dragosits M, Stadlmann J, Albiol J, Baumann K, Maurer M, Gasser B, Sauer M, Altmann F, Ferrer P, Mattanovich D, Cerdanyola B. The effect of temperature on the proteome of recombinant pichia pastoris research articles. J Proteome Res. 2009;8:1380–92.
59. Becker S, Feist AM, Mo ML, Hannum G, Palsson BØ, Herrgard MJ. Quantitative prediction of cellular metabolism with constraint-based models: the COBRA Toolbox. Nat Protoc. 2007;2(3):727–38.
60. Hyduke D, Schellenberger J, Que R, Fleming R, Thiele I, Orth J, Feist A, Zielinski D, Bordbar A, Lewis N, Rahmanian S, Kang J, Palsson B. COBRA Toolbox 2.0. Protoc Exch. 2011;[Online]. Available: http://www.nature.com/protocolexchange/protocols/2097#/related-articles.
61. Bornstein BJ, Keating SM, Jouraku A, Hucka M. LibSBML: an API library for SBML. Bioinformatics. 2008;24(6):880–1.
62. Keating SM, Bornstein BJ, Finney A, Hucka M. SBMLToolbox: an SBML toolbox for MATLAB users. Bioinformatics. 2006;22(10):1275–7.
63. van Urk H, Postma E, Scheffers W a, van Dijken JP. Glucose transport in crabtree-positive and crabtree-negative yeasts. J Gen Microbiol. 1989;135:2399–406.
64. Cárcamo M, Saa PA, Torres J, Torres S, Mandujano P, Correa JRP, Agosin E. Effective dissolved oxygen control strategy for high-cell-density cultures. IEEE Lat Am Trans. 2014;12(3):389–94.
65. Villadsen J, Patil KR. Optimal Fed-batch cultivation when mass transfer becomes limiting. Biotechnol Bioeng. 2007;98(3):706–10.
66. Cárcamo M. Producción de proteínas recombinantes en cultivos Fed-batch de *Saccharomyces cerevisiae y Escherichia coli*. Santiago: Pontificia Unversidad Católica de Chile; 2013.
67. Tolner B, Smith L, Begent RHJ, Chester K a. Production of recombinant protein in Pichia pastoris by fermentation. Nat Protoc. 2006;1(2):1006–21.
68. Marx H, Mecklenbräuker A, Gasser B, Sauer M, Mattanovich D. Directed gene copy number amplification in Pichia pastoris by vector integration into the ribosomal DNA locus. FEMS Yeast Res. 2009;9(8):1260–70.
69. Egea J, Balsa-Canto E. Dynamic optimization of nonlinear processes with an enhanced scatter search method. Ind Eng Chem Res. 2009;48(9):4388–401.

70. Balsa-Canto E, Rodriguez-Fernandez M, Banga JR. Optimal design of dynamic experiments for improved estimation of kinetic parameters of thermal degradation. J Food Eng. 2007;82(2):178–88.
71. Sacher J, Saa P, Cárcamo M, López J, Gelmi C a, Pérez-Correa R. Improved calibration of a solid substrate fermentation model. Electron J Biotechnol. 2011;14:5.
72. Sriram K, Rodriguez-Fernandez M, Doyle FJ. Modeling cortisol dynamics in the neuro-endocrine axis distinguishes normal, depression, and post-traumatic stress disorder (PTSD) in humans. PLoS Comput Biol. 2012;8:2.
73. Petersen B, Gernaey K, Vanrolleghem PA. Practical identifiability of model parameters by combined respirometric-titrimetric measurements. Water Sci Technol. 2001;43(7):347–55.
74. Landaw EM, DiStefano 3rd JJ. Multiexponential, multicompartmental, and noncompartmental modeling. II. Data analysis and statistical considerations. Am J Physiol Regul Integr Comp Physiol. 1984;246:5.
75. Sánchez BJ, Soto DC, Jorquera H, Gelmi CA, Pérez-Correa JR. HIPPO: An iterative reparametrization method for identification and calibration of dynamic bioreactor models of complex processes. Ind Eng Chem Res. 2014; 53(48):18514–25.
76. Stephens MA. EDF statistics for goodness of fit and some comparisons. J Am Stat Assoc. 1974;69(347):730–7.
77. Segrè D, Vitkup D, Church GM. Analysis of optimality in natural and perturbed metabolic networks. Proc Natl Acad Sci U S A. 2002;99(23):15112–7.
78. Rebnegger C, Graf AB, Valli M, Steiger MG, Gasser B, Maurer M, Mattanovich D. In Pichia pastoris, growth rate regulates protein synthesis and secretion, mating and stress response. Biotechnol J. 2014;9(4):511–25.
79. Boles E, Hollenberg CP. The molecular genetics of hexose transport in yeasts. FEMS Microbiol Rev. 1997;21(1):85–111.
80. Cheng H, Lv J, Wang H, Wang B, Li Z, Deng Z. Genetically engineered Pichia pastoris yeast for conversion of glucose to xylitol by a single-fermentation process. Appl Microbiol Biotechnol. 2014;98(8):3539–52.
81. Verney EB. The osmotic pressure of the proteins of human serum and plasma. J Physiol. 1926;61(3):319–28.
82. Raiford DW, Heizer EM, Miller RV, Akashi H, Raymer ML, Krane DE. Do amino acid biosynthetic costs constrain protein evolution in Saccharomyces cerevisiae? J Mol Evol. 2008;67(6):621–30.
83. West MG, Horne DW, Appling DR. Metabolic role of cytoplasmic isozymes of 5,10-methylenetetrahydrofolate dehydrogenase in Saccharomyces cerevisiae. Biochemistry. 1996;35(9):3122–32.
84. Porro D, Sauer M, Branduardi P, Mattanovich D. Recombinant protein production in yeasts. Mol Biotechnol. 2005;31(3):245–59.
85. Morales Y, Tortajada M, Picó J, Vehí J, Llaneras F. Validation of an FBA model for Pichia pastoris in chemostat cultures. BMC Syst Biol. 2014;8(1):142.

6

JuPOETs: a constrained multiobjective optimization approach to estimate biochemical model ensembles in the Julia programming language

David M. Bassen[2], Michael Vilkhovoy[1], Mason Minot[1], Jonathan T. Butcher[2] and Jeffrey D. Varner[1*]

Abstract

Background: Ensemble modeling is a promising approach for obtaining robust predictions and coarse grained population behavior in deterministic mathematical models. Ensemble approaches address model uncertainty by using parameter or model families instead of single best-fit parameters or fixed model structures. Parameter ensembles can be selected based upon simulation error, along with other criteria such as diversity or steady-state performance. Simulations using parameter ensembles can estimate confidence intervals on model variables, and robustly constrain model predictions, despite having many poorly constrained parameters.

Results: In this software note, we present a multiobjective based technique to estimate parameter or models ensembles, the Pareto Optimal Ensemble Technique in the Julia programming language (JuPOETs). JuPOETs integrates simulated annealing with Pareto optimality to estimate ensembles on or near the optimal tradeoff surface between competing training objectives. We demonstrate JuPOETs on a suite of multiobjective problems, including test functions with parameter bounds and system constraints as well as for the identification of a proof-of-concept biochemical model with four conflicting training objectives. JuPOETs identified optimal or near optimal solutions approximately six-fold faster than a corresponding implementation in Octave for the suite of test functions. For the proof-of-concept biochemical model, JuPOETs produced an ensemble of parameters that gave both the mean of the training data for conflicting data sets, while simultaneously estimating parameter sets that performed well on each of the individual objective functions.

Conclusions: JuPOETs is a promising approach for the estimation of parameter and model ensembles using multiobjective optimization. JuPOETs can be adapted to solve many problem types, including mixed binary and continuous variable types, bilevel optimization problems and constrained problems without altering the base algorithm. JuPOETs is open source, available under an MIT license, and can be installed using the Julia package manager from the JuPOETs GitHub repository

Keywords: Ensemble modeling, Multiobjective optimization, Julia

*Correspondence: jdv27@cornell.edu
[1]Department of Chemical and Biomolecular Engineering, Cornell University, 14853 Ithaca, NY, USA
Full list of author information is available at the end of the article

Background

Ensemble modeling is a promising approach for obtaining robust predictions and coarse grained population behavior in deterministic mathematical models. It is often not possible to uniquely identify all the parameters in biochemical models, even when given extensive training data [1]. Thus, despite significant advances in standardizing biochemical model identification [2], the problem of estimating model parameters from experimental data remains challenging. Ensemble approaches address parameter uncertainty in systems biology and other fields like weather prediction [3–6] by using parameter families instead of single best-fit parameter sets. Parameter families can be selected based upon simulation error, along with other criteria such as diversity or steady-state performance. Simulations using parameter ensembles can estimate confidence intervals on model variables, and robustly constrain model predictions, despite having many poorly constrained parameters [7, 8]. There are many techniques to generate parameter ensembles. Battogtokh et al., Brown et al., and later Tasseff et al. generated experimentally constrained parameter ensembles using a Metropolis-type random walk [3, 5, 9, 10]. Liao and coworkers developed methods to generate ensembles that all approach the same steady-state, for example one determined by fluxomics measurements [11]. They have used this approach for model reduction [12], strain engineering [13, 14] and to study the robustness of non-native pathways and network failure [15]. Maranas and coworkers have also applied this method to develop a comprehensive kinetic model of bacterial central carbon metabolism, including mutant data [16]. We and others have used ensemble approaches, generated using both sampling and optimization techniques, that have robustly simulated a wide variety of signal transduction processes [9, 10, 17–19], neutrophil trafficking in sepsis [20], patient specific coagulation behavior [21], uncertainty quantification in metabolic kinetic models [22] and to capture cell to cell variation [23]. Further, ensemble approaches have been used in synthetic biology to sample possible biocircuit configurations [24]. Thus, ensemble approaches are widely used to robustly simulate a variety of biochemical systems.

Identification of biochemical models requires significant training data perhaps taken from diverse sources. These real-world data sets often contain intrinsic conflicts resulting from, for example, the use of different cell lines, different measurement technologies, different reagent vendors or lots, uncontrollable experimental artifacts or general cross laboratory variability. Parameter ensembles that optimally balance these inherent conflicts lead to more robust model performance. Multiobjective optimization is an ensemble generation technique that naturally balances conflicts in noisy training data [25]. Multiobjective optimization has been used to identify signal transduction models [18, 23], for the design of synthetic circuits [24], to design the folding behaviors of novel RNAs [26], to design bioprocesses [27], and to understand bacterial adaptation [28]. Thus, it is a widely used approach for a variety of biochemical applications. Previously, we developed the Pareto Optimal Ensemble Technique (POETs) algorithm to address the challenge of competing or conflicting training objectives. POETs, which integrates simulated annealing (SA) and multiobjective optimization through the notion of Pareto rank, estimates parameter ensembles which optimally trade-off between competing (and potentially conflicting) experimental objectives [29]. However, the previous implementation of POETs, in the Octave programming language [30], suffered from poor performance and was not configurable. For example, Octave-POETs does not accommodate user definable objective functions, bounds and problem constraints, cooling schedules, different variable types e.g., a mixture of binary and continuous design variables or custom diversity generation routines. Octave-POETs was also not well integrated into a package or source code management (SCM) system. Thus, upgrades to the approach containing new features, or bug fixes were not centrally managed.

Implementation

In this software note, we present an open-source implementation of the Pareto optimal ensemble technique in the Julia programming language (JuPOETs). JuPOETs takes advantage of the unique features of Julia to address many of the shortcomings of the previous implementation. Julia is a cross-platform, high-performance programming language for technical computing that has performance comparable to C but with syntax similar to MATLAB/Octave and Python [31]. Julia also offers a sophisticated compiler, distributed parallel execution, numerical accuracy, and an extensive function library. Further, the architecture of JuPOETs takes advantage of the first-class function type in Julia allowing user definable behavior for all key aspects of the algorithm, including objective functions, custom diversity generation logic, linear/non-linear parameter constraints (and parameter bounds constraints) as well as custom cooling schedules. Julia's ability to naturally call other languages such as Python or C also allows JuPOETs to be used with models implemented in a variety of languages across many platforms. Additionally, Julia offers a built-in package manager which is directly integrated with GitHub, a popular web-based Git repository hosting service offering distributed revision control and source code management. Thus, JuPOETs can be adapted to many problem types, including mixed binary and continuous variable types, bilevel problems and constrained problems without altering the base algorithm, as was required in the previous POETs implementation.

JuPOETs optimization problem formulation

JuPOETs solves the $\mathcal{K}-$dimensional constrained multiobjective optimization problem:

$$\min_{\mathbf{p}} \begin{cases} O_1\left(\mathbf{x}(t,\mathbf{p}),\mathbf{p}\right) \\ \vdots \\ O_{\mathcal{K}}\left(\mathbf{x}(t,\mathbf{p}),\mathbf{p}\right) \end{cases} \qquad (1)$$

subject to the model equations and constraints:

$$\begin{aligned} \mathbf{f}\left(t,\mathbf{x}(t,\mathbf{p}),\dot{\mathbf{x}}(t,\mathbf{p}),\mathbf{u}(t),\mathbf{p}\right) &= \mathbf{0} \\ g_1\left(t,\mathbf{x}(t,\mathbf{p}),\mathbf{u}(t),\mathbf{p}\right) &\geq 0 \\ &\vdots \\ g_{\mathcal{C}}\left(t,\mathbf{x}(t,\mathbf{p}),\mathbf{u}(t),\mathbf{p}\right) &\geq 0 \end{aligned}$$

and parameter bound constraints:

$$\mathcal{L} \leq \mathbf{p} \leq \mathcal{U}$$

The quantity O_j denotes the jth objective function ($j = 1, 2, \ldots, \mathcal{K}$), typically the sum of squared errors for the jth data set for biochemical modeling applications. The terms $\mathbf{f}(t,\mathbf{x}(t,\mathbf{p}),\dot{\mathbf{x}}(t,\mathbf{p}),\mathbf{u}(t),\mathbf{p})$ denote the system of model equations (e.g., differential equations, differential algebraic equations or linear/non-linear algebraic equations) where $\mathbf{p}$ denotes the decision variable vector e.g., unknown model parameters ($\mathcal{D} \times 1$). In typical biochemical modeling applications, the model equations $\mathbf{f}(\cdot)$ are a system of continuous real-valued non-linear differential equations that comprise a kinetic model, but other types of models e.g., stoichiometric models are also common. The quantity t denotes time, $\mathbf{x}(t,\mathbf{p})$ denotes the model state (with an initial state $\mathbf{x}_0$), and $\mathbf{u}(t)$ denotes an input vector. The decision variables (e.g., kinetic parameters) can be subject to bounds constraints, where $\mathcal{L}$ and $\mathcal{U}$ denote the lower and upper bounds, respectively as well as $\mathcal{C}$ problem specific constraints $g_i\left(t,\mathbf{x}(t,\mathbf{p}),\mathbf{u}(t),\mathbf{p}\right), i = 1,\ldots,\mathcal{C}$. The decision variables $\mathbf{p}$ are typically real-valued kinetic constants, or metabolic fluxes in the case of stoichiometric models. However, other variables types e.g., binary or categorical decision variables can also be accommodated.

JuPOETs integrates simulated annealing (SA) [32] with Pareto ranking to estimate decision variables on or near the optimal tradeoff surface between competing objectives (Fig. 1 and Algorithm 1). A tradeoff surface defines the best possible performance for every conflicting objective, such that an increase in the performance of one objective does not decrease the performance of at least one other objective. Pareto rank is a scalar measure of distance away from the optimal tradeoff surface (low rank is near the surface, while higher ranks are progressively further away). Thus, the central idea underlying POETs is a mapping between the value of the objective vector evaluated at $\mathbf{p}_{i+1}$ (decision variable guess at iteration $i+1$) and the scalar Pareto rank (Fig. 1). Traditional simulated annealing uses a scalar performance value e.g., simulation error to make a probabilistic decision to keep or reject a set of decision variables; decision variables with better performance are always accepted, while those with worse performance are sometimes accepted depending upon a parameter called the temperature. On the other hand, JuPOETs makes this same decision using the Pareto rank instead of a single performance objective. The problem of estimating biochemical model parameters from experimental data is typically posed as an error minimization problem over continuous real-valued decision variables (model parameters) subject to the model equations. A parameter set $\mathbf{p}_{i+1}$ lies along the optimal tradeoff surface if no other parameter guess leads to decreased error for

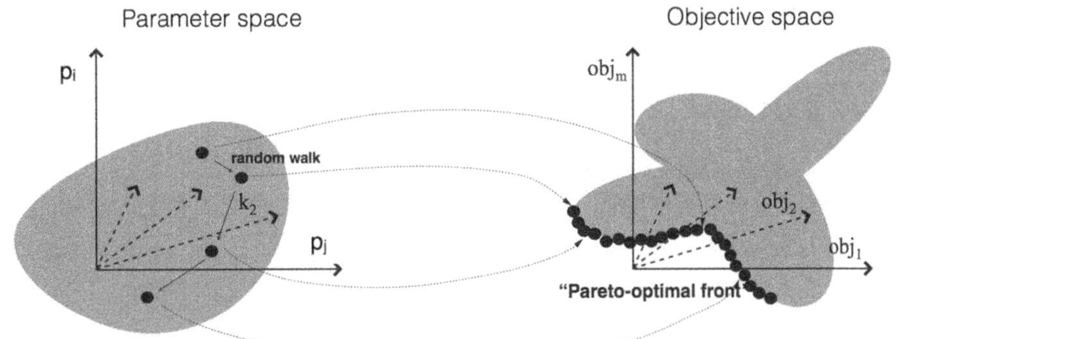

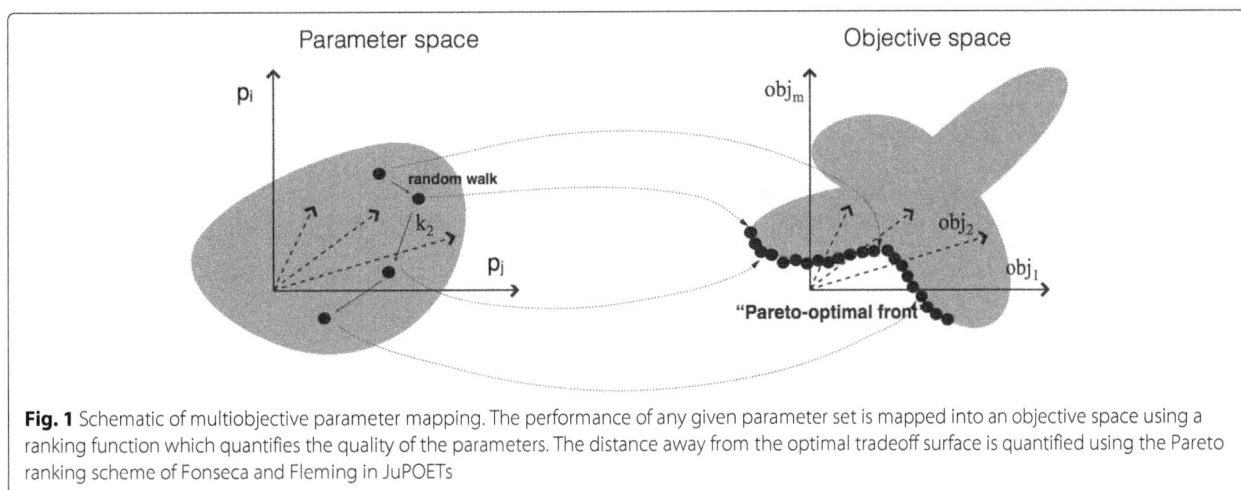

Fig. 1 Schematic of multiobjective parameter mapping. The performance of any given parameter set is mapped into an objective space using a ranking function which quantifies the quality of the parameters. The distance away from the optimal tradeoff surface is quantified using the Pareto ranking scheme of Fonseca and Fleming in JuPOETs

Algorithm 1: Pseudo-code for the JuPOETs run-loop. The user must specify the `objective` function and an initial parameter guess. The user can optionally specify the `neighbor`, `acceptance`, `cooling` and `refinement` functions (or use the default implementations). The rank archive $\mathcal{R}$, solution archive $\mathcal{S}$ and objective archive $\mathcal{O}$ are initialized from the initial guess. The initial guess (potentially following a single objective local refinement step) is perturbed in the `neighbor` function, which generates a new solution whose performance is evaluated using the user supplied `objective` function. The new solution and objective values are then added to the respective archives and ranked using the builtin `rank` function. If the new solution is accepted (based upon a probability calculated with the user supplied `acceptance` function) it is added to the solution and objective archive. This solution is then perturbed during the next iteration of the algorithm. However, if the solution is not accepted, it is removed from the archive and discarded. The temperature is adjusted using the user supplied `cooling` function after each $\mathcal{I}$ iterations. When JuPOETs terminates, the parameter solution archive $\mathcal{S}$, objective archive $\mathcal{O}$ and rank archive $\mathcal{R}$ are retuned to the caller.

input : User specified objective function, and initial guess ($\mathcal{D} \times 1$). User can also specify custom neighbor, acceptance. cooling and refinement functions or use the default functions provided.

Output: Rank archive $\mathcal{R}$ ($\mathcal{A} \times 1$), parameter solution archive $\mathcal{S}$ ($\mathcal{D} \times \mathcal{A}$) and objective archive $\mathcal{O}$ ($\mathcal{K} \times \mathcal{A}$), where $\mathcal{A}$ denotes the number of accepted solutions

initialize: $\mathcal{R}$, $\mathcal{S}$ and $\mathcal{O}$ using initial guess $\mathbf{p}_o$;
initialize: T $\leftarrow$1.0;
initialize: T_{min} $\leftarrow$1/10000;
initialize: Maximum number of steps per temperature $\mathcal{I}$;

// Call to local refinement function (single objective problem)
$\mathbf{p}_o$ $\leftarrow$ user-function:`refinement`($\mathbf{p}_o$);

while $T > T_{min}$ **do**
  i $\leftarrow$ 1;
  while $i < \mathcal{I}$ **do**
    // Generate a new parameter solution using user neighbor function
    $\mathbf{p}_{i+1}$ $\leftarrow$ user-function::`neighbor`($\mathbf{p}^*$);
    // Evaluate $\mathbf{p}_{i+1}$ using user objective function
    $\mathbf{o}_{i+1}$ $\leftarrow$ user-function::`objective`($\mathbf{p}_{i+1}$);
    Add $\mathbf{p}_{i+1}$ to solution archive $\mathcal{S}$;
    Add $\mathbf{o}_{i+1}$ to objective archive $\mathcal{O}$;
    // Calculate Pareto rank of solutions in $\mathcal{O}$ using builtin rank function
    $\mathcal{R}$ $\leftarrow$ builtin-function::`rank`($\mathcal{O}$);
    // Accept $\mathbf{p}_{i+1}$ into the archive with user defined probability
    $\mathcal{P}$ $\leftarrow$ user-function::`acceptance`($\mathcal{R}, T$);
    if $\mathcal{P}$ >*rand* **then**
      // Update the best solution with $\mathbf{p}_{i+1}$
      $\mathbf{p}^* \leftarrow \mathbf{p}_{i+1}$;
      prune $\mathcal{S}$, $\mathcal{R}$ and $\mathcal{O}$ of all solutions above a rank threshold;
    else
      Remove $\mathbf{p}_{i+1}$ from solution archive $\mathcal{S}$;
      Remove $\mathbf{o}_{i+1}$ from error archive $\mathcal{O}$;
    end
    i $\leftarrow$ i + 1;
  end
  // Update T using the user cooling function
  T $\leftarrow$ user-function::cooling(T);
end

every objective. JuPOETs calculates the performance of a candidate parameter set $\mathbf{p}_{i+1}$ by calling the user defined `objective` function; `objective` takes a parameter set as an input, evaluates the model equations, and using this solution, returns the $\mathcal{K} \times 1$ objective vector. Candidate parameter sets are generated by the user supplied `neighbor` function; the default implementation of `neighbor` is a random perturbation, however other perturbation logic can be implemented by the user. The error vector associated with $\mathbf{p}_{i+1}$ is ranked using the builtin Pareto `rank` function, by comparing the error at iteration $i + 1$ to the error archive $\mathcal{O}_i$ (all error vectors up to iteration i meeting a ranking criterion). Parameter sets on or near the optimal trade-off surface between the objectives have a rank equal to 0 (no other current parameter sets are better). These rank zero parameter sets define the Pareto optimal group for the ensemble, wherein Pareto optimality is defined as a parameter set not being dominated by any other sets within the ensemble. Sets with increasing non-zero rank are progressively further away from the optimal trade-off surface. Thus, a parameter set with a rank $= 0$ is *better* in a trade-off sense than rank > 0. We implemented the Fonseca and Fleming ranking scheme in the builtin `rank` function [33]:

$$\texttt{rank}\left(\mathcal{O}_{i+1}\left(\mathbf{p}_{i+1}\right) \mid \mathcal{O}_i\right) = r \tag{2}$$

where rank r is the number of parameter sets that dominate (are better than) parameter set $\mathbf{p}_{i+1}$, and $\mathcal{O}_{i+1}\left(\mathbf{p}_{i+1}\right)$ denotes the objective vector evaluated at $\mathbf{p}_{i+1}$. We used the Pareto rank to inform the SA calculation. The parameter set $\mathbf{p}_{i+1}$ was accepted or rejected by the SA at each iteration, by calculating an `acceptance` probability $\mathcal{P}\left(\mathbf{p}_{i+1}\right)$:

$$\mathcal{P}(\mathbf{p}_{i+1}) \equiv \exp\left\{-\texttt{rank}\left(\mathcal{O}_{i+1}\left(\mathbf{p}_{i+1}\right) \mid \mathcal{O}_i\right)/T\right\} \tag{3}$$

where T is the simulated annealing temperature; the temperature provides control over how strictly decreasing Pareto rank is enforced. As $\texttt{rank}\left(\mathcal{O}_{i+1}\left(\mathbf{p}_{i+1}\right) \mid \mathcal{O}_i\right) \rightarrow 0$, the acceptance probability moves toward one, ensuring that we explore parameter sets along the Pareto surface. Occasionally, (depending upon T) a parameter set with a high Pareto rank is accepted by the SA allowing a more diverse search of the parameter space. However, as T is reduced as a function of iteration count (using the `cooling` function), the probability of accepting a high-rank set decreases. Parameter sets could also be accepted by the SA but *not* permanently archived in $\mathcal{S}_i$, where $\mathcal{S}_i$ is the solution archive. Only parameter sets with rank less than or equal to a threshold (rank ≤ 4 by default) are included in $\mathcal{S}_i$, where the archive is re-ranked and filtered after accepting every new parameter set. Parameter bounds were implemented in the `neighbor` function as box constraints, while problem specific constraints were implemented in `objective` using a penalty method:

$$O_i + \lambda \sum_{j=1}^{\mathcal{C}} \min\left\{0, g_j\left(t, \mathbf{x}(t, \mathbf{p}), \mathbf{u}(t), \mathbf{p}\right)\right\} \qquad i = 1, \ldots, \mathcal{K} \tag{4}$$

where λ denotes the penalty parameter ($\lambda = 100$ by default). However, because both the neighbor and objective functions are user defined, different constraint implementations are easily defined.

To use JuPOETs, the user specifies the `neighbor`, `acceptance`, `cooling` and `objective` functions along with an initial decision variable guess. Default implementations of the `neighbor`, `acceptance` and `cooling` functions can be used directly, or they can be overridden by user defined logic. However, the user must provide an implementation of the `objective` function and provide an initial decision variable guess. Lastly, if the user is operating JuPOETs in hybrid mode, then a `refinement` function pointer must also be specified. Hybrid mode temporarily switches the search from a multiobjective to a single objective problem, where the sum of the objective functions can be used to update the best (or initial) parameter guess. The specific hybrid mode search logic is up to the user; by default hybrid mode is off, and the default `refinement` implementation is simply a pass through function. However, we have shown previously that POETs operated in hybrid mode (where the single objective problem used a pattern search approach) had better performance that POETs alone [29]. Thus, hybrid mode is generally recommended for most applications. In addition, there are several user configurable parameters that can be adjusted to control the performance of JuPOETs: `maximum_number_of_iterations` controls the number of iterations per temperature (default 20); `rank_cutoff` controls the upper rank bound on the solution archive (default 5); `temperature_min` controls the minimum temperature after which JuPOETs returns the error and solution archives (default 0.001); `show_trace` controls the level of output shown to the user (default true). After the completion of the run, JuPOETs returns the parameter solution archive $\mathcal{S}$, objective archive $\mathcal{O}$ and rank archive $\mathcal{R}$. The parameter solution archive $\mathcal{S}$ contains is an $\mathcal{D} \times \mathcal{A}$ array, where $\mathcal{A}$ denotes the number of solutions in the archive when JuPOETs terminated. On the other hand, the objective archive $\mathcal{O}$ is an $\mathcal{K} \times \mathcal{A}$ array containing the performance values for each objective corresponding the columns of $\mathcal{S}$. Lastly, JuPOETs returns the rank archive $\mathcal{R}$ which is an $\mathcal{A} \times 1$ array of Pareto ranks corresponding to the columns of $\mathcal{S}$. One technical note, if JuPOETs is run from multiple starting locations, and the archives from each of these runs is combined into a single collective archive, the combined parameter rank archive may become invalid. In these cases, it is required to re-rank the parameter sets using the built-in `rank` function to produce a collective parameter ranking.

Results and discussion

JuPOETs identified optimal or nearly optimal solutions significantly faster than Octave-POETs for a suite of multiobjective algebraic test problems (Table 1). The algebraic test problems were constrained non-linear functions with bound constraints and additional non-linear constraints on the decision variables in one case. The problems had up to three-dimensional continuous real-valued decision vectors, and each case had two objective functions. The wall-clock time for JuPOETs and Octave-POETs was measured for 10 independent trials for each of the test problems. The same `cooling`, `neighbor`, `acceptance`, and `objective` logic was employed between the implementations, and all other parameters were held constant. For each test function, the search domain was partitioned into 10 segments, where an initial parameter guess was drawn from each partition. The number of search steps for each temperate was $\mathcal{I}$ = 10 for all cases, and the cooling parameter was α = 0.9. On average, JuPOETs identified optimal or near optimal solutions for the suite of test problems six-fold faster (60s versus 400s) than Octave-POETs (Fig. 2). JuPOETs produced the characteristic tradeoff curves for each test problem, given both decision variable bound and problem constraints (Fig. 3). Thus, JuPOETs estimated an ensemble of solutions to constrained multiobjective algebraic test problems significantly faster than the current Octave implementation. Next, we tested JuPOETs on a proof-of-concept biochemical model identification problem.

JuPOETs estimated an ensemble of biochemical model parameters that were consistent with the mean of synthetic training data (Fig. 4). Four synthetic training data sets were generated from a prototypical biochemical network consisting of 6 metabolites and 7 reactions (Fig. 4, inset right). We considered a common case in which the same extracellular measurements of A_e, B_e, C_e and cellmass were made on four hypothetical cell types, each having the same biological connectivity but different performance. Network dynamics were modeled using the hybrid cybernetic model with elementary modes (HCM) approach of Ramkrishna and coworkers [34]. In the HCM approach, metabolic networks are first decomposed into a set of elementary modes (EMs) (chemically balanced steady-state pathways, see [35]). Dynamic combinations of elementary modes are then used to characterize network behavior. Each elementary mode is catalyzed by a pseudo enzyme; thus, each mode has both kinetic and enzyme synthesis parameters. The proof of concept network generated 6 EMs, resulting in 13 model parameters (continuos real-valued decision variables). The synthetic training data was generated by randomly varying these parameters.

The general form of the biochemical test problem was given by:

$$\min_{\mathbf{p}} (O_1, \ldots, O_{\mathcal{K}}) \tag{5}$$

subject to model and bounds constraints. We considered four training data sets ($\mathcal{K}$ = 4), each of which contained time-series measurements of A_e, B_e, C_e and cellmass. Each objective O_j, $j = 1, \ldots, \mathcal{K}$ quantified the squared difference between the simulated (x_i) and measured extracellular species abundance (y_i) in the jth data set:

$$O_j = \sum_i \sum_\tau (x_i(\tau) - y_i(\tau))^2 \qquad j = 1, \ldots, \mathcal{K} \tag{6}$$

where, i denotes the species index and τ denotes the time index. The abundance of extracellular species i (x_i), the pseudo enzyme e_l (catalyzes flux through mode l), and cellmass were governed by the model equations:

Table 1 Multi-objective optimization test problems. We tested the JuPOETs implementation on three two-dimensional test problems, with one-, two- and three-dimensional parameter vectors. Each problem had parameter bounds constraints, however, on the Binh and Korn function had additional non-linear problem constraints. For the Fonesca and Fleming problem, N = 3

Name	Dimension	Function	Domain	Constraints
Schaffer function	1	$O_1(x) = x^2$	$-10 \leq x \leq 10$	
		$O_2(x) = (x-2)^2$		
Binh and Korn function	2	$O_1(x,y) = 4x^2 + 4y^2$	$0 \leq x \leq 5$	$g_1(x,y) = (x-5)^2 + y^2 \leq 25$
		$O_2(x,y) = (x-5)^2 + (y-5)^2$	$0 \leq x \leq 3$	$g_2(x,y) = (x-8)^2 + (y+3)^2 \leq 7.7$
Fonseca and Fleming function	3	$O_1(x_i) = 1 - \exp\left(-\sum_{i=1}^{N}\left(x_i - \frac{1}{\sqrt{N}}\right)^2\right)$	$-4 \leq x_i \leq 4$	
		$O_2(x_i) = 1 - \exp\left(-\sum_{i=1}^{N}\left(x_i + \frac{1}{\sqrt{N}}\right)^2\right)$		

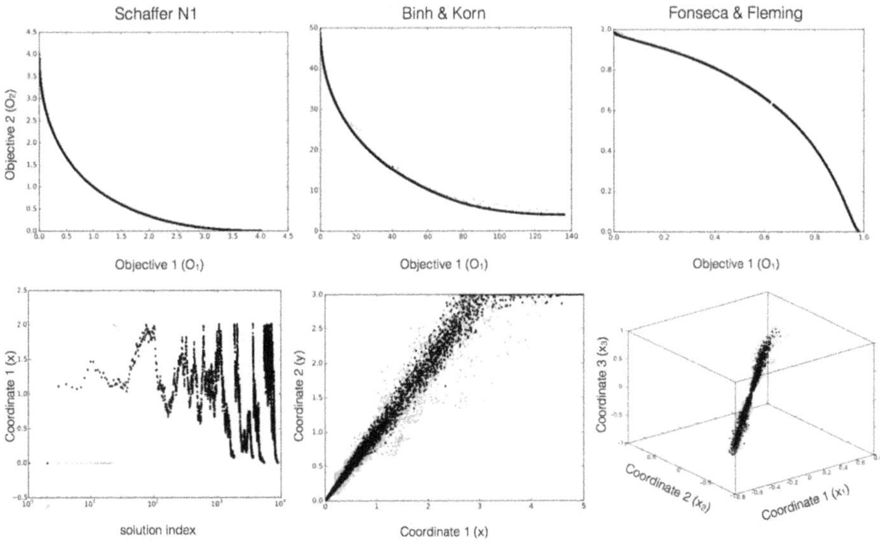

Fig. 2 The performance of JuPOETs on the multi-objective test suite. The execution time (wall-clock) for JuPOETs and POETs implemented in Octave was measured for 10 independent trials for the suite of test problems. The number of steps per temperature $\mathcal{I} = 10$, and the cooling parameter $\alpha = 0.9$ for all cases. The problem domain was partitioned into 10 equal segments, an initial guess was drawn from each segment. For each of the test functions, JuPOETs estimated solutions on (rank zero solutions, *black*) or near (*gray*) the optimal tradeoff surface, subject to bounds and problem constraints

$$
\begin{aligned}
\frac{dx_i}{dt} &= \sum_{j=1}^{\mathcal{R}} \sum_{l=1}^{\mathcal{L}} \sigma_{ij} z_{jl} q_l\left(\mathbf{e}, \mathbf{p}, \mathbf{x}\right) c \qquad i = 1, \ldots, \mathcal{M} \\
\frac{de_l}{dt} &= \alpha_l + r_{El}\left(\mathbf{p}, \mathbf{x}\right) u_l - \left(\beta_l + r_G\right) e_l \qquad l = 1, \ldots, \mathcal{L} \\
\frac{dc}{dt} &= r_G c
\end{aligned}
$$

where $\mathcal{R}$ and $\mathcal{M}$ denote the number of reactions and extracellular species in the model and $\mathcal{L}$ denotes the number of elementary modes. The quantity σ_{ij} denotes the stoichiometric coefficient for species i in reaction j and z_{jl} denotes the normalized flux for reaction j in mode l. If $\sigma_{ij} > 0$, species i is produced by reaction j; if $\sigma_{ij} < 0$,

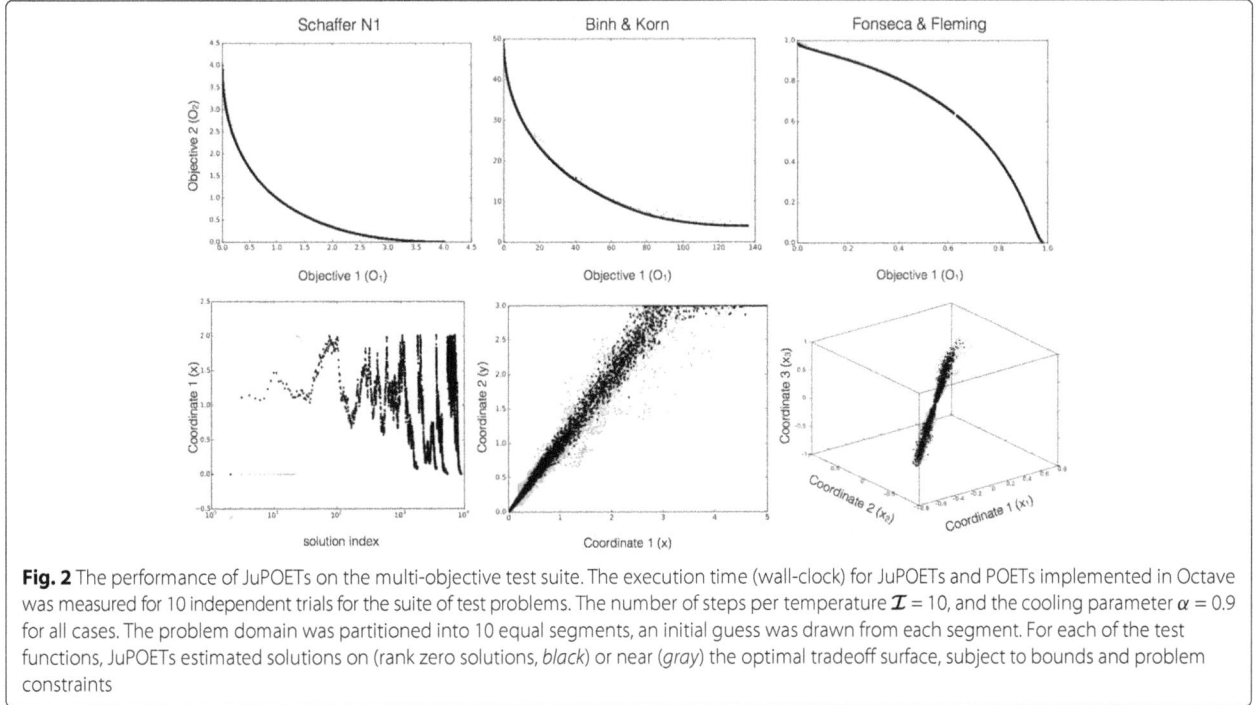

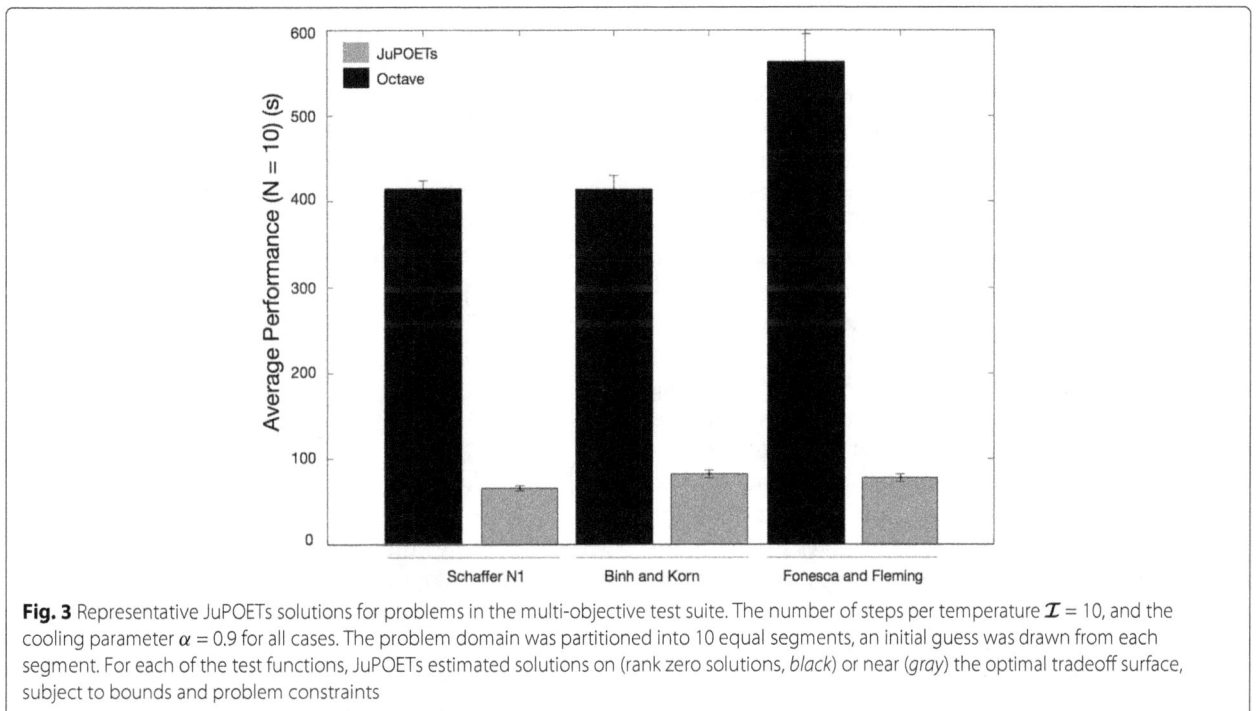

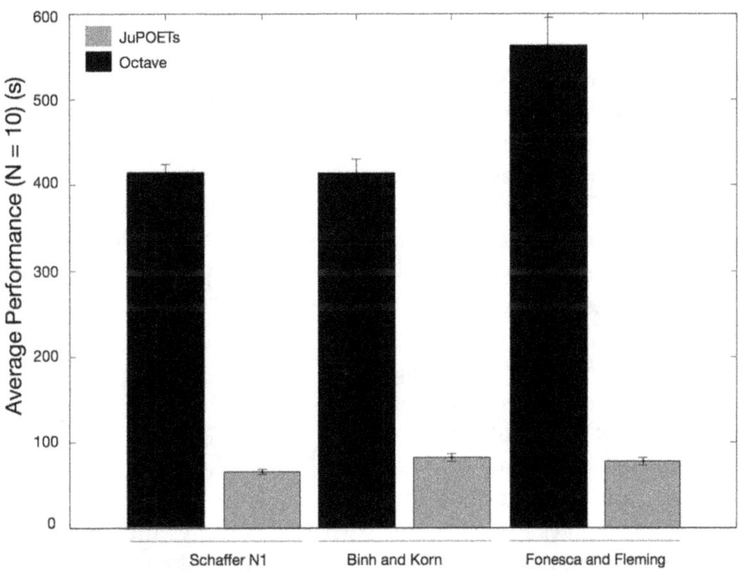

Fig. 3 Representative JuPOETs solutions for problems in the multi-objective test suite. The number of steps per temperature $\mathcal{I} = 10$, and the cooling parameter $\alpha = 0.9$ for all cases. The problem domain was partitioned into 10 equal segments, an initial guess was drawn from each segment. For each of the test functions, JuPOETs estimated solutions on (rank zero solutions, *black*) or near (*gray*) the optimal tradeoff surface, subject to bounds and problem constraints

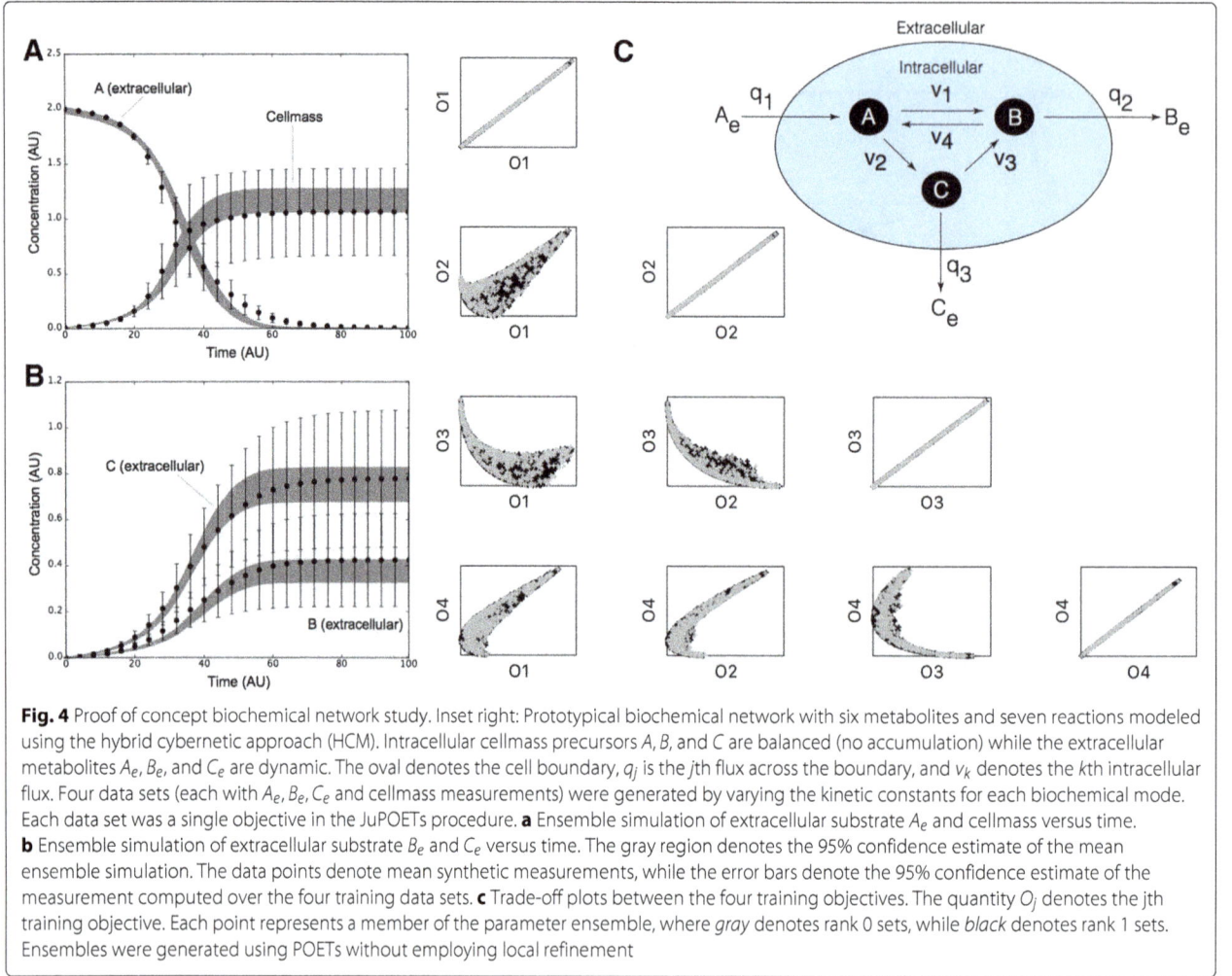

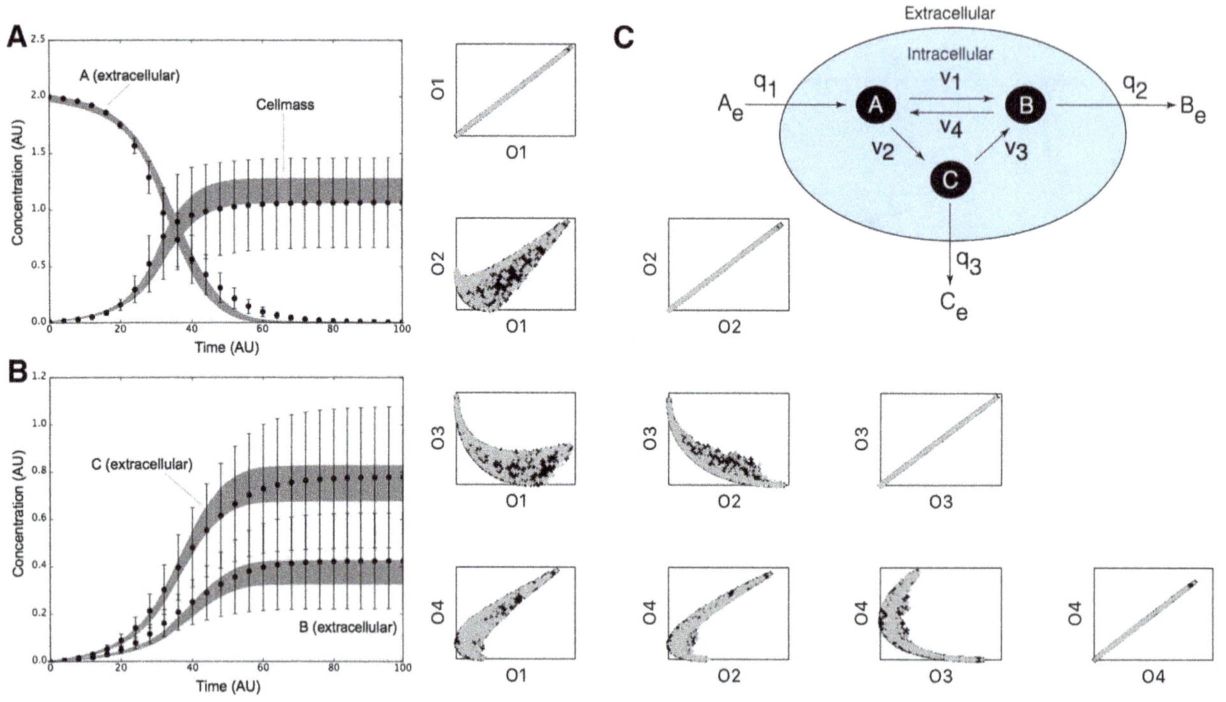

Fig. 4 Proof of concept biochemical network study. Inset right: Prototypical biochemical network with six metabolites and seven reactions modeled using the hybrid cybernetic approach (HCM). Intracellular cellmass precursors *A*, *B*, and *C* are balanced (no accumulation) while the extracellular metabolites A_e, B_e, and C_e are dynamic. The oval denotes the cell boundary, q_j is the *j*th flux across the boundary, and v_k denotes the *k*th intracellular flux. Four data sets (each with A_e, B_e, C_e and cellmass measurements) were generated by varying the kinetic constants for each biochemical mode. Each data set was a single objective in the JuPOETs procedure. **a** Ensemble simulation of extracellular substrate A_e and cellmass versus time. **b** Ensemble simulation of extracellular substrate B_e and C_e versus time. The gray region denotes the 95% confidence estimate of the mean ensemble simulation. The data points denote mean synthetic measurements, while the error bars denote the 95% confidence estimate of the measurement computed over the four training data sets. **c** Trade-off plots between the four training objectives. The quantity O_j denotes the jth training objective. Each point represents a member of the parameter ensemble, where *gray* denotes rank 0 sets, while *black* denotes rank 1 sets. Ensembles were generated using POETs without employing local refinement

species i is consumed by reaction j; if $\sigma_{ij} = 0$, species i is not connected with reaction j. Extracellular species, cellmass and pseudo-enzyme were subject to the initial conditions $\mathbf{x}(t_o) = \mathbf{x}_o$, $c(t_o) = c_o$ and $e_l = 0.5$, respectively. The term $q_l(\mathbf{e}, \mathbf{p}, \mathbf{x})$ denotes the specific uptake/secretion rate for mode l where $\mathbf{e}$ denotes the pseudo enzyme vector, $\mathbf{p}$ denotes the unknown kinetic parameter vector (decision variables), $\mathbf{x}$ denotes the extracellular species vector, and c denotes the cell mass; $q_l(\mathbf{e}, \mathbf{p}, \mathbf{x})$ is the product of a kinetic term $(\bar{q}_l)$ and a control variable governing enzyme activity. Flux through each mode was catalyzed by a pseudo enzyme e_l, synthesized at the regulated specific rate $r_{E,l}(\mathbf{p}, \mathbf{x})$, and constitutively at the rate α_l. The term u_l denotes the cybernetic variable controlling the synthesis of enzyme l. The term β_l denotes the rate constant governing non-specific enzyme degradation, and r_G denotes the specific growth rate through all modes. The specific uptake/secretion rates and the specific rate of enzyme synthesis were modeled using saturation kinetics. The specific growth rate was given by:

$$r_G = \sum_{l=1}^{\mathcal{L}} z_{\mu l} q_l(\mathbf{e}, \mathbf{p}, \mathbf{x})$$

where $z_{\mu l}$ denotes the growth flux μ through mode l. The control variables u_l and v_l, which control the synthesis and activity of each enzyme respectively, were given by:

$$u_l = \frac{z_{sl}\bar{q}_l}{\sum_{l=1}^{\mathcal{L}} z_{sl}\bar{q}_l} \tag{7}$$

and

$$v_l = \frac{z_{sl}\bar{q}_l}{\max_{l=1,\ldots,\mathcal{L}} z_{sl}\bar{q}_l} \tag{8}$$

where z_{sl} denotes the uptake flux of substrate s through mode l. Each unknown kinetic parameter was continuous and real-valued, and subject to bounds constraints: $\mathcal{L} \leq \mathbf{p} \leq \mathcal{U}$.

JuPOETs produced an ensemble of approximately $\dim \mathcal{S} \simeq 13{,}000$ parameter sets that captured the mean of the measured data sets for extracellular metabolites and cellmass (Fig. 4a and b). JuPOETs minimized the difference between the simulated and measured values for extracellular metabolites A_e, B_e, C_e and cellmass, where the residual for each data set was treated as a single objective (leading to four objectives). The 95% confidence estimate produced by the ensemble was consistent with the mean of the measured data, despite having significant uncertainty in the training data. JuPOETs produced a consensus estimate of the synthetic data by calculating optimal trade-offs between the training data sets (Fig. 4c). Multiple trade-off fronts were visible in the objective plots, for example between data set 3 (O_3) and data set 2 (O_2). Thus, without a multiobjective approach, it would be challenging to capture these data sets as fitting one leads to decreased performance on the other. However, the ensemble contained parameter sets that described each data set independently (Fig. 5). Thus, JuPOETs produced an ensemble of parameters that gave the mean of the training data for conflicting data sets, while simultaneously estimating parameter sets that performed well on each individual objective function.

Currently, JuPOETs does not consider parameter identifiability when constructing parameter ensembles. Although JuPOETs produces parameter estimates that give model performance similar to the training data, we do not have strict statistical confidence that the *true* parameter values are contained within the ensemble. However, despite this, ensembles produced by POETs can be predictive [18, 23]. Thus, JuPOETs produces a collection of parameters that are constrained by the performance of the model, and not by specific hypotheses regarding the individual values of the raw model parameters. Of course, knowledge of specific parameter values, or the relationship between parameter combinations, can be used to inform the search through either bounds or problem specific constraints (for example, as demonstrated in the first example problem).

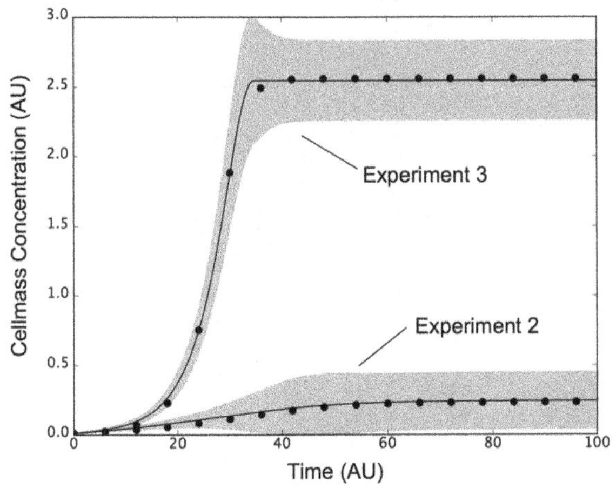

Fig. 5 Experiment to experiment variation captured by the ensemble. Cellmass measurements (*points*) versus time for experiment 2 and 3 were compared with ensemble simulations. The full ensemble was sorted by simultaneously selecting the top 25% of solutions for each objective with rank ≤ 1. The best fit solution for each objective (*line*) $\pm$ 1-standard deviation (*gray region*) for experiment 2 and 3 brackets the training data despite significant differences the training values between the two data sets

Conclusions

In this software note, we presented JuPOETs, a multiobjective technique to estimate parameter ensembles in the Julia programming language. JuPOETs is open source, and available for download under an MIT license from the JuPOETs GitHub repository at https://github.com/varnerlab/POETs.jl. We demonstrated JuPOETs on a suite of algebraic test problems, and a proof-of-concept ODE based biochemical model. While JuPOETs outperformed (and was significantly more flexible) than the previous Octave implementation, there are several areas that could be explored further. First, JuPOETs should be compared with other multiobjective evolutionary algorithms (MOEAs) to determine its relative performance on test and real world problems. Many evolutionary approaches e.g., the non-dominated sorting genetic algorithm (NSGA) family of algorithms, have been adapted to solve multiobjective problems [36, 37]. However, since there is a lack of open source Julia implementations of these alternative approaches, we did not benchmark the relative performance of JuPOETs in this note. One advantage that JuPOETs may have when compared to a strictly evolutionary approaches, is the inclusion of a local refinement step (hybrid mode), which temporarily reduces the problem to a single objective formulation. Previously, POETs run in hybrid mode led to better convergence on a proof-of-concept signal transduction model compared to the same approach without the hybrid refinement step [29]. Other hybrid multiobjective methods have also been shown to be more efficient than evolutionary approaches alone, for a variety of biochemical optimization problems [24, 38]. Thus, there are several different algorithms that we can use to benchmark, and improve the performance of JuPOETs, after we implement them in Julia. Another strategy to improve the performance of JuPOETs is to reduce the number (or cost) of function evaluations that are required to obtain optimal or near optimal solutions. For example, in many real world parameter estimation problems, the bulk of the execution time is spent evaluating the objective functions. One strategy to improve JuPOETs performance could be to optimize surrogates [39], while another would be parallel execution of the objective functions. Currently, JuPOETs serially evaluates the objective function vector. However, parallel evaluation of the objective functions e.g., using the `parallel` Julia macro or other techniques, could be implemented without significantly changing the JuPOETs run loop. Taken together, JuPOETs demonstrated improved flexibility, and performance over POETs in parameter identification and ensemble generation for multiple objectives. JuPOETs has the potential for widespread use due to the flexibility of the implementation, and the high level syntax and distribution tools native to the Julia programming language.

Acknowledgements
We gratefully acknowledge Ani Chakrabarti, Russell Gould and Kathy Rogers for their input and suggestions regarding new features to include into JuPOETs. We also gratefully acknowledge the suggestions from the anonymous reviewers to improve this manuscript and JuPOETs.

Funding
This study was supported by an award from the National Science Foundation (NSF CBET-0955172) and the National Institutes of Health (NIH HL110328) to J.B, and by a National Science Foundation Graduate Research Fellowship (DGE-1144153) to D.B. Lastly, J.V was supported by an award from the US Army and Systems Biology of Trauma Induced Coagulopathy (W911NF-10-1-0376).

Authors' contributions
JV developed the software presented in this study. MM and MV developed the proof-of-concept biochemical model. The manuscript was prepared and edited for publication by DB, JB and JV. All authors read and approved the final manuscript.

Competing interests
The authors declare that they have no competing interests.

Author details
[1]Department of Chemical and Biomolecular Engineering, Cornell University, 14853 Ithaca, NY, USA. [2]Department of Biomedical Engineering, Cornell University, 14853 Ithaca, NY, USA.

References

1. Gadkar KG, Varner J, Doyle FJ. Model identification of signal transduction networks from data using a state regulator problem. Syst Biol (Stevenage). 2005;2(1):17–30.
2. Gennemark P, Wedelin D. Benchmarks for identification of ordinary differential equations from time series data. Bioinformatics. 2009;25(6): 780–6. doi:10.1093/bioinformatics/btp050.
3. Battogtokh D, Asch DK, Case ME, Arnold J, Shüttler HB. An ensemble method for identifying regulatory circuits with special reference to the qa gene cluster of Neurospora crassa. Proc Natl Acad Sci USA. 2002;99(26): 16904–9.
4. Kuepfer L, Peter M, Sauer U, Stelling J. Ensemble modeling for analysis of cell signaling dynamics. Nat Biotechnol. 2007;25(9):1001–6. doi:10.1038/nbt1330.
5. Brown KS, Sethna JP. Statistical mechanial approaches to models with many poorly known parameters. Phys Rev E. 2003;68:021904–19.
6. Palmer TN, Shutts GJ, Hagedorn R, Doblas-Reyes FJ, Jung T, Leutbecher M. Representing model uncertainty in weather and climate prediction. Ann Rev Earth Planetary Sci. 2005;33:163–93.
7. Gutenkunst RN, Waterfall JJ, Casey FP, Brown KS, Myers CR, Sethna JP. Universally sloppy parameter sensitivities in systems biology models. PLoS Comput Biol. 2007;3(10):1871–8. doi:10.1371/journal.pcbi.0030189.
8. Song S, Varner J. Modeling and analysis of the molecular basis of pain in sensory neurons. PLoS ONE. 2009;4:6758–72.
9. Tasseff R, Nayak S, Salim S, Kaushik P, Rizvi N, Varner JD. Analysis of the molecular networks in androgen dependent and independent prostate cancer revealed fragile and robust subsystems. PLoS ONE. 2010;5(1):8864. doi:10.1371/journal.pone.0008864.
10. Tasseff R, Nayak S, Song SO, Yen A, Varner JD. Modeling and analysis of retinoic acid induced differentiation of uncommitted precursor cells. Integr Biol (Camb). 2011;3(5):578–91. doi:10.1039/c0ib00141d.
11. Tran LM, Rizk ML, Liao JC. Ensemble modeling of metabolic networks. Biophys J. 2008;95(12):5606–17. doi:10.1529/biophysj.108.135442.
12. Tan Y, Rivera JGL, Contador CA, Asenjo JA, Liao JC. Reducing the allowable kinetic space by constructing ensemble of dynamic models with the same steady-state flux. Metab Eng. 2011;13(1):60–75. doi:10.1016/j.ymben.2010.11.001.
13. Contador CA, Rizk ML, Asenjo JA, Liao JC. Ensemble modeling for strain development of l-lysine-producing escherichia coli. Metab Eng. 2009;11(4–5):221–33. doi:10.1016/j.ymben.2009.04.002.
14. Tan Y, Liao JC. Metabolic ensemble modeling for strain engineers. Biotechnol J. 2012;7(3):343–53. doi:10.1002/biot.201100186.
15. Lee Y, Lafontaine Rivera JG, Liao JC. Ensemble modeling for robustness analysis in engineering non-native metabolic pathways. Metab Eng. 2014;25:63–71. doi:10.1016/j.ymben.2014.06.006.
16. Khodayari A, Zomorrodi AR, Liao JC, Maranas CD. A kinetic model of escherichia coli core metabolism satisfying multiple sets of mutant flux data. Metab Eng. 2014;25:50–62. doi:10.1016/j.ymben.2014.05.014.
17. Luan D, Zai M, Varner JD. Computationally derived points of fragility of a human cascade are consistent with current therapeutic strategies. PLoS Comput Biol. 2007;3(7):142. doi:10.1371/journal.pcbi.0030142.
18. Song SO, Varner J. Modeling and analysis of the molecular basis of pain in sensory neurons. PLoS ONE. 2009;4(9):6758. doi:10.1371/journal.pone.0006758.
19. Nayak S, Siddiqui JK, Varner JD. Modelling and analysis of an ensemble of eukaryotic translation initiation models. IET Syst Biol. 2011;5(1):2. doi:10.1049/iet-syb.2009.0065.
20. Song SO, Song SOK, Hogg J, Peng ZY, Parker R, Kellum JA, Clermont G. Ensemble models of neutrophil trafficking in severe sepsis. PLoS Comput Biol. 2012;8(3):1002422. doi:10.1371/journal.pcbi.1002422.
21. Luan D, Szlam F, Tanaka KA, Barie PS, Varner JD. Ensembles of uncertain mathematical models can identify network response to therapeutic interventions. Mol Biosyst. 2010;6(11):2272–86. doi:10.1039/b920693k.
22. Andreozzi S, Miskovic L, Hatzimanikatis V. iSCHRUNK–in silico approach to characterization and reduction of uncertainty in the kinetic models of genome-scale metabolic networks. Metab Eng. 2016;33:158–68. doi:10.1016/j.ymben.2015.10.002.
23. Lequieu J, Chakrabarti A, Nayak S, Varner JD. Computational modeling and analysis of insulin induced eukaryotic translation initiation. PLoS Comput Biol. 2011;7(11):1002263. doi:10.1371/journal.pcbi.1002263.
24. Otero-Muras I, Banga JR. Multicriteria global optimization for biocircuit design. BMC Syst Biol. 2014;8:113. doi:10.1186/s12918-014-0113-3.
25. Handl J, Kell DB, Knowles J. Multiobjective optimization in bioinformatics and computational biology. IEEE/ACM Trans Comput Biol Bioinform. 2007;4(2):279–92. doi:10.1109/TCBB.2007.070203.
26. Taneda A. Multi-objective optimization for RNA design with multiple target secondary structures. BMC Bioinformatics. 2015;16(1):280. doi:10.1186/s12859-015-0706-x.
27. Sendin J, Otero-Muras I, Alonso AA, Banga J. Improved Optimization Methods for the Multiobjective Design of Bioprocesses. Ind Eng Chem Res. 2006;45:8594–603.
28. Angione C, Lió P. Predictive analytics of environmental adaptability in multi-omic network models. Sci Rep. 2015;5:15147. doi:10.1038/srep15147.
29. Song SO, Chakrabarti A, Varner JD. Ensembles of signal transduction models using pareto optimal ensemble techniques (poets). Biotechnol J. 2010;5(7):768–80. doi:10.1002/biot.201000059.
30. Eaton JW, Bateman D, Hauberg S. GNU octave version 3.0.1 manual: a high-level interactive language for numerical computations. North Charleston: CreateSpace Independent Publishing Platform; 2009.
31. Bezanson J, Edelman A, Karpinski S, Shah VB. Julia: A fresh approach to numerical computing. arXiv CoRR. abs/1411.1607. Ithaca: Cornell University; 2014.

32. Kirkpatrick S, Gelatt Jr CD, Vecchi MP. Optimization by simulated annealing. Science. 1983;220(4598):671–80. doi:10.1126/science.220.4598.671.
33. Fonseca CM, Fleming PJ. Genetic Algorithms for Multiobjective Optimization: Formulation, Discussion and Generalization. arXiv CoRR Publisher. In: Proceedings of the 5th International Conference on Genetic Algorithms. Ithaca: Cornell University; 1993. p. 416–23.
34. Kim J, Varner J, Ramkrishna D. A hybrid model of anaerobic e. coli gjt001: Combination of elementary flux modes and cybernetic variables. Biotechnol Prog. 2008;24(5):993–1006. doi:10.1002/btpr.73.
35. Schuster S, Fell DA, Dandekar T. A general definition of metabolic pathways useful for systematic organization and analysis of complex metabolic networks. Nat Biotechnol. 2000;18(3):326–2. doi:10.1038/73786.
36. Kalyanmoy D, Pratap A, Agarwal S, Meyarivan T. A Fast and Elitist Multiobjective Genetic Algorithm: NSGA-II. IEEE Trans Evol Comp. 2002;6: 182–97.
37. Huband S, Hingston P, Barone L, While L. A review of multiobjective test problems and a scalable test problem toolkit. IEEE Trans Evol Comp. 2006;10:477–506.
38. Sendin JOH, Otero-Muras I, Alonso AA, Banga JR. Improved optimization methods for multiple objective design of bioprocesses. Ind Eng Chem Res. 2006;45:8594–603.
39. Booker AJ, Dennis JE, Frank PD, Serafini DB, Torczon V, Trosset MW. A rigorous framework for optimization of expensive functions by surrogates. Struct Optim. 1999;17:1–13.

7

Estimating genome-wide regulatory activity from multi-omics data sets using mathematical optimization

Saskia Trescher*, Jannes Münchmeyer and Ulf Leser

Abstract

Background: Gene regulation is one of the most important cellular processes, indispensable for the adaptability of organisms and closely interlinked with several classes of pathogenesis and their progression. Elucidation of regulatory mechanisms can be approached by a multitude of experimental methods, yet integration of the resulting heterogeneous, large, and noisy data sets into comprehensive and tissue or disease-specific cellular models requires rigorous computational methods. Recently, several algorithms have been proposed which model genome-wide gene regulation as sets of (linear) equations over the activity and relationships of transcription factors, genes and other factors. Subsequent optimization finds those parameters that minimize the divergence of predicted and measured expression intensities. In various settings, these methods produced promising results in terms of estimating transcription factor activity and identifying key biomarkers for specific phenotypes. However, despite their common root in mathematical optimization, they vastly differ in the types of experimental data being integrated, the background knowledge necessary for their application, the granularity of their regulatory model, the concrete paradigm used for solving the optimization problem and the data sets used for evaluation.

Results: Here, we review five recent methods of this class in detail and compare them with respect to several key properties. Furthermore, we quantitatively compare the results of four of the presented methods based on publicly available data sets.

Conclusions: The results show that all methods seem to find biologically relevant information. However, we also observe that the mutual result overlaps are very low, which contradicts biological intuition. Our aim is to raise further awareness of the power of these methods, yet also to identify common shortcomings and necessary extensions enabling focused research on the critical points.

Keywords: Gene regulation, Regulatory network, Systems biology, Mathematical optimization

Background

Gene regulation is one of the most important biological processes in living cells. It is indispensable for adapting to changing environments, stimuli, and developmental stage and plays an essential role in the pathogenesis and course of diseases. Mechanistically, the transcription of DNA into RNA is predominantly controlled by a complex network of transcription factors (TFs) (see Fig. 1). These proteins bind to enhancer or promoter regions adjacent to the genes they regulate [1], which may enhance or inhibit the recruitment of RNA polymerase and thereby activate or repress gene transcription [2]. Gene products also can be modified post-translationally via microRNAs (miRNAs) degrading the transcript or inhibiting their translation [3]. Besides, a multitude of other mechanisms influence gene regulation, such as chromatin remodelling [4], epigenetic effects [5], and compound-building of transcription factors [2]. Distortion of regulatory processes is inflicted with various diseases [6, 7], especially with cancer [8, 9].

Due to this importance, many efforts have been devoted to the elucidation of human regulatory relationships and networks. Wide-spread experimental techniques are transcriptome measurements to quantify gene

* Correspondence: saskia.trescher@informatik.hu-berlin.de
Knowledge Management in Bioinformatics, Computer Science Department, Humboldt-Universität zu Berlin, Unter den Linden 6, 10099 Berlin, Germany

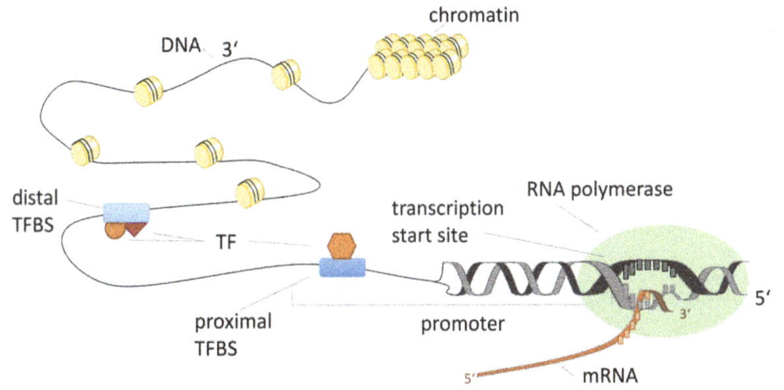

Fig. 1 Transcription of DNA into RNA. Transcription factors (TFs) bind to distal or proximal TF binding sites (TFBS) enhancing the binding of RNA polymerase and activating the transcription of DNA into RNA

and transcription factor co-expression [10], chromatin immunoprecipitation (ChIP) on chips or followed by sequencing for identifying binding patterns of specific TFs [11], and bisulfite sequencing to find epigenetic signals of regulation [12]. Many large-scale datasets of such experiments have been published and are available in public repositories such as the Gene Expression Omnibus (GEO) [13], the Cancer Genome Atlas (TCGA) [14] or the Encyclopedia of DNA Elements (ENCODE) [15]. Computational methods are also used, for instance, to identify transcription factor binding sites (TFBS) [16] or to find known TFBS within the genome (e.g., [17, 18]). Several databases have been created which store relevant information, such as lists of binding motifs (TRANSFAC [19] or JASPAR [20]) or targets of regulatory miRNAs [21].

Such measurements and predictions are used by network reconstruction algorithms to predict regulatory relationships and regulatory networks [22]. A plethora of different methods have been proposed, ranging from purely qualitative methods [23] over simple statistical approaches [24] to more advanced probabilistic frameworks [25]. Early methods were plagued by insufficient data and a general scarcity of background knowledge, which led to rather unstable results [26]. This situation has changed dramatically over the last years, as results of more and more large screens have been made publicly available [27] and also the knowledge on principal regulatory relationships has increased [28, 29]. This, in turn, has increased the interest in methods which predict genome-wide networks using a systematic, unified, mathematical framework.

Here, we review five rather recent methods and conduct a quantitative comparison of their results with the goal to identify their mutual strengths and weaknesses. They all have in common that they assume both the set of regulators (transcription factors or micro RNAs) to be known and the topology of the regulatory network to be given. By combining this background knowledge with specific omics data sets, especially transcriptome data, they try to infer the activity of regulators in a certain experimental condition or disease using mathematical optimization. All presented methods are global methods in the sense that they compute activities genome-wide (as much as represented by the underlying network), thus removing the shortcomings of local methods which ignore cross-talk between sub-models and global effects within samples. The methods predominantly produce a ranked list of regulators, sorted by their activity in a given group of samples; given that a multitude of biological influences is ignored during inference, especially kinetic and temporal effects, their goal cannot be to produce absolute snapshots of regulatory activity. We describe each method in detail and compare them with respect to the most important properties, such as the data being used, the method applied for deriving optimized activity values, or the evaluation performed to show effectiveness. We further implemented a quantitative comparison including four of the presented methods to objectively analyze their results. As contrast, we also include ARACNE [30] as sixth method; this algorithm uses only local reasoning and requires no background knowledge, but is still rather popular.

Methods

We describe in detail five methods which infer transcription factor activity from omics data sets using a background network of transcription factors and the genes they regulate. All use some form of mathematical optimization. To emphasize the common ground of these at-first-sight rather different methods, we explain their underlying models using a simple framework for defining the relationships of transcription factors and genes. This framework is presented first; it should be understood as a least common denominator, not as a proper method for network inference by itself. We then describe five recently published methods for genome-wide TF activity estimation as extensions or constraints to this general framework, namely the approach by Schacht et al. [31] (estimation of TF activity by the effect on their target genes), RACER [32], RABIT [33], ISMARA [34] and biRte [35]. Additionally, we contrast these more comprehensive methods with the local inference algorithm ARACNE [30], a popular tool for the de-novo reconstruction of gene regulatory networks. Key properties of all methods (input, mathematical model, computation, output) are summarized in Table 1.

Mathematical framework

To combine regulatory networks and quantitative omics data and to thereby deduce regulatory activity, all methods described here use a genome-wide mathematical model. Sample specific gene expression values $g_{i,s}$, derived from one biological condition, i.e., grouped into a single class, for in total G genes and S samples need to be provided as input. The background regulatory network is represented as a directed graph where the nodes designate regulators and regulated entities (mostly TFs and genes, but also miRNAs, regulatory sites, or TF complexes) and directed edges indicate a regulatory relationship between the two connected nodes, for example the influence of a TF on the expression of a gene (see Fig. 2).

We will use the variable t for regulators, i for regulated entities, and $b_{t,i}$ for the strength of an edge from a TF/miRNA t to a gene i representing, for instance, a binding affinity. As abstract framework for explaining the different methods we propose a simple linear model predicting gene expression $\widehat{g_{i,s}}$ of gene i in sample s in terms of the activity of all T transcription factors $\beta_{t,s}$, which regulate i, and the binding affinities $b_{t,i}$. In contrast to Fig. 2, where TFs can influence each other, this model ignores TF – TF relations and feedback loops:

$$\widehat{g_{i,s}} = \sum_{t=1}^{T} \beta_{t,s} b_{t,i}$$

Given this model and a set of quantitative measurements of gene expression $g_{i,s}$, the goal of the mathematical optimization is to find parameters β such that the sum of squared errors of measured vs predicted gene expression over all genes and samples is minimized using a certain norm, for example the L_2 norm:

$$\min \sum_{i=1}^{G} \sum_{s=1}^{S} \left(g_{i,s} - \widehat{g_{i,s}}\right)^2$$

Estimation of TF activity by the effect on their target genes [31]

The idea of this method is to use the expression levels of TF's target genes to infer their integrated effect (see Fig. 3). The method uses expression data and database curated TF binding information as input whereby the TF – gene network is restricted to genes regulated by more than 10 TFs and TFs with at least 5 target genes. The model is closely related to the abovementioned general framework, only adding a term for the sample specific effect of a TF. Specifically, the activity of a TF is modelled linearly by its cumulative effect on its target genes normalized by the sum of target genes or the TF's gene expression level:

$$\widehat{g_{i,s}} = c + \sum_{t} \beta_t b_{t,i} \left(\theta_{a,t} act_{t,s} + \theta_{g,t} g_{t,s}\right)$$

where $\widehat{g_{i,s}}$ denotes the predicted gene expression of gene i in sample s, c is an additive offset, β_t describes the estimated activity of TF t and $b_{t,i}$ refers to the underlying strength of the relation between TF t and gene i reflecting the binding affinity. The estimated effect of a TF in a certain sample is calculated via the switch-like term in parentheses, where either the activity definition $act_{t,s} = \frac{\sum_i b_{t,i} g_{i,s}}{\sum_i b_{t,i}}$ or the gene expression of the TF itself $g_{t,s}$ is taken into account using the restrictions $\theta_{a,t}, \theta_{g,t} \in \{0, 1\}$ and $\theta_{a,t} + \theta_{g,t} = 1$. This switch term represents a meta-parameter to find the best model and has no biological interpretation. The model outputs an activity value and the information which switch parameter is chosen for each TF of the reduced network.

During the optimization, the sum of error terms (absolute value of the difference between predicted and measured gene expression) is minimized which is achieved via mixed-integer linear programming using the Gurobi 5.5 optimizer.[1] The authors of this method state that the activity definition (see above) was used in 95% of their test cases, but the switch-like combination of both terms yielded still better optimization results. In the paper, the optimization task is greatly simplified as the model is computed for each gene separately and allows only a maximum number of 6 regulating TFs. The TF – gene network indicating the strength of a relation between a TF and a gene is created for 1120 TFs using knowledge from the commercial MetaCore™ database,[2] ChEA [36] and ENCODE [15]. Due to the restriction of the network mentioned above, the actual model is then based on 521 TFs and 636 target genes only.

Evaluation of the results was performed using expression data from 59 cell lines of the NCI-60 panel [37, 38] and from melanoma cell lines ("Mannheim cohort") [39]. A sample based leave-one-out and 10-fold cross validation of predicted and measured gene expression yielded Pearson correlation scores of about 0.6 for both

Table 1 Overview of methods for estimating regulatory activity from transcriptome data comparing input data, modelling, computational aspects and outcome variables

Method	Input	Model	Computation	Output
Approach by Schacht et al.	- mRNA expression data - TF binding information	Linear model $\widehat{g_{i,s}} = c + \sum_t \beta_t b_{t,i} (\theta_{a,t} act_{t,s} + \theta_{g,t} g_{t,s})$ with $act_{t,s} = \frac{\sum_i b_{t,i} g_{i,s}}{\sum_i b_{t,i}}, \ \theta_{a,t} + \theta_{g,t} = 1, \theta_{a,t}, \theta_{g,t} \in \{0, 1\}$	- Optimization criterion: minimize sum of absolute errors - Mixed-integer linear programming - Optimization via Gurobi 5.5	- parameter for each TF: β_t - decision for each TF if $\theta_{a,t}$ or $\theta_{g,t}$ was chosen
RACER	- mRNA expression data - copy number variation - DNA methylation - miRNA expression signals - TF binding information - miRNA target site info (*c*)	Linear models: 1) $\widehat{g_{i,s}} = c + \theta_{CNV,s} CNV_{i,s} + \theta_{DM,s} DM_{i,s} + \sum_t \beta_{t,s}\, b_{t,i} + \sum_{mi} \beta_{mi,s}\, c_{i,mi} miRNA_{mi,s}$ 2) $\widehat{g_{i,s}} = \tilde{c} + \tilde{\theta}_{i,CNV} CNV_{i,s} + \tilde{\theta}_{i,DM} DM_{i,s} + \sum_t \gamma_{i,t}\, \beta_{t,s} + \sum_{mi} \gamma_{i,mi}\, \beta_{mi,s}$	- Optimization criterion: minimize sum of squared errors with L_1 norm penalty on linear coefficients - Elastic-net regularized generalized linear models and LASSO	1) sample-specific TF and miRNA activities $\beta_{t,s}$ and $\beta_{mi,s}$ 2) TF-gene $\gamma_{i,t}$ and miRNA-gene $\gamma_{i,mi}$ interactions across all samples
RABIT	- differential mRNA expression data - somatic mutations - DNA methylation - copy number variation - TF binding info - recognition motifs for RNA-binding protein (RBP)	Linear model: $\widehat{g_t} = \sum_f \theta_f B_{f,i} + \sum_t \beta_t b_{t,i}$ With *B*: background factors (gene CNA, promoter DNA methylation, promoter degree promoter CpG content)	- Frisch-Waugh-Lovell method, select subset of significant TFs via model selection procedure and remove TFs with insignificant correlation across tumors	- regulatory activity score for each TF (t value of linear regression coefficient of *t*-test)
ISMARA	- gene expression or chromatin state measurements - annotation of promoters (number of predicted sites for motifs) - transcripts and associated promoters - miRNA target site predictions	Linear model $\widehat{g_{p,s}} = c_p + c_s + \sum_m N_{p,m}\, \beta_{m,s}$	- Optimization criterion: minimize sum of errors - Bayesian procedure, ridge regression - Gaussian prior for $\beta_{m,s}$ to avoid overfitting	- inferred motif activity profiles $\beta_{m,s}$ with set of TFs and miRNAs binding to sites of these motifs (= key regulators) - predicted target promoters, associated transcripts and genes - Network of known interactions between predicted targets and predicted regulatory interactions - enriched ontology categories
biRte	- mRNA differential expression - miRNA, TF measurements, CNV (optionally) - regulator (*R*) – target network	Likelihood model: $L_{D,\theta}(R) = p(D \vert R, \theta) = \prod_{\hat{D}} p(\hat{D} \vert R, \theta) = \prod_{\hat{D}} \prod_c \prod_i p(\hat{D}_{ic} \vert R_c, \theta)$	- data specific marginal likelihoods using estimation of hidden state variables with via MCMC - Nested effects model structure Learning to reconstruct transcriptional network	- Estimation of active regulators - Estimation of associated transcriptional network
ARACNE	- microarray expression profiles	none	- local estimation of pairwise gene expression profile mutual information	- Reconstruction of gene regulatory network

Gene expression data is named "g" with index i, estimated parameters with "β", TF binding information with "b", TFs with "t", samples with "s", miRNAs with "mi" and model constants with "c". Other variables are explained in the text

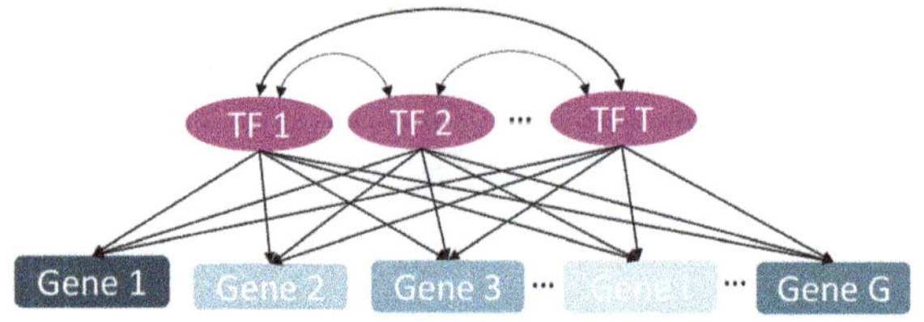

Fig. 2 General scheme of a TF – gene network where all *T* TFs are connected to each other and can regulate all of the *G* genes

data sets. A gene set enrichment analysis of the target genes for TFs modelled by the activity definition yielded 64 significantly enriched concepts including cell cycle, immune response and cell growth for the data from the NCI-60 panel. Additionally, a *t*-test was computed between melanoma and other cell lines of the NCI-60 panel to find differentially expressed genes of melanogenesis. For the resulting genes, regulation models were built and used to predict gene expression in the melanoma cell line data set yielding good prediction performances.

***RACER* [32]**

RACER (Regression Analysis of Combined Expression Regulation) aims to integrate generic cell-line data with sample-specific measurements using a two-stage regression (see Fig. 4). Firstly, sample-specific regulatory activities for TFs and miRNAs are calculated. Subsequently, general TF/miRNA – gene interactions are derived.

Compared to our general framework, RACER includes additionally miRNA binding information. It assumes a linear combination, which is not further justified, of the regulatory effects of TFs and miRNAs on mRNA level. RACER can incorporate a variety of sample specific data including mRNA and miRNA expression values, CNV and DNA methylation. Optimization is applied twice to reduce model complexity, where the method first infers sample-specific TF and miRNA activities and uses these, in a second step, to compute general TF/miRNA – gene interactions.

In the first regression step, mRNA, miRNA, CNV and DNA methylation data are used to calculate the sample specific activities:

$$\widehat{g_{i,s}} = c + \theta_{CNV,s} CNV_{i,s} + \theta_{DM,s} DM_{i,s} + \sum_t \beta_{t,s}\, b_{t,i} + \sum_{mi} \beta_{mi,s}\, c_{i,mi} miRNA_{mi,s}$$

where $\widehat{g_{i,s}}$ denotes the predicted gene expression of gene i in sample s, c is an intercept, $\beta_{t,s}$ describes the estimated activity of TF t in sample s and $b_{t,i}$ is the TF – gene binding score for TF t and gene i. The parameter $\beta_{mi,s}$ stands for the estimated activity of miRNA mi in sample s and is multiplied by $c_{i,mi}$, the number of conserved target sites on 3'UTR of the target gene i for miRNA mi, and by the expression level of miRNA mi in sample s. $\theta_{CNV,s}$ (respectively $\theta_{DM,s}$) are the regression parameters for CNV signals $CNV_{i,s}$ (respectively DNA methylation data $DM_{i,s}$). Using $\beta_{t,s}$ and $\beta_{mi,s}$ from the first regression step, TF – gene and miRNA – gene interactions across all samples are calculated in a second model:

$$\widehat{g_{i,s}} = \tilde{c} + \tilde{\theta}_{i,CNV} CNV_{i,s} + \tilde{\theta}_{i,DM} DM_{i,s} + \sum_t \gamma_{i,t}\, \beta_{t,s} + \sum_{mi} \gamma_{i,mi}\, \beta_{mi,s}$$

where the sums apply only to a number of selected TFs and miRNAs with nonzero binding signals $b_{t,i} > 0$ and conserved target sites $c_{i,mi} > 0$. The resulting parameters $\gamma_{i,t}$ and $\gamma_{i,mi}$ indicate the strength of a TF/miRNA – gene relationship across all samples. To obtain robust estimates, $\gamma_{i,mi}$ is additionally weighted by the averaged activities of the miRNA.

In each of the two regression steps, the optimization criterion is to minimize the sum of squared errors with L_1 penalty on the linear coefficients to induce a sparse solution and to set irrelevant parameters to zero after the fitting. This sparse LASSO solution is obtained through elastic-net regularized generalized linear models. A supplementary feature selection procedure comparing the full model to a restricted model leaving one TF or miRNA out provides the most predominant TF/miRNA regulators. TF binding scores are collected from the generic cell line of erythroleukemia cells K562 from ENCODE

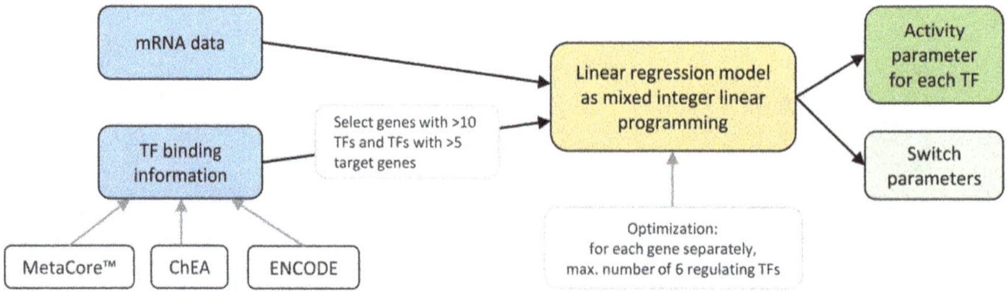

Fig. 3 Flow chart of the approach by Schacht et al. The input data sets (marked in *blue*) are partly filtered and passed to a linear regression model (*yellow*) which calculates an activity value for each TF (*green*)

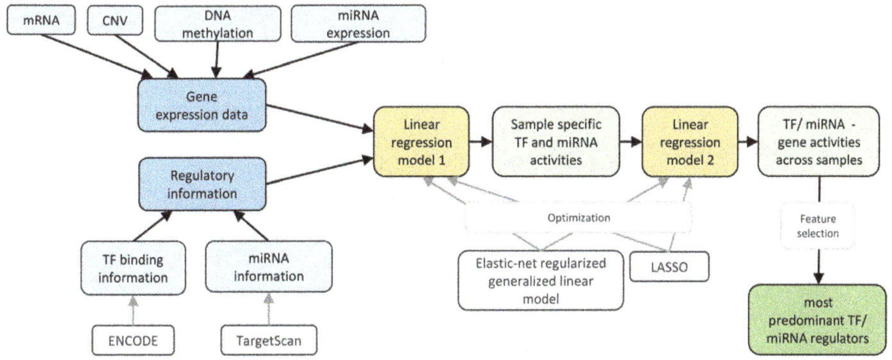

Fig. 4 Scheme of RACER method. The input data sets (marked in *blue*) are passed to a two-step linear regression model (*yellow*) which calculates sample specific activity values for each regulator and determines the most predominant regulators (*green*)

for 97 TFs and 16653 genes. Further, the number of conserved target sites on 3'UTR is taken from sequence-based information from TargetScan for 470 miRNAs and 16653 genes. The RACER method is implemented in R and publicly available under http://www.cs.utoronto.ca/~yueli/racer.html.

The method was evaluated using expression data from an acute myeloid leukemia (AML) data set from TCGA with 173 samples [40] via a sample based 10-fold cross validation on the prediction of gene expression. To assess the quality of predictions, the Spearman rank correlation was calculated resulting in a reassuring value of approximately 0.6. Further, the full model was compared to models excluding one type of the input variables. The full model performed best and a substantial reduction of Spearman correlation was observed by omitting TF regulation (20%) and DNA methylation (5%). RACER also performed with competitive accuracy in predicting known miRNA – mRNA and TF – gene relationships compared to other methods like GenMiR++ [41] or ENCODE TF binding scores [15] using e.g., validated interactions from the MirTarBase [42] and knockdown studies. The feature selection procedure revealed 18 predominant transcriptional regulators in the AML dataset. Using their associated targets, a functional enrichment analysis showed that DNA repair and the tumor necrosis factor pathway were enriched. When applying this panel to cluster patients at different cytogenetic risks, the clustering pattern of the regulatory activities was largely consistent with the risk groups. Further, a literature survey on AML showed that many TF regulators among the top predictions had a role in leukemogenesis.

RABIT [33]

Regression Analysis with Background Integration (RABIT) is a method for finding expression regulators in cancer by a large scale analysis across diverse cancer types. It integrates TF binding information with tumor profiling data to search for TFs driving tumor-specific gene expression patterns (see Fig. 5). It can be applied to predict cancer-associated RNA-binding protein (RBP) recognition motifs which are key components in the determination of miRNA function [43].

In contrast to our general framework, RABIT can, like RACER, make use of CNV and DNA methylation data additionally integrating promoter CpG content and promoter degree information (total number of ChIP-seq peaks near the gene transcription start site) and takes RBP or TF binding information as regulatory input. The computational model consists of three steps (see Fig. 5). First, RABIT tests in each tumor whether the target genes, identified by the BETA method [44], show differential expression compared to the normal controls including a control for background effects from CNVs, promoter DNA methylation, promoter CpG content and promoter degree:

$$\widehat{g}_i = \sum_f \theta_f B_{f,i} + \sum_t \beta_t b_{t,i}$$

where $\widehat{g}_i$ represents the predicted differential gene expression between tumor and normal samples in gene i, B includes values of the f different background factors for gene i, b contains RBP or TF binding information and θ and β are the respective regression parameter vectors. The regression coefficients β are estimated by minimizing the squared difference between measured and predicted gene expression. The regulatory activity score for each TF/RBP is defined by a t-value (regression coefficient divided by standard error) and its significance by the corresponding t-test. If multiple profiles exist for the same TF from different conditions or cell lines, the profile with the highest absolute value of TF regulatory activity score is selected. In a second step, a stepwise forward selection is applied to find a subset of TFs among those screened in step one optimizing the model error. Lastly, TFs with insignificant cross-tumor correlation are removed from the results.

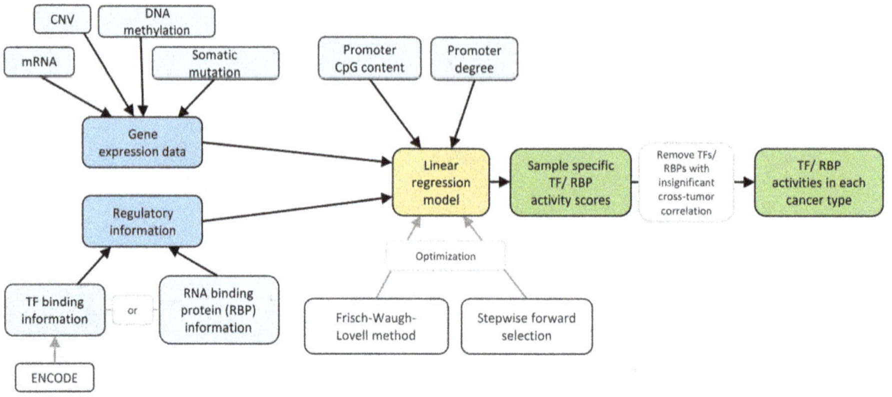

Fig. 5 Flow chart of RABIT method. The input data sets (marked in *blue*) are passed to a linear regression model (*yellow*) which calculates sample specific activity values for each regulator and determines general regulatory activities (*green*)

Computationally, the regression coefficients are calculated via the efficient Frisch-Waugh-Lovell method. TF binding information is taken from 686 TF ChIP-seq profiles from ENCODE representing 150 TFs and 90 cell types. Additionally, recognition motifs for 133 RBPs and their putative targets are collected by searching recognition motifs over the 3'UTR regions [45]. An implementation of the RABIT method can be downloaded from http://rabit.dfci.harvard.edu/download.

RABIT was applied to 7484 tumor profiles of 18 cancer types from TCGA using gene expression, somatic mutation, CNV and DNA methylation data. To systematically assess the results, the cancer relevance level of a TF was calculated as percentage of tumors with the TF target genes differentially regulated (averaged across all TCGA cancer types). A comparison to cancer gene databases, i.e., the NCI cancer gene index project [46], the Bushman Laboratory cancer driver gene list [47, 48], the COSMIC somatic mutation catalog [49] and the CCGD mouse cancer driver genes [50], showed a consistent picture. Further, RABIT's performance was compared to other regression models like LAR or LASSO where RABIT had the best classification results when classifying all TFs into three categories by NCI cancer index and achieved better cross-validation error and shorter running time. The regulatory activity of RBPs showed that some alternative splicing factors could affect tumor-specific gene expression by binding to target gene 3'UTR regions.

ISMARA [34]

In contrast to the previous three methods and to our general framework which directly scores TFs or other regulators, ISMARA (Integrated System for Motif Activity Response Analysis) infers the activity of regulatory motifs (short nucleotide sequences) and thereby indirectly deduces the effects of TFs and miRNAs (see Fig. 6). ISMARA is a web service where no parameter settings or specific processing of the input data, gene expression or ChIP-seq data are necessary. It can also be used to

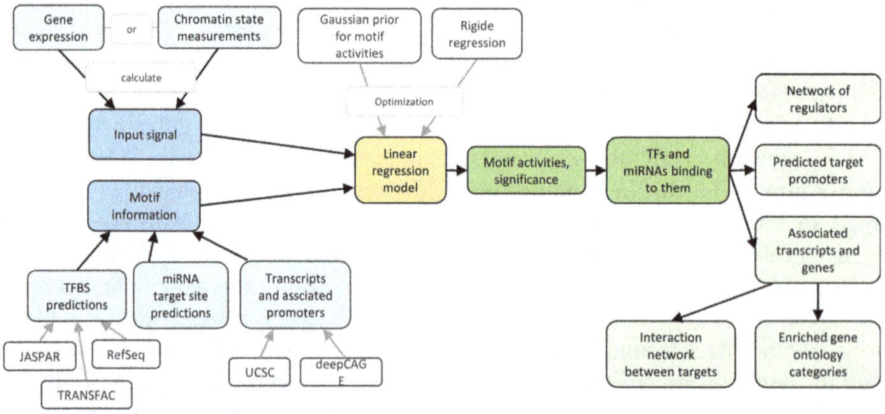

Fig. 6 ISMARA model scheme. The input data sets (marked in *blue*) are passed to a linear regression model (*yellow*) which calculates motif activities and determines associated regulators (*green*)

calculate regulatory activity differences between samples and consider replicates or data from time series.

ISMARA takes sample specific measurements and information about regulatory motifs for TFs and miRNAs into account. Based on the input of gene expression data or chromatin state measurements, the input signal is calculated for each promoter in each sample. The input signal levels are modelled linearly in terms of the binding site predictions and unknown motif activities:

$$\widehat{g_{p,s}} = c_p + c_s + \sum_m N_{p,m}\,\beta_{m,s}$$

where $\widehat{g_{p,s}}$ refers to the input signal for a promoter p in sample s, c_p *and* c_s are intercepts for each promoter and sample, $N_{p,m}$ summarizes the TF/miRNA binding site predictions (sum of the posterior probabilities of all predicted TF/miRNA binding sites for motif m in promoter p) and $\beta_{m,s}$ stands for the estimated motif activities. Like in the other presented methods, the optimization criterion is to minimize the sum of squared error terms between predicted and measured gene expression. Primarily, ISMARA provides the inferred motif activity profiles ($\beta_{m,s}$) sorted by significance and a set of TFs and miRNAs that bind to these motifs representing the key regulators. Further, a list containing their predicted target promoters, associated transcripts and genes, a network of known interaction between these targets and a list of enriched gene ontology categories is displayed. The web service ISMARA is available under http://ismara.unibas.ch.

ISMARA employs a Bayesian procedure with a Gaussian likelihood model and a Gaussian prior distribution for $\beta_{m,s}$to avoid overfitting. Information about regulatory motifs is provided via the annotation of promoters based on deep sequencing data of transcription start sites. To obtain a set of promoters and their associated transcripts, the 5' ends of mRNA mappings from UCSC genome database are clustered with the promoters. TF binding site predictions in the proximal promoter region are collected using 190 position weight matrices representing 350 TFs from JASPAR, TRANSFAC, motifs from the literature and their own analyses of ChIP-seq and ChIP-chip data. Additionally, miRNA target sites for about 100 seed families are annotated in the 3'UTRs of transcripts associated with each promoter.

For evaluation, ISMARA was applied to data from well-studied systems and results were compared to the literature. Inferred motif activities were highly reproducible and even more robust than the expression profiles from which motif activities were derived. When comparing samples from 16 human cell types (GEO accession number GSE30611) from younger and older donors, ISMARA was able to identify a key regulator of aging-related changes in expression of lysosomal genes. A joint analysis of the human GNF atlas of 79 tissues and cell lines [51] and the NCI-60 reference cancer cell lines [52] revealed that many of the top dysregulated motifs were well-known in cancer biology like HIF1A and has-miR-205 miRNA. They also suggested novel predictions for regulating TFs in innate immunity, mucociliary differentiation and cancer.

biRte [35]

BiRte (Bayesian inference of context-specific regulator activities and transcriptional networks) takes a mathematically different approach compared to the abovementioned methods integrating TF/miRNA target gene predictions with sample specific expression data into a joint probabilistic framework (see Fig. 7). Compared to our general scheme of a TF – gene network (Fig. 2), biRte takes the TF/miRNA – gene network without the interactions between regulators to estimate regulatory activities and infers the network between regulators in a second step.

BiRte takes as input differential gene expression data (mRNA), an underlying regulatory network including TF/miRNA – target gene binding information and optionally CNV data, miRNA and TF expression measurements. As opposed to our general framework, biRte defines a likelihood model for the set of active TFs/miRNAs (called regulators R which can be seen as hidden variables) based on the entire gene expression data D and certain model parameters θ:

$$\begin{aligned} L_{D,\theta}(R) &= p(D|R,\theta) = \prod_{\hat{D}} p(\hat{D}|R,\theta) \\ &= \prod_{\hat{D}}\prod_{c}\prod_{i} p(\hat{D}_{ic}|R_c,\theta) \end{aligned}$$

Here D represents the set of all available experimental data including mRNA, CNV, miRNA and TF expression data and D_{ic} refers to its ith feature measured under experimental condition c. The condition specific hidden state variables R_c are estimated with help of the Markov Chain Monte Carlo (MCMC) method where a regulator can switch from an active to an inactive state (switch) or an inactive and an active regulator exchange their activity states (swap). Thereby, the posterior probability for each regulator and condition to influence the expression of its target genes is estimated. Simultaneously, a variable selection procedure is applied to achieve sparsity of the model. The optimization goal is not, as one would expect, to return the configuration with highest posterior probability among all sampled ones but to take marginal selection frequencies during sampling into account and filter those above a defined cutoff. After the

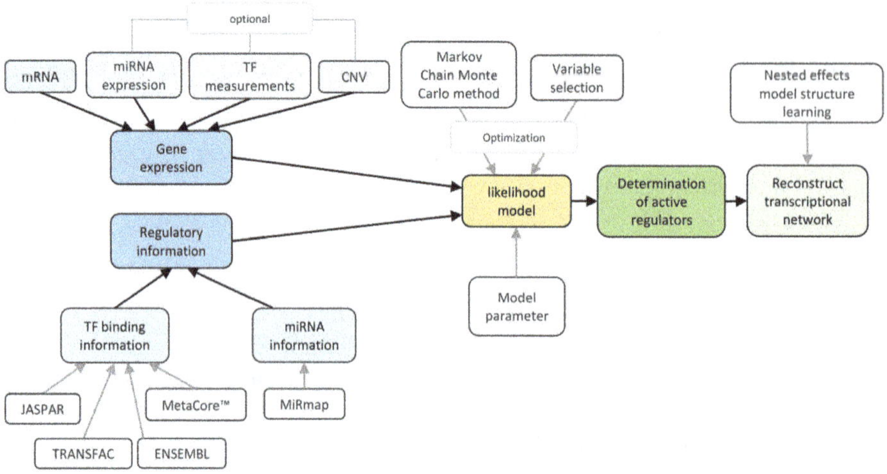

Fig. 7 Scheme of biRte method. The input data sets (marked in *blue*) are passed to a likelihood model (*yellow*) which determines active regulators (*green*)

determination of active regulators, the associated transcriptional network containing TFs and miRNAs is inferred from the observable differential expression of target genes and target gene predictions for individual regulators.

In practice, the stochastic sampling scheme based on MCMC allows swap operations only when regulators show a significant overlap of regulated targets. The variable selection procedure is implemented via a spike and slab prior [53] which can integrate prior knowledge about the activity of regulators. To infer the associated transcriptional network, Nested Effects Model (NEM) [54] structure learning is applied. An input miRNA – gene network is constructed based on MiRmap [55] for 356 miRNAs. The TF – target gene network with 344 TFs is compiled by computing TF binding affinities to promoter sequences according to the TRAP model [56] using data from ENSEMBL, TRANSFAC, JASPAR and MetaCore™. An implementation of biRte is available for R on Bioconductor under https://bioconductor.org/packages/release/bioc/html/birte.html.

Several simulations were conducted to study model behavior. On the basis of a human regulatory subnetwork and accordingly simulated expression data of 900 target genes biRte was compared to BIRTA [57], GEMULA [58] and a hypergeometric test and further to other network reconstruction algorithms like ARACNE [30], GENIE3 [59] and GeneNet [60]. BiRte performed best in regulator activity predictions including a favorable computation time and was robust against false positive and false negative target gene predictions. Additionally, biRte was applied to an E.coli growth control and to a prostate cancer data set including 44 normal and 47 cancer samples from GEO (GSE29079) with corresponding array data from 464 human miRNAs (GSE54516) and the results showed a principal agreement with the biological literature.

ARACNE [30]

We compare ARACNE (Algorithm for the Reconstruction of Accurate Cellular Networks) [30] as an established, yet local, tool for the reconstruction of gene regulatory networks to the previous five recent genome-wide approaches. The algorithm is background knowledge-free and identifies transcriptional interactions based on mutual information including non-linear and non-monotonic relationships and distinguishes between direct and indirect relationships (see Fig. 8). ARACNE is a free tool available under http://califano.c2b2.columbia.edu/aracne.

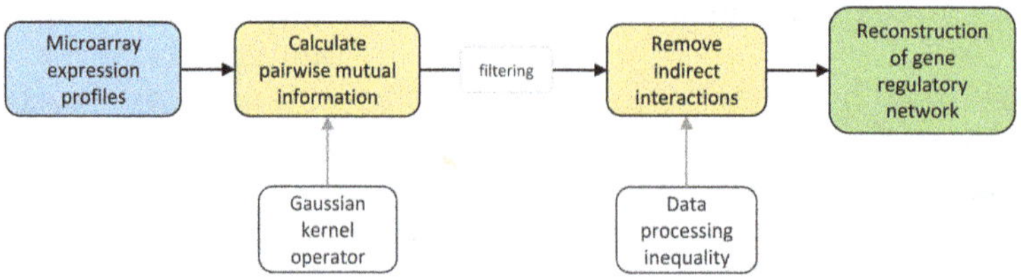

Fig. 8 ARACNE flow chart. The input data set (marked in *blue*) is used to calculate pairwise mutual information where indirect interactions are removed (*yellow*) and which allow a reconstruction of the gene regulatory network (*green*)

ARACNE uses as input only microarray expression profiles and estimates candidate interactions by calculating the pairwise gene expression profile mutual information I defined as

$$I(g_i, g_j) = I_{i,j} = S(g_i) + S(g_j) - S(g_i, g_j)$$

where S denotes the entropy. $I_{i,j}$ measures the relatedness of genes g_i and g_j and equals zero if both are independent. In a second step, the mutual information values are filtered using an appropriate threshold depending on the distribution of all mutual information values between random permutations of the original data set and indirect interactions are removed.

Computationally, a Gaussian kernel operator is used to calculate mutual information scores. In a subsequent step, the data processing inequality (DPI) [61] is applied to remove probably indirect candidate interactions. The DPI states that if the genes g_i and g_k interact only through a third gene g_j, then

$$I(g_i, g_k) \leq \min\left(I(g_i, g_j), I(g_j, g_k)\right)$$

Thus, the least of the three mutual information scores can come from indirect interactions only [30].

ARACNE's performance was evaluated on the reconstruction of realistic synthetic datasets [62] and on an expression profile dataset consisting of about 340 B lymphocytes derived from normal, tumor-related and experimentally manipulated populations [63] against Relevance Networks and Bayesian networks. Regarding the synthetic networks, ARACNE had consistently better precision and recall values compared to the two other algorithms and reached very good precision at significant recall levels. It recovers far more true connections and fewer false connections than the other methods with better performance on tree-like topologies compared to scale-free topologies. A reconstructed B-cell specific regulatory network was found to be highly enriched in known c-MYC targets where about 50% of the predicted genes to be first neighbors were reported in the literature.

Results

We described five recent methods for the genome-wide inference of regulatory activity, namely the approach by Schacht et al., RACER, RABIT, ISMARA, and biRte. They all assume the topology of the regulatory network to be known, cast activity estimation as an optimization problem regarding the difference between predicted and measured values, take different types of sample specific omics data into account, and eventually produce a list of regulators like transcription factors or miRNAs, ranked by their estimated activities in the samples under study. We also included ARACNE which is background knowledge-free and uses only local dependency measures to reconstruct a regulatory network and indirectly infer activities. All of the presented methods essentially follow the same goal, i.e., accurate ranking of regulatory activity, but differ in the types of measurements being integrated, the background knowledge necessary for their application, the complexity and refinement of the underlying model of gene regulation, and the concrete paradigm used for solving the optimization problem. Most of the methods, except for the approach by Schacht et al., are available online via a downloadable implementation, a web service, or an R package providing an operable solution for the interested user. Whereas an overview of the main features of each method ca be found in Table 1, we now first compare the algorithms regarding their general properties in a descriptive way.

The data sets used for evaluation vary between all methods. Therefore, we further implemented an evaluation framework to compare the method by Schacht et al., RACER, RABIT and biRte in an objective and quantitative way. We used experimental data of three publicly available data sets from TCGA [64] and a regulatory network as background knowledge. We first used only mRNA expression data as input to the four methods to ensure the result's comparability, whereas in a second evaluation step, also other omics data sets were included where possible. We further analyzed the relevance of regulators found by different methods using a literature search.

General properties

Experimental data types included

The methods differ in the types of measurements being integrated, which corresponds to the level of detail of their model of gene regulations. All six methods use mRNA as input. RACER, RABIT and biRte can also integrate CNV, DNA methylation, TF/miRNA expression data, or somatic mutations. ISMARA calculates an input signal from microarray, RNA-seq, or ChIP-seq data.

Additionally, all presented methods use prior knowledge about the underlying regulatory network. These networks are extracted from different data sources and pre-processed in different manners. All methods require at least knowledge about TF – gene relationships, yet RACER, biRte and ISMARA also incorporate information about miRNAs. When using RABIT, the user can choose whether to provide TF or RNA binding protein information. The approach of Schacht et al. and biRte extract regulatory information partly from the commercial MetaCore™ database, whereas the other methods use only publicly available databases, like ENCODE, JASPAR or TRANSFAC. The networks which are used for the evaluations published in the respective papers are publicly available for the case of RACER (network for 16653 genes, 97 TFs and 470 miRNAs), RABIT (predicted binding scores of 63 RBP motifs and 17463 genes) and biRte (network for E.coli including 160 TFs). Neither Schacht et al. nor ISMARA make this data available.

Mathematical models of regulatory activity

The methods use different mathematical models to infer regulatory activity. The approach by Schacht et al., RACER, RABIT and ISMARA use linear regression whereas biRte applies a probabilistic framework. ARACNE, as a local method, is based on mutual information. RACER and RABIT can be seen as extensions of the approach by Schacht et al. since they essentially use the same model structure but incorporate more input data types and more classes of regulatory information. Further, RACER applies a two-stage regression to infer regulatory activity.

Optimization frameworks

For assessing regulator activities, Schacht et al., RACER, RABIT and ISMARA minimize the sum of error terms between measured and predicted gene expression. However, the methods use rather different algorithms for solving the resulting optimization problem, and also apply different constraints to achieve model sparsity, robustness of inference, and feature selection. In the approach by Schacht et al., the regression model is computed for each gene separately and allows only a maximum number of six regulating TFs. RACER uses a LASSO approach, while ISMARA follows a Bayesian model that infers regulator activities as posterior distributions. LASSO can be interpreted as a Bayesian model using Laplacian priors instead of Gaussian priors in the regression framework obtaining point estimates of the regulatory activities and enforcing sparseness of the solution [32]. In contrast, biRte uses a likelihood model with a spike and slab prior to induce model sparsity. This approach implements a selective shrinkage of model coefficients such that estimates are less biased compared to a LASSO prior [65]. With the help of the spike and slab prior, sparsity can be controlled in a variable dependent manner allowing the inclusion of prior belief in the activity of each regulator [35].

Computed outputs

Schacht et al. and biRte determine activity of regulators over all samples at once, whereas RACER and biRte first infer sample-specific activities which are combined to cross-tumor activities only in a second optimization step. In contrast, ISMARA in first place infers motifs activity; these activities are used to deduce the effects of TFs and miRNAs by their motif binding profiles. ISMARA primarily provides sample specific TF and miRNA activity but also offers an option to group samples and compare average regulatory activity between different conditions. Like biRte and ARACNE, it also infers the network of the regulators themselves.

Methods and data sets used for evaluation

The type and extent of evaluation performed for the different methods vary greatly. They range from direct application to biological problems over the comparison of results to the biological literature to simulation studies. All methods published evaluations results on publicly available datasets, e.g., from the National Cancer Institute, TCGA or GEO, but unfortunately address different tissues and cancer types. Sample-based cross-validation is applied in the work by Schacht et al., RACER, RABIT and ISMARA. The first two of these methods use correlation coefficients between measured and predicted gene expression for assessing prediction quality. RACER, RABIT and biRte compare their results to the outcome of other algorithms and to those of restricted models, for example excluding one type of the input variables. All methods search the literature to compare their predictions to previously published studies on the respective biological question. Overall, ISMARA provide the most extensive biological evaluation using a battery of relevant use cases, whereas biRte excels in systematic simulation studies. Sadly, there are very few works which compare any of the methods presented on the same problem; the only result we are aware of compared ARACNE and biRte regarding their performance in network reconstruction on simulated data, in which biRte attained higher robustness against false positive and false negative target gene predictions [35].

Quantitative comparison

Although certain evaluation steps were carried out for all methods, results in the original papers are not comparable as they used different input datasets, different background regulatory networks, and different evaluation metrics. Therefore, in addition to the comparison of general properties of the methods, we implemented an evaluation framework using three independent and publicly available test data sets to compare the method by Schacht et al., RACER, RABIT and biRte in an objective and quantitative way. All evaluated methods were given the same regulatory network as input.

Data sets

For the evaluation we used experimental data from TCGA [64] for three cancer types: Colon adenocarcinoma (COAD), liver hepatocellular carcinoma (LIHC) and pancreatic adenocarcinoma (PAAD). For all three cancer types, mRNA expression, CNV, DNA methylation and miRNA expression data is available for primary tumor and normal tissue samples. These data sets are openly accessible via the NCI Genomic Data Commons Data Portal[3] or the NCI Genomic Data Commons Legacy Archive[4] (DNA methylation data).

For mRNA gene expression we used processed RNA-seq data in the form of FPKM (fragment per kilobase of exon per million mapped reads) values. The files included Ensembl Gene IDs which were converted to HGNC symbols using the Ensembl [66] BioMart tool[5] to match the IDs of the TF – gene network. In two cases, when multiple Ensembl Gene IDs mapped to one HGNC symbol, we chose the gene with highest log2 fold change between case and control group. miRNA expression was given as RPM (reads per million miRNA mapped) measurements. Both mRNA and miRNA data were centered using a weighted mean such that the mean of the case group equaled the negative mean of the control group, and normalized via a weighted standard deviation. CNV data was retrieved as masked copy number segment where the Y chromosome and probe sets with frequent germline copy-number variation had already been removed. Chromosomal regions were mapped to genes using the R package biomaRt [67]. If multiple records mapped to one gene, the median of the segment mean values was calculated. For DNA Methylation data we used the beta-values of Illumina Human Methylation 450 arrays as methylation scores. Multiple scores for the same gene were averaged within a sample.

We restricted our analyses to the samples for which all four input data types were available. When multiple measurements for one sample and data type were available, we used only the first one in alphabetical order of the file name. After this selection procedure, 165 samples remained for COAD, 404 for LIHC and 180 for PAAD. A list including sample and file information is available in Additional file 1.

Together with the experimental data, all evaluated methods were given the same regulatory network as input. We used a publicly available human TF – gene network [28] based on a text-mining approach and complemented it with TF – gene interactions from the public TRANSFAC[6] database [19]. This network included 2894 interactions between 429 TFs and 1218 genes. The network is provided in Additional file 2.

Evaluated methods

We conducted the quantitative comparison for the method proposed by Schacht et al., RACER, RABIT and biRte. ISMARA was not included since it is (a) only available as a web service, (b) can only be used with its own, proprietary underlying regulatory network model, and (c) requires the upload of raw data which is prohibited by TCGA's terms of use. Also ARACNE [30] was not included in the quantitative evaluation since it does not use background knowledge and we therefore consider its results as incomparable to the other methods.

- For the approach by Schacht et al. we re-implemented their method as closely as possible to the original design using Python and the Cuneiform workflow language [68, 69]. Due to the high number of integer parameters in the original method, the complexity of optimizing the whole network at once would have by far exceeded computational measures. Therefore, like in the original paper, we computed the model for each gene separately and restricted the number of regulating TFs per gene to six. We added a second step where we used these TF – gene interactions building a sub-network to optimize TF activity globally to describe the interplay of the TFs' effects on their target genes. As in the implementation of Schacht et al., we used the Gurobi Optimizer.[7]
- For RACER we used the available R scripts[8] and extracted the resulting sample-specific regulatory activities.
- RABIT published a C++ implementation which they provide on their website[9] and which we used with the FDR option set to 1. As RABIT takes differential expression into account, we used the difference of expression values between case and control group as input and ordered the TFs by t-value as proposed in the RABIT paper.
- BiRte is available as a bioconductor R package. We used R version 3.3.2 with biRte version 1.10.0 and applied the method "birteLimma" to estimate regulatory activities with the options niter and nburnin set to 10000. As biRte has a randomized component, the resulting TF activities are not exactly the same for different runs. We averaged the final activity scores over 1000 iterations of birteLimma.

For our re-implemented method by Schacht et al. and RACER we computed separate models for case and control group and ranked the TFs by their activity difference between the two groups.

To ensure the result's comparability, we first used only mRNA expression data as input to the four methods. In a second evaluation, we included also other omics data sets where possible. BiRte was evaluated on mRNA and CNV data, RABIT on mRNA, CNV and DNA methylation data, and RACER additionally used miRNA expression as input. We obtained lists with the regulators ranked according to the absolute value of their computed activity for each cancer type and method, with and without the use of additional inputs. For each cancer type we calculated the size of the overlaps in the four different results using the top 10 and top 100 regulators. The results for the top 10 regulators using either only mRNA or multiple omics data sets as input are shown in Table 2.

Only mRNA as input

When only mRNA is used as input, one TF is commonly found by the three methods RACER, RABIT and biRte in each data set, respectively: PHOX2B for COAD, EPAS1 for LIHC and ELF1 for PAAD. A literature search of these TFs and their targets revealed clear associations to the respective cancer type. The TF obtained commonly for COAD, PHOX2B, is related to TLX2, a gene which has been shown to play a role in the tumorigenesis of gastrointestinal stromal tumors [70]. EPAS1, which was found in the LIHC top 10 TFs of three methods, is linked to CXCL12, which plays an important role in metastasis formation of hepatocellular carcinoma by promoting the migration of tumor cells [71, 72]. For PAAD, three methods ranked TF ELF1 high, which is related to 14 genes in our network, inter alia to BRCA2 and LYN. Mutations in the BRCA2 gene have been implicated in pancreatic cancer susceptibility [73, 74], whereas the knockdown of LYN reduced human pancreatic cancer cell proliferation, migration, and invasion [75]. These results underline that the methods are able to find biologically relevant information about regulation processes in cancer.

Table 2 HGNC Symbols of the top 10 regulators found by each method for COAD (using 165 samples), LIHC (404 samples) and PAAD (180 samples) and the use of only mRNA data as input (left panel) and multiple input data sets (RACER: mRNA, miRNA, CNV and DNA methylation; RABIT: mRNA, CNV and DNA methylation; biRte: mRNA and CNV; right panel). TFs with equal activity values are marked with*. TFs found by several method's top 10 are marked in bold (when found by RACER, RABIT and biRte), blue (RACER and RABIT), red (RABIT and biRte) or yellow (RACER and biRte)

Data set	Only mRNA				Multi-omics		
	Schacht et al.	RACER	RABIT	biRte	RACER (mRNA, miRNA, CNV, DNA methyl.)	RABIT (mRNA, CNV, DNA methyl.)	biRte (mRNA, CNV)
COAD	INSM1	HOXA5	MYC	AHR*	MIR130A	MYC	GUCA2A
	NR0B1	SP4	KLF5	NR1I3*	MIR598	NRF1	SLC25A34
	SNAI1	MECOM	CDX2	KLF5	MIR640	KLF5	PLCD1
	FOXC1	MLXIPL	NRF1	PRDM1	MIR554	RARG	AHR
	PHOX2A	CDX2	PRDM1	CDX1	MIR921	GFI1B	FAM163B
	FOXA1	NRF1	NFYA	**PHOX2B**	MIR631	E2F1	NR1I3
	SREBF2	MYC.MAX.ZBTB17	NFKB1	ESRRA	MIR1202	CDX2	KLF4
	NR4A1	**PHOX2B**	**PHOX2B**	HOXA5	MIR548G	NFYA	TRPM6
	SNAI2	HOXA10	RARG	TCF7L2	MIR602	HOXA5	ADAMDEC1
	ARNT.HIF1A	MYC	PITX2	SOX2	MIR623	PITX2	TMIGD1
LIHC	NFIL3	GBX2	HNF4A	PHOX2A	MIR187	HNF4A	PHOX2A
	NR0B1	STAT5B	MYC	**EPAS1**	MIR892A	EGR1	EPAS1
	ELF2	POU3F1	NRF1	HNF4A	MIR638	SP1	HNF4A
	NR4A2	**EPAS1**	HNF1A	FLI1	MIR517A	NRF1	ADRA1A
	ZNF384	POU5F1	SP1	MTF1	MIR493	*DNA methyl.*	MTF1
	INSM1	ELK3	RARB	IKZF1	MIR572	MYC	IKZF1
	ATOH1	PHOX2A	MTF1	NFATC1	*CNV*	SOX10	EGR1
	SP4	FOXF2	SOX10	POU3F1	MIR192	MTF1	FLI1
	KLF11	MMP3	NR1I3	POU3F2	MIR1281	RARB	CEBPB
	POU4F1	GCM1	**EPAS1**	NFKB1	MIR1244	NR1I3	FOS
PAAD	RARB	**ELF1**	SPI1	SPI1	MIR653	*DNA methyl.*	RNU6-830P
	RBPJ	SATB1	GATA2	PRDM1	MIR552	SPI1	RN7SKP94
	USF1.USF2	IRF1	PES1	PES1	MIR381	PES1	RNA5SP60
	BARX2	STAT1.STAT2.IRF9	FOXO3	BACH1	MIR668	NFKB1.REL	SPI1
	USF2	IKZF1	**ELF1**	**ELF1**	MIR587	PURA	PHBP14
	STAT3.STAT1	NFATC2	RELA.REL	PURA	*CNV*	NFE2	TOMM22P6
	ETV4	MYF5	NFE2	TFAP2B	MIR596	ATF1	IL22
	HOXA1	GATA2	CTCF	SATB1	MIR1180	FOXO3	EEF1A1P24
	STAT4	NFYC	ATF1	NR2C2	MIR190B	NFATC2	LINC01375
	ESR1	PHOX2A	PURA	STAT1	MIR216A	IRF1	EIF4EP4

Several TFs in the top 10 are found by two of the four methods For instance, RACER and RABIT have four common top 10 TFs (CDX2, NRF1 and MYC next to PHOX2B) in the COAD data set. However, the top 10 TFs found by the method by Schacht et al. do not overlap with any top 10 TFs of the other methods in any data set. The agreement of RACER, RABIT and biRte in the top 10 TFs hints to the biological importance of the found TFs since this overlap is statistical significant as the probability of finding common TFs in three sets of ten randomly chosen ones out of 429 TFs (p-value) is below 0.006. Additionally, the methods do identify different TFs for different data sets, indicating the importance of the actual cancer specific mRNA expression values and that results are not dictated by the background network.

The results for the number of overlapping regulators in the top 100 between the four methods and the three different data sets are shown in Fig. 9. For RABIT, only 76 TFs for COAD (resp. 67 for LIHC and 57 for PAAD) could be ranked since all other TFs had an activity value equal to zero.

When looking at the overlap of three of the four methods, the number of overlapping TFs is still the highest for the triplet RACER, RABIT and biRte. For the LIHC dataset two TFs are found in the top 100 of all four methods (E2F4 and SOX10). E2F4 is a downstream target of ZBTB7, which was associated to the expression of cell cycle-associated genes in liver cancer cells [76]. Two target genes of E2F4, CDK1 and TP73 were also involved in liver cancer development [77] and proposed as prognostic marker of poor patient survival prognosis in hepatocellular carcinoma [78]. Further, epigenetic alterations of the EDNRB gene, a target of SOX10, might play an important role in the pathogenesis of hepatocellular carcinoma [79]. Even if the result of four methods finding two common TFs is not statistically significant (p-value = 0.36), their association to liver hepatocellular carcinoma shows that the methods reach their goal of identifying relevant TFs.

However, when comparing different data sets, the methods tend to rank the same TFs under the top 100 to a greater or lesser extent. For example, the overlap of all top 100 TFs of the three cancer types is only one TF for RABIT and nine TFs for biRte, but 16 TFs for the method by Schacht et al. and even 32 TFs for RACER. Therefore, the results from RABIT and biRte seem to be more cancer type specific and less dependent on the regulatory network than the results from RACER. However, we did not specifically investigate the influence of the underlying network and its topology on the results which would be an interesting point for further research.

Multi-omics data as input

When not only taking mRNA into account but also miRNA, CNV and DNA methylation, the results are more difficult to compare between the methods, since they all use a different way of combining different types of data due to their models and implementations.

We are aware of the lower level of comparability of this approach regarding the multi-omics results in contrast to a scenario, where all methods are evaluated on the same set of input data. However, we intended to use maximum set of input data for each method to cover the effect of the use of multiple omics data sets compared to only mRNA as input.

BiRte was evaluated on mRNA and CNV data, RABIT on mRNA, CNV and DNA methylation data, and RACER additionally used miRNA expression as input. Whereas RACER and RABIT considered CNV or DNA methylation data as one background factor and compute only one activity value, biRte evaluated the influence of each CNV separately.

The results (see Table 2, right panel) show that RACER exclusively ranks miRNAs high; not a single TF is found among the top 10 regulators. Also, the influence of CNVs was high in LIHC and PAAD. However, the TFs that RACER found in the top 10 when using only mRNA data as input are still ranked high in the multi-omics scenario, e. g the COAD top three TFs of the mRNA results are ranked 13th, 16th and 14th in the

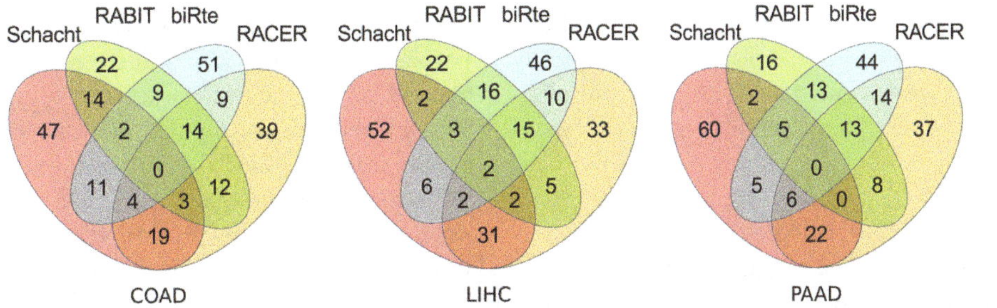

Fig. 9 Number of overlapping TFs in the top 100 of ranked TFs per method (for RABIT the overlap with the top 76/67/57 TFs (having activity > 0) in COAD/LIHC/PAAD is shown)

results of the multi-omics input. The difference of the results coming from the two input types is less for RABIT: seven TFs are still in the top 10 for COAD (8 for LIHC and 6 for PAAD) when using CNV and DNA methylation additionally to mRNA data. Therefore, the contribution of additional input data seems not to be crucial for the performance of RABIT. BiRte considers each CNV as a potential regulator which increases the total number of regulators enormously. Still, two commonly present TFs in the top 10 of the COAD data set (even six for LIHC and one for PAAD) are found by either the sole mRNA input and the multi-omics approach.

The overlap of the top 10 of RABIT and biRte in the multi omics case is considerable with three TFs in LIHC (HNF4A, EGR1 and MTF1; p-value = 0.001), and one TF in PAAD (SPI1; p-value = 0.21). Three of them (HNF4A, MTF1 and SPI1) were already found when using only mRNA data as input.

The results for the use of different input data sets show that the top ranked regulators are drastically changed when using additionally miRNA data in RACER, but change less when only CNV or DNA methylation data is provided in RABIT and biRte. However, the results from multi omics analyses are difficult to compare since the combination of input data sets is not consistent across the three different methods.

Discussion

Background networks

A crucial input to the models is the underlying regulatory network which is needed to reduce the search space for actual regulatory activity. However, the construction of comprehensive TF/miRNA – gene regulatory networks is difficult for various reasons. Firstly, a comprehensive characterization of the human regulatory repertoire is lacking since only about half of the estimated 1,500–2,000 TFs in the mammalian genome is known [80]. ChIP experiments, prone to a high false positive rate [81], were used to identify TF binding patterns but each assay is limited to the detection of one TF in one condition and therefore TF binding has not been characterized for many TFs in most cell types. Further, the local proximity of a binding site to the transcriptional start site of a gene does not automatically implicate transcriptional regulation. With regard to post-transcriptional regulators, the functions for only a few of the around 1,200 different miRNAs have been experimentally determined and current data on miRNA targets is mostly based on computational predictions [82]. Generally, the knowledge about TF and miRNA binding is scattered over the biological literature and different, partly commercial, databases, impeding the construction of comprehensive networks [28]. Therefore, any comparative evaluation of the methods presented here would have to make sure that the same background network is used for each computation. Besides, studies on the impact of network incompleteness or different error rates in networks would be important to assess the ability of the methods to cope with such common problems. Simulation studies will be vital in this regard.

The graph view on regulation

The modelling of regulatory networks as graphs, as used in all presented methods, is perhaps not the optimal representation for the underlying biological regulatory processes. A graph cannot easily account for important effects such as TF complex formation and temporal and spatial synchronization of activities. Furthermore, TF binding is affected by chromatin state and the impact of posttranslational modifications on transcriptional activity which are difficult to include in a graph view on regulation. The model's dependence on the topological structure and the robustness to changes in the underlying network have not been evaluated or discussed in any of the presented methods even if these issues are known to have a severe influence in network analysis [83].

Underlying mathematical model

Linear models, widely spread in different fields of science, provide a simple and easily understandable design but over-simplify the underlying biological processes. Nonlinear behavior, e.g., saturation effects, cannot be represented. Considering that the number of available samples is typically relatively small, the incorporation of many different data types and according parameters into the model could result in excessively complex designs prone to overfitting, but this issue lacks general awareness. Only two of the presented methods incorporate parameter priors (ISMARA and biRte), and two apply cross validation techniques to estimate prediction performance (method by Schacht et al. and RACER). Further, the effect of temporal buffering between TF binding and the actual effect on gene expression is not included in any of the methods.

Comparability

All methods produce a ranked list of regulators. Comparing these results across different methods, even when applied on the same data set and using the same background network, is difficult since no generally accepted benchmarks are available. Therefore, there currently is no objective measure to designate a best method. The closest comparable evaluation effort we are aware of is implemented in the "DREAM5 – Network Inference" challenge [84], which targets gene regulatory network reconstruction. The invited participants reverse-engineered a network from gene expression data, including a simulated network, and evaluated the results on a subset of known interactions or the known network for the in-silico case. The approach of GENIE3 [59] which trains a random forest to predict target gene expression performed best and the integration of predictions from multiple inference methods showed robust and high performance across diverse data sets. However, an extensive competitive evaluation to determine active regulators based on a given regulatory network has, to the best of our knowledge not been carried out yet.

We therefore compared the results of four methods in a quantitative way. The experimental data and the regulatory network we used as input are publicly available to ensure transparency of our results. The results suggest that the methods are able to find biologically relevant information about regulation processes in cancer. However, the result overlaps are rather low (though sometimes statistically significant). This seems surprising as all methods essentially follow the same goal, i.e., identification of the most differentially active TFs or genes. We think further research is necessary to exactly characterize the specific strengths of each method. Furthermore, we did not investigate the influence of the underlying network on the results, which is another topic for further research.

Conclusion

Despite their often rather involved procedures and models, none of the presented methods adequately reflects the biological reality of regulatory activity in cells. A specific disease phenotype is rarely caused by a single gene but rather a product of the interplay of genetic variability, epigenetic modifications and post-transcriptional regulation mechanisms [85]. The presented methods ignore a multitude of such factors like the effects of chromatin state and alternative splicing, nonlinear relationships between regulatory activity and gene expression, or kinetic and temporal effects. Furthermore, TFs themselves regulate the expression of other TFs forming feedback loops which are not considered in any of the presented methods. Nevertheless, the methods apparently are able to detect strong signals and produced promising results in terms of ranking transcription factors by their activity and are thus valuable tools for identifying biomarkers for specific phenotypes.

Endnotes

[1]http://www.gurobi.com/products/gurobi-optimizer

[2]http://lsresearch.thomsonreuters.com/pages/solutions/1/metacore

[3]https://gdc-portal.nci.nih.gov

[4]https://gdc-portal.nci.nih.gov/legacy-archive

[5]http://www.ensembl.org/biomart/martview, release 87

[6]http://www.gene-regulation.com/pub/databases.html, release 7.0

[7]version 6.04, available under a free academic license

[8]http://www.cs.utoronto.ca/~yueli/racer.html (accessed 17 October 2016)

[9]http://rabit.dfci.harvard.edu (accessed 05 February 2016)

Additional files

Additional file 1: Lists information about the samples and files from TCGA included in our quantitative evaluation for all three cancer types (COAD, LIHC and PAAD).

Additional file 2: Includes an adjacency list of the connected nodes of the TF - gene network. The list includes three columns ("TF", "gene", "edge") where each row indicates an association with the value of "edge" between a TF and a gene. Complexes of TFs are indicated with a separating "." between their components.

Acknowledgements

We thank Dr. Holger Fröhlich, the author of biRte, for his help in the usage of biRte with multiple omics data sets as input and Christopher Schiefer for his contribution to the re-implementation of the method proposed by Schacht et al. We acknowledge the advice of Prof. Dr. Erik van Nimwegen concerning ISMARA. The results in this work are in part based upon data generated by the TCGA Research Network.

Funding

We would like to acknowledge the funding provided to S.T. and J.M. from the Berlin School of Integrative Oncology (BSIO, Graduate School 1091) which is supported by the German Research Foundation (Deutsche Forschungsgemeinschaft, DFG) in the framework of the Excellence Initiative of the German federal and state governments.

Authors' contributions

ST performed literature research, quantitative comparisons and drafted the manuscript with the help of JM and UL. All authors read and approved the final manuscript.

Authors' information

Not applicable.

Competing interests

The authors declare that they have no competing interests.

References

1. Lemon B, Tjian R. Orchestrated response: a symphony of transcription factors for gene control. Genes Dev. 2000;14(20):2551–69.
2. Spitz F, Furlong EE. Transcription factors: from enhancer binding to developmental control. Nat Rev Genet. 2012;13(9):613–26.
3. Guo H, Ingolia NT, Weissman JS, Bartel DP. Mammalian microRNAs predominantly act to decrease target mRNA levels. Nature. 2010; 466(7308):835–40.
4. Clapier CR, Cairns BR. The biology of chromatin remodeling complexes. Annu Rev Biochem. 2009;78:273–304.
5. Jaenisch R, Bird A. Epigenetic regulation of gene expression: how the genome integrates intrinsic and environmental signals. Nat Genet. 2003; 33(Suppl):245–54.
6. Gong X, Jia P, Zhao Z. Investigating microRNA-transcription factor mediated regulatory network in glioblastoma. 2010 IEEE International Conference on Bioinformatics and Biomedicine Workshops; 2010. p. 258–63.

7. Jiang Q, Wang Y, Hao Y, Juan L, Teng M, Zhang X, Li M, Wang G, Liu Y. miR2Disease: a manually curated database for microRNA deregulation in human disease. Nucleic Acids Res. 2009;37:98–104.
8. Mayo MW, Baldwin AS. The transcription factor NF-kappaB: control of oncogenesis and cancer therapy resistance. Biochim Biophys Acta. 2000; 1470(2):M55–62.
9. Esquela-Kerscher A, Slack FJ. Oncomirs - microRNAs with a role in cancer. Nat Rev Cancer. 2006;6(4):259–69.
10. Allocco DJ, Kohane IS, Butte AJ. Quantifying the relationship between co-expression, co-regulation and gene function. BMC Bioinformatics. 2004;25:5–18.
11. Johnson DS, Mortazavi A, Myers RM, Wold B. Genome-wide mapping of in vivo protein-DNA interactions. Science. 2007;316(5830):1497–502.
12. Lou S, Lee H-M, Qin H, Li J-W, Gao Z, Liu X, Chan LL, Lam V, So W-Y, Wang Y, Lok S, Wang J, Ma RC, Tsui SK, Chan J, Chan T-F, Yip KY. Whole-genome bisulfite sequencing of multiple individuals reveals complementary roles of promoter and gene body methylation in transcriptional regulation. Genome Biol. 2014;15(7):408.
13. Edgar R, Domrachev M, Lash AE. Gene expression omnibus: NCBI gene expression and hybridization array data repository. Nucleic Acids Res. 2002; 30(1):207–10.
14. The Cancer Genome Atlas Research Network. Comprehensive genomic characterization defines human glioblastoma genes and core pathways. Nature. 2008;455(7216):1061–8.
15. Gerstein MB, Kundaje A, Hariharan M, Landt SG, Yan K-K, Cheng C, Mu XJ, Khurana E, Rozowsky J, Alexander R, Min R, Alves P, Abyzov A, Addleman N, Bhardwaj N, Boyle AP, Cayting P, Charos A, Chen DZ, Cheng Y, Clarke D, Eastman C, Euskirchen G, Frietze S, Fu Y, Gertz J, Grubert F, Harmanci A, Jain P, Kasowski M, Lacroute P, Leng J, Lian J, Monahan H, O'Geen H, Ouyang Z, Partridge EC, Patacsil D, Pauli F, Raha D, Ramirez L, Reddy TE, Reed B, Shi M, Slifer T, Wang J, Wu L, Yang X, Yip KY, Zilberman-Schapira G, Batzoglou S, Sidow A, Farnham PJ, Myers RM, Weissman SM, Snyder M. Architecture of the human regulatory network derived from ENCODE data. Nature. 2012; 489(7414):91–100.
16. Tompa M, Li N, Bailey TL, Church GM, De Moor B, Eskin E, Favorov AV, Frith MC, Fu Y, Kent WJ, Makeev VJ, Mironov AA, Noble WS, Pavesi G, Pesole G, Régnier M, Simonis N, Sinha S, Thijs G, van Helden J, Vandenbogaert M, Weng Z, Workman C, Ye C, Zhu Z. Assessing computational tools for the discovery of transcription factor binding sites. Nat Biotechnol. 2005; 23(1):137–44.
17. Elemento O, Tavazoie S. Fast and systematic genome-wide discovery of conserved regulatory elements using a non-alignment based approach. Genome Biol. 2005;6:R18.
18. Ernst J, Plasterer HL, Simon I, Bar-Joseph Z. Integrating multiple evidence sources to predict transcription factor binding in the human genome. Genome Res. 2010;20(4):526–36.
19. Wingender E, Dietze P, Karas H, Knüppel R. TRANSFAC: a database on transcription factors and their DNA binding sites. Nucleic Acids Res. 1996; 24(1):238–41.
20. Sandelin A, Alkema W, Engström P, Wasserman WW, Lenhard B. JASPAR: an open-access database for eukaryotic transcription factor binding profiles. Nucleic Acids Res. 2004;32:D91–4.
21. Griffiths-Jones S, Grocock RJ, van Dongen S, Bateman A, Enright AJ. miRBase: microRNA sequences, targets and gene nomenclature. Nucleic Acids Res. 2006;34:D140–4.
22. Hecker M, Lambeck S, Toepfer S, van Someren E, Guthke R. Gene regulatory network inference: data integration in dynamic models-a review. Biosystems. 2009;96(1):86–103.
23. Liang S, Fuhrman S, Somogyi R. Reveal, a general reverse engineering algorithm for inference of genetic network architectures. Pacific Symp Biocomput. 1998;18–29.
24. Bansal M, Belcastro V, Ambesi-Impiombato A, di Bernardo D. How to infer gene networks from expression profiles. Mol Syst Biol. 2007;3:78.
25. Li P, Zhang C, Perkins EJ, Gong P, Deng Y. Comparison of probabilistic Boolean network and dynamic Bayesian network approaches for inferring gene regulatory networks. BMC Bioinformatics. 2007;8 Suppl 7:S13.
26. Markowetz F, Spang R. Inferring cellular networks-a review. BMC Bioinformatics. 2007;8 Suppl 6:S5.
27. Rung J, Brazma A. Reuse of public genome-wide gene expression data. Nat Rev Genet. 2013;14:89–99.
28. Thomas P, Durek P, Solt I, Klinger B, Witzel F, Schulthess P, Mayer Y, Tikk D, Blüthgen N, Leser U. Computer-assisted curation of a human regulatory core network from the biological literature. Bioinformatics. 2015;31(8):1258–66.
29. Krämer A, Green J, Pollard J, Tugendreich S. Causal analysis approaches in ingenuity pathway analysis. Bioinformatics. 2014;30(4):523–30.
30. Margolin AA, Nemenman I, Basso K, Wiggins C, Stolovitzky G, Dalla Favera R, Califano A. ARACNE: an algorithm for the reconstruction of gene regulatory networks in a mammalian cellular context. BMC Bioinformatics. 2006;7 Suppl 1:S7.
31. Schacht T, Oswald M, Eils R, Eichmüller SB, König R. Estimating the activity of transcription factors by the effect on their target genes. Bioinformatics. 2014;30(17):i401–7.
32. Li Y, Liang M, Zhang Z. Regression analysis of combined gene expression regulation in acute myeloid leukemia. PLoS Comput Biol. 2014;10(10):e1003908.
33. Jiang P, Freedman ML, Liu JS, Liu XS. Inference of transcriptional regulation in cancers. Proc Natl Acad Sci. 2015;112(25):7731–6.
34. Balwierz PJ, Pachkov M, Arnold P, Gruber AJ, Mihaela Z, van Nimwegen E. ISMARA: automated modeling of genomic signals as a democracy of regulatory motifs. Genome Res. 2014;24(5):869–84.
35. Fröhlich H. biRte: Bayesian inference of context-specific regulator activities and transcriptional networks. Bioinformatics. 2015;31(20):3290–8.
36. Lachmann A, Xu H, Krishnan J, Berger SI, Mazloom AR, Ma'ayan A. ChEA: transcription factor regulation inferred from integrating genome-wide ChIP-X experiments. Bioinformatics. 2010;26(19):2438–44.
37. Liu H, D'Andrade P, Fulmer-Smentek S, Lorenzi P, Kohn KW, Weinstein JN, Pommier Y, Reinhold WC. MRNA and microRNA expression profiles of the NCI-60 integrated with drug activities. Mol Cancer Ther. 2010;9(5):1080–91.
38. Shoemaker RH. The NCI60 human tumour cell line anticancer drug screen. Nat Rev Cancer. 2006;6(10):813–23.
39. Hoek KS, Schlegel NC, Brafford P, Sucker A, Ugurel S, Kumar R, Weber BL, Nathanson KL, Phillips DJ, Herlyn M, Schadendorf D, Dummer R. Metastatic potential of melanomas defined by specific gene expression profiles with no BRAF signature. Pigment Cell Res. 2006;19(4):290–302.
40. The Cancer Genome Atlas Research Network. Genomic and epigenomic landscapes of adult de novo acute myeloid leukemia. N Engl J Med. 2013; 368(22):2059–74.
41. Huang JC, Babak T, Corson TW, Chua G, Khan S, Gallie BL, Hughes TR, Blencowe BJ, Frey BJ, Morris QD. Using expression profiling data to identify human microRNA targets. Nat Methods. 2007;4(12):1045–9.
42. Hsu SD, Lin FM, Wu WY, Liang C, Huang WC, Chan WL, Tsai WT, Chen GZ, Lee CJ, Chiu CM, Chien CH, Wu MC, Huang CY, Tsou AP, Huang HD. MiRTarBase: a database curates experimentally validated microRNA-target interactions. Nucleic Acids Res. 2011;39:D163–9.
43. van Kouwenhove M, Kedde M, Agami R. MicroRNA regulation by RNA-binding proteins and its implications for cancer. Nat Rev Cancer. 2011;11(9):644–56.
44. Wang S, Sun H, Ma J, Zang C, Wang C, Wang J, Tang Q, Meyer CA, Zhang Y, Liu XS. Target analysis by integration of transcriptome and ChIP-seq data with BETA. Nat Protoc. 2013;8(12):2502–15.
45. Ray D, Kazan H, Cook KB, Weirauch MT, Najafabad HS, Gueroussov S, Albu M, Zheng H, Yang A, Na H, Irimia M, Matzat LH, Dale RK, Smith SA, Yarosh C, Kelly SM, Nabet B, Mecenas D, Li W, Laishram RS, Qiao M, Lipshitz HD, Piano F, Corbett AH, Carstens RP, Frey BJ, Anderson RA, Lynch KW, Penalva LO, Lei EP, Fraser AG, Blencowe BJ, Morris QD, Hughes TR. A compendium of RNA-binding motifs for decoding gene regulation. Nature. 2013;499(7457):172–7.
46. National Cancer Institute Wiki. Cancer gene index End user documentation. 2014. Available: https://wiki.nci.nih.gov/x/hC5yAQ. [Accessed 14 Jul 2016].
47. Sadelain M, Papapetrou EP, Bushman FD. Safe harbours for the integration of new DNA in the human genome. Nat Rev Cancer. 2012;12(1):51–8.
48. Vogelstein B, Papadopoulos N, Velculescu VE, Zhou S, Diaz Jr LA, Kinzler KW. Cancer genome landscapes. Science. 2013;339(6127):1546–58.
49. Futreal PA, Coin L, Marshall M, Down T, Hubbard T, Wooster R, Rahman N, Stratton MR. A census of human cancer genes. Nat Rev Cancer. 2004;4(3):177–83.
50. Abbott KL, Nyre ET, Abrahante J, Ho YY, Vogel RI, Starr TK. The candidate cancer gene database: a database of cancer driver genes from forward genetic screens in mice. Nucleic Acids Res. 2015;43:D844–8.
51. Su A, Wiltshire T, Batalov S, Lapp H, Ching KA, Block D, Zhang J, Soden R, Hayakawa M, Kreiman G, Cooke MP, Walker JR, Hogenesch JB. A gene atlas of the mouse and human protein encoding transcriptomes. Proc Natl Acad Sci. 2004;101(16):6062–7.

52. Ross DT, Scherf U, Eisen MB, Perou CM, Rees C, Spellman P, Iyer V, Jeffrey SS, Van de Rijn M, Waltham M, Pergamenschikov A, Lee JC, Lashkari D, Shalon D, Myers TG, Weinstein JN, Botstein D, Brown PO. Systematic variation in gene expression patterns in human cancer cell lines. Nat Genet. 2000;24(3): 227–35.
53. George EI, Mcculloch RE. Approaches for bayesian variable selection. Stat Sin. 1997;7:339–73.
54. Markowetz F, Kostka D, Troyanskaya OG, Spang R. Nested effects models for high-dimensional phenotyping screens. Bioinformatics. 2007;23(13):i305–12.
55. Vejnar CE, Zdobnov EM. MiRmap: comprehensive prediction of microRNA target repression strength. Nucleic Acids Res. 2012;40(22):11673–83.
56. Roider HG, Kanhere A, Manke T, Vingron M. Predicting transcription factor affinities to DNA from a biophysical model. Bioinformatics. 2007;23(2): 134–41.
57. Zacher B, Abnaof K, Gade S, Younesi E, Tresch A, Fröhlich H. Joint bayesian inference of condition-specific miRNA and transcription factor activities from combined gene and microRNA expression data. Bioinformatics. 2012; 28(13):1714–20.
58. Geeven G, van Kesteren RE, Smit AB, de Gunst MC. Identification of context-specific gene regulatory networks with GEMULA-gene expression modeling using LAsso. Bioinformatics. 2012;28(2):214–21.
59. Huynh-Thu VA, Irrthum A, Wehenkel L, Geurts P. Inferring regulatory networks from expression data using tree-based methods. PLoS One. 2010; 5(9):e12776.
60. Opgen-Rhein R, Strimmer K. From correlation to causation networks: a simple approximate learning algorithm and its application to high-dimensional plant gene expression data. BMC Syst Biol. 2007;1:37.
61. Cover T, Thomas J. Elements of Information Theory. New York: Wiley; 1991.
62. Mendes P, Sha W, Ye K. Artificial gene networks for objective comparison of analysis algorithms. Bioinformatics. 2003;19 suppl 2:ii122–9.
63. Klein U, Tu Y, Stolovitzky GA, Mattioli M, Cattoretti G, Husson H, Freedman A, Inghirami G, Cro L, Baldini L, Neri A, Califano A, Dalla-Favera R. Gene expression profiling of B cell chronic lymphocytic leukemia reveals a homogeneous phenotype related to memory B cells. J Exp Med. 2001; 194(11):1625–38.
64. Weinstein JN, Collisson EA, Mills GB, Shaw KRM, Ozenberger BA, Ellrott K, Shmulevich I, Sander C, Stuart JM. The cancer genome atlas Pan-cancer analysis project. Nat Genet. 2013;45(10):1113–20.
65. Hernández-Lobato D, Hernández-Lobato JM, Suárez A. Expectation propagation for microarray data classification. Pattern Recognit Lett. 2010;31(12):1618–26.
66. Yates A, Akanni W, Amode MR, Barrell D, Billis K, Carvalho-Silva D, Cummins C, Clapham P, Fitzgerald S, Gil L, Girón CG, Gordon L, Hourlier T, Hunt SE, Janacek SH, Johnson N, Juettemann T, Keenan S, Lavidas I, Martin FJ, Maurel T, McLaren W, Murphy DN, Nag R, Nuhn M, Parker A, Patricio M, Pignatelli M, Rahtz M, Riat HS, Sheppard D, Taylor K, Thormann A, Vullo A, Wilder SP, Zadissa A, Birney E, Harrow J, Muffato M, Perry E, Ruffier M, Spudich G, Trevanion SJ, Cunningham F, Aken BL, Zerbino DR, Flicek P. Ensembl 2016. Nucleic Acids Res. 2016;44(D1):D710–6.
67. Durinck S, Spellman PT, Birney E, Huber W. Mapping identifiers for the integration of genomic datasets with the R/Bioconductor package biomaRt. Nat Protoc. 2009;4(8):1184–91.
68. Brandt J, Bux M, Leser U. Cuneiform: a functional language for large scale scientific data analysis. Proc Work EDBT/ICDT. 2015;1330:17–26.
69. Bux M, Brandt J, Lipka C, Hakimazadeh K, Dowling J, Leser U. SAASFEE: scalable scientific workflow execution engine. Very Large Data Bases. 2015; 8(12):1892–5.
70. Naumov VA, Generozov EV, Zaharjevskaya NB, Matushkina DS, Larin AK, Chernyshov SV, Alekseev MV, Shelygin YA, Govorun VM. Genome-scale analysis of DNA methylation in colorectal cancer using infinium human methylation 450 bead chips. Epigenetics. 2013;8(9):921–34.
71. Liu H, Pan Z, Li A, Fu S, Lei Y, Sun H, Wu M, Zhou W. Roles of chemokine receptor 4 (CXCR4) and chemokine ligand 12 (CXCL12) in metastasis of hepatocellular carcinoma cells. Cell Mol Immunol. 2008;5(5):373–8.
72. Rubie C, Frick VO, Wagner M, Weber C, Kruse B, Kempf K, König J, Rau B, Schilling M. Chemokine expression in hepatocellular carcinoma versus colorectal liver metastases. World J Gastroenterol. 2006;12(41):6627–33.
73. Couch FJ, Johnson MR, Rabe KG, Brune K, de Andrade M, Goggins M, Rothenmund H, Gallinger S, Klein A, Petersen GM, Hruban RH. The prevalence of BRCA2 mutations in familial pancreatic cancer. Cancer Epidemiol Biomarkers Prev. 2007;16(2):342–6.
74. Greer JB, Whitcomb DC. Role of BRCA1 and BRCA2 mutations in pancreatic cancer. Gut. 2007;56(5):601–5.
75. Je DW, O YM, Ji YG, Cho Y, Lee DH. The inhibition of SRC family kinase suppresses pancreatic cancer cell proliferation, migration, and invasion. Pancreas. 2014;43(5):768–76.
76. Yang X, Zu X, Tang J, Xiong W, Zhang Y, Liu F, Jiang Y. Zbtb7 suppresses the expression of CDK2 and E2F4 in liver cancer cells: implications for the role of Zbtb7 in cell cycle regulation. Mol Med Rep. 2012;5(6):1475–80.
77. Bisteau X, Caldez MJ, Kaldis P. The complex relationship between liver cancer and the cell cycle: a story of multiple regulations. Cancers. 2014;6(1):79–111.
78. Stiewe T, Tuve S, Peter M, Tannapfel A, Elmaagacli AH, Pützer BM. Quantitative TP73 transcript analysis in hepatocellular carcinomas. Clin Cancer Res. 2004;10(2):626–33.
79. Hsu LS, Lee HC, Chau GY, Yin PH, Chi CW, Lui WY. Aberrant methylation of EDNRB and p16 genes in hepatocellular carcinoma (HCC) in Taiwan. Oncol Rep. 2006;15(2):507–11.
80. Vaquerizas JM, Kummerfeld SK, Teichmann SA, Luscombe NM. A census of human transcription factors: function, expression and evolution. Nat Rev Genet. 2009;10(4):252–63.
81. Pickrell JK, Gaffney DJ, Gilad Y, Pritchard JK. False positive peaks in ChIP-seq and other sequencing-based functional assays caused by unannotated high copy number regions. Bioinformatics. 2011;27(15):2144–6.
82. Rajewsky N. microRNA target predictions in animals. Nat Genet. 2006; 38(Suppl):S8–13.
83. Luscombe NM, Babu MM, Yu H, Snyder M, Teichmann SA, Gerstein M. Genomic analysis of regulatory network dynamics reveals large topological changes. Nature. 2004;431:308–12.
84. Marbach D, Costello JC, Küffner R, Vega NM, Prill RJ, Camacho DM, Allison KR, Consortium TD, Kellis M, Collins JJ, Stolovitzky G. Wisdom of crowds for robust gene network inference. Nat Methods. 2012;9:796–804.
85. Davidsen PK, Turan N, Egginton S, Falciani F. Multi-level functional genomics data integration as a tool for understanding physiology: a network perspective. J Appl Physiol. 2016;120(3):297–309.

8

A new efficient approach to fit stochastic models on the basis of high-throughput experimental data using a model of IRF7 gene expression as case study

Luis U. Aguilera[1], Christoph Zimmer[2] and Ursula Kummer[1*]

Abstract

Background: Mathematical models are used to gain an integrative understanding of biochemical processes and networks. Commonly the models are based on deterministic ordinary differential equations. When molecular counts are low, stochastic formalisms like Monte Carlo simulations are more appropriate and well established. However, compared to the wealth of computational methods used to fit and analyze deterministic models, there is only little available to quantify the exactness of the fit of stochastic models compared to experimental data or to analyze different aspects of the modeling results.

Results: Here, we developed a method to fit stochastic simulations to experimental high-throughput data, meaning data that exhibits distributions. The method uses a comparison of the probability density functions that are computed based on Monte Carlo simulations and the experimental data. Multiple parameter values are iteratively evaluated using optimization routines. The method improves its performance by selecting parameters values after comparing the similitude between the deterministic stability of the system and the modes in the experimental data distribution. As a case study we fitted a model of the IRF7 gene expression circuit to time-course experimental data obtained by flow cytometry. IRF7 shows bimodal dynamics upon IFN stimulation. This dynamics occurs due to the switching between active and basal states of the IRF7 promoter. However, the exact molecular mechanisms responsible for the bimodality of IRF7 is not fully understood.

Conclusions: Our results allow us to conclude that the activation of the IRF7 promoter by the combination of IRF7 and ISGF3 is sufficient to explain the observed bimodal dynamics.

Keywords: IRF7, IFN, Stochastic models, Parameter estimation

Background

Computer models contribute to the integrative understanding of complex molecular processes in the cell. The most commonly used approach is deterministic modeling based on ordinary differential equations (ODEs). When the studied system comprises species with a low molecular count, stochastic formalisms, e.g. Gillespie's algorithm that simulates trajectories and uses discrete molecule numbers are more appropriate [1]. In recent years, the use of stochastic models has substantially increased [2].

Considering stochasticity in biological systems has changed the quantitative and qualitative understanding obtained by previous deterministic models [3]. Examples of biological phenomena discovered by stochastic modeling include gene expression in burst-like patterns [4], productive or latent cell decision after HIV-infection [5], and the presence of oscillatory behavior induced by noise [6]. However, in contrast to the plethora of methods used to fit and analyze deterministic models [7–10], there are only very limited sets of methods available to do the same with stochastic models.

*Correspondence: ursula.kummer@bioquant.uni-heidelberg.de
[1]Department of Modeling of Biological Processes, COS Heidelberg / Bioquant, Heidelberg University, Im Neuenheimer Feld 267, 69120 Heidelberg, Germany
Full list of author information is available at the end of the article

Thus, only recently, methods for parameter fitting of stochastic models have been developed and are still under development. Parameter estimation methods for stochastic models have been so far designed for single time course data [11–16]. For experiments leading to data distributions, they include the moment closure [17–19], and the comparison of experimental and simulation distributions [20–22]. Despite the current efforts, there are still major problems that limit the full applicability of those methods. The main drawback is the high computational execution time. To solve this problem novel efficient strategies to fit stochastic models are needed.

In the following we introduce a new approach that is a variant of existing methods based on the comparison of distributions. Differently to the existing methods we use the experimental data to define a condition that the model must fulfill in the deterministic regime. This condition tests, if the deterministic steady states are in the close proximity to the modes of the experimental distribution. Only if the model fulfills this condition, the parameter set is evaluated in the stochastic regime. In this way, the algorithm directs the evaluation of parameter sets towards regions in parameter space with high potential to reproduce the experimental data. Additionally, this method applies to non-linear models and complex experimental data distributions.

To test the potential of this new method, we selected an open scientific question and real experimental data. We investigated the molecular mechanisms responsible for the experimentally observed bimodal dynamics of IRF7 expression after Interferon (IFN) stimulation.

It is well documented that in isogenic cell populations not all cells produce an antiviral response when infected by a virus [23]. The mechanisms behind this heterogeneity have been related to stochastic events in the signaling pathways responsible to elicit the antiviral response [24–26]. IFN is a cytokine that activates the JAK-STAT signaling pathway in virus-infected cells. The JAK-STAT signaling pathway induces the translocation of ISGF3 (a transcription factor) into the nucleus to directly activate the transcription of a set of several hundred IFN-stimulated genes (ISGs) [27]. IRF7 is an ISG with a central role in the immune response [28, 29]. Recently, it was observed that after IFN stimulation murine fibroblasts show a switch-like pattern of IRF7 expression, which is reflected at the population level as bimodality [30].

We developed a model to describe the observed bimodal dynamics of IRF7 expression after IFN stimulation. We hypothesize that the binding of IRF7 and ISGF3 to the IRF7 promoter is the mechanism responsible for this bimodal behavior. Our simulation results reproduced IRF7 switch-like dynamics at single cell level, and bimodality was achieved at the population level. Hereby, we used the newly developed method to fit the model and correctly reproduce the experimental data. Our results allow us to conclude that the IRF7 promoter activation by the combination of IRF7 dimers and ISGF3 is sufficient to quantitatively explain the observed IRF7 bimodal dynamics.

Methods

Experimental data

Published experimental data were produced by Rand et al. [30]. The data described the expression of IRF7 after IFN-β stimulation in a population of murine NIH3T3 fibroblasts. Rand's experiments were done in the following way: First, cells were transfected with a BAC (Bacterial Artificial Chromosome) containing IRF7 and reporter mCherry genes fused, subsequently cultures were treated with different concentrations of murine IFN-β. For illustrative purposes we selected the treatment with 150U of IFN-β, a concentration where bimodality is prevalent. Then, fluorescence in the cultures was monitored using flow cytometry at different time points during 48 hours.

Numerical methods

In the deterministic regime the model was simulated using symbolic methods in Matlab [31] and/or using the LSODA algorithm in COPASI 4.11 [32]. Stability was calculated by making the right-hand side of the differential equations equal to zero and subsequently by finding the roots of the system using function *solve* in Matlab. Eigenvalues were calculated using function *eig* in Matlab. For solving the model under stochastic dynamics we used the Gibson and Bruck algorithm [33] coded in COPASI. The random search and genetic algorithm were coded in Matlab. Raw experimental data was analyzed using the function *FCS data reader* coded in Matlab [34]. The modes in the distributions were calculated using the function *PeakFinder* coded in Matlab [35]. The source code of the project can be accessed via: https://sourceforge.net/projects/irf7-bimodaldynamics/.

Results

New algorithm to fit stochastic biological models

Comparing experimental and simulation distributions

The measurements of fluorescence were made comparable with the corresponding observable chemical species S^o in the model by a function h that maps state $S^o(t_i)$ to observation $S^\dagger(t_i)$ as follows:

$$S^\dagger(t_i) = h(S^o(t_i)), \tag{1}$$

at each measurement time point $t_i, i = 1, \ldots, n$.

Subsequently, a process to compare distributions was developed based on [20]. First, considering a specific set of parameter values $\theta = \{\theta_1, \ldots, \theta_d\}$, we performed ns repetitions of the stochastic simulations $\boldsymbol{s}(t_i) =$

$\{s_1(t_i), \ldots, s_{ns}(t_i)\}$. The total of those stochastic simulations were used to build histograms with a fixed number of bins L for each t_i. Subsequently, the (discrete) probability density function (PDF) of the simulations $P_s(\boldsymbol{s}(t_i), \theta, b_l)$ were computed using the center of each bin b_l, for $l = 1, \ldots, L$, and the normalized form of the histograms. On the other hand, having nm repetitions of single cell experimental data $\boldsymbol{m}(t_i) = \{m_1(t_i), \ldots, m_{nm}(t_i)\}$ histograms and PDFs $P_e(\boldsymbol{m}(t_i), b_l)$ were built. Even though, complex formulas can be use to calculate the number of bins in an histogram, here L was kept constant for P_s and P_e and was calculated as $L = \sqrt{nm}$, this is a simple approach used by default by most programming languages (i.e. Matlab). Finally, an approach using the difference of squares was used as a metric to calculate the distance between $P_s(\boldsymbol{s}(t_i), \theta, b_l)$ and $P_e(\boldsymbol{m}(t_i), b_l)$, as follows:

$$F(\theta, \boldsymbol{m}) = \sum_{i=1}^{n} \sum_{l=1}^{L} (P_e(\boldsymbol{m}(t_i), b_l) - P_s(\boldsymbol{s}(t_i), \theta, b_l))^2. \quad (2)$$

Algorithm implementation

The objective function $F(\theta, \boldsymbol{m})$ can be used to calculate a parameter estimate

$$\hat{\theta} = \text{argmin}_{\theta} F(\theta, \boldsymbol{m}). \quad (3)$$

As optimization usually requires plenty of function evaluations and each evaluation of F requires ns stochastic simulations, a direct optimization strategy is computationally unfeasible even for simple models. Therefore, we introduce a new strategy that selects good candidate parameters and only performs the stochastic simulations for these parameters.

In large systems where fluctuations can be discarded, the stochastic system can be reduced to the deterministic one [36, 37]. For this reason, in most cases the deterministic dynamics can be associated with a measure of central tendency in the PDFs obtained after the stochastic simulations. Using this reasoning we will introduce the strategy to efficiently estimate parameters for stochastic models making use of deterministic dynamics as an initial indicator.

Assuming that the experimental data is in equilibrium at the last measurement point t_n, we used the modes from $P_e(\boldsymbol{m}(t_n), b_l)$ as a central measure of tendency, denoting the modes by $\boldsymbol{\alpha}(t_n) = (\alpha_1(t_n), \ldots, \alpha_q(t_n))$, with q being the total number of modes.

Then, using the ODE version of the model ($\dot{\boldsymbol{X}}(\theta, t)$) we determine its stability. Here we tested whether the system has stable steady states and denote them with $\boldsymbol{X}^*(\theta) = (X_1^*(\theta), \ldots, X_{ss}^*(\theta))$, being ss the number of stable steady states. If the system has no stable steady state, it holds $ss = 0$ and the vector $\boldsymbol{X}^*(\theta)$ is empty. The calculation of steady states can either be performed by solving $\dot{\boldsymbol{Y}}(\theta) = 0$ where $\dot{\boldsymbol{Y}}$ is the right hand side of the ODE system with $\boldsymbol{X}(\theta, t)$ replaced by $\boldsymbol{Y}(\theta)$. Solving this equation can be impossible for large systems. Alternatively, one can numerically solve the ODE systems until the flux is zero for all components. Varying the initial conditions leads to the different steady states. The stability of the system was calculated after determining the sign of the eigenvalues ($\lambda < 0$), for stable steady states [38].

Having $\boldsymbol{X}^*(\theta)$ and $\boldsymbol{\alpha}(t_n)$ we introduce the *deterministic precondition*:

$$\beta_k^{low} X_k^*(\theta) \le \alpha_k(t_n) \le \beta_k^{up} X_k^*(\theta), \quad \text{for } 1 \le k \le q, \text{ and } 0 \le \beta_k^{low} \le 1 \text{ and } 1 \le \beta_k^{up}. \quad (4)$$

Equation (4) means that the number of modes equals the number of stable steady states: $ss = q$, and that each of the stable steady states is close to its corresponding mode. Only, if this is fulfilled, we calculate stochastic simulations and evaluate the objective function $F(\theta, \boldsymbol{m})$. If the precondition is not fulfilled, we do not carry out stochastic simulations. In this case, we assign a high (bad) objective function value to direct the parameter search towards parameters that pass the deterministic precondition. A graphical representation of the deterministic precondition is given in Fig. 1.

At this point, it is important to notice that the deterministic precondition is not stating that the stable steady states form the ODE system are equal to the mean obtained by solving the stochastic system. Rather, the deterministic precondition tests if the deterministic steady state lies around a certain range.

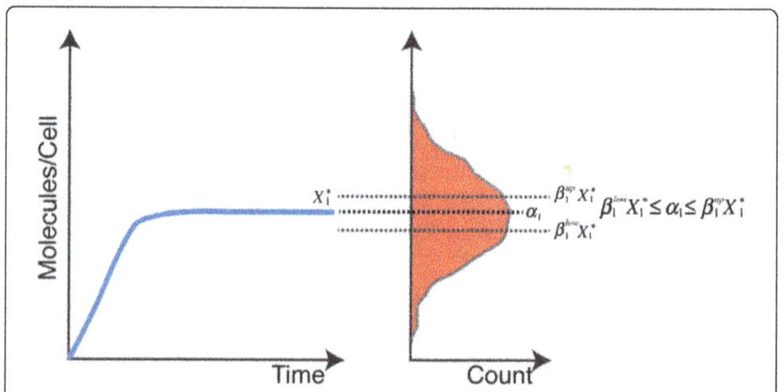

Fig. 1 The deterministic precondition. Left side the deterministic dynamics of the system can be seen, here the observable variable evolves to a single stable steady state $X_1^*(\theta)$. This steady state can be calculated by finding the point in the system where the net flux is equal to zero $\dot{\mathbf{X}}(\theta) = 0$. At the right hand side of the plot, the PDF of the experimental data is shown. This PDF is rotated and the x-axis shows the count and the y-axis shows the bin values with units of Molecules/Cell. From both graphs it can be seen that the deterministic precondition is valid when $X_1^*(\theta)$ is inside the range defined by $\beta_1^{low} X_1^*(\theta)$ and $\beta_1^{up} X_1^*(\theta)$

$$F_{\text{cond}}(\theta, \boldsymbol{m}) = \begin{cases} F(\theta, \boldsymbol{m}), \ \beta_k^{low} X_k^*(\theta) \leq \alpha_k(t_n) \leq \beta_k^{up} X_k^*(\theta) - \text{"Condition passed"} \\ \infty, \qquad \text{else} - \text{"Condition not passed"} \end{cases} \tag{5}$$

Our extended objective function with the deterministic check reads as:
and the estimate is defined as:

$$\hat{\theta}_{\text{cond}} = \text{argmin}_{\theta} F_{\text{cond}}(\theta, m). \tag{6}$$

As previously discussed, our method is based on the central assumption that the modes in the experimental data are related to the stable steady states in the deterministic mathematical model. If this assumption should fail, either of the following could happen: a) The method accepts a parameter that passed the deterministic precondition but does not show a good agreement of the distributions. If it does not show good agreement, the objective function $F(\theta, \boldsymbol{m})$ will show a high value which means that this parameter does not lead to a good fit. Therefore, this case leads to a loss of computational time, but not to wrong results and it overall is still less expensive than evaluating every parameter set. b) We reject a parameter that fails the deterministic precondition but would have lead to a good fit. In this case, we lose indeed a good parameter. This shows that our method is obviously a heuristic strategy that does not guarantee to find the exact global minimum, but this is true for almost any optimization strategy. The algorithmic steps are depicted in Fig. 2.

Parameter estimation methods

The deterministic precondition was first implemented using a random search algorithm. Random search is a global optimization strategy that tests random combinations of parameter values. The successful output in this method is dependent on the total number of evaluated parameter sets [39]. The more evaluated parameter values, the higher the probability to find the global minimum. The pseudo-code for the random search is given in Algorithm 1.

The second implementation of the deterministic precondition was using the more directed Genetic Algorithm. Genetic Algorithm mimics evolution and is based on reproduction and selection. This algorithm is made of a population of individuals (parameter sets), and each contains a genome that is defined by the number of parameters to optimize. The individuals are ranked after solving the objective function, and a population of parental individuals is selected according to an elitism rate (ϵ). New individuals (offspring) are generated by pairing and recombining the parental genomes (cross-over). Variability is introduced in the population by adding mutations in the new individuals according to a given mutation rate (μ). By the continuous process of selecting the best parameters after each generation, the algorithm evolves towards the regions in the parameter space that reduce the

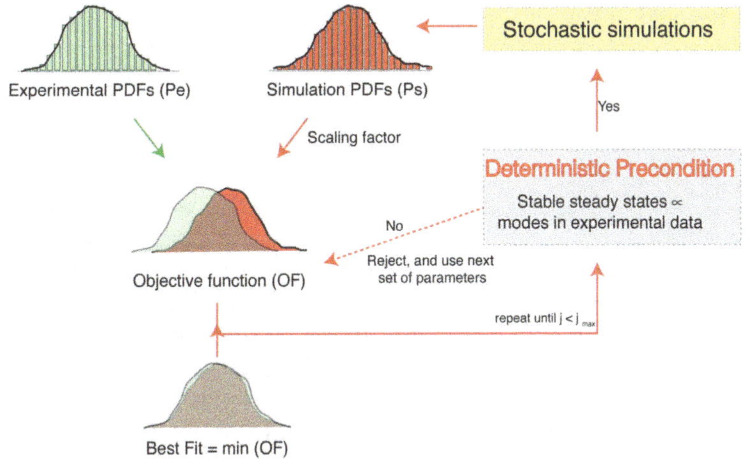

Fig. 2 Algorithm to fit stochastic models to experimental data. The algorithm solves the model in the deterministic and stochastic regimes. A condition observed in the experimental data is defined. A set of parameter values is evaluated in the deterministic regime to test if the model reproduces this condition. If the condition is met the stochastic simulation is performed. Otherwise, the parameter values are rejected. The PDF obtained from the experimental data is compared with the PDF obtained after running the stochastic simulations. This comparison is made using a difference of squares as an objective function. This process is repeated until evaluating a total number of parameter sets or after a termination criterion is met. The parameter set that best reproduces the experimental data is given by the minimum value obtained after the iterative evaluation of the objective function

Algorithm 1: Random search with the deterministic precondition

Data: High-throughput data PDFs. Biochemical model.
Define: Ranges for the parameter values.
Number of parameters sets to test is *N*.
Result: Parameter values that best reproduce the experimental data.

```
Generate matrix of random parameter values.
for j = 1 : N do
    Assign the j^th parameter set in the model ;
    Run deterministic simulations ;
    Test deterministic precondition ;
    if deterministic precondition is true then
        Run stochastic simulations ;
        Objective Function (OF) = F(θ, m) ;
    else
        Reject the j^th parameter set;
        Set OF = ∞ ;
    end
end
```

values in the objective function. The pseudo-code for the Genetic Algorithm is given in Additional file 1.

Identifying parameters for a constitutive gene expression circuit with in-silico data

To prove the functionality and benefits of the new proposed algorithm, we first applied it to a simple example where the parameters are known a priori. The model describes the stochastic dynamics of two variables, protein and mRNA of a gene with constitutive expression [40], a graphical representation of the constitutive gene expression circuit is given in Fig. 3-a. The model is described by the following reactions:

$$\varnothing \xrightarrow{\theta_1} mRNA, \tag{7}$$

$$mRNA \xrightarrow{\theta_2} \varnothing, \tag{8}$$

$$mRNA \xrightarrow{\theta_3} Protein, \tag{9}$$

$$Protein \xrightarrow{\theta_4} \varnothing. \tag{10}$$

To estimate the parameter values for this model we carried out the following procedure: First, we generated the in-silico data by selecting the following true parameter values $\theta^o = (5, 0.03, 0.1, 0.03)$ and running 10000 independent stochastic simulations from which PDFs were built. Four observable time points were defined at 50 min, 100 min, 150 min and 200 min (see Fig. 3b). From the PDFs a mode was obtained at $\alpha_1(t_n) = [543]$ a.u.(arbitrary units), where $t_n = 200$ min. Then, we defined the deterministic precondition using Eq. (4), and assigning minimum and maximum acceptance ranges by setting $\beta_1^{low} = 0.95$ and $\beta_1^{up} = 1.05$, respectively. A deeper analysis of the stability of the constitutive gene expression circuit is given in Additional file 2. To build simulated PDFs we used 1000 repetitions of the stochastic model, this number was empirically calculated according to Additional file 2. For illustrative purposes, we assumed that the values for the parameters responsible for the mRNA transcription and protein translation (θ_1, θ_3) were unknown, and the new algorithm was used to estimate those parameters. The deterministic precondition was applied using the RS strategy obtaining that the algorithm only evaluates stochastic dynamics in 3.1% of the tested parameter values, reducing in this way the total simulation time. Additionally, the complete algorithm was repeated 1000 times and histograms of the estimated parameter were computed to determine whether they are close to $\theta^{(0)}$. As can be observed in Fig. 3-c the deterministic precondition reduces the evaluation of different parameters under stochastic dynamics by selecting only those parameters that are in a well-defined region in the proximity of $\theta^{(0)}$. For this model, the main benefits of using the deterministic precondition was the reduction of the number of parameters evaluated under stochastic dynamics. This rejection of parameters was made in an area outside the true parameter values, and hence no difference in accuracy is expected in comparison with a method without using the deterministic precondition. A complete description of the analysis of the performance, accuracy and error for this example is given in Additional file 2. The model for the constitutive gene expression circuit can be obtained from BioModels database under reference MODEL1608100000.

Mathematical model for IRF7 expression dynamics

Subsequently we applied our algorithm to a real problem with flow cytometry data. Here we studied the dynamics of murine IRF7 gene expression upon IFN stimulation. For this reason, we developed a model that comprises known key components and feedback mechanisms. The overall system describes the active IRF7 promoter (*Pa*) by the binding of IRF7 dimer and ISGF3 to the DNA binding sites ISRE and IRFE, respectively [41], the transcription and translation of IRF7, and its subsequent phosphorylation and dimerization. IRF7 protein binding to the IRFE binding site in the promoter results in the production of more IRF7 protein, constituting a positive feedback loop [42]. A graphical representation of the IRF7 gene expression dynamics is given in Fig. 4.

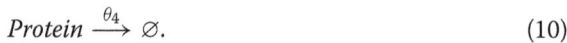

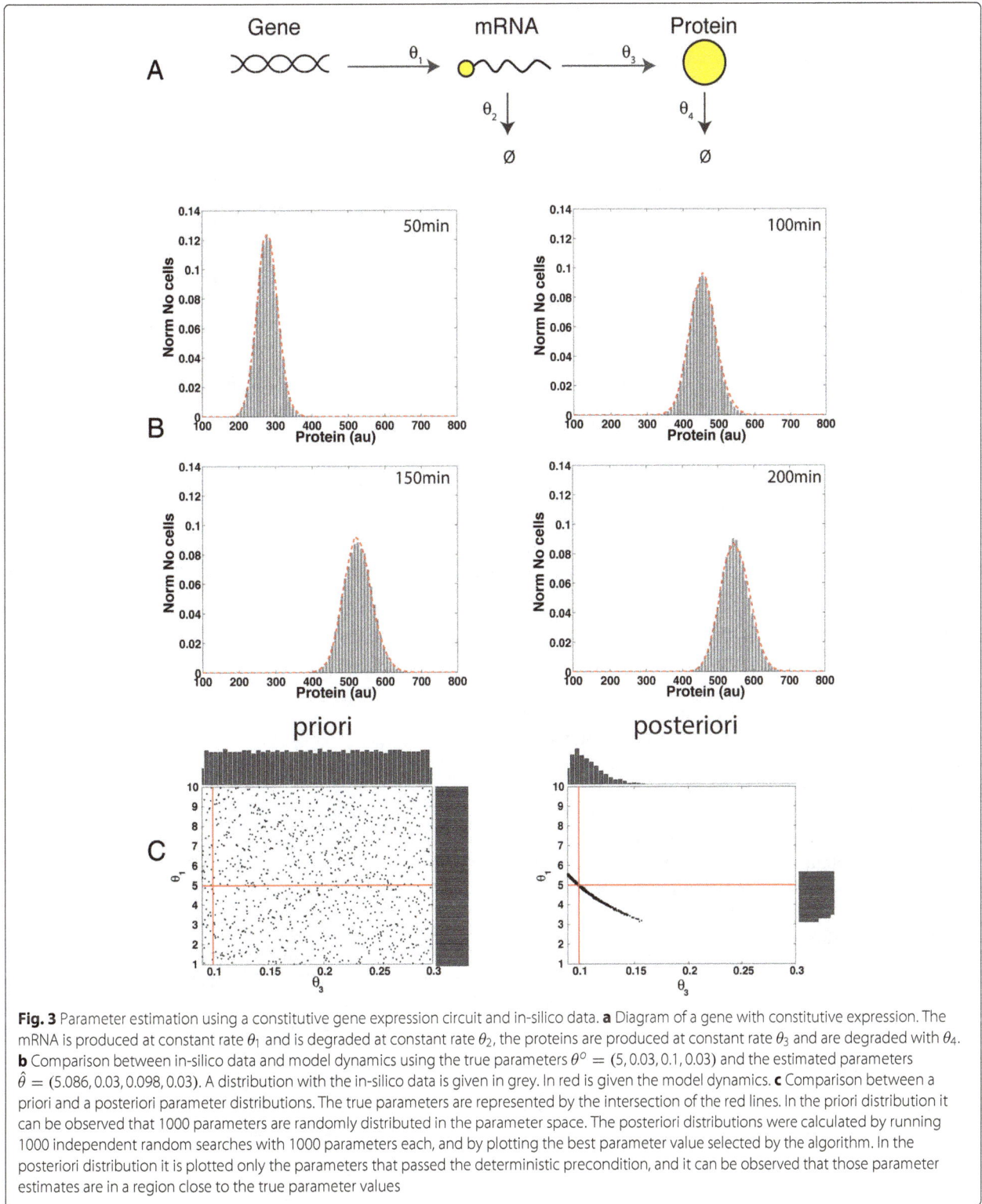

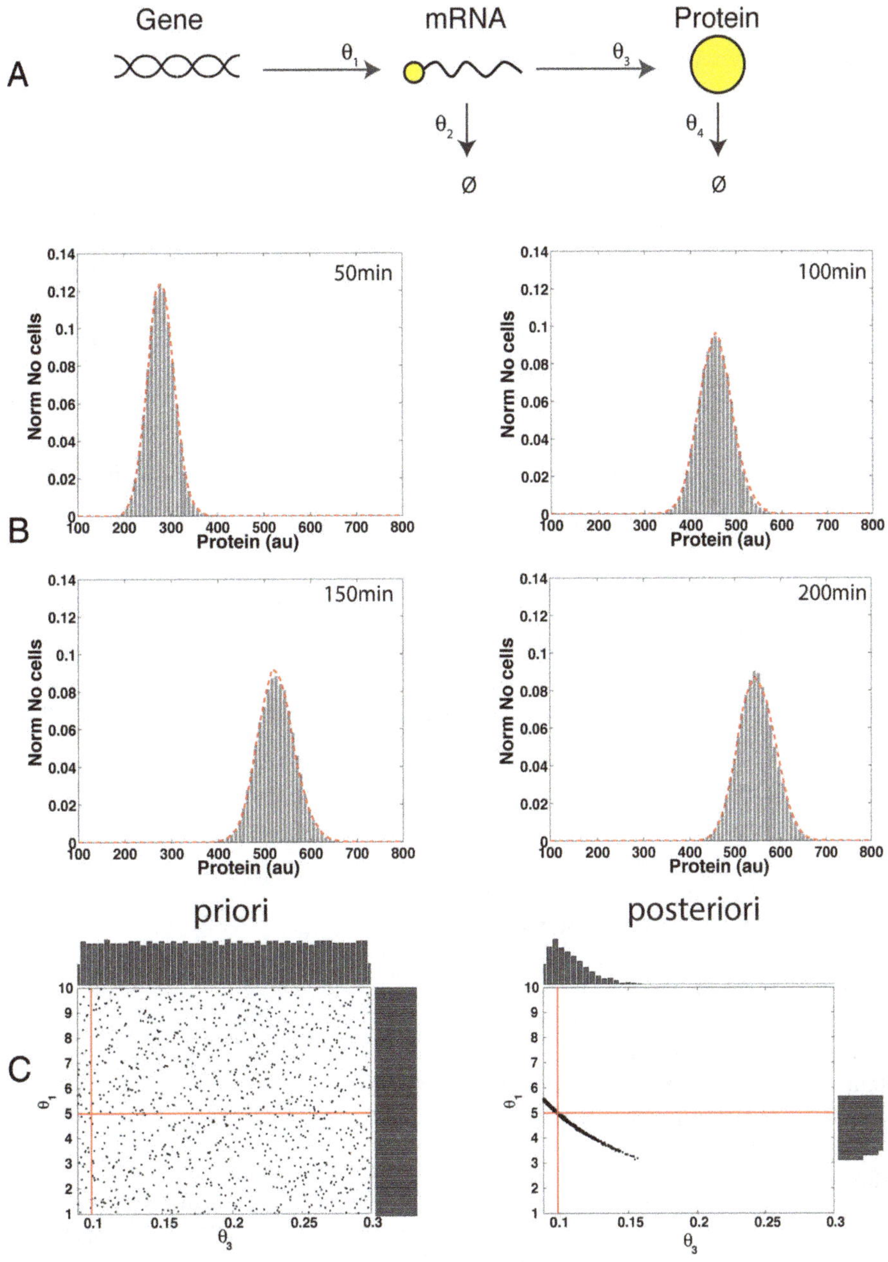

Fig. 3 Parameter estimation using a constitutive gene expression circuit and in-silico data. **a** Diagram of a gene with constitutive expression. The mRNA is produced at constant rate θ_1 and is degraded at constant rate θ_2, the proteins are produced at constant rate θ_3 and are degraded with θ_4. **b** Comparison between in-silico data and model dynamics using the true parameters $\theta^O = (5, 0.03, 0.1, 0.03)$ and the estimated parameters $\hat{\theta} = (5.086, 0.03, 0.098, 0.03)$. A distribution with the in-silico data is given in grey. In red is given the model dynamics. **c** Comparison between a priori and a posteriori parameter distributions. The true parameters are represented by the intersection of the red lines. In the priori distribution it can be observed that 1000 parameters are randomly distributed in the parameter space. The posteriori distributions were calculated by running 1000 independent random searches with 1000 parameters each, and by plotting the best parameter value selected by the algorithm. In the posteriori distribution it is plotted only the parameters that passed the deterministic precondition, and it can be observed that those parameter estimates are in a region close to the true parameter values

Chemical reactions

The developed model consists of 7 species and 13 reactions (reactions (11) to (23)). In the model, reaction (11) describes the IFN activation of the JAK-STAT signaling pathway. For the description of this reaction and according to [30] a saturable function was used. Downstream reactions of the pathway were lumped in reaction (12), so that the same variable was used to describe the output of the JAK-STAT signaling pathway, namely ISGF3, in this highly simplified model. Reaction (13) describes the

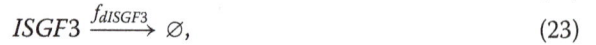

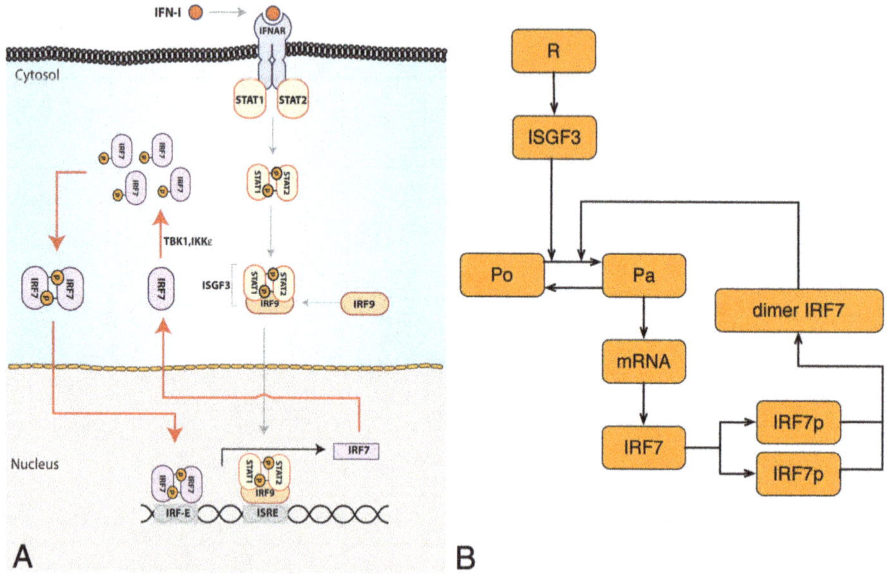

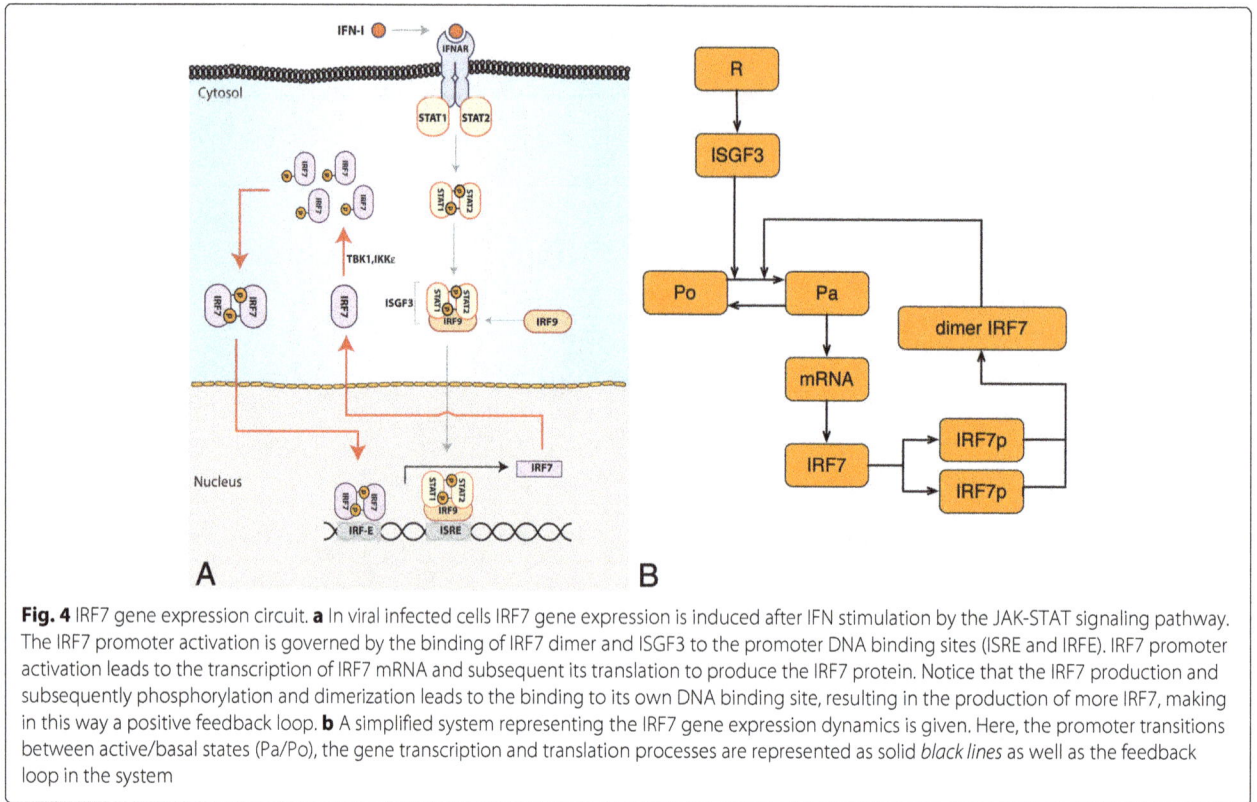

Fig. 4 IRF7 gene expression circuit. **a** In viral infected cells IRF7 gene expression is induced after IFN stimulation by the JAK-STAT signaling pathway. The IRF7 promoter activation is governed by the binding of IRF7 dimer and ISGF3 to the promoter DNA binding sites (ISRE and IRFE). IRF7 promoter activation leads to the transcription of IRF7 mRNA and subsequent its translation to produce the IRF7 protein. Notice that the IRF7 production and subsequently phosphorylation and dimerization leads to the binding to its own DNA binding site, resulting in the production of more IRF7, making in this way a positive feedback loop. **b** A simplified system representing the IRF7 gene expression dynamics is given. Here, the promoter transitions between active/basal states (Pa/Po), the gene transcription and translation processes are represented as solid *black lines* as well as the feedback loop in the system

IRF7 promoter activation by the binding of IRF7 dimer and ISGF3 to ISRE and IRFE, respectively. Subsequently, we incorporated IRF7 mRNA transcription by the active promoter, and to a lesser extent by the basal promoter state, as reactions (14) and (15), respectively. Reaction (16) considers the translation of IRF7 mRNA to produce IRF7 protein. IRF7 protein is phosphorylated in reaction (17). Two phosphorylated IRF7 proteins form a IRF7 dimer in reaction (18). Finally, IRF7 promoter inactivation, degradation of mRNAs, ISGF3, IFN and IRF7 proteins are represented by reactions (19) to (23), respectively:

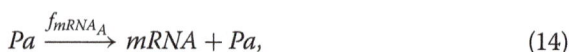

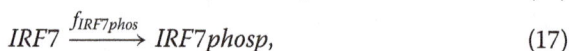

$$IFN \xrightarrow{f_{IFN}} 2IFN, \tag{11}$$

$$IFN \xrightarrow{f_{ISGF3}} ISGF3 + IFN, \tag{12}$$

$$\varnothing \xrightarrow{f_{Pa}} Pa, \tag{13}$$

$$Pa \xrightarrow{f_{mRNA_A}} mRNA + Pa, \tag{14}$$

$$\varnothing \xrightarrow{f_{mRNA_B}} mRNA, \tag{15}$$

$$mRNA \xrightarrow{f_{IRF7}} IRF7 + mRNA, \tag{16}$$

$$IRF7 \xrightarrow{f_{IRF7phos}} IRF7phosp, \tag{17}$$

$$2 \cdot IRF7phosp \xrightarrow{f_{IRF7dimer}} IRF7dimer, \tag{18}$$

$$Pa \xrightarrow{f_{dPa}} \varnothing, \tag{19}$$

$$IFN \xrightarrow{f_{dIFN}} \varnothing, \tag{20}$$

$$mRNA \xrightarrow{f_{dmRNA}} \varnothing, \tag{21}$$

$$IRF7dimer \xrightarrow{f_{dIRF7dimer}} \varnothing, \tag{22}$$

$$ISGF3 \xrightarrow{f_{dISGF3}} \varnothing, \tag{23}$$

The reaction rates are given in Table 1.

Mathematical equations

To evaluate the deterministic precondition we consider the corresponding ODEs (Eq. (24) to (30)):

$$\frac{d\,IFN}{dt} = f_{IFN} - f_{dIFN} \tag{24}$$

$$\frac{d\,ISGF3}{dt} = f_{ISGF3} - f_{dISGF3} \tag{25}$$

$$\frac{d\,P_a}{dt} = f_{P_a} - f_{dP_a} \tag{26}$$

$$\frac{d\,mRNA}{dt} = f_{mRNA_A} + f_{mRNA_B} - f_{dmRNA} \tag{27}$$

$$\frac{d\,IRF7}{dt} = f_{IRF7} - f_{IRF7phosp} \tag{28}$$

$$\frac{d\,IRF7phosp}{dt} = f_{IRF7phosp} - 2 \cdot f_{IRF7dimer} \tag{29}$$

$$\frac{d\,IRF7dimer}{dt} = f_{IRF7dimer} - f_{dIRF7dimer} \tag{30}$$

Table 1 Reaction rates considered in the model

Name	Definition
f_{IFN}	$V_{IFN}\left(\frac{IFN^n}{k_{IFN}^n+IFN^n}\right)$
f_{dIFN}	$k_{dIFN}\cdot IFN$
f_{ISGF3}	$k_{ISGF3}\cdot IFN$
f_{dISGF3}	$k_{dISGF3}\cdot ISGF3$
f_{P_a}	$k_{on}\left(\frac{ISGF3\cdot IRF7dimer}{k_{al3}\cdot k_{al7}+k_{al3}\cdot ISGF3+k_{al7}\cdot IRF7dimer+ISGF3\cdot IRF7dimer}\right)(1-P_a)$
f_{dP_a}	$k_{off}\cdot P_a$
f_{mRNA_A}	$k_{Active}\cdot P_a$
f_{mRNA_B}	$k_{Basal}(1-P_a)$
f_{dmRNA}	$k_{dmRNA}\cdot mRNA$
f_{IRF7}	$k_{IRF7}\cdot mRNA$
$f_{IRF7phosp}$	$k_{dIRF7}\cdot IRF7$
$f_{IRF7dimer}$	$k_{IRF7dimer}\cdot IRF7phosp\cdot IRF7phosp$
$f_{dIRF7dimer}$	$k_{dIRF7dimer}\cdot IRF7dimer$

The model for the IRF7 circuit can be obtained from BioModels database under reference MODEL1608100001.

Fitting the stochastic IRF7 gene expression model to experimental data

All parameters of the above described model for the IRF7 gene circuit were fitted by using experimental data of IRF protein expression. In the model, a global quantity to describe the different forms of the IRF7 protein was defined as follows:

$$IRFtotal(t) = IRF7(t) + IRF7phosp(t) + IRF7dimer(t), \tag{31}$$

and this variable was mapped to $IRF7^{\dagger}(t_i)$ flow cytometry measurements as follows:

$$IRF7^{\dagger}(t_i) = \varphi IRF7total(t_i), \tag{32}$$

where φ is a scaling factor.

Using the experimental data, PDFs were build and the modes in the distributions were determined using the PeakFinder function [35] obtaining two elements of $\alpha(t_n) = [77, 1000]$ a.u. (arbitrary units), where $t_n = 48$ h. φ relates the values of fluorescence with the molecular count described by the mathematical model. Unfortunately, no calibration curve is provided with the data to calculate this parameter. For this reason, multiple values were tested for φ obtaining consistent results, for illustrative purposes we report $\varphi = 1$. We defined the deterministic precondition using Eq. (4), and assigning minimum and maximum acceptance ranges by setting $\beta_k^{low} = 0.95$ and $\beta_k^{up} = 1.05$, for $k = 1, 2$. The allowed ranges for the parameter values are given in Table 2. The deterministic precondition was introduced in two different optimization strategies, random search, and genetic algorithms. In both optimization strategies 1000 realizations of the stochastic simulations were performed if the parameter set fulfilled the deterministic precondition.

Using the random search strategy, we tested 10000 parameter sets from which less than the 1% passed the deterministic precondition and were stochastically evaluated. The reduction of parameter values allowed us to efficiently find a set of parameter values that reproduced the experimental data (the fitting for the random search algorithm is given in Additional file 3). A complete analysis of the performance of the use of the deterministic precondition with the random search algorithm is given in Additional file 4. To test the reproducibility of the estimated parameter values, the method was repeated 100 times obtaining well-defined parameter distributions that show the predominant values, see Additional file 5.

Then, we implemented the deterministic precondition using a genetic algorithm with adaptive population size. Here we implemented an initial population of 3000 individuals for the first generation, and 5 subsequent generations with 20 individuals. Notice, that a large initial population of parameters was needed by the expected high rejection percentage of parameters by the deterministic precondition during the first generation. As parameters for the algorithm we used $\epsilon = 0.4$ and $\mu = 0.2$. Our simulation results showed a constant decrease in the value of the objective function value during the generations, which indicates progress during fitting. By the use of the deterministic precondition 99% of the parameters were rejected in the first generation and in the subsequent generations around 30% of the parameter values were rejected, improving in this way the efficiency of the algorithm (see Fig. 5). The comparison between the experimental data and the model simulation distributions is given in Fig. 6 showing a high degree of agreement. In addition, to check the validity of the methodology further we fit the model to two additional flow cytometry measurements that describe the stimulation of the cell culture with 100U and 250U of IFN. The respective results are given in Additional file 6. Complete analysis of the performance of the use of the deterministic precondition in the genetic algorithm is given in Additional file 4.

IRF7 temporal dynamics

The simple model of IRF7 gene expression described above is sufficient to explain IRF7 bimodality. Using the optimized parameter values given in Table 2 and the initial conditions given in Table 3, the stochastic temporal dynamics of the model were simulated and are given in Fig. 7. In Fig. 7-a it can be seen that the IFN concentration evolves to a steady state. Subsequently, the first affected variable is ISGF3 that equally evolves towards

Table 2 Description of the parameter values

Name	Description	Range	Nominal	Units
V_{IFN}	Maximum activation rate of the IFN pathway	[2.8, 11.2]	6.135	(Molecules/Cell)*min
n	Hill coefficient	-	2	Dimensionless
k_{IFN}	Saturation constant for the IFN pathway	[0.0022, 0.0088]	0.0055	Molecules/Cell
k_{dIFN}	Decay rate of the IFN pathway	[0.0232, 0.0926]	0.0492	min^{-1}
k_{ISGF3}	Constant for ISGF3 production	[0.00012, 0.00048]	0.0003	min^{-1}
k_{dISGF3}	Decay rate of the ISGF3	[0.00068, 0.00272]	0.0017	min^{-1}
k_{on}	Promoter activation	[184.55, 738.2]	522.59	min^{-1}
k_{aI3}	Constant of promoter activation by IFN	[7681.6, 30727]	22687.02	Molecules/Cell
k_{aI7}	Constant of promoter activation by IRF7	[13399, 53597]	35281.99	Molecules/Cell
k_{off}	Promoter inactivation	[0.00044, 0.00176]	0.0013	min^{-1}
k_{Active}	IRF7 transcription rate by active promoter	[0.5402, 2.161]	1.144	min^{-1}
$^{a}k_{Basal}$	IRF7 basal transcription rate	[0.0312, 0.125]	0.0861	min^{-1}
$^{b}k_{dmRNA}$	Decay rate of mRNA	[0.029, 0.116]	0.0715	min^{-1}
$^{c}k_{IRF7}$	Translation rate of IRF7	[14, 56]	43.867	min^{-1}
k_{dIRF7}	Rate of IRF7 phosphorylation	[1.540, 6.160]	3.877	min^{-1}
$k_{IRF7dimer}$	Rate of IRF7 dimerization	[0.235, 0.94]	0.602	(Cell/Molecules)/min
$k_{dIRF7dimer}$	Decay rate of IRF7 dimers	[0.209, 0.836]	0.439	min^{-1}
φ	Scaling factor	-	1	Cell/Molecules

[a] k_{Basal} was calculated to be at least one order of magnitude smaller than k_{Active}
[b] Degradation rates for the mRNA were calculated assuming a mRNA half-life in the order of minutes [50]
[c] Based on the average translation rate in NIH3T3 cells [49]

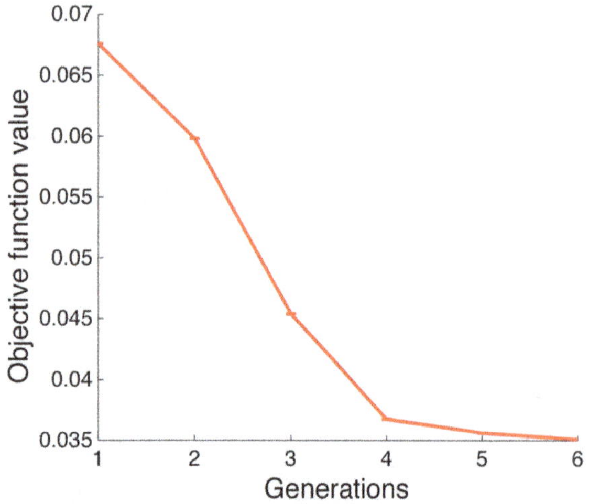

Fig. 5 Genetic Algorithm performance. The deterministic precondition was introduced in a genetic algorithm strategy. The genetic algorithm was implemented using an adaptive population size using a population of 3000 individuals for the first generation, and a population of 20 individuals for the subsequent generations. As algorithm parameters we used $\mu = 0.2$, and $\epsilon = 0.4$. In the plot a decrease in the objective function value during the generations in the GA is shown

a stable steady state, see Fig. 7-b. Figure 7-c shows the IRF7 promoter dynamics exhibiting transitions between active/basal states. This promoter state transition is a characteristic of genes with regulated expression [43]. Subsequently, IRF7 mRNA expression displays a pattern that is affected by this stochastic switching between two possible states, one with basal expression and the other with active expression, see Fig. 7-d. For single-cell trajectories, IRF7 protein expression shows a switch-like expression. For the whole population of those trajectories bimodality is observed (Fig. 7-e), the same stands for the different forms of the IRF7 protein, phosphorylated (Fig. 7-f) dimer (Fig. 7-g), and the total amount of IRF7 proteins (Fig. 7-h). The system's steady states are given in Table 4.

Discussion

The promise of systems biology is to achieve a quantitative understanding of the molecular processes in the cell with the aid of computational models. However, a bottleneck is the availability of reliable parameter values needed in those models. Often, it is very difficult or even impossible to measure all of these parameters. For deterministic models, this problem has been well tackled by the development of efficient methods for parameter estimation. Contrarily, for stochastic dynamics the landscape

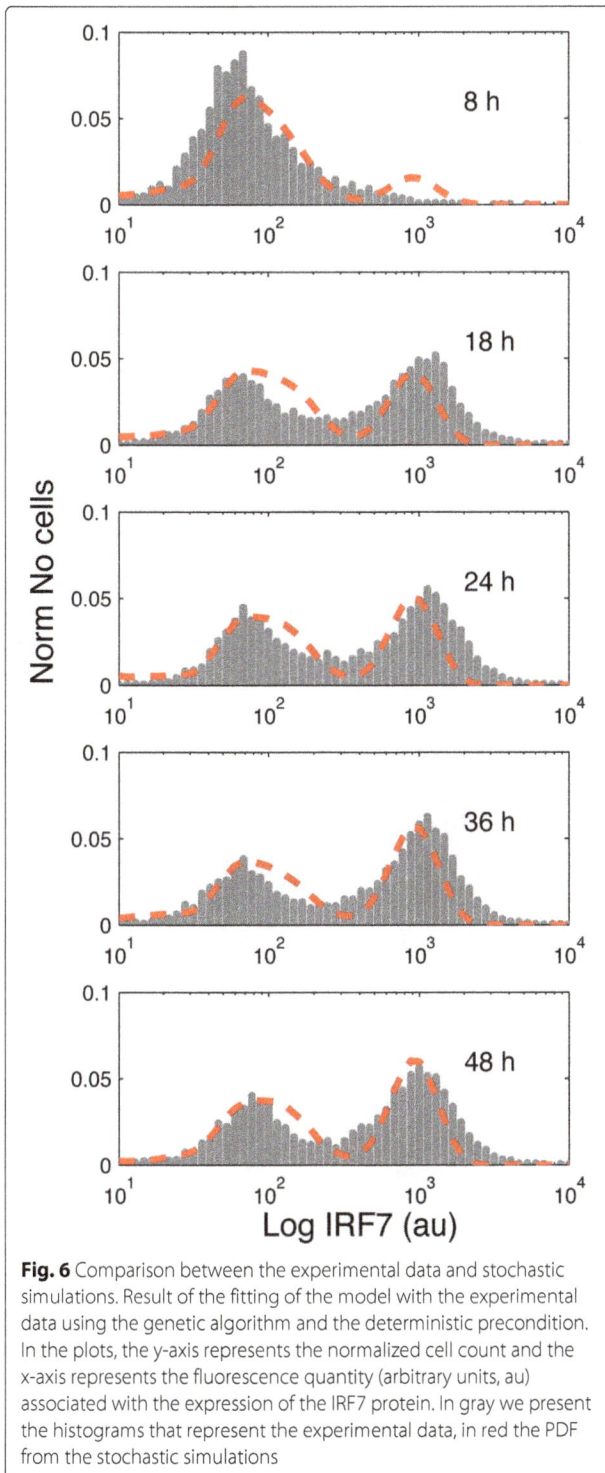

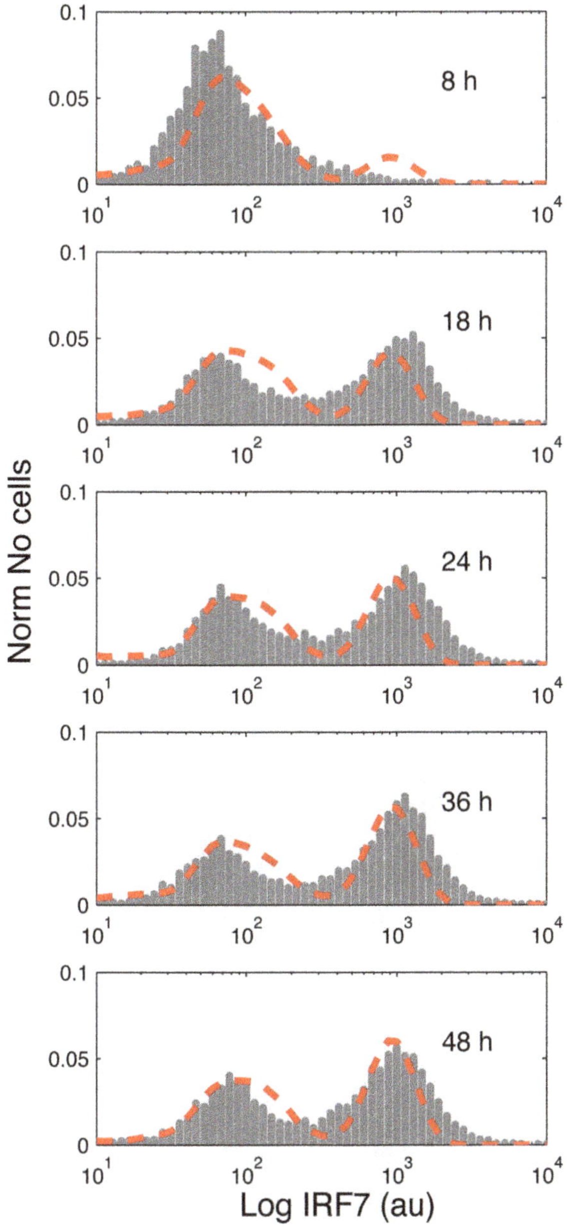

Fig. 6 Comparison between the experimental data and stochastic simulations. Result of the fitting of the model with the experimental data using the genetic algorithm and the deterministic precondition. In the plots, the y-axis represents the normalized cell count and the x-axis represents the fluorescence quantity (arbitrary units, au) associated with the expression of the IRF7 protein. In gray we present the histograms that represent the experimental data, in red the PDF from the stochastic simulations

of methods for parameter estimation is poorer and still under development. For this reason, innovative and efficient algorithms are needed to fit and validate realistic stochastic models.

During the last years important advances have been achieved in the field of parameter inference for stochastic models. On one hand, strategies that involve mathematical procedures of moment estimation have been suggested [17–19]. On the other, Bayesian methods that test multiple parameters values to find approximated solutions that represent the data have been successfully implemented [20]. Currently, the main problems observed in most methods include the computational cost, the accuracy of the obtained solution, and its potential to be implemented in large and non-linear systems. Different strategies have been suggested to improve the accuracy and alleviate the computational performance [21, 22].

Table 3 System's initial conditions

Variable	Initial condition (molecules/cell)
IFN	150
ISGF3	1
Pa	0
mRNA	1
IRF7	1
IRF7phosp	0
IRF7dimer	0

In our method, we tackled the computational cost by introducing a deterministic precondition that works as a by-pass in the algorithm avoiding large amounts of unnecessary stochastic simulations. The concept of reducing the parameter space by introducing a precondition defined by the experimental data has been suggested previously by Hori et al. [21]. In their method, a small order linear model is used to optimize the experimental data by finding the root of a Lyapunov equation. In contrast, we use a deterministic precondition. Both approaches have advantages and disadvantages. Thus, Hori's method is constrained by the need to find the Lyapunov equation. Our method is based on steady state calculation and Monte Carlo simulations that are standard and well-known methodologies used in systems biology.

Recently a new algorithm developed by Lillacci et al., has been shown to significantly reduce the computational cost and achieve a high accuracy fit for stochastic models and flow cytometry data [22]. This method uses the Kolmogorov distance as a metric to calculate the difference between model simulation and experimental data. By choosing this metric, it is possible to estimate the minimal number of simulations needed to compare experimental and model simulations under a certain tolerance value. This estimated value decreases as the number of experimental data increases. In typical flow cytometry experiments the number of measured cells is in the order of tens of thousands and this number in Lillacci's Algorithm is translated in a reduction of at least one order of magnitude in the number of required stochastic simulations. This algorithm has been applied to a model with

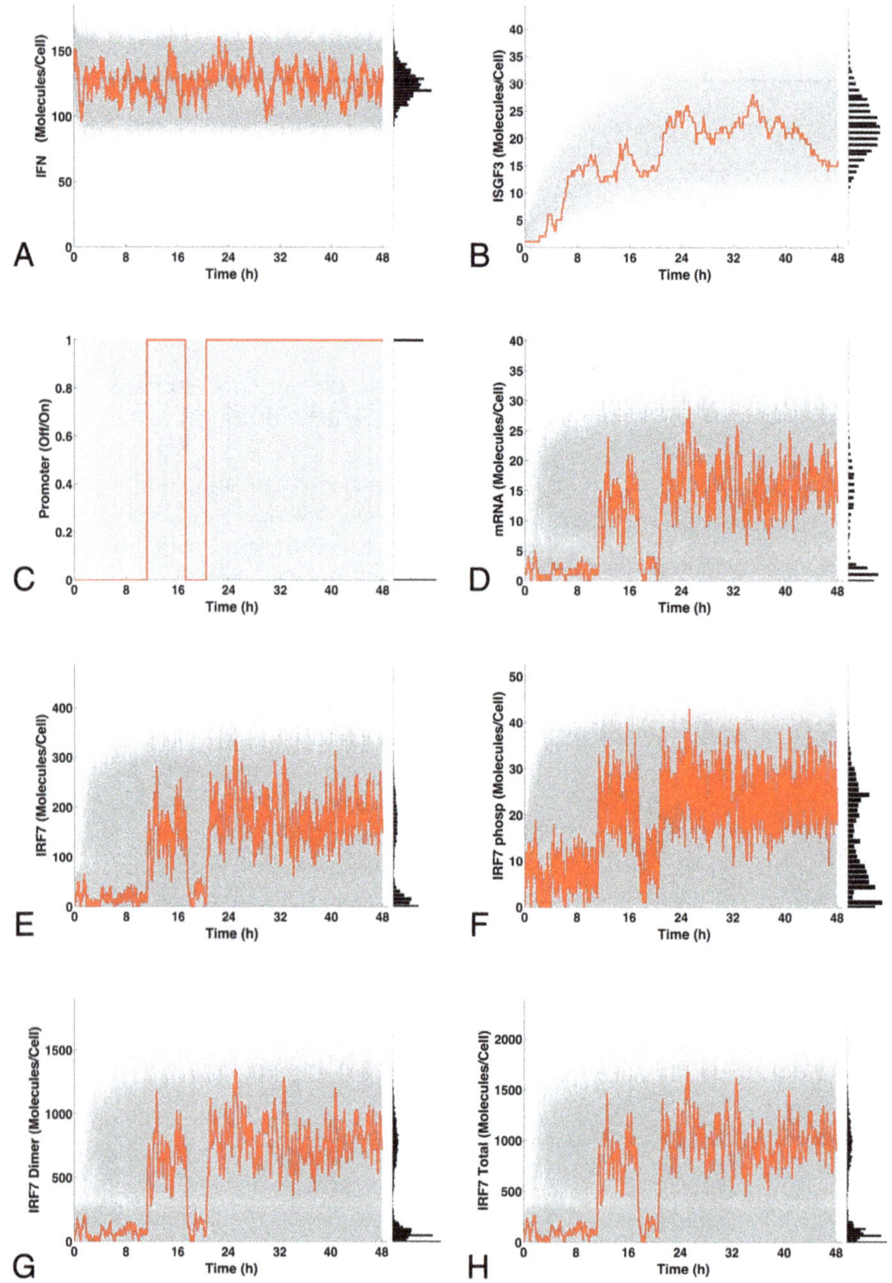

Fig. 7 Time courses in the IRF7 circuit after IFN stimulation. The temporal dynamics of the promoter, mRNA and IRF7 protein dynamics were obtained after stochastic simulations. A representative trajectory that presents a single cell dynamics is given by the red lines. The stochastic simulations were repeated 1000 times using the same initial condition obtaining the histograms that represent the cell population. **a** Time courses of IFN showing the evolution to a steady state, the same is observed in **b** for the ISGF3. **c** IRF7 promoter shows the transition between the two possible states Off/On. **d** IRF7 mRNA showing a basal expression state and a state of active expression. Bimodality is observed in **e** for the IRF7 protein, the same bimodal behavior was observed in **f** for the IRF7 phosphorylated protein, in **g** for the IRF7 protein dimer, and in **h** for all forms of the IRF7 protein

18 reactions and 20 free parameters obtaining a good fit to flow cytometry experimental data that reproduces bimodal dynamics. Comparing our method with Lillacci's algorithm it is important to point out that both methods tackle the computational cost issue in two fundamentally different ways. Lillacci's algorithm minimizes the number of needed stochastic simulations, whereas our algorithm minimizes the number of parameters sets evaluated with stochastic dynamics. A powerful new algorithm may be the result of combining both methodologies.

It is well known from non-linear dynamics that the model architecture and parameter values determine the

Table 4 System's steady states

Variable	1st sss (eigenvalue)	2nd sss (eigenvalue)
IFN	0 (-15.95)	124.70 (-0.0017)
ISGF3	0 (-3.87)	22.00 (-0.049)
Pa	0 (-0.44)	0.99 (-0.072)
mRNA	1.20 (-0.071)	15.93 (-0.32)
IRF7	13.62 (-0.049)	180.32 (-0.44)
IRF7phosp	6.62 (-0.0017)	24.09 (-3.87)
IRF7dimer	60.19 (-0.0013)	796.69 (-58.03)
IRF7total	80.43	954.01

behavior of the system. Hence, more complex stable dynamics such as limit cycles, and higher-multistability (a system with more than two stable steady states) can be obtained. Unstable systems are usually not of interest in the biological context. Our method was designed and tested for monostable and bistable systems, all other cases being rejected by the deterministic precondition. The case of limit cycles can be better approached by existing parameter estimation methods for single time-course data [12–16]. The case of systems with higher-multistability can in principle be treated by our method, but the application is limited by the costly need to find multiple stable steady-states in the system. Finally, it is relevant to mention that comparing stable states of the ODE model with modes of the measured flow cytometry distribution in some models can present some inherent problems. Bimodal fluorescence distributions may have no relation to bistability, e.g. when the systems has large differences in transitions rates for the promoter states [44].

Our method was implemented using two well-established optimization strategies, random search and genetic algorithms. Using a random search algorithm we observed that less than 1% of the population of parameters are subject to stochastic evaluations, this reduction in the tested parameters allowed us to explore a larger proportion of the parameter space, which is especially relevant for systems with multiple unknown parameters and no initial parameter guesses. On the other hand, using genetic algorithms we achieved a convergence to a minimal value in the optimization function after a few generations (see Fig. 5). Additionally, during each generation in the genetic algorithm a large percentage of the total evaluated parameters is rejected by the deterministic precondition. This percentage is dependent on the parameter of the algorithm (rate of elitism and mutation rate). In both strategies the introduction of the deterministic precondition significantly improved the parameter estimation process. Moreover, the implementation of deterministic precondition is not restricted to random search and genetic algorithms, it can also be implemented in other optimization algorithms, e.g. particle swarm or simulated annealing. A complete analysis of the performance of using the deterministic condition is given in Additional file 4.

The accuracy of the method is given by the agreement between experimental data and simulations. As can be observed in Fig. 6 a good fit was found, albeit not a perfect one. We observed that even increasing the number of tested parameter values during the random search and/or increasing the number of generations during the genetic algorithm did not improve the fit. For this reason, we consider that the differences between the model and the experimental data might be explained by the fact that our system is a highly simplified model that was built by lumping some steps in the biological system. Our team is working to fit more complex models in a future publication. To test the reproducibility of the obtained parameter values, the method was repeated 100 times and distributions of the obtained parameters were built. As can be observed in Additional file 5, when a parameter is identifiable this is reflected in a narrow distribution, on the other hand, when a parameter is non-identifiable this is reflected in a wide distribution. In the example given by reactions (11) to (23) we observed that most parameters are not identifiable. Thus, non-identifiability is expected in the parameters contained in reactions (11) and (20). Those reactions have opposite effects on the IFN dynamics. For this reason, multiple combinations of values in parameters V_{IFN}, k_{IFN} and k_{dIFN} have similar effects on the overall system dynamics. Contrarily, the parameters involved in the production of the mRNA (k_{Active} and k_{Basal}) are better confined.

Rand et al. described bimodal gene expression of IRF7 after IFN stimulation. This phenomenon was explained at the cell population level by the effects of the IFN paracrine response [30]. Here we observed that bimodality also can be explained in absence of paracrine response and only taking into account the molecular mechanism responsible for the IRF7 promoter dynamics. Bistability in gene-expression circuits is commonly associated with the switching between active/basal states in the gene promoter. In most cases, DNA cooperativity in the promoter is the basis of promoter-state switching [45]. However, for type-I IFN responses a promoter activation in a cooperative manner has recently been discarded [46]. Taking the recent literature into account, we developed a model on the basis of a positive feedback loop circuit comprising the independent activation of ISRE and IRFE elements by one ISGF3 molecule and one IRF7 dimer, respectively (see Fig. 4). This model has the needed and sufficient elements to sustain a bistable system (that is a positive feedback loop and non-linear dynamics in the reaction rates) [47, 48]. Additionally, multiple model structures testing different biological scenarios including: additional feedback loops, additional intermediate elements, and promoter cooperative activation were tested, obtaining that the presented model reproduces the experimental data best. Our simulation results agree with the experimental data obtained from flow cytometry [30]. Additional characteristics observed in the experimental data are also reproduced by the model, such as the basal expression of IRF7 without IFN stimulation [29], see Additional file 7. The number of molecules in the active state of the system (given by the obtained 2nd steady state, see Table 4) agrees with the order of magnitude reported for mammal mRNAs (average = 17 Molecules/Cell with a range between 1 to 200 Molecules/Cell) and the range of protein concentration (average = 50,000 Molecules/Cell with a range between 100 to 10^8 Molecules/Cell) [49].

Conclusion

Here, we present a method to fit stochastic models to experimental data. The method is based on the comparison of distributions. The central idea of the method is to use a deterministic precondition that is defined by the experimental data as a filter avoiding large amounts of costly stochastic simulations. Using this idea, the number of parameters evaluated under stochastic dynamics is reduced, resulting in a significant improvement in the performance of the algorithm. As a case study, we used a model of IRF7 gene expression investigating the origin of bimodality in its dynamics upon IFN stimulation. Our results allowed us to conclude that a circuit with IRF7 promoter activation by one IRF7 dimer and one ISGF3 molecule is sufficient to explain the observed bimodal dynamics.

Additional files

Additional file 1: Pseudocode for the genetic algorithm with the deterministic precondition. Additional file with the pseudocode for the genetic algorithm.

Additional file 2: Identifying parameters for a constitutive gene expression circuit with in-silico data. Additional file with an example where the method is applied to a simulation case with synthetic data. The file contains **Figures A1-A3**, and **Tables A1** and **A2**.

Additional file 3: Random search strategy. Additional file with **Figure A4**. Results obtained by using the random search strategy.

Additional file 4: Algorithm's speed and performance analysis. Additional file with an analysis of the algorithm's speed and performance. The file contains **Figure A5**, and **Tables A3** and **A4**.

Additional file 5: Testing the reproducibility of the selected parameter values. Additional file containing **Figure A6** that shows the parameter distributions.

Additional file 6: Fitting the model to different doses of IFN. Additional file with figure A7 and A8 that shows the fits of the model to flow cytometry data describing the stimulation of cells with 100U and 250U of IFN.

Additional file 7: Model dynamics without IFN stimulation. Additional file with **Figure A9** that shows the time courses for the model in absence of IFN stimulation.

Abbreviations

au: Arbitrary units; BAC: Bacterial artificial chromosome; DP: Deterministic precondition; GA: Genetic algorithm; IFN: Interferon; ISG: Interferon stimulated genes; ODE: Ordinary differential equation; PDF: Probability density function; RS: Random search; sss: Stable steady state

Acknowledgements

We gratefully thank Dr. Frank T. Bergmann for providing a modified version of Copasi that is used as part of this project. In the same way, we express our sincere thanks to Dr. Sven Sahle for his helpful comments helping us to increase the algorithm performance by suggesting changes in the data acquisition and analysis. We sincerely thank Dr. Ulfert Rand and Dr. Mario Köster for providing the experimental data.

Funding

This research was supported by Consejo Nacional de Ciencia y Tecnología (CONACYT) under grant 263795 in the program "Estancias postdoctorales al extranjero para la consolidación de grupos de investigación convocatoria 2015" to LUA. CZ was supported by BIOMS. The project was also supported by BMBF (Immunoquant).

Authors' contributions

LUA, UK and CZ conceived and designed the research. LUA wrote the manuscript, designed and performed the simulations. UK suggested extensions and modifications of the research. UK and CZ participated in discussions and revised the manuscript critically. All authors read and approved the final manuscript.

Competing interests

The authors declare that they have no competing interests.

Author details

[1]Department of Modeling of Biological Processes, COS Heidelberg / Bioquant, Heidelberg University, Im Neuenheimer Feld 267, 69120 Heidelberg, Germany. [2]BIOMS (Center for Modeling and Simulation in the Biosciences), Heidelberg University, Im Neuenheimer Feld 267, 69120 Heidelberg, Germany.

References

1. Gillespie DT. A general method for numerically simulating the stochastic time evolution of coupled chemical reactions. J Comput Phys. 1976;22(4): 403–34. doi:10.1016/0021-9991(76)90041-3.
2. Adrews SS, Dinh T, Arkin AP. Stochastic models of biological processes. Encycl Complex Syst Sci. 20098730–49. doi:10.1007/978-0-387-30440-3_524.
3. Raj A, van Oudenaarden A. Nature, nurture, or chance: stochastic gene expression and its consequences. Cell. 2008;135(2):216–26. doi:10.1016/j.cell.2008.09.050.
4. Golding I, Paulsson J, Zawilski SM, Cox EC. Real-time kinetics of gene activity in individual bacteria. Cell. 2005;123(6):1025–36. doi:10.1016/j.cell.2005.09.031.
5. Weinberger LS, Burnett JC, Toettcher JE, Arkin AP, Schaffer DV. Stochastic gene expression in a lentiviral positive-feedback loop: HIV-1 Tat fluctuations drive phenotypic diversity,. Cell. 2005;122(2):169–82. doi:10.1016/j.cell.2005.06.006.

6. Samoilov M, Plyasunov S, Arkin AP. Stochastic amplification and signaling in enzymatic futile cycles through noise-induced bistability with oscillations,. Proc Natl Acad Sci U S A. 2005;102(7):2310–5. doi:10.1073/pnas.0406841102.
7. Moles CG, Mendes P, Banga JR. Parameter estimation in biochemical pathways: a comparison of global optimization methods. Genome Res. 20032467–74. doi:10.1101/gr.1262503.
8. Rodriguez-Fernandez M, Egea JA, Banga JR. Novel metaheuristic for parameter estimation in nonlinear dynamic biological systems. BMC Bioinformatics. 2006483. doi:10.1186/1471-2105-7-483.
9. Bock HG, Kostina E, Schlöder JP. Numerical Methods for Parameter Estimation in Nonlinear Differential Algebraic Equations. GAMM-Mitteilungen. 2007;30(2):376–408. doi:10.1002/gamm.200790024.
10. Marquardt DW. An algorithm for least-squares estimation of nonlinear parameters. J Soc Ind Appl Math. 1963;11(2):431–41. doi:10.1137/0111030.
11. Zimmer C. Reconstructing the hidden states in time course data of stochastic models. Math Biosci. 2015;269:117–29. doi:10.1016/j.mbs.2015.08.015.
12. Zimmer C, Sahle S. Deterministic inference for stochastic systems using multiple shooting and a linear noise approximation for the transition probabilities. IET Syst Biol. 20151–12. doi:10.1049/iet-syb.2014.0020.
13. Komorowski M, Costa MJ, Rand DA, Stumpf MPH. Sensitivity, robustness, and identifiability in stochastic chemical kinetics models. Proc Natl Acad Sci. 2011;108(21):8645–50. doi:10.1073/pnas.1015814108.
14. Golightly A, Wilkinson DJ. Bayesian inference for stochastic kinetic models using a diffusion approximation. Biometrics. 2005;61(3):781–8. doi:10.1111/j.1541-0420.2005.00345.x.
15. Tian T, Xu S, Gao J, Burrage K. Simulated maximum likelihood method for estimating kinetic rates in gene expression. Bioinformatics. 2007;23(1): 84–91. doi:10.1093/bioinformatics/btl552.
16. Boys RJ, Wilkinson DJ, Kirkwood TBL. Bayesian inference for a discretely observed stochastic kinetic model. Stat Comput. 2008;18(2):125–35. doi:10.1007/s11222-007-9043-x.
17. Zechner C, Ruess J, Krenn P, Pelet S, Peter M, Lygeros J, Koeppl H. Moment-based inference predicts bimodality in transient gene expression. Proc Natl Acad Sci. 2012;109(21):8340–345. doi:10.1073/pnas.1200161109.
18. Kügler P. Moment fitting for parameter inference in repeatedly and partially observed stochastic biological models. PLoS ONE. 2012;7(8):1–15. doi:10.1371/journal.pone.0043001.
19. Miller-Jensen K, Skupsky R, Shah PS, Arkin AP, Schaffer DV. Genetic selection for context-dependent stochastic phenotypes: Sp1 and TATA mutations increase phenotypic noise in HIV-1 gene expression. PLoS Comput Biol. 2013;9(7):. doi:10.1371/journal.pcbi.1003135.
20. Poovathingal SK, Gunawan R. Global parameter estimation methods for stochastic biochemical systems,. BMC Bioinformatics. 2010;11:414. doi:10.1186/1471-2105-11-414.
21. Hori Y, Khammash MH, Hara S. Efficient parameter identification for stochastic biochemical networks using a reduced-order realization. In: Proceedings of the European Control Conference 2013. Zurich: European Control Conference, ECC; 2013. p. 4154–49.
22. Lillacci G, Khammash M. The signal within the noise: efficient inference of stochastic gene regulation models using fluorescence histograms and stochastic simulations. Bioinformatics. 2013;29(18):2311–9. doi:10.1093/bioinformatics/btt380.
23. Ivashkiv LB, Donlin LT. Regulation of type I interferon responses,. Nat Rev Immunol. 2014;14(1):36–49. doi:10.1038/nri3581.
24. Hu J, Sealfon SC, Hayot F, Jayaprakash C, Kumar M, Pendleton AC, Ganee A, Fernandez-Sesma A, Moran TM, Wetmur JG. Chromosome-specific and noisy IFNB1 transcription in individual virus-infected human primary dendritic cells. Nucleic acids Res. 2007;35(15):5232–41. doi:10.1093/nar/gkm557.
25. Zhao M, Zhang J, Phatnani H, Scheu S, Maniatis T. Stochastic expression of the interferon-β gene. PLoS Biol. 2012;10(1):1001249. doi:10.1371/journal.pbio.1001249.
26. Levin D, Harari D, Schreiber G. Stochastic receptor expression determines cell fate upon interferon treatment. Mol Cell Biol. 2011;31(16): 3252–66. doi:10.1128/MCB.05251-11.
27. Schneider WM, Chevillotte MD, Rice CM. Interferon-stimulated genes: a complex web of host defenses. Annu Rev Immunol. 2014;32:513–45. doi:10.1146/annurev-immunol-032713-120231.
28. Vasquez K, Sigrist K, Kucherlapati R, Demant P, Dietrich WF, Agoulnik S, Plus S. IRF-7 is the master regulator of. Nature. 2005;434(April):772–7. doi:10.1038/nature03419.1.
29. Honda K, Taniguchi T. IRFs: master regulators of signalling by Toll-like receptors and cytosolic pattern-recognition receptors. Nat Rev Immunol. 2006;6(9):644–58. doi:10.1038/nri1900.
30. Rand U, Rinas M, Schwerk J, Nöhren G, Linnes M, Kröger A, Flossdorf M, Kály-Kullai K, Hauser H, Höfer T, Köster M. Multi-layered stochasticity and paracrine signal propagation shape the type-I interferon response. Mol Syst Biol. 2012;584:584. doi:10.1038/msb.2012.17.
31. MATLAB version (R2015a); 2015.
32. Hoops S, Sahle S, Gauges R, Lee C, Pahle J, Simus N, Singhal M, Xu L, Mendes P, Kummer U. COPASI–a COmplex PAthway SImulator. Bioinformatics (Oxford, England). 2006;22(24):3067–74. doi:10.1093/bioinformatics/btl485.
33. Gibson MA, Bruck J. Efficient exact stochastic simulation of chemical systems with many species and many channels. J Phys Chem A. 2000;104(9):1876–89. doi:10.1021/jp993732q.
34. Balkay L. FCS data reader. MATLAB Central File Exchange. 2006.
35. Yoder N. peakfinder. MATLAB Central File Exchange. 2009.
36. Gillespie DT. The chemical langevin equation. J Chem Phys. 2000;113(1): 297–306. doi:10.1063/1.481811.
37. Gillespie DT. Stochastic simulation of chemical kinetics. Annu Rev Phys Chem. 2007;58:35–55. doi:10.1146/annurev.physchem.58.032806.104637.
38. Strogatz SH. Nonlinear dynamics and chaos with applications to physics, biology, chemistry, and engineering. New York: Westview Press; 2000. doi:10.1137/1037077.
39. Brownlee J. Clever algorithms nature-inspired programming recipes. Melbourne: Lulu Enterprises; 2011.
40. Legewie S, Herzel H, Westerhoff HV, Blüthgen N. Recurrent design patterns in the feedback regulation of the mammalian signalling network. Mol Syst Biol. 2008;4(1):190. doi:10.1038/msb.2008.29.
41. Ning S, Huye LE, Pagano JS. Regulation of the transcriptional activity of the IRF7 promoter by a pathway independent of interferon signaling. J Biol Chem. 2005;280(13):12262–70. doi:10.1074/jbc.M404260200.
42. Litvak V, Ratushny AV, Lampano AE, Schmitz F, Huang AC, Raman A, Rust AG, Bergthaler A, Aitchison JD, Aderem A. A FOXO3-IRF7 gene regulatory circuit limits inflammatory sequelae of antiviral responses. Nature. 2012;490(7420):421–5. doi:10.1038/nature11428.
43. Kaufmann DE, Walker BD. Treatment interruption to boost specific HIV immunity in acute infection. Curr Opin HIV AIDS. 2007;2(1):21–5.
44. Dobrzyński M, Nguyen LK, Birtwistle MR, von Kriegsheim A, Fernández AB, Cheong A, Kolch W, Kholodenko BN. Nonlinear signalling networks and cell-to-cell variability transform external signals into broadly distributed or bimodal responses. J R Soc Interface. 2014;11(98):20140383. doi:10.1098/rsif.2014.0383.
45. Kaern M, Elston TC, Blake WJ, Collins JJ. Stochasticity in gene expression: from theories to phenotypes. Nat Rev Genet. 2005;6(6):451–64. doi:10.1038/nrg1615.
46. Begitt A, Droescher M, Meyer T, Schmid CD, Baker M, Antunes F, Knobeloch KP, Owen MR, Naumann R, Decker T, Vinkemeier U. STAT1-cooperative DNA binding distinguishes type 1 from type 2 interferon signaling. Nat Immunol. 2014;15(2):168–76. doi:10.1038/ni.2794.
47. Cherry JL, Adler FR. How to make a biological switch. J Theor Biol. 2000;203(2):117–33. doi:10.1006/jtbi.2000.1068.
48. Craciun G, Tang Y, Feinberg M. Understanding bistability in complex enzyme-driven reaction networks. Proc Natl Acad Sci. 2006;103(23): 8697–702. doi:10.1073/pnas.0602767103.
49. Schwanhäusser B, Busse D, Li N, Dittmar G, Schuchhardt J, Wolf J, Chen W, Selbach M. Global quantification of mammalian gene expression control. Nature. 2011;473(7347):337–42. doi:10.1038/nature10098.
50. Yang E, van Nimwegen E, Zavolan M, Rajewsky N, Schroeder M, Magnasco M, Darnell JE. Decay rates of human mrnas: correlation with functional characteristics and sequence attributes. Genome Res. 2003;13(8):1863–72. doi:10.1101/gr.1272403.

9

SBpipe: a collection of pipelines for automating repetitive simulation and analysis tasks

Piero Dalle Pezze* and Nicolas Le Novère

Abstract

Background: The rapid growth of the number of mathematical models in Systems Biology fostered the development of many tools to simulate and analyse them. The reliability and precision of these tasks often depend on multiple repetitions and they can be optimised if executed as pipelines. In addition, new formal analyses can be performed on these repeat sequences, revealing important insights about the accuracy of model predictions.

Results: Here we introduce SBpipe, an open source software tool for automating repetitive tasks in model building and simulation. Using basic YAML configuration files, SBpipe builds a sequence of repeated model simulations or parameter estimations, performs analyses from this generated sequence, and finally generates a LaTeX/PDF report. The parameter estimation pipeline offers analyses of parameter profile likelihood and parameter correlation using samples from the computed estimates. Specific pipelines for scanning of one or two model parameters at the same time are also provided. Pipelines can run on multicore computers, Sun Grid Engine (SGE), or Load Sharing Facility (LSF) clusters, speeding up the processes of model building and simulation. SBpipe can execute models implemented in COPASI, Python or coded in any other programming language using Python as a wrapper module. Future support for other software simulators can be dynamically added without affecting the current implementation.

Conclusions: SBpipe allows users to automatically repeat the tasks of model simulation and parameter estimation, and extract robustness information from these repeat sequences in a solid and consistent manner, facilitating model development and analysis. The source code and documentation of this project are freely available at the web site: https://pdp10.github.io/sbpipe/.

Keywords: Pipeline, Modelling, Simulation, Parameter estimation

Background

The range of software tools developed by the Systems Biology community has grown considerably in the last few years, in particular aimed at supporting mathematical modelling of biological networks. The development of a mathematical model typically comprises successive phases: design, parameterisation, simulation and testing. Model design is the phase where the core of the problem to investigate is summarised using a mathematical formalism. Once designed, the model parameters need to be calibrated, for example using some experimental data. After this stage, the model is used for generating predictions which are then tested experimentally. Depending on the outcome, a model can be refined in order to improve or correct its prediction.

Many tools already exist to generate, simulate and analyse mathematical models [1, 2]. Although these tools provide modellers with key functionalities for model parameter estimation and simulation, it has become clear that the accuracy of these tasks depends on multiple repetitions. Furthermore, the analysis of this batch of repeats can reveal important insights regarding the model itself and the data used for calibration. Therefore, it is useful to repeat tasks such as parameter estimation or stochastic simulation, collect statistics and visualise these results.

SBpipe is an open source software tool which provides modellers with a collection of pipelines for model development and simulation. A pipeline for parameter estimation allows users to repeat a model calibration

*Correspondence: piero.dallepezze@babraham.ac.uk
The Babraham Institute, Babraham Campus, Cambridge CB22 3AT, UK

many times on a multicore machine or a computer cluster. The generated fit sequence is then analysed, and information about the profile likelihood from parameter estimation samples is represented graphically and textually. Support for model simulation is also provided with pipelines for time course model simulation, as well as single and double parameter scans.

Implementation

SBpipe is an open source software package developed with the Python [3] and R [4] programming languages. Python is the main programming language connecting all the package components, whereas R is used for generating statistics and plots. The use of this statistics-dedicated programming language for analysing the results allows users to run the provided R scripts independently of SBpipe using an R environment. This can be convenient if further data analyses are needed or plots need to be annotated or edited.

Pipelines in SBpipe are configured using YAML configuration files. This allows modellers to easily edit their tasks manually or programmatically if needed. Examples of configuration files can be found within the main package in the folder

```
tests/insulin_receptor/
```

In order to maintain a flexible and extendible design, SBpipe abstracts the concepts of simulator and pipeline. The class `Simul` is a generic simulator interface used by the pipelines in SBpipe. This mechanism uncouples pipelines from simulators which can therefore be configured in each pipeline configuration file. Currently, the available simulators are `Copasi` and `Python`. These simulators process models developed in COPASI [5] and models coded in Python, respectively.

SBpipe passes the report file name as an input argument to the latter. The Python program is then responsible for generating a report file containing the simulation (or parameter estimation) results. Python can also be used as a wrapper module for running models coded in any programming language. Rather than coding a model itself, the Python file can call an external program containing the model. This Python wrapper must forward the report file name to this external program which becomes responsible of generating the report file. With this simple approach, users can run their existing models using customised command options or any program library they need. The `tests/` folder contains examples of models coded in R, Octave, or Java programming languages, and executed using basic Python module wrappers. The supplied R models depend on the packages minpack.lm, deSolve, and sde, whereas the supplied Java model requires a JVM. Dependencies for these additional models must be installed separately.

The class `Pipeline` represents a generic pipeline, which is extended by each SBpipe pipeline. The following pipelines are currently available:

- `simulate`: deterministic or stochastic time course stimulation;
- `single_param_scan`: scan a model parameter;
- `double_param_scan`: scan two model parameters;
- `param_estim`: model parameter estimation including sampling of the parameter likelihood.

An SBpipe pipeline performs three tasks: data generation, data analysis, and report generation. The first task loads and runs a simulator at runtime and organises the generated data. The second task computes statistics and plots from these data. Finally, the third task generates a LaTeX/PDF report containing the computed plots. Because of the interdependency between these tasks, their execution is sequential. However, users can select the tasks to run in the pipeline configuration file. A typical scenario requiring a task to be turned off would be the analysis of data previously generated data using different configuration thresholds. In this case, the data generation task can be disabled to prevent SBpipe from re-running the simulations.

Pipelines for parameter estimation or stochastic model simulation can be computationally intensive. SBpipe allows users to generate repeats of model simulation or parameter estimation in parallel. In a configuration file, users can select the number of repeats, and whether the jobs should be executed locally using Python multiprocessing or in a computer cluster. In this case, SBpipe supports the cluster types Sun Grid Engine (SGE) and Load Sharing Facility (LSF).

The project is available on the GitHub repository. Numerous test cases are also provided within the package. Every time the source code is updated online, these tests are automatically executed by Travis.CI, a GitHub application for continuous integration service. For standard users, these tests are useful examples of how to configure SBpipe. User and developer documentations for this project are available online and within the project folder.

Results

To demonstrate SBPIPE functions we will use a minimal model of insulin receptor (IR). This IR model is a module of a more complex Insulin/TOR model [6] (Biomodels database [7] id: BIOMD0000000581). This choice enables users to quickly reproduce the results shown in this article using the SBpipe test suite and to present the results in the most compact manner. This model describes the activation of the insulin receptor upon insulin stimulation. In the presence of insulin, the insulin receptor beta (*IRβ*) is phosphorylated on Y1164. The phosphorylated receptor is then dephosphorylated and enters in a refractory state. This latter state is used to introduce a delay in the system succintly representing receptor internalisation, degradation and synthesis, thus reducing the number of model parameters. Finally, from this refractory state the receptor can become functional again. Details of the model are provided in Additional file 1: Table S1, Figure S1. The generic pipeline work flow is shown for the parameter estimation pipeline in Fig. 1a. To illustrate how SBpipe can reveal parameter identifiability issues from multiple parameter estimations, two fit sequences are independently generated using sufficient and insufficient data sets (Additional file 1: Tables S2–S4). For each group, SBpipe generates $N = 1000$ independent parameter estimations using Particle Swarm optimisation algorithm [8] as implemented in COPASI. These calibrations are then processed in the data analysis task. Although SBpipe does not contain a pipeline for computing identifiability analysis directly, the parameter estimation pipeline can help identify issues in parameter estimation by projecting the estimates for each parameter. This analysis uses not only the best fit of each of the N estimations, but also the sub-optimal fits. As these fits represent samples of the parameter space, they can reveal a *sampled profile likelihood estimation (PLE)* for each estimated parameter. For direct methods calculating model parameter profile likelihoods using COPASI, see [9] or https://pypi.python.org/pypi/PyCoTools.

Results of estimation tasks using data sets presented in Table S2A and Table S2B are shown in the *Identifiable* or *Non-identifiable* columns of Fig. 1, respectively. The *Identifiable* column shows how the parameter $k1$ presents clear confidence intervals at 66%, 95%, and 99% percents of confidence levels (CL). The *Non-identifiable* column shows how the same parameter is practically non-identifiable to the right of the confidence interval. Parameter distributions and correlations are also computed for the best fits, and for the fits with objective values lesser than a confidence level of 95%. For the complete results generated by this pipeline, see Additional file 1: Tables S2–S4, Figures S2–S8.

Results generated by the time-course simulation pipeline are shown in Fig. 1b. Deterministic and stochastic model simulations are illustrated for the phosphorylated state of the IR species. For deterministic simulation, time courses of model variables are simply plotted. For stochastic simulations, SBpipe can represent time courses with mean (black line), the 95% confidence intervals of the mean (cyan bars), and one standard deviation (blue bars). The second panel in Fig. 1b show this plot using a sequence of 40 independent stochastic simulations. If available, data corresponding to model variables can easily be added to the plot by specifying the data set file name in the configuration file. For the complete results, see Additional file 1: Figures S9–S10.

Figure 1c shows the results from the single parameter scan pipeline. Simulations are ran with values of the parameter `k1` within the 95% confidence interval as determined by the parameter estimation using the data with a sufficient number of data points. If needed, differential scales can also be configured in order to discriminate protein levels. This is particularly useful if a simulated protein knockdown (or overexpression) is investigated. For the complete results, see Additional file 1: Figures S11–S12.

Results generated by the double parameter scan pipeline are shown in Fig. 1d. In this analysis two model parameters are scanned simultaneously and these data are reported for each time point separately. For instance, it can be useful for revealing combinatorial effects of two drugs affecting a timecourse. For the complete results, see Additional file 1: Figures S13–S15. An example of this analysis can be found in [10], where it was applied for exploring the combination of mTOR and ROS treatments in a cellular senescence model.

Discussion

SBpipe is a software tool which allows modellers to automatically repeat certain tasks in model development and analysis, such as parameter estimation and simulation, and obtain additional information about the robustness of the model. Its use should increase productivity and the confidence in the results obtained with the model.

Parameter estimation from experimental data is a challenging task which can easily produce unreliable results due to local minima, parameter non-identifiability, or inadequate optimisation algorithm configuration. From the generation and analysis of a fit sequence, SBpipe can reveal crucial insights about a model structure, the reliability of each parameter, as well as indications about the sufficiency and quality of the experimental data used to calibrate the model. This knowledge is required for assessing whether parameters are well defined and the overall model predictions are reliable.

Several software tools exist to automate aspects model building and simulation tasks, and a comprehensive review of these packages is beyond the scope of this article. Some of these comprehensive packages such as AMIGO2 [11] and SBPOP [12] rely on proprietary software (e.g. Matlab). Condor-COPASI [13] is an example of open source alternative. This server-based software tool integrates COPASI with Condor, a high-throughput computing environment. It allows COPASI users to run and analyse models on a Condor pool. SBpipe distinguishes from Condor-COPASI for three main reasons: 1) although COPASI models are supported, users can run repeated model parameter estimations and simulations using any

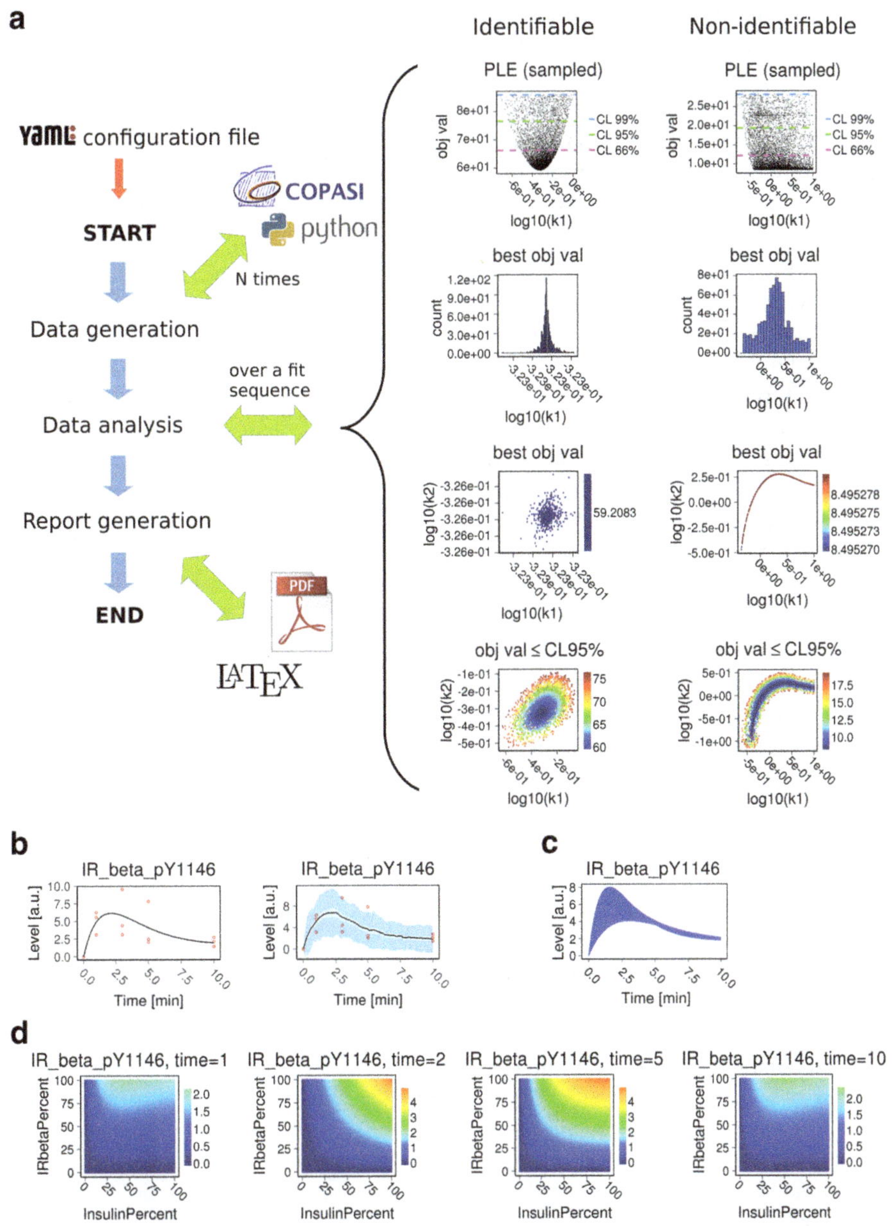

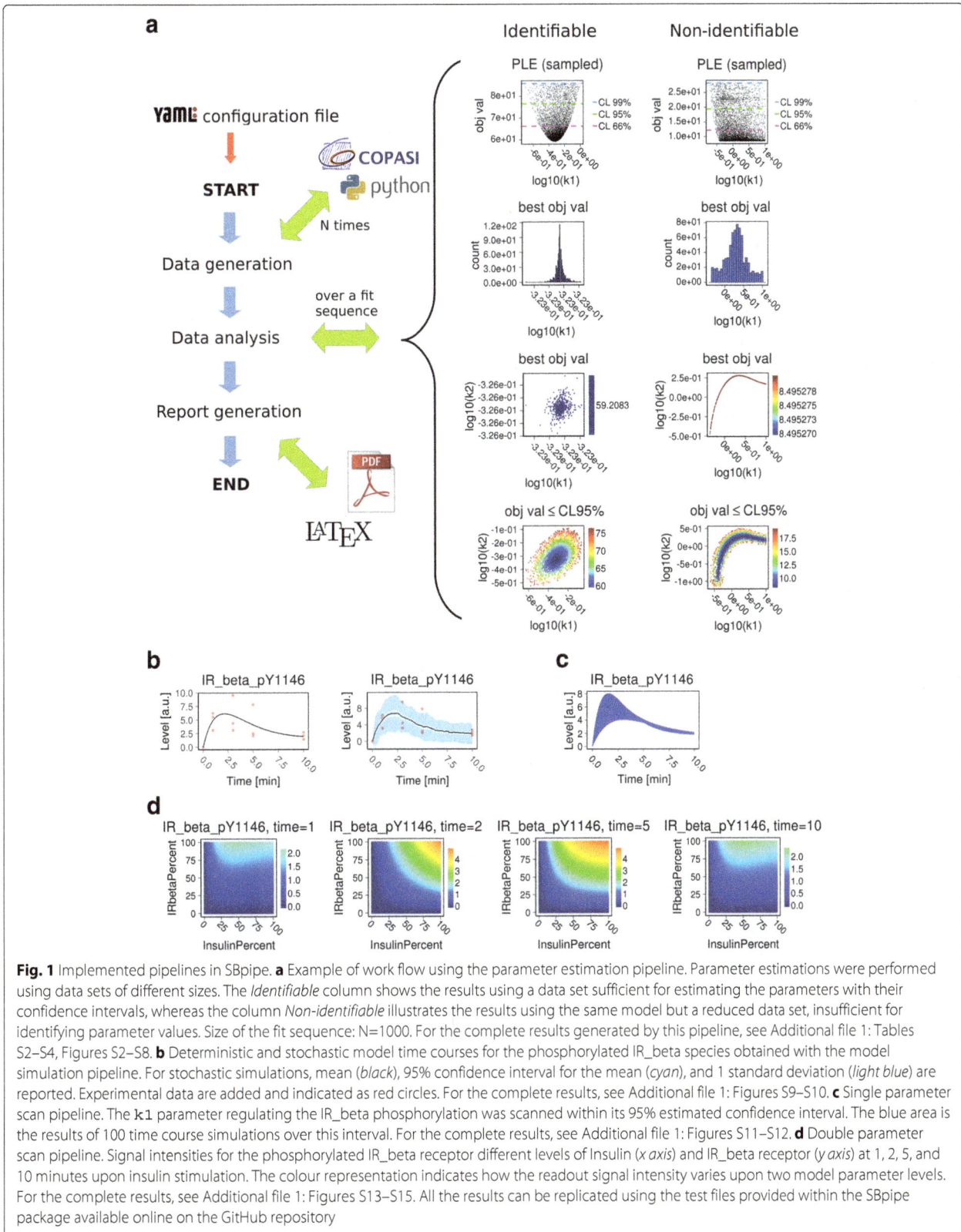

Fig. 1 Implemented pipelines in SBpipe. **a** Example of work flow using the parameter estimation pipeline. Parameter estimations were performed using data sets of different sizes. The *Identifiable* column shows the results using a data set sufficient for estimating the parameters with their confidence intervals, whereas the column *Non-identifiable* illustrates the results using the same model but a reduced data set, insufficient for identifying parameter values. Size of the fit sequence: N=1000. For the complete results generated by this pipeline, see Additional file 1: Tables S2–S4, Figures S2–S8. **b** Deterministic and stochastic model time courses for the phosphorylated IR_beta species obtained with the model simulation pipeline. For stochastic simulations, mean (*black*), 95% confidence interval for the mean (*cyan*), and 1 standard deviation (*light blue*) are reported. Experimental data are added and indicated as red circles. For the complete results, see Additional file 1: Figures S9–S10. **c** Single parameter scan pipeline. The `k1` parameter regulating the IR_beta phosphorylation was scanned within its 95% estimated confidence interval. The blue area is the results of 100 time course simulations over this interval. For the complete results, see Additional file 1: Figures S11–S12. **d** Double parameter scan pipeline. Signal intensities for the phosphorylated IR_beta receptor different levels of Insulin (*x axis*) and IR_beta receptor (*y axis*) at 1, 2, 5, and 10 minutes upon insulin stimulation. The colour representation indicates how the readout signal intensity varies upon two model parameter levels. For the complete results, see Additional file 1: Figures S13–S15. All the results can be replicated using the test files provided within the SBpipe package available online on the GitHub repository

other software or programming library; 2) it is a client-based software tools and therefore it does not require cluster administration; 3) SBpipe can also run locally via multithreading, which is ideal for preliminary testing of the most suitable algorithms for parameter estimation and simulation, before starting intensive jobs on a cluster.

SBpipe requires some familiarity with command line tools, although no programming skill is needed when COPASI models are used. Users only need to create a configuration file and run it using a simple command set. Users with a background in programming languages can also benefit from SBpipe functionalities using mathematical models coded with their preferred language if needed. In contrast to standard pipeline frameworks, SBpipe does not currently offer support for dependency management at coding level and reentrancy at execution level. The former is defined as a way to precisely define the dependency order of functions. The latter is the capacity of a program to continue from the last interrupted task. Although many pipeline frameworks are available for bioinformatics, the definition of a clear and spread standard specifying how pipelines can be configured is still limited in our opinion. In the future we hope to also use a pipeline framework as an additional way to run SBpipe tasks. Benefitting of dependency declaration and execution reentrancy would in particular be beneficial for running SBpipe on clusters or on the cloud.

From an implementation standpoint, SBpipe design is sufficiently generic to permit rapid extension of new pipelines. With this solid but flexible design, SBpipe aims to encourage the development of pipelines for systems modelling into a single community activity.

Conclusions

SBpipe is a novel open source software that enables systems biology modellers to simulate models, scan and estimate model parameters in a large scale. Novel analyses from multiple repeats are also computed via publication quality plots and tables. This project permits to increase productivity and reliability in model building and simulation.

Abbreviations

CL: Confidence level; IR: Insulin receptor; LSF: Load sharing facility; PLE: Profile likehood estimation; SGE: Sun grid engine

Acknowledgements

We acknowledge Dr Lu Li, Dr An Nguyen, and Dr Pınar Pir for helpful feedback.

Funding

This work was funded by British BBSRC (BBS/E/B/000C0419).

Authors' contributions

PDP and NLN conceived and designed the project; PDP implemented the software. PDP and NLN wrote the manuscript. Both authors read and approved the final manuscript.

Competing interests

The authors declare that they have no competing interests.

References

1. Ghosh S, Matsuoka Y, Asai Y, Hsin KY, Kitano H. Software for systems biology: from tools to integrated platforms. Nat Rev Genet. 2011;12: 821–32.
2. Le Novère N. Quantitative and logic modelling of molecular and gene networks. Nat Rev Genet. 2015;16:146–58.
3. van Rossum G. Python tutorial, Technical Report CS-R9526. Amsterdam: Centrum voor Wiskunde en Informatica (CWI); 1995.
4. R Development Core Team. R: A Language and Environment for Statistical Computing: Vienna; 2008. ISBN 3-900051-07-0.
5. Hoops S, Sahle S, Gauges R, Lee C, Pahle J, Simus N, et al. COPASI - a COmplex PAthway SImulator. Bioinformatics. 2006;22(24):3067–74.
6. Dalle Pezze P, Sonntag A, Thien A, Prentzell M, Gödel M, Fischer S, et al. A Dynamic Network Model of mTOR Signaling Reveals TSC-Independent mTORC2 Regulation. Sci Signal. 2012;5(217):ra25.
7. Le Novère N, Bornstein B, Broicher A, Courtot M, Donizelli M, Dharuri H, et al. BioModels Database: a free, centralized database of curated, published, quantitative kinetic models of biochemical and cellular systems. Nucleic Acids Res. 2006;34(Database issue):D689–91.
8. Kennedy J, Eberhart R. Particle Swarm Optimization. In: Proceedings of the Fourth IEEE International Conference on Neural Networks (Perth, Australia). Perth: IEEE; 1995. p. 1942–1948.
9. Schaber J. Easy parameter identifiability analysis with COPASI. Biosystems. 2012;110(3):183–5.
10. Dalle Pezze P, Nelson G, Otten E, Korolchuk V, Kirkwood T, von Zglinicki T, et al. Dynamic Modelling of Pathways to Cellular Senescence Reveals Strategies for Targeted Interventions. PLOS Comput Biol. 2014;10(8):1–20.
11. Balsa-Canto E, Henriques D, Gábor A, Banga JR. AMIGO2, a toolbox for dynamic modeling, optimization and control in systems biology. Bioinformatics. 2016;32(21):3357.
12. Schmidt H, Jirstrand M. Systems Biology Toolbox for MATLAB: a computational platform for research in systems biology. Bioinformatics. 2006;22(4):514.
13. Kent E, Hoops S, Mendes P. Condor-COPASI: high-throughput computing for biochemical networks. BMC Syst Biol. 2012;6(91). http://bmcsystbiol.biomedcentral.com/articles/10.1186/1752-0509-6-91.

10

A combined model reduction algorithm for controlled biochemical systems

Thomas J. Snowden[1,2], Piet H. van der Graaf[3,2] and Marcus J. Tindall[1,4*]

Abstract

Background: Systems Biology continues to produce increasingly large models of complex biochemical reaction networks. In applications requiring, for example, parameter estimation, the use of agent-based modelling approaches, or real-time simulation, this growing model complexity can present a significant hurdle. Often, however, not all portions of a model are of equal interest in a given setting. In such situations methods of model reduction offer one possible approach for addressing the issue of complexity by seeking to eliminate those portions of a pathway that can be shown to have the least effect upon the properties of interest.

Methods: In this paper a model reduction algorithm bringing together the complementary aspects of proper lumping and empirical balanced truncation is presented. Additional contributions include the development of a criterion for the selection of state-variable elimination via conservation analysis and use of an 'averaged' lumping inverse. This combined algorithm is highly automatable and of particular applicability in the context of 'controlled' biochemical networks.

Results: The algorithm is demonstrated here via application to two examples; an 11 dimensional model of bacterial chemotaxis in *Escherichia coli* and a 99 dimensional model of extracellular regulatory kinase activation (ERK) mediated via the epidermal growth factor (EGF) and nerve growth factor (NGF) receptor pathways. In the case of the chemotaxis model the algorithm was able to reduce the model to 2 state-variables producing a maximal relative error between the dynamics of the original and reduced models of only 2.8% whilst yielding a 26 fold speed up in simulation time. For the ERK activation model the algorithm was able to reduce the system to 7 state-variables, incurring a maximal relative error of 4.8%, and producing an approximately 10 fold speed up in the rate of simulation. Indices of controllability and observability are additionally developed and demonstrated throughout the paper. These provide insight into the relative importance of individual reactants in mediating a biochemical system's input-output response even for highly complex networks.

Conclusions: Through application, this paper demonstrates that combined model reduction methods can produce a significant simplification of complex Systems Biology models whilst retaining a high degree of predictive accuracy. In particular, it is shown that by combining the methods of proper lumping and empirical balanced truncation it is often possible to produce more accurate reductions than can be obtained by the use of either method in isolation.

Keywords: Model reduction, Lumping, Empirical balanced truncation, Controlled systems

*Correspondence: m.tindall@reading.ac.uk
[1]Department of Mathematics and Statistics, University of Reading, RG6 6AX, Reading, UK
[4]The Institute for Cardiovascular and Metabolic Research (ICMR), University of Reading, RG6 6AX Reading, UK
Full list of author information is available at the end of the article

Background

The field of Systems Biology has seen a considerable increase in both the number of models created and their complexity across the past decade. The BioModels Database, which acts as an open repository for Systems Biology models, saw the number of models it stores increase approximately ten-fold between 2005 and 2010, with the average number of reactions per model having nearly tripled in the same period [1]. This increase in complexity, specifically in the number of species or reactions modelled by each system, has become a defining characteristic of research in this area.

Such systems are typically developed by bringing together biochemical and physiological knowledge to inform highly detailed mechanistic models of biological networks (e.g. signalling pathways, protein-protein interactions, and genetic cascades). Mathematically, these networks are typically modelled via high-dimensional systems of stiff, nonlinear ordinary differential equations (ODEs).

This model complexity, however, can present a number of issues with regards to their use and analysis. For example parameter estimation techniques and real-time numerical simulation can both be difficult to perform for high dimensional or overparameterised systems. Even the basic intuitiveness of a system can potentially be obscured by its complexity. Additionally, complexity of this form is often associated with the 'curse of dimensionality', whereby the data that can be obtained for such systems in practice are sparse relative to the volume of the state and parameter spaces.

Model reduction techniques [2] offer one possible approach to easing complexity. A method of model reduction here refers to any method designed to construct a lower order representation (either in terms of the number of state variables or parameters) of a model with which some set of the original dynamical behaviour can be satisfactorily approximated.

A range of model reduction methods exist in the literature, many of which have previously been applied to models of biochemical reaction networks. The most commonly applied methods are based upon time-scale separation. These simplify a system by exploiting the wide ranges in reaction rates and equilibration speeds typical of biochemical reaction networks. These approaches include variants of the quasi-steady state approximation [3–9], variants of the rapid equilibrium approximation [10–15], the intrinsic low dimensional manifold method [16–20], and computational singular perturbation [21–24]. Beyond time-scale exploitation, sensitivity analysis can also be used to guide model reduction by identifying and eliminating those portions of a network least responsible for the dynamical behaviour of interest [25–29]. Optimisation based methods [30–35] seek to evaluate a range of possible reduced models under a given metric of reduction accuracy before returning the best available option. Lumping methods, meanwhile, reduce a network by treating groups of state-variables as a single dynamical, 'lumped' variable [36–41]. Additionally, there exist a range of singular value decomposition (SVD) reduction methods based upon the matrix decomposition of the same name. These exploit the property that SVD can be used to give lower rank approximations of matrices. Such methods have seen limited application to biochemical systems, but a number of publications employing a particular variant known as balanced truncation can be found in the literature [42–44].

Each model reduction approach has advantages and disadvantages in the reduction of large scale models of biochemical reaction networks. There is no 'one size fits all' method of reduction; the best method available depends inextricably on the properties of the model and the aims of reduction.

Systems Biology models, such as those this paper seeks to reduce, usually possess a number of mathematical properties that can influence the suitability of specific reduction methodologies: most notably, such models are often stiff, nonlinear and of extremely high dimensionality (e.g. containing 10s or even 100s of state-variables and parameters). Stiffness, in this context, refers to a parameter dependent property of a system of differential equations whereby their numerical integration can require taking a step-length that is excessively short relative to the exact solution's smoothness in a given interval. This has relatively severe implications for simulating such models, as stiffness is associated with issues of numerical stability and computational processing time. In the case of large-scale Systems Biology models, stiffness is typically a consequence of reactions in the system evolving and equilibrating at greatly different timescales as compared with one another. Meanwhile, the nonlinearity of these models implies that a number of analytical methodologies will not be applicable. Often linearisation is used in this context, but in many such cases this incurs a prohibitively large error. Finally, the high dimensionality of such systems can also present issues of combinatorial complexity or an excessive computational burden for some methods of mathematical analysis.

This paper specifically seeks to address the topic of controlled biochemical reaction networks. Here a controlled network refers to any model for which the dynamics are influenced by the concentration of a particular reactant that can be considered as an input into the network, within which some given combinations of the reactants can be considered as outputs of the system. In the context of Systems Biology this includes, for example, models of receptor signalling pathways where the concentration of an extracellular ligand may be seen as an input controlling the pathway. The concentration of some subset of the intracellular signalling species may also be considered an output that is directly observed or inferred from some measure of the cellular response. The recently emerging field of Quantitative Systems Pharmacology [45, 46], which proposes to mechanistically model the effects of drug action from the genetic scale upwards is particularly amenable to such a formulation.

Here we develop a model reduction algorithm focused on maintaining the input-output relationship of a controlled biochemical reaction network. The algorithm combines several approaches including conservation analysis, proper lumping and empirical balanced truncation. For controlled systems with a specified output, empirical balanced truncation is designed to give a reduction that accurately maintains the input-output relationship. Unfortunately, due to the need to repeatedly simulate the system under a range of perturbed conditions, empirical balanced truncation can be highly sensitive to model stiffness. Hence, given the stiffness that is commonly associated with such systems, in the combined algorithm proper lumping is employed as a preconditioner to enable the application of empirical balanced truncation whilst retaining an accurate reduced model.

Our paper is structured as follows: we will first outline the general model reduction problem and then proceed to provide an overview of conservation analysis, empirical truncation, and proper lumping. A detailed account of how these methods can be brought together to obtain more accurate reductions than can be obtained via application of any single method in isolation will then be given. Finally, we demonstrate the algorithm via application to two examples: an 11 dimensional model of bacterial chemotactic signalling in *Escherichia coli* [47] and a 99 dimensional model of extracellular signal-regulated kinase (ERK) phosphorylation mediated via the epidermal growth factor (EGF) and nerve growth factor (NGF) receptor pathways [48]. Results are compared and a number of enhancements to the core methods are discussed.

Problem outline

The models we seek to address are comprised of systems of coupled, nonlinear ODEs. These are formulated as initial value problems and can be expressed by a control affine, state-space representation such that

$$\dot{\boldsymbol{x}}(t) = \boldsymbol{f}(\boldsymbol{x}(t)) + \sum_{i=1}^{l} \boldsymbol{g}_i(\boldsymbol{x}(t))u_i(t), \tag{1a}$$

$$\boldsymbol{y}(t) = \boldsymbol{h}(\boldsymbol{x}(t)), \tag{1b}$$

with initial conditions $\boldsymbol{x}(0) = \boldsymbol{x}_0$ and where the over-dot represents the time derivative, such that $\dot{\boldsymbol{x}} = \frac{\mathrm{d}x}{\mathrm{d}t}$. Here the state variables $\boldsymbol{x}(t) \in \mathbb{R}^n$ typically represent the time-varying concentrations of the modelled species, $\boldsymbol{u}(t) \in \mathbb{R}^l$ (such that $u_i(t) \in \boldsymbol{u}(t)$) represent the input variables, and $\boldsymbol{y} \in \mathbb{R}^p$ represent the output variables. Here $\boldsymbol{f}(\boldsymbol{x}(t))$ is the set of functions describing the dynamical interaction between the individual reactants, each set of functions $\boldsymbol{g}_i(\boldsymbol{x}(t))$ describes the dynamical behaviour of the reactants interacting with each of the inputs, and $\boldsymbol{h}(\boldsymbol{x}(t))$ describes the combinations of the reactant concentrations corresponding to each of the outputs.

We seek a reduced model of the form

$$\dot{\tilde{\boldsymbol{x}}}(t) = \tilde{\boldsymbol{f}}(\tilde{\boldsymbol{x}}(t)) + \sum_{i=1}^{l} \tilde{\boldsymbol{g}}_i(\tilde{\boldsymbol{x}}(t))u_i(t), \tag{2a}$$

$$\bar{\boldsymbol{y}}(t) = \tilde{\boldsymbol{h}}(\tilde{\boldsymbol{x}}(t)), \tag{2b}$$

where $\tilde{\boldsymbol{x}} \in \mathbb{R}^{\tilde{n}}$ represents a reduced set of state variables (such that $\tilde{n} < n$) and for which, given a set of inputs $\boldsymbol{u}(t)$, the reduced set of outputs $\bar{\boldsymbol{y}}(t)$ represents an approximation of the original set $\boldsymbol{y}(t)$. Similarly to $\boldsymbol{f}$, $\boldsymbol{g}$, and $\boldsymbol{h}$ in the unreduced model, $\tilde{\boldsymbol{f}}(\tilde{\boldsymbol{x}}(t))$ and $\tilde{\boldsymbol{g}}_i(\tilde{\boldsymbol{x}}(t))$ are sets of functions describing the dynamical effects of interactions between the reduced state-variables and inputs. Meanwhile $\tilde{\boldsymbol{h}}(\tilde{\boldsymbol{x}}(t))$ approximately maps the reduced state-variables to the outputs.

The accuracy of a reduced model in capturing the dynamics of the original can be defined in a number of ways to suit the needs of the modeller. The most common approaches, however, are based upon the instantaneous error between the outputs of the two systems

$$\boldsymbol{\epsilon}(t) = \left|\boldsymbol{y}(t) - \bar{\boldsymbol{y}}(t)\right|. \tag{3}$$

The most common metrics are the L^2-norm $\|\boldsymbol{\epsilon}(t)\|_2 = \left(\int \boldsymbol{\epsilon}(t)^2\, dt\right)^{1/2}$ or the ∞-norm $\|\boldsymbol{\epsilon}(t)\|_\infty = \sup\{\boldsymbol{\epsilon}(t)\}$. Throughout this paper we employ a form of maximal relative error E as the metric that we aim to minimise, such that

$$\epsilon_i \in \boldsymbol{\epsilon}: \ \epsilon_i(t) = \frac{\|y_i(t) - \bar{y}_i(t)\|}{y_i(t)} \quad \text{and} \quad E = \|\boldsymbol{\epsilon}(t)\|_\infty. \tag{4}$$

Here, the relative error is selected such reduced models can be compared fairly for a range of different perturbation magnitudes applied both to the inputs and the initial condition of the state-variables.

Methods

Our combined model reduction algorithm is designed to bring together the methods of nondimensionalisation, conservation analysis, proper lumping and empirical balanced truncation. At its core the method employs proper lumping as a preconditioner (to reduce model stiffness) prior to the application of empirical balanced truncation. In this section we briefly review the variants of the methods employed before providing a more detailed account of the algorithm.

Nondimensionalisation

Nondimensionalisation refers to the process of rescaling the variables in a system such that the physical units (typically units of concentration and time) are removed from the model [49]. There are number of purposes for nondimensionalisation in the analysis of biochemical systems — primary amongst these is its use in accessing characteristic or intrinsic properties of the reaction network. Usually these are ratios of kinetic rate constants and conserved values that enable greater intuition into how the parameterisation of a model governs its behaviour.

This yields a nondimensionalised parameter set $\tilde{\boldsymbol{p}}$ with entries representing ratios of the original parameters $\boldsymbol{p}$. Often, nondimensionalisation can result in a reduction in the dimension of the new parameter set $\tilde{\boldsymbol{p}}$ by finding ratios that are fixed to 1 irrespective of the original parameterisation. This does not, however, result in a reduction in the number of modelled reactions or reactants and hence does not reduce complexity as previously defined. Additionally, the dimensionless parameters may lose their innate biological meaning as the ratios they represent may not always hold particular biological significance.

There is usually a large possible number of combinations of the original parameters that could be used to yield these nondimensional ratios - for example, it is common to rescale time relative to a single kinetic rate parameter, amongst which a great number of choices may exist. Whilst there is no single method to determine a 'best' or 'optimal' nondimensionalisation, in the case where the system is fully parameterised and parameter values are fixed, selecting a nondimensionalised parameter set $\tilde{\boldsymbol{p}}$ with entries all of a similar order will typically improve the numerical properties of the model for computational issues such as rounding error. To achieve this, the combined model reduction algorithm randomly samples 50 possible parameter combinations as nondimensionalsations and selects the one that minimises σ, where

$$\sigma = \log_{10}\left(\frac{\max(\tilde{\boldsymbol{p}})}{\min(\tilde{\boldsymbol{p}})}\right). \tag{5}$$

Conservation analysis

It is common in models of biochemical reaction networks for the total concentration of certain subsets of the species to remain constant at all times independent of the model's specific parameterisation [50]. Such subsets are commonly referred to as conserved moieties. Replacement of state-variables via the algebraic exploitation of conservation relations is a common first step in the analysis of biochemical reaction networks. Eliminating one of the conserved state-variables for each of the conservation relations can be used to yield a simplified realisation of the system.

In principle, all such conservation relations for a given biochemical reaction network can be determined by finding the linear dependencies in the associated stoichiometry matrix of the system. Mathematically this relies upon computing the left null space of the stoichiometry matrix for which a number of well-established methods exist. A review of a range of such techniques, including Gaussian elimination and singular value decomposition, can be found in Sauro and Ingalls [51]. For smaller scale systems such methods will usually find all available conservation relations, but for higher dimensional systems difficulties may occur. In particular, it is often necessary to select more stable computational methods to avoid missing conservation relations or finding false ones due to numerical error. A particularly stable method based upon the construction of a QR decomposition via Householder reflections has been developed by Vallbhajosyula et al. [52]. We therefore employ a form of this approach here, a more detailed mathematical treatment of which can be found in Section 1.2 of the Additional file 1.

This form of analysis leads to a simplified realisation of the system under which it is possible to obtain nonsingular Jacobians for given states of the model. As such it can be seen as a first step in model reduction schemes involving numerical methods.

Proper lumping

A lumping can potentially refer to any direct mapping $L : \mathbb{R}^n \to \mathbb{R}^{\tilde{n}}$ of the original state variables $\boldsymbol{x}(t) \in \mathbb{R}^n$ to a reduced set $\tilde{\boldsymbol{x}}(t) \in \mathbb{R}^{\tilde{n}}$ where $\tilde{n} < n$. Here we limit ourselves to linear lumping, such that we have a projection of the form

$$\tilde{\boldsymbol{x}}(t) = L\boldsymbol{x}(t), \tag{6}$$

and, additionally, proper lumping such that the projection L becomes a matrix $L \in \{0,1\}^{\tilde{n}\times n}$ where each column is pairwise orthogonal. This implies that each of the original state-variables corresponds to, at most, one of the lumped state-variables in the reduced model as is depicted in Fig. 1(a).

The dynamics of the reduced variables $\tilde{\boldsymbol{x}}(t)$ are obtained via application of the Galerkin projection [53] (a detailed account of which can be found in the Section 1.1 of the Additional file 1) to the original system (2), such that

$$\dot{\tilde{\boldsymbol{x}}}(t) = L\boldsymbol{f}(\bar{L}\tilde{\boldsymbol{x}}(t)) + \sum_{i=1}^{l} L\boldsymbol{g}_i(\bar{L}\tilde{\boldsymbol{x}}(t))u_i(t),$$

$$\text{with } \tilde{\boldsymbol{x}}(0) = \tilde{\boldsymbol{x}}_0 = L\boldsymbol{x}_0, \tag{7a}$$

$$\tilde{\boldsymbol{y}}(t) = \boldsymbol{h}(\bar{L}\tilde{\boldsymbol{x}}(t)), \tag{7b}$$

where $\bar{L} \in \mathbb{R}^{n\times\tilde{n}}$ represents a generalised inverse of L such that $L\bar{L} = I_{\tilde{n}}$ (the $\tilde{n}\times\tilde{n}$ identity matrix). An approximation for the original state variables from the reduced variables can therefore be computed as

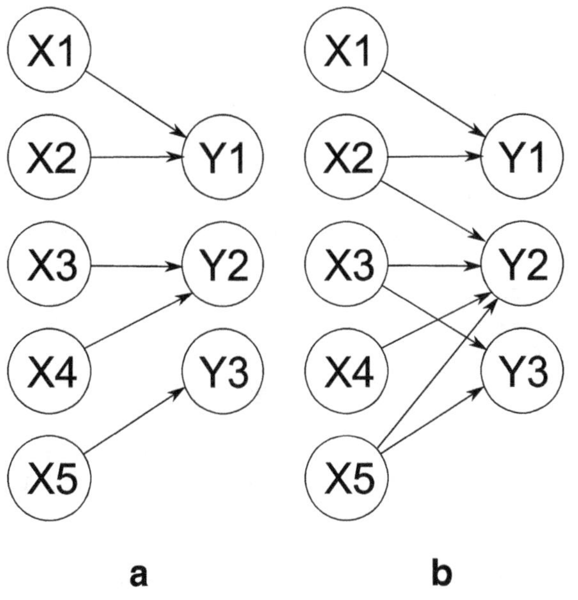

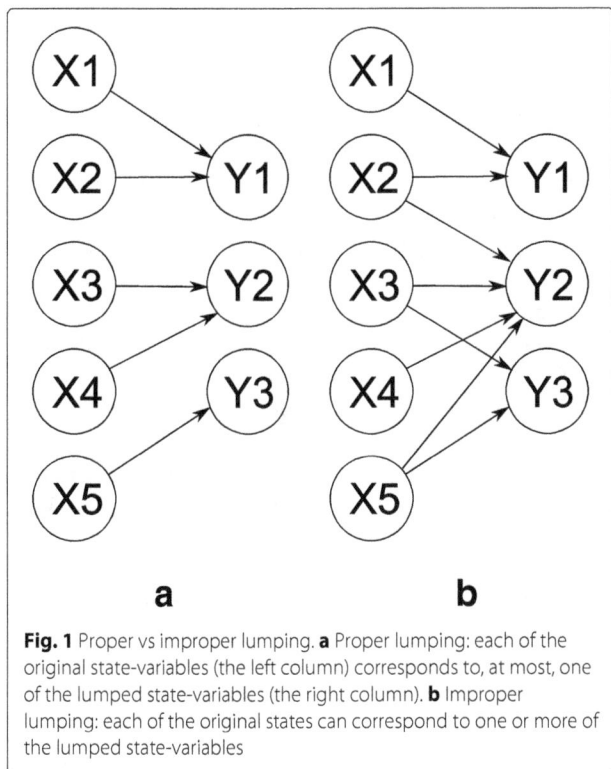

Fig. 1 Proper vs improper lumping. **a** Proper lumping: each of the original state-variables (the left column) corresponds to, at most, one of the lumped state-variables (the right column). **b** Improper lumping: each of the original states can correspond to one or more of the lumped state-variables

$$x(t) \approx \bar{L}\tilde{x}(t). \tag{8}$$

Finding an optimal lumping

There exist a number approaches for selecting an appropriate lumping matrix L for producing a system of given reduced dimensionality $\tilde{n}$. Here we choose to employ the scheme described by Dokoumetzidis and Aarons [37]. This algorithm runs an exhaustive search of possible lumping matrices to determine which produces the lowest error between simulation of the original model and the reduced model. To speed up this process, it is assumed (from justifications given in the original paper) that the lowest error k dimensional reduction obtained via lumping of an n dimensional system can also be found as the optimal lumping of two states in the $k+1$ dimensional reduction. This yields a 'forward selection' strategy, where 2 of the state-variables are lumped at each step, which greatly decreases the combinatorial burden of possible lumping matrices that must be evaluated.

Lumping and stiffness

Stiffness here is defined as the ratio between the largest and smallest eigenvalues of the Jacobian matrix of the system evaluated at its unperturbed initial condition, such that

$$\chi = \frac{\lambda\left(J|_{x=x_0}\right)_{\max}}{\lambda\left(J|_{x=x_0}\right)_{\min}}, \quad x_0 \in \mathbb{R}^n. \tag{9}$$

A high stiffness coefficient implies that parts of the system are evolving at significantly greater rates than others which can lead to traditional methods of simulation being numerically unstable.

The lumping algorithm of Dokoumetzidis and Aarons [37] will tend to sum together those state-variables that interact on faster timescales than their neighbours, and hence rapidly reach a form of proportional equilibrium. As a result the reduced model will tend to contain a lower range of timescales and a lower stiffness coefficient with every additional dimension eliminated. Furthermore, proper lumping, which creates reduced state-variables as straightforward sums of the originals, retains a degree of a biological interpretability that may not hold for alternative, coordinate transforming methods of model reduction.

Empirical balanced truncation

Our approach to empirical balanced truncation is based upon the procedure developed by Hahn and Edgar [54]. The aim is to construct two covariance matrices, known as the empirical controllability and observability Gramians, via repeated simulations of the system under perturbations of the input and the initial conditions respectively. The controllability Gramian provides information on how changes in the input will alter the state of the system. The observability Gramian provides information on the magnitude of the output any given initial condition of the system can produce. As these Gramians are positive semi-definite matrices, one good way to interpret them is as ellipsoids in the phase-space of the model [55]. The ellipsoid of the controllability Gramian represents the set of states that can be reached for a given magnitude of input, the ellipsoid of the observability Gramian describes the set of initial states that can produce an output of a given magnitude. A detailed mathematical account of the computation of empirical Gramians can be found in Section 1.3 of the Additional file 1.

We have modified the empirical balanced truncation procedure of Hahn and Edgar to deal with a system where state-variables have been eliminated via conservation analysis. In particular the conserved totals should be treated as functions of the initial conditions of the system and be altered in accordance with the perturbations used to create the observability Gramian. This allows us to perturb the initial conditions of the system without risking violation of conservation.

Once the Gramians have been computed the aim is to construct a balancing transformation of the state variables which also acts to equalise and diagonalise the Gramians. As the Gramians are now diagonal, each of the associated state-variables is therefore orthogonal in terms of their contribution to the input-output relationship. Hence, under this transformation, the state-variables which contribute least can be truncated without influencing the remaining terms.

The balancing transformation can be obtained in a numerically stable way following the approach of Laub et al. [56]. Firstly, perform a Cholesky factorisation of both the controllability Gramian $\mathcal{P}$ and the observability Gramian $\mathcal{Q}$ to obtain

$$\mathcal{P} = L^{\mathsf{T}}L \quad \text{and} \quad \mathcal{Q} = R^{\mathsf{T}}R,$$

where L and R represent the upper triangular factors of the Gramians. Now, take a singular value decomposition of the newly formed matrix LR^{T} and select a reduced dimensionality $\tilde{n}$ of the new model to obtain

$$LR^{\mathsf{T}} = (U_1 \; U_2)\begin{pmatrix} \Sigma_1 & 0 \\ 0 & \Sigma_2 \end{pmatrix}\begin{pmatrix} V_1^{\mathsf{T}} \\ V_2^{\mathsf{T}} \end{pmatrix},$$

where U_1 is an $n \times \tilde{n}$ matrix, Σ_1 is an $\tilde{n} \times \tilde{n}$ matrix (of the form $\Sigma_1 = \mathrm{diag}(\sigma_1^2, \ldots, \sigma_{\tilde{n}}^2)$) and V_1^{T} is a $\tilde{n} \times n$ matrix. Finally, set

$$T_1 = \Sigma_1^{-\frac{1}{2}} V_1^{\mathsf{T}} R \quad \text{and} \quad S_1 = L^{\mathsf{T}} U_1 \Sigma_1^{-\frac{1}{2}}. \tag{10}$$

Given the state-variable projection $\tilde{\boldsymbol{x}} = T_1\boldsymbol{x}$ and the particular generalised inverse leading to the approximation $\boldsymbol{x} \approx S_1\tilde{\boldsymbol{x}}$, we construct the reduced dynamics for this system again via application of the Galerkin projection to the original system (2), such that

$$\dot{\tilde{\boldsymbol{x}}}(t) = T_1\boldsymbol{f}(S_1\tilde{\boldsymbol{x}}(t)) + \sum_{i=1}^{l} T_1\boldsymbol{g}_i(S_1\tilde{\boldsymbol{x}}(t))u_i(t), \quad T_1\boldsymbol{x}(0) = \tilde{\boldsymbol{x}}_0, \tag{11a}$$

$$\tilde{\boldsymbol{y}}(t) = \boldsymbol{h}(S_1\tilde{\boldsymbol{x}}(t)). \tag{11b}$$

This reduced system will not only feature fewer state-variables, but will also typically be faster to simulate and contain fewer parameters.

Stiffness and empirical balanced truncation

The construction of empirical Gramians via repeated simulation of the system under perturbations of input and initial conditions can be sensitive to numerical error. Where balanced truncation is applied using Gramians with a high degree of associated error, blow-up problems can occur for simulations of the reduced system. This accumulation of numerical error is particularly likely to occur for systems with a high-stiffness coefficient ($\chi \gg 1$).

Given the stiffness reducing property of lumping, however, and its retention of some biological meaning it can potentially be treated as a preconditioning step - enabling the application of more numerically sensitive methods (such as empirical balanced truncation) to previously lumped systems.

The combined model reduction algorithm

Here a combined, automatable algorithm for model reduction bringing together the methods of nondimensionalisation, conservation analysis, proper lumping and empirical balanced truncation is introduced. A high-level overview of the algorithm's steps is shown in Fig. 2.

The algorithm is designed to be automatically applicable to models given in Systems Biology Markup Language (SBML) form — a standardised format for the representation, storage and easy communication of Systems Biology models. Many such models of this form can commonly be obtained from online repositories. Publicly open databases, such as the previously mentioned BioModels Database, contain thousands of such models enabling researchers to share their work in a more accessible way.

As preliminary steps, nondimensionalisation and conservation analysis are applied to the model. Nondimensionalisation is applied to improve numerical accuracy by reducing the range of parameters accounted for in the model. Conservation analysis is then applied in order to obtain a simplified realisation of the system and remove any associated singularities from the system's associated Jacobian.

At the core of the algorithm, however, are the methods of proper lumping and empirical balanced truncation. Theoretically, empirical balanced truncation should produce lower output error reductions across a range of inputs than lumping. However, due to the need to construct accurate covariance matrices of data from repeated numerical simulations, empirical balanced truncation can fail when applied to highly stiff systems. Conversely, lumping strips the model of some of its stiffness for each reduced dimension. In the combined algorithm, therefore, the complimentary aspects of proper lumping and empirical balanced truncation are exploited. Lumping is used to reduce the system until the stiffness coefficient $\chi_{\tilde{n}}$ of the reduced model is within a satisfactory range $\chi_{\tilde{n}} < \chi_c$, for χ_c some pre-chosen critical stiffness value (from numerical experimentation with example models we set this to be 250 in the automated algorithm). Empirical balanced truncation is then employed to obtain further model reduction whilst maintaining a good error bound between the outputs of the original and reduced models.

The algorithm as presented will proceed until the reduced system exceeds the maximum tolerated error, here defined to be 5%. The algorithm then returns the lowest dimensional reduced system that meets this constraint. It is not possible to know a priori what degree of reduction will be attainable by the algorithm, and the actual reduction achieved is both model structure and parameterisation dependent. Application of the combined

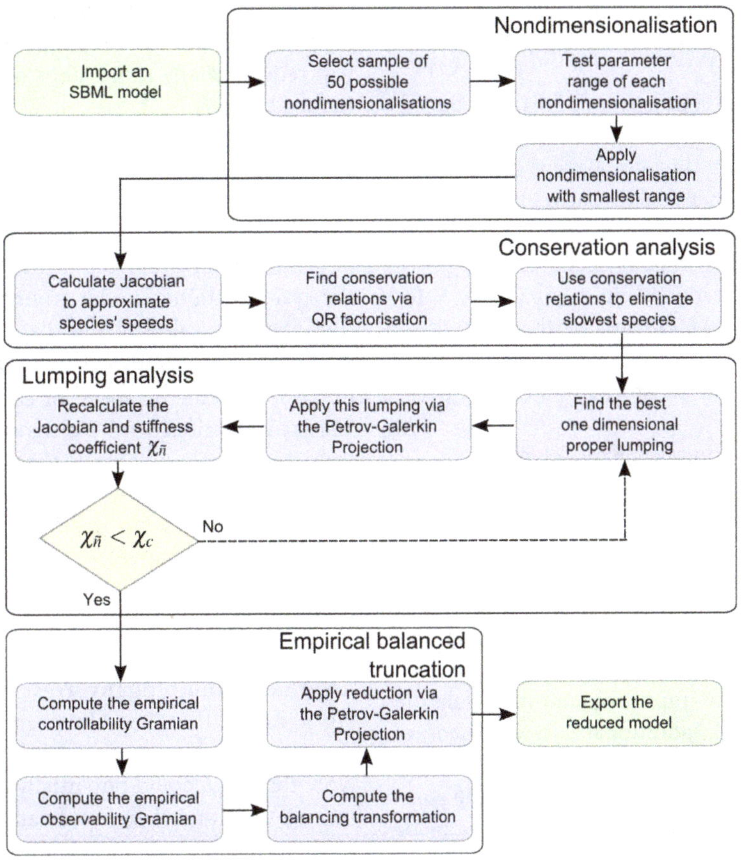

Fig. 2 A flow chart giving an overview of the order of steps performed in the combined algorithm. This chart only represents the high-level actions of the algorithm with many of the blocks representing subroutines that are somewhat more nuanced in practice. A detailed account of all steps can be found in the Additional file 1

reduction algorithm does require that the model it is applied to is fully parameterised prior to reduction. As such, the reduced model is only guaranteed to be accurate locally to a point in parameter space. For small perturbations to the initial parameterisation it is likely that the reduced model will remain accurate, however for larger deviations it may not be suitable for describing the overall input-output behaviour of the system.

Combined algorithm implementation

The main steps of the combined model reduction algorithm are as follows.

1. Import the model of interest with a set of predefined state-variables ($\boldsymbol{x}(t) \in \mathbb{R}^n$), reactions, and rate parameters into the algorithm.
2. Request the user to define the set of inputs ($\boldsymbol{u}$) and the set of outputs ($\boldsymbol{y} = h(\boldsymbol{x}(t))$) in which they are interested.
3. Randomly select a sample of 50 possible nondimensionalisations under the original parameterisation of the model.
4. Calculate the range of the new parameter set for each nondimensionalisation.
5. Select and apply the nondimensionalisation sampled with the smallest range of orders of magnitude of parameters to help avoid rounding-off and truncation errors.
6. Compute the conservation relationships using a standard QR factorization method via Householder reflections (outlined, for example, in [51]).
7. Prioritise which state-variables to replace in the dynamical system via exploitation of the conservation relations.
8. Compute the linear, proper lumping matrix for all lumpable pairs of state-variables remaining in the system. Expressed alternatively, compute the set of all possible proper lumping matrices that will reduce the number of state-variables by one.
9. Compute the lumping inverse.
10. Simulate each lumped system and compute the associated maximal relative output error.

11. Select the lumping matrix L and associated inverse that produces the lowest error and apply this reduction.
12. Calculate the stiffness coefficient $\chi_{\tilde{n}}$ for this reduced system.
13. Return to step (8). Exit the loop of steps (8)-(13) when the lumping either violates the maximum tolerated error (typically set to 5%) or reduces the stiffness coefficient to within an acceptable range $\chi_{\tilde{n}} < \chi_c$ or when the reduced dimensionality $\tilde{n} = 1$.
14. Compute the empirical Gramians for the lumped system.
15. Compute the balancing transformation for the given Gramians.
16. Apply the transformation to the model via the Galerkin projection and truncate state-variables until the maximum acceptable error is reached.
17. Produce the symbolic form of the reduced model and terminate.

The main algorithm was implemented in a commercial software package (Matlab 2013b, MathWorks Inc., Natick, MA). To enable the importing and manipulation of models stored in SBML, additional use was made of two open-source toolboxes - libSBML [57] and SBtoolbox2 [58]. Note that a more detailed account of the steps and implementation of the combined algorithm can be found in Section 3 of the Additional file 1.

Algorithm enhancements

The algorithm additionally features two specific enhancements to the combined methodologies that will be outlined in the following section. The first of these addresses the question of how to select which state-variables should be explicitly eliminated from the system of ODEs via the exploitation of conservation relations. The second addresses the question of which of the possible generalised inverses should be used to reduce the system via lumping with the Galerkin projection.

Selection of state-variables for elimination via conservation analysis: Typically, the most accurate possible proper lumping of two state-variables in a model will occur between those two that most rapidly equilibrate with respect to the remainder of the system. In short, if two variables quickly reach a point where their proportional concentration with respect to each other is approximately constant, they can be lumped accurately. If a state-variable from this 'best' lumped pair has been replaced by an algebraic equation from the conservation relations, then it will necessarily be unavailable for lumping in the combined algorithm. Another concern is that the choice of state-variables eliminated will have an effect upon the reduction of stiffness attainable via the application of lumping; if the fastest interacting state-variables are not represented explicitly the stiffness will require many more steps of lumping in order to be reduced.

To avoid this issue the combined algorithm initially 'speed' ranks the state-variables in the model such that the slowest variables can then be selected for elimination. This speed ranking is achieved via numerical calculation of the Jacobian evaluated at the steady-state $\boldsymbol{x} = \boldsymbol{x}^*_{\boldsymbol{u}_0}$ of the system attained under an unperturbed input $\boldsymbol{u} = \boldsymbol{u}_0$. The absolute value of the diagonal entries of the Jacobian $J|_{\boldsymbol{x}=\boldsymbol{x}^*_{\boldsymbol{u}_0}}$ are then used as an approximate metric of the initial time-scale or speed of each of the state-variables. This is equivalent to the outgoing rate of concentration from each state-variable at the unperturbed steady-state of the network. Those state-variables corresponding to the largest values are deemed fast, whilst those corresponding to the smallest values are deemed slow and prioritised for elimination via application of conservation analysis.

Alternative lumping inverses: In order to apply a lumping L under the Galerkin projection, some generalised right-inverse $\bar{L}$ of L such that $L\bar{L} = I_{\tilde{n}}$ is required. Note that, as $\bar{L}$ could potentially be any generalised right-inverse of L, and there exists an infinite number of ways to construct such a matrix. However, the particular choice made will have an effect on the maximal relative output error incurred by the reduced model as defined by Eq. (4). Some of the alternatives are evaluated here.

In the original Wei and Kuo papers [59, 60] they suggest selecting the $\bar{L}$ that reconstructs the steady-state of the system such that $\boldsymbol{x}(t) = \bar{L}\tilde{\boldsymbol{x}}(t)$ for $t \to \infty$. This can be constructed as follows. Let

$$X := \text{diag}\left\{\boldsymbol{x}^*\right\},$$

where $\boldsymbol{x}^*$ represents the steady-state of the system such that $\lim_{t\to+\infty} \boldsymbol{x}(t) = \boldsymbol{x}^*$, then

$$\bar{L} = XL^\top \left(LXL^\top\right)^{-1}. \tag{12}$$

As, in the case of a controlled biochemical network, the steady-state of the system is necessarily dependent on the value of the inputs $\boldsymbol{u}(t)$. We here seek to maintain the generality of the reduced model by employing the steady-state attained under an unperturbed input $\boldsymbol{u} = \boldsymbol{u}_0$, such that $\boldsymbol{x} = \boldsymbol{x}^*_{\boldsymbol{u}_0}$.

The Dokoumetzidis and Aarons [37] paper, which introduces the proper lumping algorithm that the scheme presented here is based upon, follows the work of Li and Rabbitz [61]. They suggested the somewhat simpler approach of using the Moore-Penrose inverse of L such that $\bar{L} = L^+$, which can similarly be computed as

$$L^+ = L^\top \left(LL^\top\right)^{-1}. \tag{13}$$

Typically the approach of reconstructing the steady-state values will lead to a better approximation of the original dynamics. In the case where a lumping sums a group of species all with a steady-state value of zero, however, this will lead to a singularity and thus the Moore-Penrose approach is to be preferred.

We thus consider an alternative to the above methodologies; we design an inverse $\hat{L}$ designed to reconstruct the proportionally average value of the states-variables for all time-points with non-zero values. To understand this alternative lumping inverse, first observe that all lumping matrices L (and inverses $\bar{L}$) can be expressed as the composition of sequential lumping matrices of two state-variables. For example

$$\begin{bmatrix} 1 & 1 & 1 & 0 \\ 0 & 0 & 0 & 1 \end{bmatrix} = \begin{bmatrix} 1 & 1 & 0 \\ 0 & 0 & 1 \end{bmatrix} \begin{bmatrix} 1 & 1 & 0 & 0 \\ 0 & 0 & 1 & 0 \\ 0 & 0 & 0 & 1 \end{bmatrix},$$

which demonstrates that a lumping of 3 state-variables can be achieved via two sequential lumpings of two state-variables. The novel inverse $\hat{L}$ is constructed for the two state-variable lumping case and can be generalised to any lumping via this process of sequential composition. Hence consider the general case seeking to lump the state-variables $x_h(t)$ and $x_k(t)$ (with $h < k$), where the corresponding lumping matrix $L = \{l_{ij}\}$ has the entries

$$l_{ij} = \begin{cases} 1 \ \text{for } i = j \text{ with } i < k, \\ 1 \ \text{for } i = j - 1 \text{ with } i \geq k, \\ 1 \ \text{for } i = h, j = k, \\ 0 \ \text{otherwise.} \end{cases} \tag{14}$$

The lumping inverse $\hat{L} = \{\hat{l}_{ij}\}$ is then based upon calculating the average proportion of each of the lumped state-variables $x_h(t)$ and $x_k(t)$ during the interval $0 \leq t \leq T$; here, T is defined as some time point where the system can be assumed to have approximately reached steady-state, such that

$$\|x(T) - x^*\| \ll 1.$$

Hence each element of the inverse is constructed as follows

$$\hat{l}_{ij} = \begin{cases} 1 & \text{for } i = j \text{ with } i < h, \\ 1 & \text{for } i = j \text{ with } i > h \text{ and } j < k, \\ 1 & \text{for } i = j + 1 \text{ with } i \geq k, \\ \frac{1}{T}\int_0^T \frac{x_i(t)}{x_i(t)+x_j(t)} dt & \text{for } i = h, j = h, \\ \frac{1}{T}\int_0^T \frac{x_j(t)}{x_i(t)+x_j(t)} dt & \text{for } i = k, j = h, \\ 0 & \text{otherwise.} \end{cases} \tag{15}$$

In the case where the time $T \to \infty$ and either or both state-variables have a non-zero steady-state this reduces to the steady-state reconstructing lumping inverse defined by Eq. (12). For the zero steady-state situation, however, it avoids the issue of numerical singularities.

Indices of controllability and observability

One of the benefits of the application of empirical balanced truncation is its potential use in obtaining metrics of observability and controllability for the individual state-variables of nonlinear systems. Given the typically nonlinear nature of cell signalling models, this potentially allows a framework for accessing indices of the controllability of each state-variable via receptor activation or suspension, the observability of each state-variable in influencing the output of interest, and the contribution of each state-variable to the overall input-output relationship of the model.

The interpretation of standard Gramians as ellipsoids describing the controllability and observability of the directions in phase-space [55] can also be extended to the interpretation of empirical Gramians. Hence it is possible to employ these matrices to obtain indices of controllability and observability for individual state-variables in nonlinear systems. Given this, we define the following indices.

Observability index: The observability index of the i-th state-variable $\nu_{oi} \in \boldsymbol{\nu}_o$ is defined as

$$\nu_{oi} := \frac{\sqrt{e_i^{\mathsf{T}} \mathcal{Q} e_i}}{\max\left\{\sqrt{e_i^{\mathsf{T}} \mathcal{Q} e_i}\right\}},$$

where $0 \leq \nu_{oi} \leq 1$ and e_i represents the ith unit vector. Note the index has been rescaled relative to the maximally observable state-variable, such that each represents the comparable influence of the associated state-variable. The greater the value the more influence the state-variable has on the observed output. A value of zero indicates the state-variable has no effect on the output of interest.

Controllability index: The controllability index of the i-th state-variable $\nu_{ci} \in \boldsymbol{\nu}_c$ is defined as

$$\nu_{ci} := \frac{\sqrt{e_i^{\mathsf{T}} \mathcal{P} e_i}}{\max\left\{\sqrt{e_i^{\mathsf{T}} \mathcal{P} e_i}\right\}},$$

with $0 \leq \nu_{ci} \leq 1$. Again, the indices here are rescaled relative to their maximal value such that they are more comparable to each other and to other indexes. The greater the value the more controllable the state-variable is via the input. A value of zero indicates that the input has no effect on the corresponding state-variable.

Input-output importance index: The input-output index $\boldsymbol{\nu}$ is then defined as the element-wise product of the observability and controllability indices rescaled proportionate to the maximal value. Hence the index corresponding to the i-th state-variable $\nu_i \in \boldsymbol{\nu}$ is given by

$$\nu_i := \frac{\nu_{oi}\nu_{ci}}{\max\{\nu_{oi}\nu_{ci}\}},$$

with $0 \leq \nu_i \leq 1$. This therefore provides an overall metric of the corresponding state-variable's influence on the overall input-output relationship of the system.

Results and discussion

In this section two examples are employed to demonstrate the application of the combined model reduction algorithm, the enhancements made to the base methods, and the calculation of the indices of controllability and observability. The first of these systems is a relatively simple (11 dimensional) model of bacterial chemotaxis in *Escherichia coli* - the modest scope of this model allows the application of our methods to be more easily intuited. The second is a significantly more substantial model (99 dimensional) describing the mediation of ERK activation via both the EGF and NGF receptor pathways. Through application to a model of this size, our methods demonstrate their potential usefulness in analysing models that might be inscrutable under traditional approaches.

A model of bacterial chemotaxis

We have applied our combined model reduction methodology to a model of chemotactic signalling in *E. coli* as detailed in Tindall et al. [47] and summarised in Section 2 of the Additional file 1. This is a modest 11 dimensional model detailing a system of 12 biochemical reactions. This serves as a reasonable starting example as the model is intuitively tractable, but also of a sufficient size and complexity to be meaningfully reduced by the combined algorithm.

E. coli is a common, rod shaped, gram-negative bacterium often used as a model organism in biological studies due to both the large body of existing literature characterising its behaviour as well as the relative ease and inexpensiveness in its growth and experimental manipulation. There are many strains of *E. coli* present in nature, but the model discussed here pertains specifically to those strains that exhibit a chemotactic response. Chemotaxis is the process by which a cell senses an environmental chemical gradient and biases its movement towards those regions most suitable for growth and reproduction. In the model presented here, this process involves the transmembrane receptors on the surface of the bacterium sensing the local concentrations of an attractant or repellent; a decrease in attractant or an increase in repellent will cause the receptors to activate a signalling pathway inside the cell resulting in an increase of the intracellular concentration of the phosphorylated chemotactic CheY protein, referred to here as CheY_P. This concentration, in turn, modulates the flagellum's movement, resulting in a change of direction for the cell either towards attractants or away from repellents.

Chemotaxis represents a good example model to work with as its attractant-receptor behaviour represents a controlled system, and it is highly amenable to the input-output problem formulation that the combined algorithm seeks to address. For our analysis, the external concentration of some chemotactic attractant was treated as the input into the system and the total concentration of CheY_P (in both free and complex forms). Hence,

$$y(t) = [\,\text{CheY}_P] + [\,\text{CheA}\cdot\text{CheY}_P] + [\,\text{CheY}_P\cdot\text{CheZ}]\,. \tag{16}$$

The model has a stiffness coefficient at the initial condition of the system of 958.3, which is relatively high. Values for these initial conditions can be found in Section 2 of the Additional file 1.

Reduction

We sought to compare the reduction of this example via lumping and empirical balanced truncation alone, and our combined algorithm.

In respect of the combined algorithm, the process of reduction began by nondimensionalising the variables of the system - specifically seeking to rescale the initial conditions and coefficients in the model such that they spanned the fewest orders of magnitude with the aim of avoiding possible issues of numerical truncation. The conservation relations for the system were then calculated; as 4 conservation relations were found, this lead to a system of 7 ODEs that exactly replicated the original system's dynamics.

The algorithm then lumped the state-variables until the stiffness coefficient was less than 250. In our example this occurred at 5 state-variables. Finally, empirical balanced truncation was applied to the 5 dimensional reduced model. In this case, the empirical Gramians were constructed using data from 100 distinct simulations covering perturbations to both the models input parameter $u(t)$ and the initial conditions $\boldsymbol{x}_0$. In both cases, perturbations were uniformly sampled from 0.2 to 1.8 times each parameter's original, unperturbed value. Using the balancing transformation and straightforward truncation the associated error for each possible dimensionality of reduction (between four and one) was calculated. A significantly more detailed account of reducing the bacterial chemotaxis model can be found in Section 2.1 of the Additional file 1.

The results in Table 1 along with Figs. 3 and 4 clearly demonstrate the combined algorithm produces more accurate reductions than those of either method in isolation. At the 2-dimensional reduction level, the combined algorithm exhibited an approximately 78% improvement in reduction error in comparison to the use of lumping alone. Additionally, this reduced model produced a significant speed-up in simulation time. In the case of 100 repeated simulations over a 3 second period under the introduction of $10\mu M$ concentration of attractant ligand at $t = 0$, the original 11 dimensional system required 0.2594 seconds on average to be simulated, whilst the 2 dimensional reduced model was solved in 0.0101 seconds.

Selection of species eliminated via conservation analysis

As was previously discussed, the selection of species eliminated via conservation analysis can have a substantial effect on model reduction error. The combined algorithm hence applies a speed-ranking step designed to estimate which of the biochemical reactants are likely to be most readily lumped.

To demonstrate the use of such an approach we have calculated the associated error and stiffness coefficients incurred under lumping for the chemotaxis model where the species eliminated via conservation were selected either naively or via the speed-ranking approach. These results can be found in Table 2. In the naive case the algorithm selected species CheA, $\text{CheA}_P \cdot \text{CheY}$, CheZ and CheB to be eliminated via conservation. Whilst in the selective case the algorithm selected species CheA, CheA·CheY, CheZ, and CheB_P. These results demonstrate that carefully considering which state-variables should be replaced via conservation relations can greatly improve the overall error incurred via lumping and yield a better reduction in model stiffness.

Table 1 Error results for the nonlinear methods of model reduction applied to the *E. coli* chemotaxis model

Number of dimensions	Empirical balanced truncation	Lumping	Combined method
6	1.396%	0.15%	-
5	1.655%	0.51%	-
4	#	0.54%	0.51%
3	#	4.77%	2.00%
2	#	12.88%	2.84%
1	#	75.56%	21.86%

Note: '#' implies Matlab could not numerically simulate this reduction using ode15s due to stiffness. '-' implies the reduction error at this point was equal to the lumping error. The errors stated represent the maximal relative error between the original and reduced systems when simulated to steady-state under the introduction of $10\mu M$ concentration of attractant ligand at $t = 0$

Analysis of alternative lumping inverses

The second enhancement to our combined method concerns the choice of lumping inverse used during model reduction. This topic does not seem to have been well explored in the literature, but as will be demonstrated here this choice can have a sizeable effect on the overall accuracy of a reduced model. In the case of the bacterial chemotaxis model, results comparing the three approaches for selecting a matrix inverse can be found in Table 3. In the case of the average inverse $\hat{L}$ we have set $T = 5$ seconds.

These results show that, whilst the Moore-Penrose inverse performs worse than the others, both the steady-state and averaged inverses can produce very good results. Given that the optimal approach can vary between lumping steps, the combined algorithm has been designed in such a way as to trial both the average and steady-state inverse at each step before selecting the most accurate. The overall inverse is then returned as the composition of the sequentially best inverses for each dimensionality of lumping.

Indices of controllability and observability

In the 5 dimensional lumped model created via the algorithm, only one lumped state-variable was created. This variable represents the sum of species Y_P, $\text{A} \cdot \text{Y}_P$, and $\text{A}_P \cdot \text{Y}_P$ which can be thought of as representing the phosphotransfer chain between species i.e. A and Y.

Calculating empirical Gramians for this lumped system yields the indices given in Table 4 (note that indices here can only be explicitly calculated for the retained species that were not eliminated via exploitation of conservation relations). Here the lumped variable, $\text{Y}_P + \text{A} \cdot \text{Y}_P + \text{A}_P \cdot \text{Y}_P$, is shown to be the most easily controlled, the most easily observed, and in total the most responsible for carrying the input-output signal from change in attractant concentration through to the phosphorylation of species Y and hence chemotaxis. Additionally, this set of species is readily lumpable, having such a low associated error cost and resulting in a large reduction in model stiffness. This suggests that the phosphotransfer process occurs significantly faster than the remainder of the network and hence equilibrates quickly. This finding concurs with the known biology, and demonstrates that $\text{Y}_P + \text{A} \cdot \text{Y}_P + \text{A}_P \cdot \text{Y}_P$ is more significant in the process of chemotactic signalling than the other species described in this model. Also noteworthy is the importance of species A·Y which agrees with other work (i.e. Tindall et al. [47]). Finally, the extremely low observability and controllability of CheB suggests that the overall concentration of it has an almost negligible response to the change of extracellular attractant and very little effect on the output in comparison to the remainder of the network. This also makes sense biologically, as CheB is functionally involved in the process of adaptation

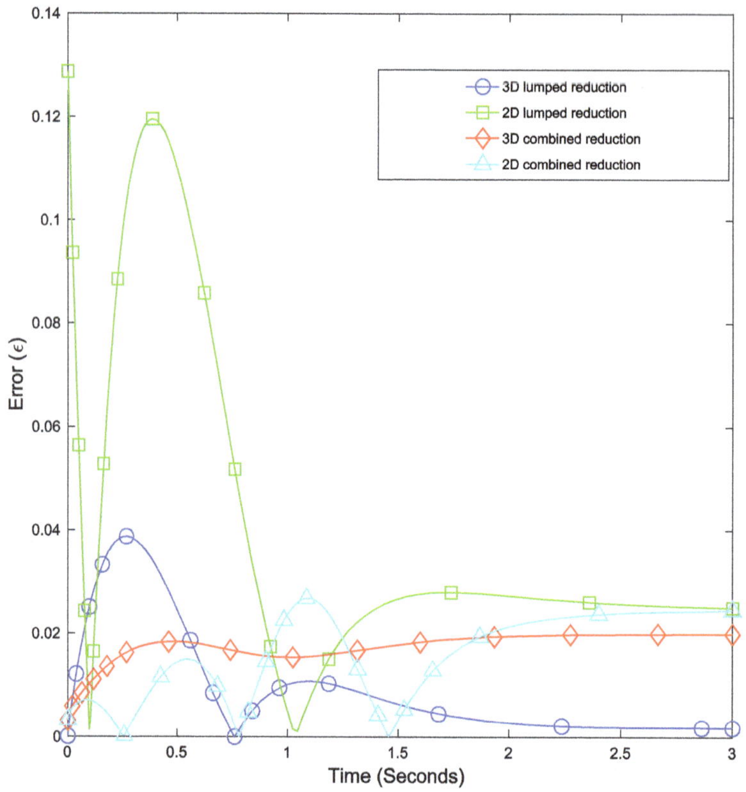

Fig. 3 Relative error between the original and reduced models of *E. coli* chemotaxis. These systems were simulated to steady-state under the introduction of $10\mu M$ concentration of extracellular attractant at $t = 0$. Figure depicts time varying errors incurred for the 3 and 2 dimensional reduced models under lumping applied in isolation and via the combined reduction algorithm

(how the cell steadily adjusts to the concentration of the chemotactic attractant), as opposed to being directly involved in the actual chemotactic flagellar response.

A model of extracellular regulatory kinase activation

The second example provided here is a model of ERK phosphorylation mediated via the EGF and NGF receptor pathways that was originally detailed in Sasagawa et al. [48]. This biological system commonly arises in the study of cancer and pain, and remains an area of ongoing clinical significance. The SBML representation employed here of this model including the parameterisation and initial conditions is available at www.ebi.ac.uk/biomodels-main/BIOMD0000000049.

This is a relatively large biochemical model describing 150 reactions and 99 species. The model additionally integrates two receptor pathways (EGFR and NGFR) allowing exploration as to how they interact. Due to its size and clinical relevance, this model therefore represents a prime candidate for the application of model reduction techniques. Although fully parameterised systems of this scope remain somewhat uncommon, their occurrence in the literature is increasing; primarily a result of increases in data and knowledge of cellular systems at finer spatial scales. However, even with such data available, approximations may still be required to model parameters particularly in cases where the model acts as a representation of more complex underlying biochemical mechanisms. We have thus employed this system as an example to demonstrate that our methodology remains valid for systems of varying complexity and size, assuming all parameter values are known.

A full description of the model and its parameterisation can be found in Sasagawa et al. [48]. A block schematic diagram of the system is given in Fig. 5; the blocks in this diagram represent various 'submodules' of the system each containing a number of reactants and the reactions that link them. We made one minor modification to the original model; the state-variable representing total concentration of proteasome was altered to have a rate of depletion such that the model was asymptotically stable. Without this proteasome would accumulate indefinitely and cause the system to be unstable.

The analysis performed here treats only one of the pathways as salient, such that the rate of EGF binding represents the only input under consideration. Additionally, the

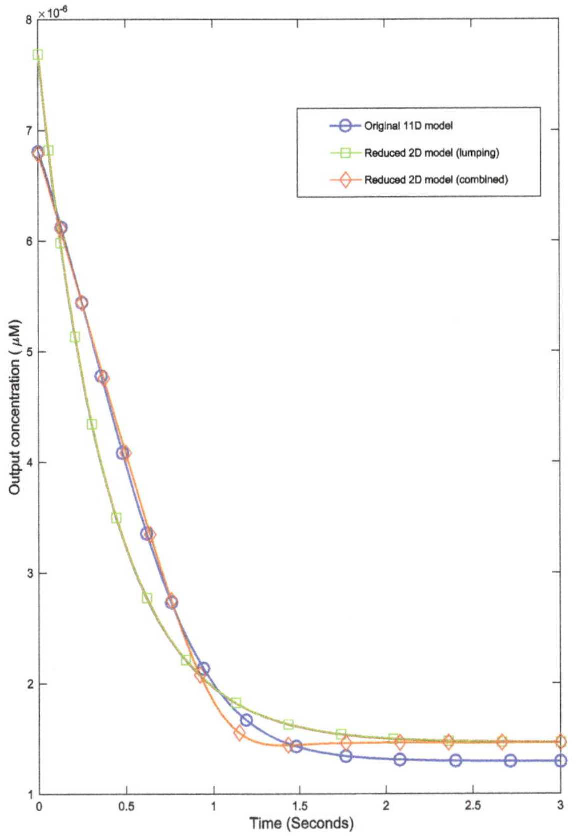

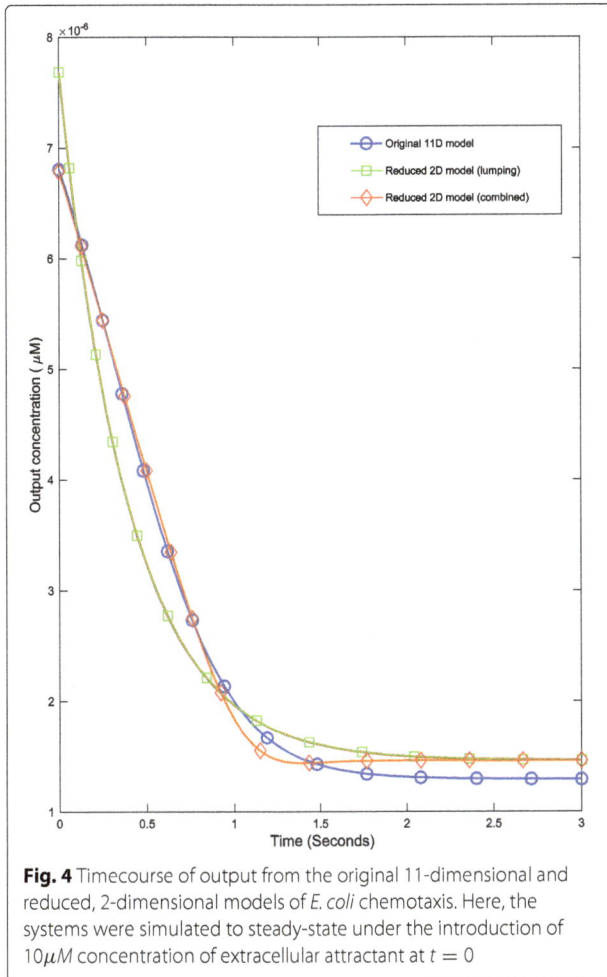

Fig. 4 Timecourse of output from the original 11-dimensional and reduced, 2-dimensional models of *E. coli* chemotaxis. Here, the systems were simulated to steady-state under the introduction of $10\mu M$ concentration of extracellular attractant at $t = 0$

Table 2 Comparison of reduction error and stiffness coefficients at each level of dimensional reduction for the *E coli* signalling model under different approaches to conservation analysis using the combined algorithm

Dimensions	Naive elimination		Selective elimination	
	Lumping error (ε)	Stiffness	Lumping error (ε)	Stiffness
6	1.37%	876.73	0.15%	794.91
5	2.45%	904.80	0.51%	56.22
4	4.8%	907.59	0.54%	60.16
3	74.18%	542.24	4.77%	58.95
2	339.99%	99.32	12.88%	25.95
1	424.33%	1	75.56%	1

This table compares the effect of either naively or rigorously selecting which species to eliminate via the application of conservation relations. The errors stated represent the maximal relative error between the original and reduced systems when simulated to steady-state under the introduction of $10\mu M$ concentration of attractant ligand at $t = 0$

Table 3 Comparison of maximal relative error under differing lumping inverses for the reduction of the *E coli* model via lumping in isolation

Dimensions	Moore-Penrose	Steady-state	Average
6	6.67%	0.15%	0.18%
5	10.58%	0.55%	0.51%
4	13.99%	0.54%	1.26%
3	26.93%	4.78%	4.77%
2	78.36%	13.45%	12.88%
1	70.92%	75.56%	82.34%

Each row represents a further level of dimensional reduction for the model, whilst the columns represent the different methods of lumping inverse. The 'Moore-Penrose' column contains values where the lumping inverse $\bar{L}$ is the Moore-Penrose or pseudoinverse of the lumping matrix L. The 'steady-state' column contains values where the lumping inverse $\bar{L}$ is selected to reconstruct the unperturbed steady-state values of the original system such that $\boldsymbol{x}^*_{u_0} = \bar{L}\tilde{\boldsymbol{x}}^*_{u_0}$. The errors stated represent the maximal relative error between the original and reduced systems when simulated to steady-state under the introduction of $10\mu M$ concentration of attractant ligand at $t = 0$

total concentration of phosphorylated ERK (in complex with other species or in isolation) is regarded as the output, such that

$$y=[\text{ppERK}]+[\text{dppERK}]+[\text{ppERK.MKP3}]+[\text{dppERK.MKP3}].$$

Additionally, the system employs initial conditions such that it begins at a non-zero steady-state with no input into the model (i.e. at the natural rate of EGF binding). Note that we chose to look at the EGF receptor - ERK activation pathway in particular as its dynamical behaviour exhibits an adaptive response [48]. The retention of this nonlinear behaviour in a reduced model serves as a good demonstration of the combined algorithm's strengths and particular applicability in the field.

Conservation analysis was performed via QR factorisation and from which 23 states could be eliminated via the speed-ranking method. Exploitation of these conservation relations resulted in a 76 dimensional model which was then further reduced via the combined algorithm, results of which can be seen in Table 5 and Fig. 6.

The original model had an extremely high stiffness coefficient at the initial condition of the system of 42658. Via lumping this could be reduced to 235.7 for the 11 dimensional lumped reduction enabling application of empirical balanced truncation. In this case, the empirical Gramians were constructed using data from 50 distinct simulations covering perturbations to both the models input parameter $u(t)$ and the initial conditions $\boldsymbol{x}_0$. In both cases, perturbations were uniformly sampled from 0.4 to 1.6 times each parameter's original, unperturbed value.

The results shown in Table 5 highlight the extent of the reductions that can be obtained via the combined method

Table 4 Controllability and observability index values for the model of chemotactic signalling in *E. Coli*

Species	Controllability	Observability	Input-output
CheA · CheY	0.865	0.881	0.762
$CheA_P$	0.846	0.497	0.421
$CheY_P$ + CheA · $CheY_P$ + $CheA_P$ · CheY	1	1	1
$CheY_P$ · CheZ	0.343	0.703	0.241
$CheB_P$	0.004	0.006	2×10^{-5}

with very little associated error. In particular the combined method has again demonstrated that it can produce better reductions than either method in isolation. The 7 dimensional reduced model had only a 4.77% associated maximal relative error as compared to the original model when simulated from steady-state under a 50% inhibition of EGF receptor binding. This is equivalent to an approximately 68% improvement in error over the 7 dimensional reduction achieved by lumping in isolation. Repeated simulation revealed that the original model had an average simulation time of 1.357 seconds, whilst the reduced 7 dimensional model required only 0.144 seconds.

Note that Matlab files regarding the reduction of the ERK activation model can be found online [62].

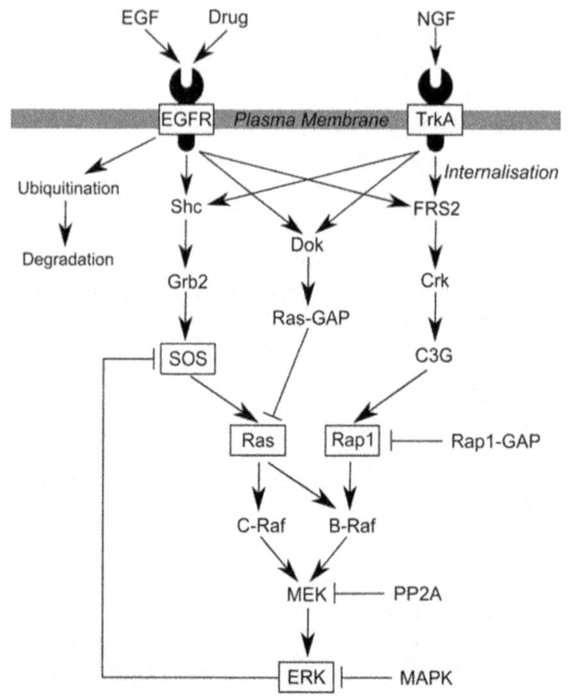

Fig. 5 Reproduction of the block schematic depiction of the ERK activation model due to Sasagawa et al. [48]

Indices of controllability and observability

Computing the controllability and observability indices for the 11-dimensional lumped version of the ERK activation model yields the results given in Table 6. In comparison to the results for the *E. coli* chemotactic signalling model, the results here are significantly more difficult to intuit. This is primarily due to the fact that the lumped variables often include species from highly disconnected areas of the original network, making it difficult to interpret their biological significance.

Despite this, however, it is still possible to obtain some insight from the indices calculated. The most controllable

Table 5 Error results comparing the application of empirical balanced truncation, lumping, and the combined method of model reduction to the ERK activation model

Dimension	EBT error	Lumping error	Stiffness	Combined error
75	0.76%	≈ 0%*	42658	–
50	#	0.01%	42633	–
25	#	0.52%	10664	–
15	#	1.26%	7934	–
14	#	2.21%	7934	–
13	#	2.29%	7934	–
12	#	1.21%	1591	–
11	#	3.07%	236	–
10	#	6.02%	264	2.84%
9	#	10.96%	211	4.02%
8	#	13.12%	43	4.32%
7	#	14.18%	42	4.77%
6	#	29.53%	44	13.08%
5	#	39.03%	45	20.81%
4	#	46.47%	212	31.09%
3	#	54.67%	42	34.58%
2	#	53.52%	18	41.10%
1	#	55.73%	1	50.46%

Note: '#' implies Matlab could not numerically simulate this reduction using ode15s due to issues associated with stiffness. '-' implies the reduction error at this point was equal to the lumping error. The errors stated represent the maximal relative error between the original and reduced systems when simulated to steady-state under the introduction of agonist increasing the rate of EGF receptor binding by 50% at $t = 0$. * implies that the error was within the numerical tolerance of the simulator

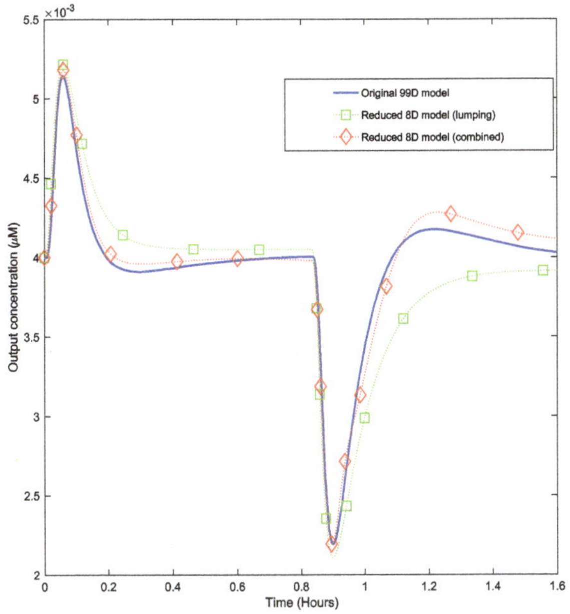

Fig. 6 Timecourses of the output from the original 99-dimensional and the reduced 8-dimensional systems. This plot emphasises the fact that the reduced model is designed to remain valid for any reasonable change in input. The system starts by being affected by an agonist that increases the rate of EGF binding by 25% for 50 minutes, at this point the input flips to an antagonist decreasing the rate of EGF binding by 50% and runs for the same time period. At any given time point the error between the original and reduced model exceeds no more than 5%

and observable lumped state-variable, $x_6(t)$, for example, primarily contains the concentration of singly phosphorylated mitogen/extracellular signal-regulated kinase (MEK), in isolation and in its various complex forms. This concurs with the known biology as phosphorylated MEK represents the point in the pathway that all possible routes of activation are directed towards before the phosphorylation of ERK occurs. The second most important state-variable, $x_4(t)$, is somewhat more difficult to parse, but is most distinguishable from the other state-variables by its inclusion of the SOS protein and associated complexes and species along this pathway. This, again, concurs with the known biology as SOS represents the most responsive branch of the pathway to EGFR binding that is described by the model. Similarly, the most unimportant state-variable, $x_3(t)$, can be seen as disproportionately representing the FRS2-C3G branch of the pathway, which is significantly more responsive to binding of the NGF receptor than that of EGF.

Overall the indices do seem to automatically provide a degree of model intuition that is often not possible for such large systems. As such they represent one possible solution to a core issue in the study of complex models. Due to the nature of the lumping algorithm, however, they are somewhat limited in their usefulness. An alternative approach, which we would hope to explore in future work, would be to constrain the lumping algorithm so as to obtain more readily interpreted lumped variables prior to the calculation of indices.

Table 6 Controllability and observability index values for the model of ERK activation controlled via the EGFR pathway

Lumped state-variable	Controllability index	Observability index	Input-output index
$x_1(t)$	0.3022	0.0176	0.0053
$x_2(t)$	0.0201	0.0153	0.0003
$x_3(t)$	0.0189	0.0009	1.681×10^{-5}
$x_4(t)$	0.8547	0.2752	0.2352
$x_5(t)$	0.6194	0.0079	0.0049
$x_6(t)$	1	1	1
$x_7(t)$	0.1412	0.1043	0.0147
$x_8(t)$	0.1378	0.0018	0.0003
$x_9(t)$	0.1539	0.0164	0.0025
$x_{10}(t)$	0.0473	0.0092	0.0004
$x_{11}(t)$	0.3312	0.0442	0.0146

Conclusions

In this paper a combined model reduction algorithm incorporating conservation analysis, proper lumping and empirical balanced truncation for the reduction of high dimensional systems of nonlinear ODEs has been presented. This algorithm was designed for specific application to controlled models of biochemical systems. Such models typically have a number of associated properties, including nonlinearity, stiffness and high dimensionality, which any model reduction approach must address. Under the combined algorithm conservation analysis is first used to obtain a simplified realisation of the system, proper lumping is then used to reduce the model whilst also reducing the stiffness coefficient and finally empirical balanced truncation is employed to achieve a more accurate and lower dimensional reduction than could be obtained by further lumping. The algorithm has additionally been designed to be highly automatable; it is implemented so as to take models stored in the SBML format and return highly accurate reduced systems without significant user input.

Whilst the algorithm has broad applicability, crucially, it focuses on the reduction of controlled biochemical systems that are amenable to an input-output formulation. In reducing a model the algorithm seeks primarily to maintain the input-output response profile, such that the reduced model can accurately predict the output of the original model for a wide range of possible inputs. Such an approach may be particularly applicable to the newly emerging field of Quantitative Systems

Pharmacology, which aims to mechanistically describe the effects of drug administration across multiple scales of action. Here the preservation of dose-response behaviour is of primary importance and tallies well with the algorithm's emphasis on maintaining a model's input-output relationship.

One of the difficulties with reduced models obtained under this approach is that the meaning of the dynamical equations is necessarily obfuscated by the coordinate transformations that have been applied. The combined algorithm starts with a fully mechanistic 'white box' model of a signalling pathway, whereby all of the biological reactions and reactants have a one-to-one correspondence with terms in the system of ODEs. The algorithm then produces a reduced 'grey-box' model where this biological interpretation of the governing equations is now hidden by the transformations applied. Computationally, however, the reduced model remains an accurate approximation to the original. The reduced state-variables can still be mapped back to the original biological species, thus allowing mechanistic insight to be gained as opposed to solely empirical observation. Hence, whilst the reduced system may no longer serve as a directly interpretable description of the original biological system, it still retains a high degree of utility for the analysis of the model's overall input-output relationship and its mechanistic underpinnings.

The algorithm in the form presented here assumes that the model in question is initially fully parameterised. It is however possible to extend the algorithm such that the reduced models constructed are guaranteed to remain valid across a range of parameterisations by additionally evaluating perturbations to the parameter values whilst constructing the lumping matrices and the empirical Gramians. Hence, at each step the possible reduced models are tested not only for perturbations to the original input and the initial conditions, but also for perturbations to the parameterisation of the system. To maintain validity across a defined range of values in parameter space, sampling approaches such as latin hypercube sampling would likely work well, but such ideas are not explored in this initial study.

The algorithm was demonstrated via application to two systems; an 11 dimensional model of bacterial chemotactic signalling in *E. coli* and a 99 dimensional model of ERK activation mediated via the EGF and NGF receptor pathways. The results for both models highlight the extent of reduction that can be obtained with very little associated error and therefore the potential usefulness of these model reduction methods in the analysis of Systems Biology models.

From the results it can be seen that both models began with a high stiffness coefficient. As a result, the application of empirical balanced truncation to the unreduced systems yielded higher error than lumping alone and numerically failed after truncating only a small proportion of the state-variables. Crucially, the proper lumping scheme was able to act as a good pre-conditioner and efficiently stripped stiffness from the systems, hence closing eigengaps in the associated Jacobians. Via subsequent application of empirical balanced truncation the combined algorithm was able to produce significantly better reductions than could be obtained by any of the individual methods in isolation. Furthermore, these reduced systems exhibited a significant speed-up in simulation time. In the case of the *E. coli* model we observed a 96.1% speed-up in simulation time whilst only incurring a 2.8% error. The ERK activation model demonstrated a 89.4% speed-up in simulation time in exchange for a 4.77% maximal relative error. For applications such as parameter optimisation or agent based modelling approaches, where an extremely large number of simulations may be required, speed-up is a substantial benefit. This speed-up can be attributed to two main factors - the reduction in the number of state variables present in each system and the reduction in their overall stiffness coefficients.

There are still a number of open questions with regards to this work, in particular to what extent does a model reduction (specifically, under a coordinate transformation of the state variables) remain accurate if the parameters in the original system are altered? Additionally, are there reduced models that remain valid over a larger range of parameterisations than others? If now, instead of considering the rate parameters to represent a set of fixed values, we consider the parameters to represent a distribution of physically acceptable values how does this affect our selection of the best reduction? Understanding these questions is a necessary next step in further growing the practical applicability of model reduction for biochemical systems.

The paper also introduced the concept of empirical indices of controllability and observability for biochemical systems; as was demonstrated these indices can be used to provide substantial insight into models of even a very large size. This usefulness, however, is dependent upon the biological interpretation of the specific lumping used to precondition the system. Developing a lumping algorithm specifically for the calculation of these indices represents a reasonable extension of this work.

The combined model reduction algorithm has demonstrated good results for the reduction of controlled, nonlinear, stiff, high-dimensional models of biochemical reaction networks. Reduced systems produced under this approach maintain a high degree of input-output accuracy and see a significant speed-up in simulation time. These results justify research into the combining of complementary reduction methodologies and highlight the use of empirical balanced truncation, which had not previously seen application in the reduction of biochemical systems.

Abbreviations
EGF: Epidermal growth factor; ERK: Extracellular regulatory kinase; MEK: Mitogen/extracellular signal-regulated kinase; NGF: Nerve growth factor; ODE: Ordinary differential equations; SBML: Systems Biology Markup Language; SVD: Singular value decomposition

Acknowledgements
None.

Funding
TS was funded for this research through an Engineering and Physical Sciences Research Council (EPSRC) Collaborative Awards in Science and Engineering (CASE) studentship in conjunction with Pfizer Global Research and Development. MT was in receipt of a Research Councils UK (RCUK) Fellowship.

Authors' contributions
PHvdG initially proposed the study of model reduction for Systems Biology models. MT was the primary supervisor of the study with input from PHvdG. TS designed and implemented the combined algorithm, and performed the calculations for both case studies. TS drafted the manuscript with input from MT. All authors read and approved the final manuscript.

Competing interests
The authors declare that they have no competing interests.

Author details
[1]Department of Mathematics and Statistics, University of Reading, RG6 6AX, Reading, UK. [2]Certara QSP, University of Kent Innovation Centre, CT2 7FG Canterbury, UK. [3]Leiden Academic Centre for Drug Research, Universiteit Leiden, NL-2333 CC Leiden, Netherlands. [4]The Institute for Cardiovascular and Metabolic Research (ICMR), University of Reading, RG6 6AX Reading, UK.

References

1. Li C, Donizelli M, Rodriguez N, Dharuri H, Endler L, Chelliah V, Li L, He E, Henry A, Stefan MI, Snoep JL, Hucka M, Le Novere N, Laibe C. BioModels Database: An enhanced, curated and annotated resource for published quantitative kinetic models. BMC Syst Biol. 2010;4:92.
2. Okino M, Mavrovouniotis M. Simplification of mathematical models of chemical reaction systems. Chem Rev. 1998;98(2):391–408.
3. Petrov V, Nikolova E, Wolkenhauer O. Reduction of nonlinear dynamic systems with an application to signal transduction pathways. Syst Biol IET. 2007;1(1):2–9.
4. Schneider KR, Wilhelm T. Model reduction by extended quasi-steady-state approximation. J Math Biol. 2000;40(5):443–50.
5. Vejchodský T, Erban R, Maini PK. Reduction of chemical systems by delayed quasi-steady state assumptions. 2014. arXiv preprint arXiv:1406.4424.
6. Vejchodský T. Accurate reduction of a model of circadian rhythms by delayed quasi steady state assumptions. 2013. arXiv preprint arXiv:1312.2825.
7. Choi J, Yang KW, Lee TY, Lee SY. New time-scale criteria for model simplification of bio-reaction systems. BMC bioinformatics. 2008;9(1):338.
8. West S, Bridge LJ, White MR, Paszek P, Biktashev VN. A method of speed coefficients for biochemical model reduction applied to the NF-κB system. J Math Biol. 2015;70(3):591–620. arXiv preprint arXiv:1403.1610.
9. Härdin HM, Zagaris A, Krab K, Westerhoff HV. Simplified yet highly accurate enzyme kinetics for cases of low substrate concentrations. FEBS J. 2009;276(19):5491–506.
10. Vora N, Daoutidis P. Nonlinear model reduction of chemical reaction systems. AIChE J. 2001;47(10):2320–32.
11. Gerdtzen ZP, Daoutidis P, Hu WS. Nonlinear model reduction for energy metabolism in Saccharomyces cerevisiae. In: American Control Conference, 2002. Proceedings of the 2002; 2002. p. 2867–72. IEEE.
12. Gerdtzen ZP, Daoutidis P, Hu WS. Non-linear reduction for kinetic models of metabolic reaction networks. Metab Eng. 2004;6(2):140–54.
13. Prescott TP, Papachristodoulou A. Layering in networks: The case of biochemical systems. In: American Control Conference (ACC), 2013; 2013. p. 4544–49. IEEE.
14. Prescott TP, Papachristodoulou A. Layered decomposition for the model order reduction of timescale separated biochemical reaction networks. J Theor Biol. 2014;356:113–22.
15. Sivakumar H, Hespanha JP. Towards modularity in biological networks while avoiding retroactivity. In: American Control Conference (ACC), 2013; 2013. p. 4550–556. IEEE.
16. Maas U, Pope SB. Simplifying chemical kinetics: intrinsic low-dimensional manifolds in composition space. Combustion Flame. 1992;88(3):239–64.
17. Vallabhajosyula RR, Sauro HM. Complexity reduction of biochemical networks. In: Simulation Conference, 2006. WSC 06. Proceedings of the Winter; 2006. p. 1690–1697. IEEE.
18. Zobeley J, Lebiedz D, Kammerer J, Ishmurzin A, Kummer U. A new time-dependent complexity reduction method for biochemical systems. In: Transactions on Computational Systems Biology I. Berlin, Heidelberg: Springer; 2005. p. 90–110.
19. Surovtsova I, Zobeley J. Focusing on dynamic dimension reduction for biochemical reaction systems. Understanding Exploiting Syst Biol Biomed Bioprocesses. 2006;31:31–46.
20. Surovtsova I, Simus N, Lorenz T, König A, Sahle S, Kummer U. Accessible methods for the dynamic time-scale decomposition of biochemical systems. Bioinformatics. 2009;25(21):2816–23.
21. Kourdis PD, Goussis DA, Steuer R. Physical understanding via reduction of complex multiscale models: glycolysis in Saccharomyces cerevisiae. In: BioInformatics and BioEngineering, 2008. BIBE 2008. 8th IEEE International Conference On; 2008. p. 1–6. IEEE.
22. Kourdis PD, Steuer R, Goussis DA. Physical understanding of complex multiscale biochemical models via algorithmic simplification: Glycolysis in Saccharomyces cerevisiae. Physica D: Nonlinear Phenomena. 2010;239(18):1798–817.
23. Kourdis PD, Palasantza AG, Goussis DA. Algorithmic asymptotic analysis of the NF-κB signaling system. Comput Math Appl. 2013;65(10):1516–34.
24. Surovtsova I, Simus N, Hübner K, Sahle S, Kummer U. Simplification of biochemical models: a general approach based on the analysis of the impact of individual species and reactions on the systems dynamics. BMC Syst Biol. 2012;6(1):14.
25. Degenring D, Froemel C, Dikta G, Takors R. Sensitivity analysis for the reduction of complex metabolism models. J Process Control. 2004;14(7):729–45.
26. Liu G, Swihart MT, Neelamegham S. Sensitivity, principal component and flux analysis applied to signal transduction: the case of epidermal growth factor mediated signaling. Bioinformatics. 2005;21(7):1194–202.
27. Smets I, Bernaerts K, Sun J, Marchal K, Vanderleyden J, Van Impe J. Sensitivity function-based model reduction: a bacterial gene expression case study. Biotech Bioeng. 2002;80(2):195–200.
28. Apri M, de Gee M, Molenaar J. Complexity reduction preserving dynamical behavior of biochemical networks. J Theor Biol. 2012;304:16–26.
29. Maurya M, Bornheimer S, Venkatasubramanian V, Subramaniam S. Reduced-order modelling of biochemical networks: application to the GTPase-cycle signalling module. IEE Proc Syst Biol. 2005;152(4):229–42.

30. Maurya MR, Scott JB, Venkatasubramanian V, Subramaniam S. Model-reduction by simultaneous determination of network topology and parameters: Application to modules in biochemical networks. In: 2005 Annual Meeting AIChE; 2005.
31. Maurya M, Bornheimer S, Venkatasubramanian V, Subramaniam S. Mixed-integer nonlinear optimisation approach to coarse-graining biochemical networks. IET Syst Biol. 2009;3(1):24–39.
32. Hangos KM, Gábor A, Szederkényi G. Model reduction in bio-chemical reaction networks with michaelis-menten kinetics. In: European Control Conference (ECC), July 17-19 2013, Zurich; 2013. p. 4478–483.
33. Taylor SR, Petzold LR, et al. Oscillator model reduction preserving the phase response: application to the circadian clock. Biophys J. 2008;95(4): 1658–73.
34. Anderson J, Chang YC, Papachristodoulou A. Model decomposition and reduction tools for large-scale networks in systems biology. Automatica. 2011;47(6):1165–74.
35. Prescott TP, Papachristodoulou A. Guaranteed error bounds for structured complexity reduction of biochemical networks. J Theor Biol. 2012;304:172–82.
36. Danø S, Madsen MF, Schmidt H, Cedersund G. Reduction of a biochemical model with preservation of its basic dynamic properties. FEBS J. 2006;273(21):4862–77.
37. Dokoumetzidis A, Aarons L. Proper lumping in systems biology models. IET Syst Biol. 2009;3(1):40–51.
38. Gulati A, Isbister G, Duffull S. Scale reduction of a systems coagulation model with an application to modeling pharmacokinetic–pharmacodynamic data. CPT: Pharmacometrics Syst Pharmacol. 2014;3(1):90.
39. Koschorreck M, Conzelmann H, Ebert S, Ederer M, Gilles ED. Reduced modeling of signal transduction–a modular approach. BMC Bioinformatics. 2007;8(1):336.
40. Sunnåker M, Schmidt H, Jirstrand M, Cedersund G. Zooming of states and parameters using a lumping approach including back-translation. BMC Syst Biol. 2010;4(1):28.
41. Sunnåker M, Cedersund G, Jirstrand M. A method for zooming of nonlinear models of biochemical systems. BMC Syst Biol. 2011;5(1):140.
42. Liebermeister W, Baur U, Klipp E. Biochemical network models simplified by balanced truncation. FEBS J. 2005;272(16):4034–43.
43. Härdin H, van Schuppen J. System reduction of nonlinear positive systems by linearization and truncation. Positive Syst. 2006;431–38.
44. Sootla A, Anderson J. On projection-based model reduction of biochemical networks – Part I: The deterministic case. Los Angeles: IEEE 53rd Annual Conference on Decision and Control (CDC); 2014. arXiv preprint arXiv:1403.3579.
45. van der Graaf PH, Benson N. Systems pharmacology: bridging systems biology and pharmacokinetics-pharmacodynamics (PKPD) in drug discovery and development. Pharm Res. 2011;28(7):1460–4.
46. Sorger PK, Allerheiligen SR, Abernethy DR, Altman RB, Brouwer KL, Califano A, D'Argenio DZ, Iyengar R, Jusko WJ, Lalonde R, et al. Quantitative and systems pharmacology in the post-genomic era: new approaches to discovering drugs and understanding therapeutic mechanisms. In: An NIH White Paper by the QSP Workshop Group-October; 2011.
47. Tindall M, Porter S, Wadhams G, Maini P, Armitage J. Spatiotemporal modelling of CheY complexes in *Escherichia coli* chemotaxis. Progress Biophys Mol Biol. 2009;100(1):40–6.
48. Sasagawa S, Ozaki YI, Fujita K, Kuroda S. Prediction and validation of the distinct dynamics of transient and sustained ERK activation. Nat Cell Biol. 2005;7(4):365–73.
49. Murray JD. Mathematical Biology I: An Introduction. New York: Springer; 2002.
50. Klipp E, Liebermeister W, Wierling C, Kowald A, Lehrach H, Herwig R. Systems Biology. Oxford: Wiley-Blackwell; 2013.
51. Sauro HM, Ingalls B. Conservation analysis in biochemical networks: computational issues for software writers. Biophys Chem. 2004;109(1): 1–15.
52. Vallabhajosyula RR, Chickarmane V, Sauro HM. Conservation analysis of large biochemical networks. Bioinformatics. 2006;22(3):346–53.
53. Antoulas AC. Approximation of Large-scale Dynamical Systems. Advances in Design and Control. Philadelphia: Society for Industrial and Applied Mathematics; 2005.
54. Hahn J, Edgar TF. An improved method for nonlinear model reduction using balancing of empirical Gramians. Comput Chem Eng. 2002;26(10): 1379–97.
55. Dullerud GE, Paganini F, Vol. 6. A Course in Robust Control Theory. New York: Springer; 2000.
56. Laub AJ, Heath MT, Paige CC, Ward RC. Computation of system balancing transformations and other applications of simultaneous diagonalization algorithms. Autom Control IEEE Trans. 1987;32(2):115–22.
57. Bornstein BJ, Keating SM, Jouraku A, Hucka M. LibSBML: an API library for SBML. Bioinformatics. 2008;24(6):880–1.
58. Schmidt H, Jirstrand M. Systems biology toolbox for matlab: a computational platform for research in systems biology. Bioinformatics. 2006;22(4):514–5.
59. Wei J, Kuo JC. Lumping analysis in monomolecular reaction systems. analysis of the exactly lumpable system. Ind Eng Chem Fundam. 1969;8(1):114–23.
60. Kuo JC, Wei J. Lumping analysis in monomolecular reaction systems. analysis of approximately lumpable system. Ind Eng Chem Fundam. 1969;8(1):124–33.
61. Li G, Rabitz H. A general analysis of approximate lumping in chemical kinetics. Chem Eng Sci. 1990;45(4):977–1002.
62. Snowden TJ, van der Graaf PH, Tindall MJ. Reduction results for a mathematical model of ERK activation (Matlab Files). 2016. doi:10.5281/zenodo.192503 https://doi.org/10.5281/zenodo.192503. Accessed 12 Apr 2016.

11

Theoretical model of mitotic spindle microtubule growth for FRAP curve interpretation

Leonid V. Omelyanchuk[1,2] and Alina F. Munzarova[1,2*]

Abstract

Background: Spindle FRAP curves depend on the kinetic parameters of microtubule polymerization and depolymerization. The empirical FRAP curve proposed earlier permits determination of only one such dynamic parameter, commonly called the "tubulin turnover". The aim of our study was to build a FRAP curve based on an already known kinetic model of microtubule growth.

Results: A numerical expression that describes the distribution of polymerizing and depolymerizing microtubule ends as a function of four kinetic parameters is presented. In addition, a theoretical FRAP curve for the metaphase spindle is constructed using previously published dynamic parameters.

Conclusion: The numerical expression we elaborated can replace the empirical FRAP curve described earlier for a spindle comprising fluorescently marked microtubules. The curve we generated fits well the experimental data.

Keywords: Mitotic spindle, FRAP (fluorescence recovery after photobleaching), Microtubules, Fluorescently marked tubulin, Growing/shrinking microtubule ends, PDE (partial differential equation)

Background

Fluorescence recovery after photobleaching (FRAP) was first introduced in 1974 [1] and is a widely used method to study turnover, transport, diffusion and interaction among biological molecules in living specimens. The use of FRAP has been facilitated by the current availability of microscopes equipped with a laser scanning device. The emergence of fluorescent protein labeling with the green fluorescent protein (GFP) and its spectral variants has greatly enhanced FRAP application. In a typical FRAP experiment, a GFP-labeled structure is rapidly and irreversibly photobleached with a high intensity laser, and fluorescence recovery is recorded as a function of time. The fluorescent molecules then diffuse into the irradiated region, while the non-fluorescent ones diffuse into the unbleached area until equilibrium is reached. The analysis of fluorescence recovery curves yields the diffusion coefficient and the fraction of free, transiently bound and immobilized molecules.

For a quantitative description of fluorescence recovery dynamics in FRAP experiments, several theoretical models have been proposed [2]. These include (1) the Pure-Diffusion Dominant Model that considers the recovery rate for weakly bound or free fluorescent molecules, and is defined exclusively by their diffusion; (2) the Effective Diffusion Model, which describes the recovery kinetics of fluorescent molecules that bind tightly the bleached structure, and is also largely defined by diffusion; (3) the Reaction Dominant Model, where diffusion is very fast and molecules rapidly equilibrate after the bleach; and (4) the Diffusion Phase-binding Phase approximation used whenever the contributions of diffusion and binding are coupled. In another early study, *Salmon* et al. [3] used the FRAP technique to trace the behavior of fluorescently labeled bovine tubulin injected in sea urchin eggs undergoing the first mitotic division. The authors provided evidence that the use of

* Correspondence: alina.munzarova@mcb.nsc.ru
[1]Institute of Molecular and Cellular Biology, Novosibirsk, Russia
[2]Novosibirsk State University, Novosibirsk, Russia

exogenous tubulin accurately represents the in vivo situation, and showed that FRAP data were best fitted by a negative-going exponential function, and that fluorescence recovery had a half-time of approximately 20 s. Taking into account that tubulin dimers diffused back into the bleached area within 1 s, the authors concluded that diffusion was not a limiting factor and that the Reaction Dominant Model was correctly interpreting their FRAP results.

One of the first theoretical description of microtubule growth and degradation [4] was based on data on microtubule dynamics obtained from in vitro experiments [5, 6]. Several kinetic models describing microtubule (MT) growth were proposed [4], and one of them, usually referred as the Hill's model [4], is still widely used. According to this model, MT growth begins with the recruitment of tubulin dimers (present at the concentration C_0) at a centrosome-associated nucleation site, followed by the polymerization of additional dimers at a rate constant J_1 (growth phase) or depolymerization at a rate constant J_2 (shrinking phase). At any length, the MTs may switch from the growth mode to a shrinking mode with a constant rate k_1, or from a shrinking mode to a growth mode with a constant rate k_2. A graphical representation of the Hill's model is reported in Fig. 1. This kinetic model was later adapted to describe the polymerization/depolymerization kinetics of microtubules in *Xenopus* egg extracts and to analyze how cyclin A and cyclin B could affect this process [7]. In one of the kinetic regimes (bounded state), the experimental data were nicely fitting the model, and all of the constants could be defined. In addition to the bounded state where MTs are on average disassembling and J_1k_2- J_2k_1 is negative, the authors also described an unbounded state (with J_1k- $J_2k' > 0$) where MTs are on average growing (our k_1 and k_2 are k' and k of *Hill* [4], respectively) [7].

The Hill's model (Fig. 1) was applied to FRAP-based studies on pre-anaphase B/metaphase spindles of *Drosophila* syncytial embryos expressing GFP-tubulin [8]. The authors concluded that both bounded and unbounded regimes are inadequate to describe the observed FRAP dynamics. Data analysis indicated that tubulin dimers turn over almost entirely during a single cycle of MT shortening and growth, and consequently the recovery time does not depend on the size or position of the bleached region along the metaphase spindle. The recovery halftime ($T_{1/2}$) after photobleaching is then simply the halftime of this cycle (i.e., $1/k_1$ or $1/k_2$ in terms of the model depicted in Fig. 1). This critical analysis served as a starting point for us to find other solutions to the kinetic scheme shown in Fig. 1.

Numerical modeling of a FRAP experiment [8] shows that the parameters that describe MT turnover can be determined from the modeling data. However, this approach is based on the mechanical model of Brust-Mascher et al. [9] and requires equations to be solved for the entire spindle, necessitating the knowledge of multiple mechanical parameters. As mentioned above, another approach to analyze MT dynamics is based on chemical kinetics description [4]. This approach operates with a smaller number of dynamic parameters and was

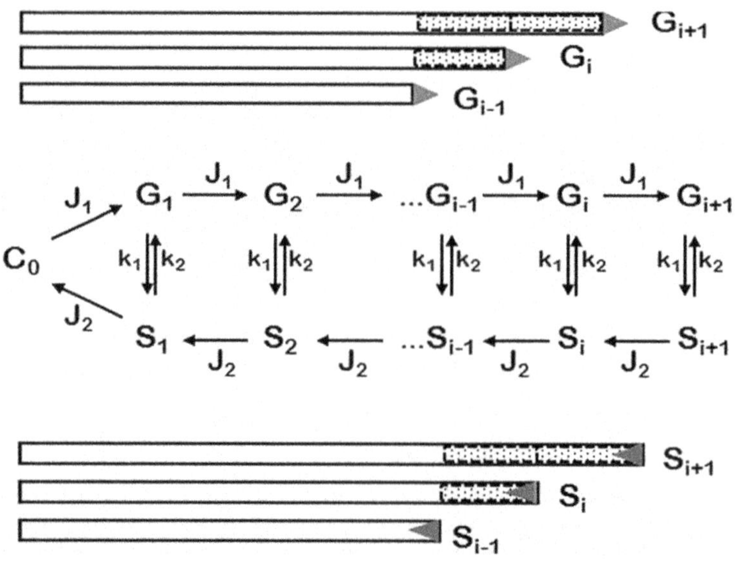

Fig. 1 Kinetic model of MT growth adapted from Hill, 1984 [4]. G_i and S_i are the concentrations of growing and shrinking microtubules containing *i* number of tubulin dimers. C_0 indicates MTs with zero length. J_1 and J_2 are the kinetic constants of growth and shrinking, respectively; k_1, k_2 are the constants for rates of growth-to-shrinking and shrinking-to-growth transitions. Growing and shrinking MTs are depicted as *bars*, and tubulin dimers are shown as *shaded bars*

verified by in vitro microtubule growth experiments. One of the objectives of the present study is developing a chemical-kinetic description for FRAP experiments on metaphase spindles. Another objective is the analysis of the dependence of spindle FRAP on parameters describing microtubule end dynamics. FRAP in a steady-state spindle has a simple relationship with the number of MT growing and shrinking ends within the photobleached area. The dynamics of these ends both with and without specific marks are not easy to evaluate, but can be inferred from the solution of system (2) for elementary intervals. Thus, our aim is describing the limit transition from the discrete Hill's equations [4] to a continuous equation, and understanding how the kinetic constants of the Hill equation compare to those in the continuous equations. We found a solution for the Cauchy problem for partial differential equation (PDE) (2) for elementary intervals and used this expression to solve the equation for fluorescence recovery in a steady state spindle. Finally, we applied our model to the experimental data [8] and performed a quantitative analysis of FRAP recovery time.

Results

Limit passage to the continuous model

Since we were not able to solve discrete equations describing MT behavior with time, we are using a continuous model. We denote the MT concentration found in the growth phase as $G_i(t)$, where i is the number of polymerized tubulin dimers. Similarly, the concentration of microtubules in the shrinking phase is denoted as $S_i(t)$. Then, the kinetic equations for the concentration of tubulin dimer appear as:

$$\begin{cases} \frac{\partial}{\partial t}G_i(t) = -J_1(G_i(t)-G_{i-1}(t)) - k_1G_i(t) + k_2S_i(t) \\ \frac{\partial}{\partial t}S_i(t) = -J_2(S_i(t)-S_{i+1}(t)) - k_2S_i(t) + k_1G_i(t) \end{cases} \quad (1)$$

We can rewrite equation (1) as follows, considering Δx is a small coordinate step:

$$\begin{cases} \frac{\partial}{\partial t}G_i(t) = -J_1\frac{\Delta x}{\Delta x}(G_i(t)-G_{i-1}(t)) - k_1G_i(t) + k_2S_i(t) \\ \frac{\partial}{\partial t}S_i(t) = -J_2\frac{\Delta x}{\Delta x}(S_i(t)-S_{i+1}(t)) - k_2S_i(t) + k_1G_i(t) \end{cases}$$

The values $v_1 = J_1\Delta x$ and $v_2 = J_2\Delta x$ have a dimension of *cm/sec*, which correspond to the linear polymerization and depolymerization speeds of a single tubulin dimer. If Δx is negligible compared to the microtubule size, the above equations can be transformed into PDEs, where t is a time and x is a coordinate:

$$\begin{cases} \frac{\partial}{\partial t}G(x,t) = -v_1\left[\frac{\partial}{\partial x}G(x,t)\right] - k_1G(x,t) + k_2S(x,t) \\ \frac{\partial}{\partial t}S(x,t) = v_2\left[\frac{\partial}{\partial x}S(x,t)\right] - k_2S(x,t) + k_1G(x,t) \end{cases} \quad (2)$$

As reported below, this transformation will help us to find a mathematical solution for MT behavior.

Stationary regime and bounded state

Let us solve the equations for the stationary regime, i.e., when the process does not depend on time, namely when $t = 0$. In this case, $G(x)$ and $S(x)$ denote the distribution of growing and shrinking ends over the coordinate x, which corresponds to the current length of MTs.

$$\begin{cases} v_1\left[\frac{\partial}{\partial x}G(x)\right] = -k_1G(x) + k_2S(x) \\ v_2\left[\frac{\partial}{\partial x}S(x)\right] = -k_1G(x) + k_2S(x) \end{cases}$$

The discriminant of this system,

$$D = \left(\frac{k_1}{v_1} - \frac{k_2}{v_2}\right)^2$$

is either positive or zero. Then a characteristic equation:

$$\lambda^2 - \left(\frac{k_2}{v_2} - \frac{k_1}{v_1}\right)\lambda = 0$$

has two roots,

$$\lambda_1 = 0, \quad \lambda_2 = \frac{k_2v_1 - k_1v_2}{v_1v_2} = -\alpha$$

Substituting the experimental values of k_1; k_2; v_1; v_2 from reference [10] into the expression for λ_2 results in a positive α. The general solution of the system (3) is

$$\begin{cases} G(x) = C_1\frac{k_2}{v_1} + C_2\frac{k_2}{v_1}e^{-\alpha x} \\ S(x) = C_1\frac{k_1}{v_1} + C_2\frac{k_2}{v_2}e^{-\alpha x} \end{cases}$$

where C_1 and C_2 are arbitrary constants. If at $x = 0$; $G(x) = G_0$, and upon x tending to infinity $G(x) = G_\infty$, then

$$\begin{cases} G(x) = G_\infty + (G_0 - G_\infty)e^{-\alpha x} \\ S(x) = \frac{k_1}{k_2}G_\infty + (G_0 - G_\infty)\frac{v_1}{v_2}e^{-\alpha x} \end{cases}$$

Using the experimentally observed values of dynamic parameters [10] and arbitrarily assigning $G_0 = 20$ and $G_\infty = 30$, the system can be visualized as shown in Fig. 2.

If x is measured in μm, the $Gk_1 = Sk_2$ ratio holds true for a wide range of large x values, whereas in the case of small

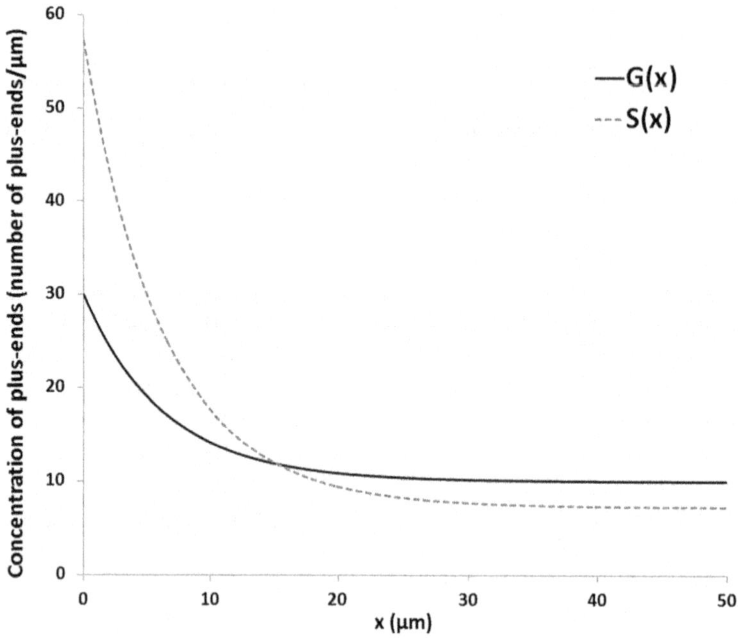

Fig. 2 Concentrations of growing and shrinking microtubule ends as a function of coordinate x for a model case

x values, which is most pertinent to the experimental situation, an alternative ratio should be applied. A negative-going exponential function for microtubule length (at $G_\infty = 0$) observed for the discrete form of equations (1) in the stationary regime (at $v_2k_1 > v_1k_2$) was discussed earlier [7] and was successfully used to interpret the MT length distribution in experiments in vitro. Based on these experimental data [7] and given the finite number of tubulin molecules in the cell, $G_\infty > 0$ is not a viable option and so G_∞ should be set as 0. In this case, a simple ratio $G(x)v_1 = S(x)v_2$ is true and corresponds to a situation where the spindle remains unchanged with time, and the numbers of growing and shrinking microtubules are equal.

Analytical solution of the equations for elementary intervals

The system of continuous equations (2) was first published by *Dogterom and Leibler* [11]; the authors stated that it could be solved analytically with the border conditions set at $t = 0$, and with all MTs having zero length. However, they did not publish the solution. In the following section we find this solution.

The solution refers to a FRAP experiment that involves photobleaching of fluorescently labeled MTs in a rectangular region of width L perpendicular to the spindle axis, followed by the analysis of the dynamics of fluorescence recovery in the bleached region. In a real experiment both G (Growing) and S (Shrinking) MT ends are distributed uniformly across this region at the initial time-point after bleaching. Let us consider the elementary interval Δ of L, which initially contains G_0 and S_0 unmarked ends (in the stationary case $G_0v_1 = S_0v_2$; v_1 and v_2 are the velocities of growth and shrinking, respectively). The G_0 ends will start to produce a marked tubulin track and will move towards the spindle equator from the elementary interval, while the S_0 ends will move to the opposite direction. Some G_0 ends will then turn into S ends (catastrophe) and some S_0 ends will turn into G ends (rescue). To describe the behavior of G_0 and S_0 ends it is necessary to solve the Cauchy problem for (2). We thus substituted the x and t variables with α and β, and converted the functions $G(x; t)$ and $S(x; t)$ into $u(\alpha;\beta)$ and $v(\alpha;\beta)$:

$$\begin{cases} \alpha = x - v_1 t \\ \beta = x + v_2 t \end{cases} \quad \begin{cases} G(x,t) = u(\alpha,\beta)e^{-a\beta+b\alpha} \\ S(x,t) = v(\alpha,\beta)e^{-a\beta+b\alpha} \end{cases} \quad \begin{cases} a = \dfrac{k_1}{v_1+v_2} \\ b = \dfrac{k_2}{v_1+v_2} \end{cases}$$

We obtain the equation system for the functions $u(\alpha; \beta)$ and $v(\alpha; \beta)$:

$$\frac{\partial}{\partial\beta}u(\alpha,\beta) = bv(\alpha,\beta)$$
$$\frac{\partial}{\partial\alpha}v(\alpha,\beta) = bv(\alpha,\beta)$$

The elimination of $v(\alpha; \beta)$ gives us the following equation for $u(\alpha; \beta)$:

$$\frac{\partial}{\partial\beta}\left(\frac{\partial}{\partial\beta}u(\alpha,\beta)\right) + a\,b\,u(\alpha,\beta) = 0 \tag{3}$$

With $u(\alpha; \beta)$ known, one can determine $v(\alpha; \beta)$ from the first equation of the system. If $u(\alpha;\beta)_{\alpha=\beta} = u_0(x)$ and $v(\alpha;\beta)_{\alpha=\beta} = v_0(x)$ are known, we can supply equation (3) with the boundary condition:

$$u(\alpha,\beta)|_{\alpha=\beta} = u_0(x)$$

$$\frac{\partial}{\partial\beta}u(\alpha,\beta)|_{\alpha=\beta} = bv_0(x)$$

The solution of this equation, already presented in ref [12], is expressed as:

$$u(\alpha,\beta) = u_0(\alpha) + b\int_{\alpha}^{\beta} I_0\left[2\sqrt{ab}\sqrt{(\xi-\alpha)(\beta-\xi)}\right]v_0(\xi)d\xi + \sqrt{ab}\int_{\alpha}^{\beta} I_1\left[2\sqrt{ab}\sqrt{(\xi-\alpha)(\beta-\xi)}\right]\sqrt{\frac{\beta-\xi}{\xi-\alpha}}u_0(\xi)d\xi \tag{4}$$

By reverting to variables x and t and functions $G(x;\, t)$ and $S(x;\, t)$ and by denoting their values at $t = 0$ as $G_0(x)$ and $S_0(x)$, we obtain:

$$G(x,t) = = G_0(x-v_1t)e^{-k_1t} + \frac{k_2}{v_1+v_2}e^{\frac{-(k_1v_2+k_2v_1)}{v_1+v_2}t} \int_{x-v_1t}^{x+v_2t} I_0\left(2\frac{\sqrt{k_1k_2}\sqrt{(\xi-x+v_1t)(x+v_2t-\xi)}}{v_1+v_2}\right) S_0(\xi)e^{\frac{(k_2-k_1)(x-\xi)}{v_1+v_2}}d\xi + \frac{\sqrt{k_1k_2}}{v_1+v_2}e^{\frac{-(k_1v_2+k_2v_1)}{v_1+v_2}t} \int_{x-v_1t}^{x+v_2t} I_1\left(2\frac{\sqrt{k_1k_2}\sqrt{(\xi-x+v_1t)(x+v_2t-\xi)}}{v_1+v_2}\right) G_0(\xi)\cdot\cdot e^{\frac{(k_2-k_1)(x-\xi)}{v_1+v_2}}\sqrt{\frac{(x+v_2t-\xi)}{(\xi-x+v_1t)}}d\xi$$

$$S(x,t) = = S_0(x+v_2t)e^{-k_2t} + \frac{k_1}{v_1+v_2}e^{\frac{-(k_1v_2+k_2v_1)}{v_1+v_2}t} \int_{x-v_1t}^{x+v_2t} I_0\left(2\frac{\sqrt{k_1k_2}\sqrt{(\xi-x+v_1t)(x+v_2t-\xi)}}{v_1+v_2}\right) e^{\frac{(k_2-k_1)(x-\xi)}{v_1+v_2}}G_0(\xi)d\xi + \frac{\sqrt{k_1k_2}}{v_1+v_2}e^{\frac{-(k_1v_2+k_2v_1)}{v_1+v_2}t} \int_{x-v_1t}^{x+v_2t} I_1\left(2\frac{\sqrt{k_1k_2}\sqrt{(\xi-x+v_1t)(x+v_2t-\xi)}}{v_1+v_2}\right) \cdot e^{\frac{(k_2-k_1)(x-\xi)}{v_1+v_2}}\sqrt{\frac{(\xi-x+v_1t)}{(x+v_2t-\xi)}}S_0(\xi)d\xi \tag{5}$$

Next, we consider $G_0(x) = G_{00}Im(x)$ and $S_0(x) = S_{00}Im(x) = (v_1/v_2)G_{00}Im(x)$ as short pulses along the x axis having the width Δ, which is small compared to the interval L, where

$$Im(x) = \begin{cases} 1, & if\ 0 < x < \Delta; \\ 0, & otherwise \end{cases},$$

G_{00} is a linear concentration of G ends at the initial time-point. Numerical calculations based on various values of $v_1;\, v_2;\, k_1;\, k_2$ show that the formulae (5) indeed represent the solutions of equation (2). It must be noted that the pair of functions:

$$\begin{cases} G(x,t) = e^{\frac{k_2(x+v_2t)}{v_1+v_2} - \frac{k_1(x-v_1t)}{v_1+v_2}} \\ S(x,t) = e^{\frac{k_2(x+v_2t)}{v_1+v_2} - \frac{k_1(x-v_1t)}{v_1+v_2}} \end{cases}$$

also represent a particular solution of equation (2). By using this solution at $t = 0$ in (4), we confirmed that the downstream substitution of the expressions obtained in (2) results in equality. Since the expression (5) cannot be easily processed numerically, we approximated the functions $G(x;\, t)$ and $S(x;\, t)$ by the simpler formulae $GF(x;\, t)$ and $SF(x;\, t)$. These formulae represent an expansion of expression (5), as follows:

$$GF(x,t) == \begin{cases} e^{\frac{k_2-k_1}{v_1+v_2}x + \frac{-(k_1v_2+k_2v_1)}{v_1+v_2}t} G_{00}\Delta \left[\frac{k_2v_1}{(v_1+v_2)v_2}I_0\left[2\frac{\sqrt{k_1k_2}\sqrt{(-x+v_1t)(x+v_2t)}}{v_1+v_2}\right] + \frac{\sqrt{k_1k_2}}{(v_1+v_2)}I_1\left[2\frac{\sqrt{k_1k_2}\sqrt{(-x+v_1t)(x+v_2t)}}{v_1+v_2}\right]\sqrt{\frac{(x+v_2t)}{(-x+v_1t)}}\right], & if -v_2t+\Delta < x < v_1tG_{00}Im \\ (x-v_1t)e^{-k_1t}, & if\ v_1t \le x \le v_1t+\Delta 0, \quad otherwise \end{cases}$$

$$SF(x,t) == \begin{cases} e^{\frac{k_2-k_1}{v_1+v_2}x + \frac{-(k_1v_2+k_2v_1)}{v_1+v_2}t} G_{00}\Delta \left[\frac{k_1}{(v_1+v_2)}I_0\left[2\frac{\sqrt{k_1k_2}\sqrt{(-x+v_1t)(x+v_2t)}}{v_1+v_2}\right] + \frac{\sqrt{k_1k_2}}{(v_1+v_2)}\frac{v_1}{v_2}I_1\left[2\frac{\sqrt{k_1k_2}\sqrt{(-x+v_1t)(x+v_2t)}}{v_1+v_2}\right]\sqrt{\frac{(-x+v_1t)}{(x+v_2t)}}\right], & if -v_2t+\Delta < x < v_1tG_{00}\frac{v_1}{v_2}Im \\ (x+v_2t)e^{-k_2t}, & if\ -v_2t \le x \le -v_2t+\Delta 0, \quad otherwise \end{cases} \tag{6}$$

We are now trying to determine the evolution of short pulses initially situated within a small Δ interval. A short pulse is concentration of growing (G) or shrinking (S) ends of x length in a small Δ interval at a given time. The analysis for short pulses located close to the origin according to equation (2) is shown in Fig. 3. The short pulses become exponentially weaker with time: the short pulse G moves right at a velocity v_1 and its height decreases with time at a rate constant k_1. The short pulse S moves left at a velocity v_2, and its height decreases exponentially with time at a rate constant k_2. We note that the asymptotic approximation we used matches the exact solution fairly well. It only differs from the exact solution at the top of the short pulse. However, taking into account that the short pulse width Δ is small, this does not result in a significant distortion of the solution.

In conclusion, we found the solution of the system of continuous equations (2). An important contribution of this solution is the possibility to differentiate between labeled and unlabeled S ends. As can be seen in Fig. 3, the S ends on the left of the $x = 0$ coordinate point are unlabeled, whereas those on the right are labeled.

Solution for finite x coordinate intervals

As mentioned, we found a solution for (2) for evolution of concentration "pulse" Δ of G and S ends initially located at $x = 0$. In this case, after time t, G and S ends at the point r located to the right of $x = 0$ gave a marked tubulin track with length r, while the ends on the left did not leave marked tracks. Thus, a positive r can be considered as the length of a marked track. In this section we are constructing the solution where G and S ends are initially distributed over the 0-L interval of x coordinate (in a real photobleaching experiment Δ could not be negligible compared to the spindle length) and are then moving in different directions according to (2). This can be done if the solution (6) is convolved with a rectangle over x ($Rect(x) = 1$ if $0 < x < L$ and $Rect(x) = 0$ otherwise). Since for the following analysis we need only the degrading S ends that have no tracks, the auxiliary integration variable xx must be negative and vary from $-L$ to 0. However, when integrating over the initial position of the impulse at 0-Δ, we need to move the limits of xx to $-L + \Delta$, and consider that the Δ interval introduces a small error (Δ is small). The concentration $S_{tail}(x,t)$ of S ends without tracks (integrated over initial position of the pulse) at the coordinate point x and at the time t is:

$$S_{tail}(x,t) = \frac{1}{L}\int_{-L+\Delta}^{\Delta} Rect(x)\, SF(xx+x,t)\, dxx$$

The total number of degrading S ends without a tubulin label could be determined by the integration of $S_{tail}(x,t)$ over x. With $t = 0$, the total number of S ends is $G_{00}\ \Delta v_1/v_2$ (at the stationary spindle $G_{00}\ v_1 = S_{00}\ v_2$, see above); therefore, the fraction $S_n(t)$ of S ends (without a label) among all S ends is:

$$S_n(t) = \frac{v_2}{v_1 G_{00}\Delta}\int_0^L S_{tail}(x,t)dx$$

This expression permits construction of a solution for a stationary spindle.

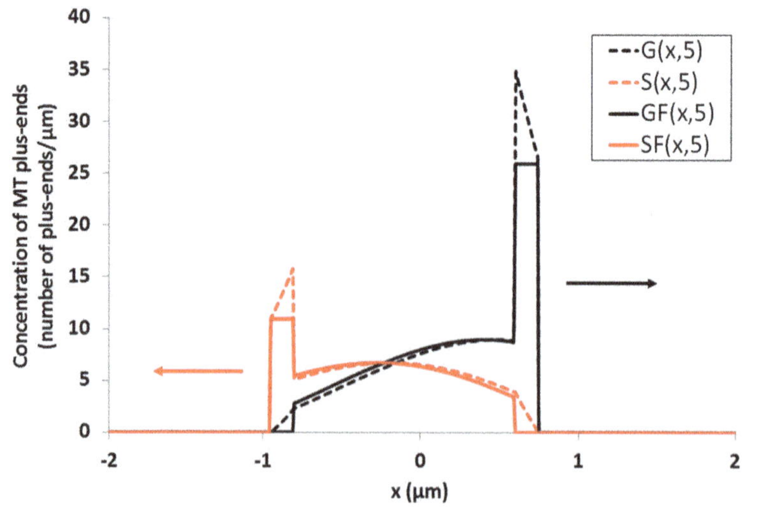

Fig. 3 The behavior of the solution of (5) and the approximate solution of (6) at t = 5 are depicted as *dashed* and *continuous lines*, respectively; *black* and *red* limes refer to growing and shrinking MTs, respectively. The parameters used are $v_1 = 0.12$ μm/s, $v_2 = 0.19$ μm/s, $k_1 = 0.27$ 1/s, $k_2 = 0.35$ 1/s, $G_{00} = 100$. The length of the photobleached area is L = 3 μm, and the short pulse width $\Delta = 0.15$ μm. *Arrows* denote the direction of the G(x; t) and S(x; t) front movements

FRAP in a stationary spindle

The essential changes in the mitotic spindle structure occur during prometaphase and at the metaphase-to-anaphase transition. During metaphase the spindle shape is relatively stable, so that the stationary state appears a good approximation.

In FRAP experiments on mitotic spindles containing fluorescent tubulin, a rectangular area perpendicular to the spindle axis is photobleached, and fluorescence recovery in this area is recorded [8]. Let us consider a stationary case (see above) where:

$$G(x) = G_0 \, e^{-\alpha x}$$

$$S(x) = G_0 \, \frac{v_1}{v_2} e^{-\alpha x}$$

In this case, the concentration of G ends in the photobleached area is constant with all G ends incorporating marked tubulin dimers. At any given time, the concentration of S ends is also constant, but S ends can be of two types: S_n ends that degrade releasing of non-marked photobleached tubulin, and S_m ends that shrink after a short growth phase and therefore degrade releasing fluorescent tubulin.

The fluorescence recovery detection area is usually small compared to the spindle length, and we can neglect concentration changes within this zone. Therefore, we can use G_0 and $S_0 = G_0 \, v_1/v_2$ instead of $G(x)$ and $S(x)$; because the spindle is stationary, G_0 and S_0 do not depend on time. If large intervals are considered, the equation would need only to account for the exponential dependence of $G(x)$ and $S(x)$ over coordinate x.

Growing G_0 ends incorporate marked tubulin at a speed of v_1, while the fraction of degrading $S_m(t)$ ends release marked tubulin at a speed v_2. If $M(t)$ is the length of marked MTs within the bleached area, its time derivative would be:

$$\frac{d}{dt}M(t) = v_1 G_0 - v_2 S_m(t)$$

By introducing the fraction of marked tubulin $Lb(t) = M(t)/E$, where E is the total length of marked MTs within the zone before photobleaching, and by accounting for $G_0 v_1 = S_0 v_2$ and $S_m(t) = S_0(1-S_n(t))$, we obtain:

$$\frac{d}{dt}Lb(t) = \frac{v_2 S_0}{E} S_n(t)$$

Integration leads to:

$$Lb(t) - Lb(t_0) = \frac{v_2 S_0}{E} \int_{t_0}^{t} S_n(t)dt$$

Before photobleaching, most G and S ends within the L interval have marked tubulin at their ends and are related by $G_0 \, v_1 = S_m \, v_2$. After fluorescence recovery (when the S_n ends have disappeared), the $G_0 \, v_1 = S_m \, v_2$ condition is again present. In such cases, the track length for both G_0 and S_m ends before and after fluorescence recovery would be $L/2$. Fluorescence recovery normalization to the level of fluorescence before bleaching is the accepted mode of experimental data analysis. In our model, the fluorescence levels before bleaching and after full recovery coincide. Thus, we can normalize the theoretical dependence fluorescence level after full recovery:

$$\frac{Lb(t) - Lb(t_0)}{Lb(\infty) - Lb(t_0)} = \frac{\int_{t_0}^{t} S_n(t)dt}{\int_{t_0}^{\infty} S_n(t)dt} \tag{7}$$

(7) is the final dependence that can be used to fit the experimental spindle FRAP curves. It is clear that (7) does not depend on $v_2 S_0/E$.

Numerical implementation

A pre-anaphase-B FRAP curve (near the equator) for *Drosophila* syncytial embryo mitosis was obtained by *Cheerambathur* et al. [8], who also proposed a theoretical model for the FRAP curve and determined the dynamic parameters of the G and S ends. Approximate parameter values are $v_1 = v_2 = 0.35$ *μm/s*, $k_1 = 0.2$ *1/s* and $k_2 = 0.25$ *1/s*. The width of the photobleached area near the equator was *2.2 μm*. They [8] also reported the speed of EB1 labeled ends (*0.25 μm/s*) in their experimental system. In our model, like in the theoretical model of [8]), v_1 and v_2 are the speeds in the photobleached area, which is slowly moving to the pole at the speed of the flux (*0.05 μm/s* according to [8]). Thus, we calculated *0.25-0.05 μm/s* for v_1, considered $k_1 = 0.2$ *1/s*, and performed minimization of the root-mean-square deviation of our theoretical curve (7) from the experimental one [8]. The simple gradient descent method was used to find a local minimum near the dynamic parameters of [8]. The theoretical and experimental curves are shown in Fig. 4; the optimal parameters are $v_1 = 0.2$ *μm/s*, $v_2 = 0.08$ *μm/s*, $k_1 = 0.2$ *1/s*, and $k_2 = 0.4$ *1/s*.

In our model, we assume that full fluorescence recovery occurs within the bleached area, regardless its position along the spindle. However, there is experimental evidence that fluorescence recovery is higher near the spindle equator than at the poles [8, 13, 14]. It should be also mentioned that in the stationary case, G and S ends are distributed over the half-spindle in an exponential manner (with exponent index $-\alpha x$), with the maximum near the pole and the minimum near the equator. Thus, photobleaching near the poles will not only affect MT ends situated in the bleached area but also the ends of "MT fragments" detached from the poles; fluorescence of these

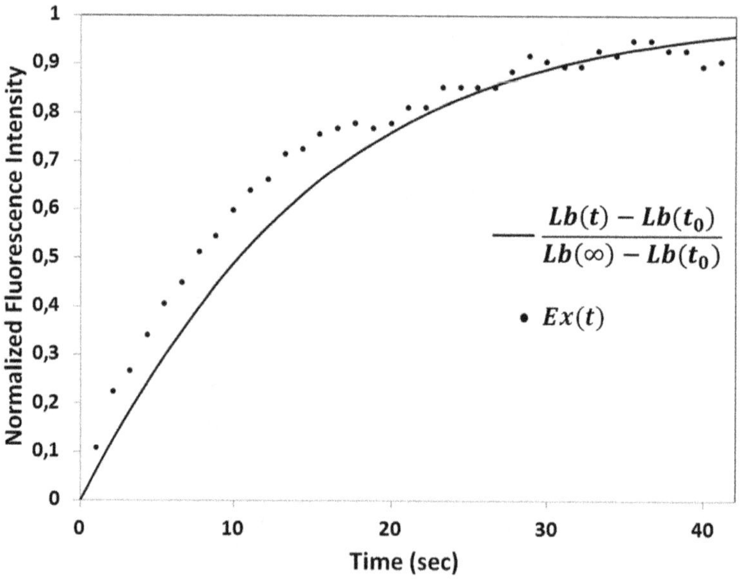

Fig. 4 Experimental curve for pre-anaphase B [8] and its theoretical approximation (reproduced with permission of the authors [8]). Ex(t) are the corrected experimental points from [8], i.e., the fluorescence level detected just after photobleaching was subtracted from the experimental values; experimental points were then normalized to the fluorescence level before photobleaching. The theoretical photobleaching curve (*continuous line*) was calculated for various values of v_1, v_2, k_1, and k_2 using Mathcad-14 software

"MT fragments" will not be restored during the experiment, lowering the overall level of fluorescence recovery. If the bleached area is near the equator, this effect would be small.

Discussion

The FRAP technique, although widely used for studying the dynamics of spindle MT behavior, has still some limitations. These limitations are due to the fact that the kinetic constant value measured in the experiments (referred to as the "rate of synthetic processes in the spindle") has no direct link to the dynamic parameters of MTs. The work of *Cheerambathur* et al. [8] partially addressed this issue by linking this parameter with k_1 and k_2, but the authors have not solved the problem completely. This is because modeling of the fluorescence recovery curve proposed earlier by *Brust-Mascher* et al. [9] requires the mechanical equations for the entire spindle to be solved, which in turn requires the knowledge of multiple mechanical parameters. Here, we provided an alternative model for the interpretation of the fluorescence recovery curve. Previously published reports describing MT polymerization and depolymerization dynamics have served as the starting point for our analysis. Based on these studies, we analytically solved the PDE, which describes the dynamics and transitions between the states of MT ends. We then used this solution to solve the equation describing fluorescence recovery in a steady-state spindle and, finally, we obtained a theoretical dependence for the fluorescence recovery curve. Our study provides a model for the FRAP recovery curve if all four parameters are known. We also made a model calculation to determine how lowering of each parameter could affect $T_{1/2}$ FRAP time. As shown in Table 1, lowering of v_1 or k_2 leads to an increase in $T_{1/2}$, while decrease of v_2 or k_1 decreases $T_{1/2}$.

Some mutant proteins decrease the rate of the process in which they are involved compared to their wild type counterparts, so we made an attempt to analyze published FRAP data for three mutant *Drosophila* mitotic proteins. Mini-spindles (Msps), a member of the XMAP215/TOG protein family, concentrates not only at the centrosomes but also at the MT plus-end [14]. Based on their own data and pre-existing data *Buster* et al. [14] concluded that Msps positively regulates transition from pausing to MT growth state. Although our model does not include MT pausing, a decreased transition to

Table 1 Alterations of MT growth/shrinking parameters affect $T_{1/2}$ FRAP times for half spindles (calculations according to our model)

v_1 (μm/s)	v_2 (μm/s)	k_1 (1/s)	k_2 (1/s)	$T_{1/2}$ (sec)
0.25	0.09	0.21	0.17	**31**
$\underline{0.24}$	0.09	0.21	0.17	**33**
0.25	$\underline{0.09}$	0.21	0.17	25
0.25	0.09	$\underline{0.20}$	0.17	28
0.25	0.09	0.21	$\underline{0.16}$	**36**

The lower parameter is underlined. Increased and decreased FRAP times are indicated by bold large numbers and small numbers, respectively growth means that the constant k_2 is decreased. Table 1 shows that the decrease in k_2 results in an increase in the FRAP time, which is what has been in fact observed by *Buster* et al. [14]. A second example is concerned with the Mast/Orbit protein, a CLASP family protein that is found at microtubule plus-ends near the kinetochore [15, 16]. RNAi-mediated depletion of Mast/Orbit leads to MT flux inhibition [14], consistent with the finding that Mast is involved in the control of microtubule polymerization [17]. Thus, Mast affects v_1 speed in our model and a mutation in *Mast* would decrease v_1. The v_1 decrease (Table 1) decreases FRAP time, according to the observations of *Buster* et al. [14]. In a third example we consider the Eb1 protein, a well-known microtubule plus-end marker that increases MT rescue frequency while decreasing pause [14]. Loss of Eb1 would therefore decrease the k_2 parameter, which would lead to an increase in FRAP time as has been in fact observed [14].

The empirical FRAP curve of *Salmon* et al. [3] permits determination of the dynamic parameter commonly called "tubulin turnover". Here, we have shown that the spindle FRAP curve depends on four kinetic parameters of MT end polymerization and depolymerization. Using the results of *Cheerambathur* et al. [8], we showed that the solution we found can adequately approximate our experimental FRAP curve for metaphase spindles. In addition, we posited that differences in fluorescence recovery between the poles and the equator of metaphase spindles, which are commonly observed in FRAP experiments, could be explained by the exponential distribution of *G* and *S* ends in the half-spindle as proposed in our model.

Conclusions

Here, we provided an alternative model for the interpretation of the fluorescence recovery curve. Based on the previously published reports describing MT polymerization and depolymerization dynamics, we analytically solved the PDE, which describes the dynamics and transitions between the states of MT ends. We then used this solution to solve the equation describing fluorescence recovery in a steady-state spindle and, finally, we obtained a theoretical dependence for the fluorescence recovery curve. A numerical expression that describes the distribution of polymerizing and depolymerizing microtubule ends as a function of four kinetic parameters is presented. We also made a model calculation to determine how lowering of each parameter could affect $T_{1/2}$ FRAP time. As shown in this study, the lowering of v_1 or k_2 leads to an increase in $T_{1/2}$, while the decrease of v_2 or k_1 decreases $T_{1/2}$. The numerical expression we elaborated can replace the empirical FRAP curve described earlier for a spindle comprising fluorescently marked microtubules. The curve we generated fits well the experimental data.

Abbreviations

FRAP: Fluorescence recovery after photobleaching; GFP: Green fluorescent protein; MT: Microtubule; PDE: Partial differential equation

Acknowledgements

After acceptance of this paper, Professor Leonid Omelyanchuk passed away just before his sixtieth birthday. He was a valuable and highly respected researcher. Prof. Omelyanchuk provided important contributions to the development cytophotometry methods and nanotubes-based biosensors. He was also highly appreciated of his studies of the mechanisms of cell division. Leonid was a very sincere and kind-hearted person, a caring husband, a good father and a dearly loving grandfather.
Leonid's memory will be forever in the hearts of his colleagues of the Institute of Molecular and Cell Biology and the Institute of Cytology and Genetics of the Siberian Branch of Russian Academy of Sciences. AFM will greatly miss him as a fine human being and a mentor.
The authors express gratitude to M.A.Yurkin, who helped us at the initial stage of the work, T.Y. Mikhailova, who found the solution (5) for equations (2) and made the expansion (6), representing the asymptotic behavior of (5) at a small Δ. We wish to thank M. Gatti and A.V. Pindyurin for critical revision and major correction of the manuscript, and A.A. Gorchakov, N.A Bulgakova and J. Saul for final reading of the text.

Declarations

This article has been published as part of BMC Systems Biology Vol 11 Suppl 1, 2017: Selected articles from BGRS\SB-2016: systems biology. The full contents of the supplement are available online at http://bmcsystbiol.biomedcentral.com/articles/supplements/volume-11-supplement-1.

Funding

This work was supported by the grant from the Ministry of Education and Science of Russian Federation #14.Z50.31.0005. Publication costs were paid by the authors.

Authors' contributions

LVO and AFM conceptualized the research project. LVO found a way to solve the non-stationary equation describing spindle FRAP curve. AFM explored the effect of gene silencing on the behavior of fluorescence recovery after photobleaching curve in the framework of our model. All authors read and edited the paper. Both authors read and approved the final manuscript.

Competing interests

The authors declare that they have no competing interests.

References

1. Peters R, Peters J, Tews KH, Bähr W. A microfluorimetric study of translational diffusion in erythrocyte membranes. Biochim Biophys Acta. 1974;367:282–94.
2. Sprague BL, Pego RL, Stavreva D, McNally JG. Analysis of binding reactions by fluorescence recovery after photobleaching. Biophys J. 2004;86:3473–95.
3. Salmon ED, Leslie RJ, Saxton WM, Karow ML, McIntosh JR. Spindle microtubule dynamics in sea urchin embryos: analysis using a fluorescein-labeled tubulin and measurements of fluorescence redistribution after laser photobleaching. J Cell Biol. 1984;99:2165–74.
4. Hill TL. Introductory analysis of the GTP-cap phase-change kinetics at the end of a microtubule. Proc Natl Acad Sci U S A. 1984;81:6728–32.

5. Mitchison T, Kirschner M. Microtubule assembly nucleated by isolated centrosomes. Nature. 1984;312:232–7.
6. Mitchison T, Kirschner M. Dynamic instability of microtubule growth. Nature. 1984;312:237–42.
7. Verde F, Dogterom M, Stelzer E, Karsenti E, Leibler S. Control of microtubule dynamics and length by cyclin A- and cyclin B-dependant kinases in xenophobus egg extracts. J Cell Biol. 1992;118:1097–108.
8. Cheerambathur DK, Civelekoglu-Scholey G, Brust-Mascher I, Sommi P, Mogilner A, Scholey JM. Quantitative analysis of an anaphase B switch: Predicted role for a microtubule catastrophe gradient. J Cell Biol. 2007; 177:995–1004.
9. Brust-Mascher I, Civelekoglu-Scholey G, Kwon M, Mogilner A, Scholey JM. Model for anaphase B: role of three mitotic motors in a switch from poleward flux to spindle elongation. Proc Natl Acad Sci U S A. 2004; 101:15938–43.
10. Rogers SL, Rogers GC, Sharp DJ, Vale RD. Drosophila EB1 is important for proper assembly, dynamics, and positioning of the mitotic spindle. J Cell Biol. 2002;158:873–84.
11. Dogterom M, Leibler S. Physical aspects of the growth and regulation of microtubule structure. Phys Rev Lett. 1993;70:1347–50.
12. Koshlyakov NS, Gliner EB, Smirnov MM. Uravneniya v chastnih proizvodnikh matematicheskoi physiki. In: Suhodskiy AM, editor. Moscow: Vischaia shkola; 1970. p. 91–2.
13. Goshima G, Mayer M, Zhang N, Stuurman N, Vale RD. Augmin: a protein complex required for centrosome-independent microtubule generation within the spindle. J Cell Biol. 2008;181:421–9.
14. Buster D, Zhang D, Sharp D. Poleward tubulin flux in spindles: regulation and function in mitotic cells. Mol Biol Cell. 2007;18:3094–104.
15. Lemos CL, Sampaio P, Maiato H, Costa M, Omelyanchuk LV, Liberal V, et al. Mast, a conserved microtubule-associated protein required for bipolar mitotic spindle organization. EMBO J. 2000;19:3668–82.
16. Inoue YH. Orbit, a novel microtubule-associated protein essential for mitosis in drosophila melanogaster. J Cell Biol. 2000;149:153–66.
17. Maiato H, Khodjakov A, Rieder CL. Drosophila CLASP is required for the incorporation of microtubule subunits into fluxing kinetochore fibres. Nat Cell Biol. 2005;7:42–7.

12

Prior knowledge guided active modules identification: an integrated multi-objective approach

Weiqi Chen[1], Jing Liu[2] and Shan He[1*]

Abstract

Background: Active module, defined as an area in biological network that shows striking changes in molecular activity or phenotypic signatures, is important to reveal dynamic and process-specific information that is correlated with cellular or disease states.

Methods: A prior information guided active module identification approach is proposed to detect modules that are both active and enriched by prior knowledge. We formulate the active module identification problem as a multi-objective optimisation problem, which consists two conflicting objective functions of maximising the coverage of known biological pathways and the activity of the active module simultaneously. Network is constructed from protein-protein interaction database. A beta-uniform-mixture model is used to estimate the distribution of *p*-values and generate scores for activity measurement from microarray data. A multi-objective evolutionary algorithm is used to search for Pareto optimal solutions. We also incorporate a novel constraints based on algebraic connectivity to ensure the connectedness of the identified active modules.

Results: Application of proposed algorithm on a small yeast molecular network shows that it can identify modules with high activities and with more cross-talk nodes between related functional groups. The Pareto solutions generated by the algorithm provides solutions with different trade-off between prior knowledge and novel information from data. The approach is then applied on microarray data from diclofenac-treated yeast cells to build network and identify modules to elucidate the molecular mechanisms of diclofenac toxicity and resistance. Gene ontology analysis is applied to the identified modules for biological interpretation.

Conclusions: Integrating knowledge of functional groups into the identification of active module is an effective method and provides a flexible control of balance between pure data-driven method and prior information guidance.

Keywords: Prior knowlege, Multi-objective evolutionary algorithm, Active module identification

*Correspondence: s.he@cs.bham.ac.uk
[1]School of Computer Science, University of Birmingham, Edgbaston, B15 2TT Birmingham, UK
Full list of author information is available at the end of the article

Background

With the development of high-throughput data collection technologies, vast amounts of omics data that cover different species and different levels of biological activities have accumulated exponentially. These varied omics data, including the genome sequencing data (genomics), genome-wide expression profiles (transcriptomics), and protein abundances data (proteomics), provide valuable information concerning the intrinsic mechanisms underlining biological processes. With the accumulation of large datasets, one of the most essential challenges for researchers is that how to properly interpret these data. Take gene expression data analysis as an example, methods have evolved from the simple single or multivariate statistical analysis, e.g., calculation of fold-change, identification of differential expressed genes, to integrated approaches that integrate prior knowledge and different datasets [1]. As a research field driven by those integrated approaches, network biology has gained popularity recently years.

Network biology offers a highly abstract model of networks to characterize various levels of biological systems and provides insights into those system by taking advantages of network theory [2, 3]. Although currently it's not able to fully capture the diversity and dynamics of complex biological system [4], it is still one of the most promising and fast developing research area in modern biology. Many studies have been performed on the construction of networks from biological systems and the structural and functional features that may respond to related biological information. Network construction methods are varied from calculating pair-wise correlation coefficient of expression data (correlation network [5]), filtering from existing interaction database (protein-protein interaction network [6–9]), or integrated approaches based on both expression data and metabolic models (tissue specific metabolic network [10]). Structural features includes degree distribution, clustering coefficient, scale-free property [11], modularity [12] and network robustness [13]. One of the most studied features is modular structure.

Modular structure is one of the essential characteristics that reveal information about the relationship and interaction among components in the network. In biological networks, modules are considered as the functional units of cellular process and organization [14]. Varied definitions of module have been proposed and numerous methods have been developed to identify those modules [15, 16], all aiming to reveal essential biological mechanisms [17, 18]. Among them, active module detection is a successfully applied integrative approach. Active module is a densely connected area in network that shows striking changes in molecular activity or phenotypic signatures, which is often associated with a given cellular response. Active module is expected to reveal dynamic and process-specific information that is correlated with cellular or disease states.

A typical active module detection algorithm takes gene expression data, calculates statistical values indicating differential expression level, and searches in corresponding network to identify modules inside which gene activity changes significantly. The jActiveModule [19] method proposed by Ideker in 2002 is considered as the first to formulate active module detection into an optimization problem. It uses the standard normal inverse of single gene's p-value to measure the activity of one gene, aggregates the node scores for a given module with adjustment and background correction, and finally searches for high-scoring modules via simulated annealing. Many following methods adopt this framework of significant-area-search method. One representative research for identifying condition-responsive protein-protein interaction module used edge-based scoring method [6]. There are also formulations that combine both node and edge score [7, 9]. As the problem of finding the maximal-scoring connected subgraph is NP-hard (non-deterministic polynomial-time hard) [19], heuristic algorithms are broadly used, e.g. simulated annealing [19], greedy search [20], and evolutionary algorithm [8, 21]. Exact approaches via integer linear programming are also developed [22].

In this paper we propose a novel multi-objective active module identification algorithm. We first formulate the active module identification problem as a multi-objective problem, which not only maximises the activity score as defined by Dittrich and Klau [22] but also maximises the prior knowledge contained in the active module. The intuition behind this multi-objective formulation is to use prior knowledge to guide the search of novel information from data, i.e., active modules. The Pareto solutions from this multi-objective optimisation problem are then the optimal trade-off between known knowledge and novel information.

In order to solve this multi-objectie problem, we proposed a modified multi-objective evolutionary algorithm. One of the important details omitted in many papers of active module identification is how to ensure the connectedness of the solutions. Without this connectedness constraint, the optimal solution is trivial, i.e., the top genes with largest node scores. In order to ensure the connectedness of the identified active modules, we design a novel constraints based on algebraic connectivity. The algorithm is applied to a small molecular interaction network that was used by Ideker [19] and then applied to a large Protein-Protein Interaction (PPI) network constructed from microarray data on drug toxicity and resistance.

Methods

Problem formulation

The network G is represented as $G = (V, E)$ with $p_v \in (0, 1)$ for $v \in V$ where V is the set of nodes, E the set of edges, and p_v the assigned p-value from differential expression analysis for each node v. In the proposed algorithm there are two objectives and one constraint for a given module A:

- Active module score S_A indicating significant changes in gene expression for a given module, to be maximized during search.
- KEGG (Kyoto Encyclopedia of Genes and Genomes) pathway coverage score R_A for the number of covered metabolic pathway by genes in module, to be maximized.
- Algebraic connectivity to check whether a given subgraph is connected or not, used as a constraint to ensure connectedness.

Active module score

Microarray analysis studies showed that expression data can be effectively estimated by many mixture-model methods that divide genes into two or more groups, one group contains genes that are differentially expressed, and other(s) not differentially expressed [1]. Among those many methods, Pounds and Morris proposed a beta-uniform mixture (BUM) model that very accurately describes the distribution of a large set of p-values produced from an microarray experiment [23]. The BUM model considers the distribution of p-values as a mixture of a special case of beta distribution ($b = 1$) and a uniform(0, 1) distribution, with a mixture parameter λ. The p-values under the null hypothesis are assumed to have a uniform distribution. Under the alternative hypothesis the distribution of p-values will have a high density for small p-values and can be described by $B(a, 1)$.

A general beta distribution $B(a, b)$ is given by

$$f(x) = \frac{\Gamma(a+b)}{\Gamma(a)\Gamma(b)} x^{a-1}(1-x)^{b-1} \tag{1}$$

where $\Gamma(.)$ denotes the gamma function. As $\Gamma(1) = 1$, the probability density function of BUM model is then reduced to

$$f(x|a, \lambda) = \lambda + (1-\lambda)ax^{a-1} \tag{2}$$

for $0 < x \leq 1$, $0 < \lambda < 1$ and $0 < a < 1$. Given a set of p-values the two parameters of BUM distribution λ and a can be calculated by maximum likelihood estimation.

Following the idea of Dittrich and Klau [22] to decompose signal component from background noise, an additive score to measure the significance of gene's differential expression is calculated as

$$\begin{aligned} S^{FDR}(x) &= \log \frac{B(a,1)(x)}{B(a,1)(\tau)} \\ &= \log \frac{ax^{a-1}}{a\tau^{a-1}} \\ &= (a-1)(\log x - \log \tau) \end{aligned} \tag{3}$$

where τ is a threshold to determine the significance of a p-value. In order to control the estimated upper bound of the false discovery rate (FDR) introduced by Benjamini and Hochberg [24], τ could then be selected to ensure that $FDR \leq \alpha$ for some predefined α using the following equation

$$\tau = \left(\frac{\hat{\pi} - \alpha\lambda}{\alpha(1-\lambda)} \right)^{\frac{1}{(a-1)}} \tag{4}$$

where $\hat{\pi} = \lambda + (1-\lambda)a$, meaning the maximum proportion of the set of p-values that could arise from the null hypothesis.

After assigning score to each of the genes, the overall score for a given module A is then the summation of all genes' scores in it, given by

$$S_A = \sum_{x \in A} S^{FDR}(x) \tag{5}$$

KEGG pathway coverage

KEGG is an integrated database of high level functions and utilities of biological systems [25]. KEGG PATHWAY is a collection of manually drawn pathway maps representing the knowledge on the molecular interaction and reaction networks. Mapping of pathway information mainly relies on molecular datasets, especially large-scale datasets such as genomics, transcriptomics, proteomics, and metabolomics. Genes involved in the same KEGG pathway are considered as functionally related to each other. In the experiment KEGG pathway coverage score R_A is formulated as the second objective to measure the enrichment of functional groups in a given module A.

The KEGG pathway information is retrieved from the KEGG REST-style entry for *Saccharomyces cerevisiae* (yeast) [26]. Each entry of the mapping data records one gene and its corresponding pathway. The records are then split into different groups labeled by the pathways. For the i-th pathway, V_i stands for the set of genes it contains. Given a module A with V_A as the set of genes, its KEGG pathway cover rate R_i over the i-th pathway is calculated as

$$R_i = \frac{|V_i \cap V_A|}{|V_i|} \tag{6}$$

meaning the percentage this pathway is covered by given module. The cover rate R_i is then compared with a threshold R_{ratio} to determine whether this pathway can be considered as enriched in the given module. The threshold shall be selected carefully. A too high value of R_{ratio} leads to a tiny group of connected pathways genes with positive active module score as the search could not expand to other area under such stringent condition. On the contrary, a very low R_{ratio} could not reflect the meaning for the second objective. In practice, R_{ratio} is set to a series of values for preliminary experiment. The results are analyzed and compared to decide a suitable value. The total enriched pathway count R_A is given by

$$R_A = |\{R_i | R_i > R_{ratio}\}|, i \in P \tag{7}$$

where P stands for total number of pathways.

Algebraic connectivity

The algebraic connectivity of a graph G, denoted as $\alpha(G)$, is the second-smallest eigenvalue of the Laplacian matrix of G. It serves as a good parameter to measure how well a graph is connected. $\alpha(G)$ is greater than zero if and only if G is a connected graph.

The Laplacian matrix L of a simple graph G is calculated as

$$L = D - A \tag{8}$$

where D is the degree matrix and A the adjacency matrix. The eigenvector ν of the square matrix L is the non-zero vector that satisfies

$$L\nu = \lambda\nu \tag{9}$$

λ is a scalar known as the eigenvalue associated with the eigenvector ν. Algebraic connectivity $\alpha(G)$ is the second smallest eigenvalue of the Laplacian matrix L.

Multi-objective optimization algorithm

In order to perform multi-objective optimization to maximize both active module score and KEGG pathway coverage score simultaneously, a multi-objective evolutionary algorithm modified from NSGA-II (non-dominated sorting genetic algorithm II, see [27]) is applied as search strategy for module detection.

Solution representation

A solution is represented as a binary vector of length $|V|$, where $|V|$ is the size of network, i.e. total number of nodes. Adding or deleting a node is performed through simply flip one bit of the vector at corresponding site.

Fitness function

Active module score S_A and KEGG coverage score R_A are used as two objectives. As the implementation of the algorithm is aimed at minimization both objectives, scores calculated from above equations would be given an extra negative sign.

Initialization

The search starts by randomly initializing a group of small cores in network. Nodes with high $S^{FDR}(x)$ scores are selected as seeds of potential modules to begin with. Number of seed nodes is decided by the population parameter for evolutionary algorithm. For instance, if population is set to 50, nodes with top 50 $S^{FDR}(x)$ scores are selected as seeds. In the case when the population size exceeds network size, every node will be selected as a seed. In initialization stage, neighboring nodes of a seed with positive scores are added to the module in which the seed represents.

Parent selection

Binary tournament selection is applied for selecting parents to reproduce. In some cases when the population converges too fast, this step is skipped to decrease selection pressure, thus the whole population would be used for reproduction.

Reproduction

Single point crossover is applied to selected parents. Mutation is performed by adding neighboring nodes with positive $S^{FDR}(x)$ score or in a pathway into current module each time. Offspring generated is added to parental population to form a combined population with twice the size, waiting to be sorted and selected.

Clearing procedure

An extra clearing procedure is applied after reproduction and before non-dominated sorting. The step is introduced because in practise the algorithm tends to generate a number of replicated solutions when converging towards global optima. However, considering the natural property of our optimization problem, it is reasonable to obtain multiple optima, both those global on the non-dominated Pareto front and those dominated local optima, each representing the most significantly changed modules and modules that do not change that significantly, but still worth looking into. This procedure, inspired and simplified from Petrowski [28], detects replicated solution groups, preserves one copy, and resets all other individuals as infeasible solutions which will soon be eliminated after soring and replacement.

Sorting and replacement

The algorithm uses fast non-dominated soring and crowding distance assignment as detailed in Ref [27] to generate new population from the combined population efficiently and preserve solution diversity.

Constraint handling

To ensure the connectivity of detected module after crossover, algebraic connectivity $\alpha(G)$ is used as a constraint. Solution with non-positive algebraic connectivity violates the constraint, indicating itself a disconnected subgraph and thus an infeasible solution. Replicated solutions are also marked infeasible in the clearing procedure.

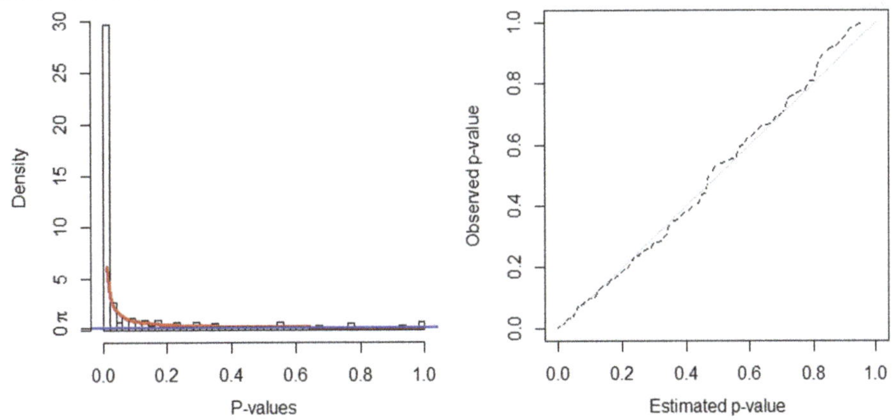

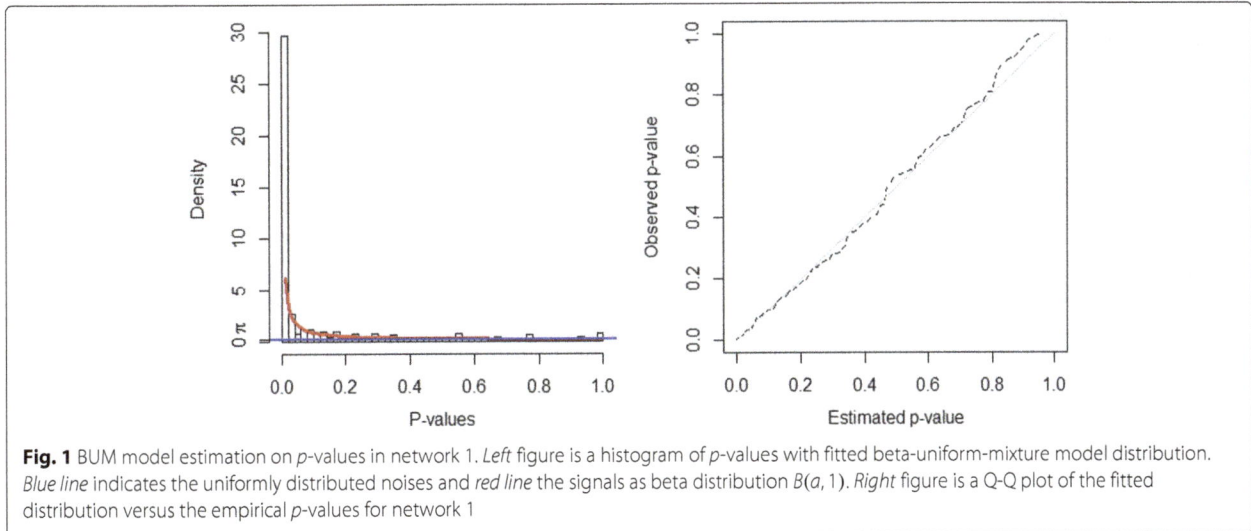

Fig. 1 BUM model estimation on *p*-values in network 1. *Left* figure is a histogram of *p*-values with fitted beta-uniform-mixture model distribution. *Blue line* indicates the uniformly distributed noises and *red line* the signals as beta distribution $B(a, 1)$. *Right* figure is a Q-Q plot of the fitted distribution versus the empirical *p*-values for network 1

Infeasible solutions are dominated by all feasible solutions.

Network construction

Network 1: a small molecular interaction network on galactose utilization pathway

A small molecular interaction network once used by Ideker [19] is used as a test network. The molecular interaction networks visualization software Cytoscape [29] provides jActiveModule as a plugin to find expression activated modules. The tutorial in Cytoscape App Store [30] provides samples data consists of a network file as a model of the galactose utilization pathway in yeast and a companion expression file contains *p*-values to describe the significance of each observed change in expression. *p*-values under condition labeled as *GAL80R* are extracted and overlaid to network file, resulting in a network with 330 genes.

Network 2: yeast drug reaction network constructed from differential analysis and interactome mapping

Gene expression data on yeast's reaction to diclofenac is downloaded from GEO (NCBI Gene Expression Omnibus) database [31]. Diclofenac is a widely used analgesic drug that can cause serious adverse drug reactions [32]. Yeast is used as model eukaryote to capture the cellular changes under the treatment of diclofenac. The data provides the microarray expression for diclofenac-treated yeast cells and control cells, each with 5 replicates. Differential expression analysis between diclofenac-treated group and control group is performed using the online tool GEO2R [33], with *p*-value adjustment set to Benjamini and Hochberg false discovery rate control. After deleting genes with adjusted *p*-value larger than 0.05, a set of differentially expressed genes is generated for interactome mapping.

Protein-protein interaction data is download from BioGRID [34], an integrated and up-to-date public database that archives and disseminates genetic and protein interaction data from model organisms and humans. To be specific, the downloaded file is BIOGRID-ORGANISM-LATEST.tab2.zip that separates interactions into distinct files based on Organism and was released on June 30, 2016. File for interactions of *Saccharomyces cerevisiae* is extracted for use. As the whole interaction data contains tens of thousands of proteins and millions of interaction records, a considerable amount of proteins have no corresponding records in given expression data or show no differential expression. Those proteins shall be excluded from the final network in order to avoid the waste of both computational resource and analysis attention. According to the filtering method applied by Muraro and Simmons [8], interactions containing at least one differentially expressed gene are selected as an attempt to include indirect interactions. The resulting network concerning yeast cellular reaction to diclofenac consists of 1803 nodes and 3356 edges.

Table 1 Parameters for experiment networks

Parameters	Network 1	Network 2
nodes	330	1803
edges	359	3356
a	0.113	0.280
λ	9.07×10^{-2}	0.168
α (FDR)	1×10^{-4}	1×10^{-4}
τ	1.76×10^{-4}	7.71×10^{-6}
R_{ratio}	0.6	0.8

Result and Discussion

Analysis of network 1

To estimate distribution for p-values, the parameters of BUM model a and λ are estimated by R package BioNet [35]. Figure 1 shows the fitted model. As the majority of genes in yeast network have a very significant p-value, threshold τ is calculated at an extremely stringent FDR level as an attempt to control the size of detected module. Parameter details are shown in Table 1.

As a benchmark, the jActiveModule method is applied to the network via Cytoscape plugin, generating 5 active modules by default. Figure 2 gives a visualization of the network by Cytoscape, with detected active modules mapped on it. To understand the biological function of modules, gene ontology (GO) annotation for biological process is applied to modules by enrichment analysis tools provided on Gene Ontology Consortium [36]. The tool only asks for a submission of gene list, GO type (biological process, molecular function, cellular component) and species. The results is shown in Table 2. Among the 5 modules, 3 modules are enriched in the GO term galactose catabolic process via UDP-galactose with p-values

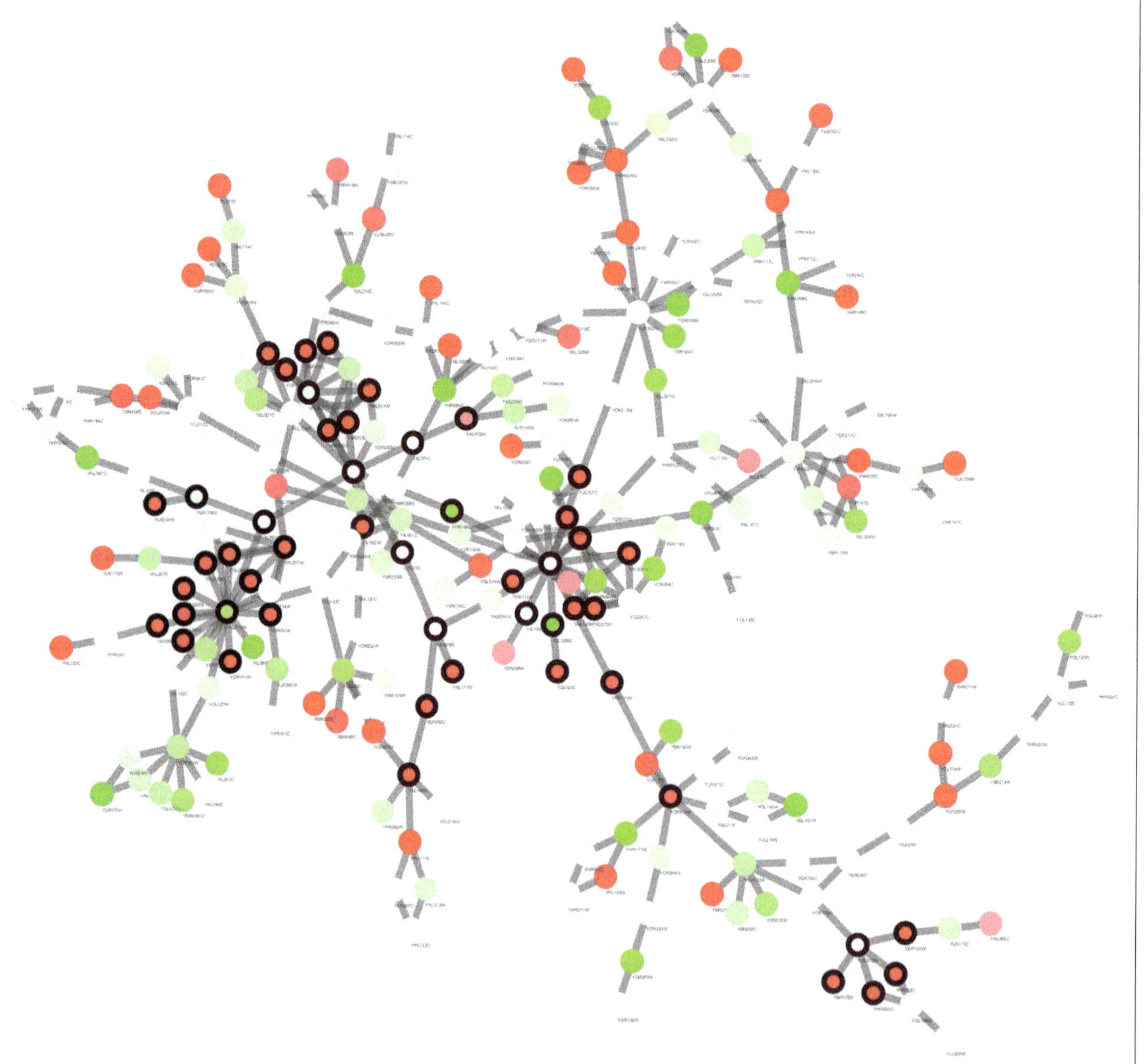

Fig. 2 Network 1 with active modules detected by jActiveModule. Each node denotes for one gene. *Node color* is a continuous mapping of the p-value generated from differential expression analysis. *Red color* indicates a significant change with small p-value and green color means no significant difference. The point where color will change between *red* and *green* is set to the threshold $\tau = 1.76 \times 10^{-4}$ that is used as a parameter for the proposed algorithm. Color of nodes near the changing point is white. Modules identified by jActiveModule are highlighted with *black node* border. Modules may overlap with each other

Table 2 Gene ontology results of modules detected by jActiveModule in network 1

Module	Size	S_A	R_A	Typical GO terms	p-value
1	26	250.39	1	galactose catabolic process via UDP-galactose	3.42×10^{-04}
				glycolytic fermentation to ethanol	2.72×10^{-03}
				amino acid catabolic process to alcohol via Ehrlich pathway	1.25×10^{-02}
2	5	58.21	0	response to heat	2.16×10^{-03}
3	16	270.79	2	galactose catabolic process via UDP-galactose	4.85×10^{-05}
4	18	169.89	2	galactose catabolic process via UDP-galactose	1.15×10^{-04}
				cellular carbohydrate metabolic process	3.27×10^{-02}
5	4	37.05	0	None	Not available

S_A and R_A are the objective functions of active module score and KEGG pathway coverage score, respectively. The values are calculated by the proposed objective functions using the same parameters setting as the proposed algorithm. $\tau = 1.76 \times 10^{-4}$ and $R_{ratio} = 0.6$

from 4.85×10^{-05} to 3.42×10^{-04}. Other 2 modules are too tiny to have accurate explanation.

The proposed algorithm is applied to network 1 with threshold $R_{ratio} = 0.6$ for KEGG pathway coverage score, resulting in a set of 13 Pareto solutions. As a feature for multi-objective optimization, all the modules in the same Pareto front are equally good. No one out performs another. In order to show the difference of those modules in trade-offs between two objectives, we selected 3 modules from the 13 Pareto solutions:

- Module 1: the extreme point on the Pareto front with maximum active module score $S_A = 393.41$.

Table 3 Gene ontology results of 3 modules on Pareto front detected by the proposed algorithm in network 1

Module	Size	S_A	R_A	Typical GO terms	p-value
1	65	393.41	9	galactose catabolic process via UDP-galactose	5.15×10^{-03}
				negative regulation of mating-type specific transcription from RNA polymerase II promoter	1.21×10^{-02}
				glycolytic fermentation to ethanol	4.05×10^{-02}
				pheromone-dependent signal transduction involved in conjugation with cellular fusion	6.39×10^{-03}
				cellular carbohydrate metabolic process	4.16×10^{-02}
2	92	268.96	19	negative regulation of mating-type specific transcription from RNA polymerase II promoter	4.67×10^{-04}
				galactose catabolic process via UDP-galactose	1.63×10^{-02}
				regulation of transcription during mitosis	7.19×10^{-03}
				gluconeogenesis	1.84×10^{-04}
				glycolytic process	2.87×10^{-02}
				pyruvate metabolic process	4.20×10^{-02}
				response to pheromone involved in conjugation with cellular fusion	3.93×10^{-06}
3	126	181.3	25	negative regulation of mating-type specific transcription from RNA polymerase II promoter	1.80×10^{-03}
				galactose catabolic process via UDP-galactose	4.48×10^{-02}
				C-terminal protein lipidation	1.62×10^{-02}
				gluconeogenesis	1.36×10^{-03}
				ADP metabolic process	2.47×10^{-04}
				pyruvate metabolic process	7.73×10^{-05}
				response to pheromone involved in conjugation with cellular fusion	1.47×10^{-02}
				ribonucleoprotein complex assembly	5.31×10^{-03}

Module 1 is the extreme point with maximized active score S_A. Module 2 is a balanced solution between S_A and R_A. Module 3 is the other extreme point with maximized pathway coverage score R_A

- Module 2: at the knee point of the Pareto front, which represents the optimal trade-off between active score ($S_A = 268.96$) and KEGG pathway coverage score ($R_A = 19$)
- Module 3: the extreme point on Pareto front with maximum KEGG pathway coverage $R_A = 25$.

GO analysis for biological process is performed on the three modules. The results together with the objective function values are tabulated in Table 3. We also visualize Modules 1 and 2 in Figs. 3 and 4, respectively.

By comparing the results in Table 3 with those in Table 2, we found that Module 1 identified by the proposed algorithm have better active module score (S_A) and KEGG pathway coverage score (R_A) than all the modules found by jActiveModule algorithm. Such results indicate that by incorporating the prior knowledge, we can guide the algorithm to search areas in the network with more significant activity.

From these two figures and Table 3, we found that compared with jActiveModule that searches for small and separated modules, the proposed algorithm tends to identify

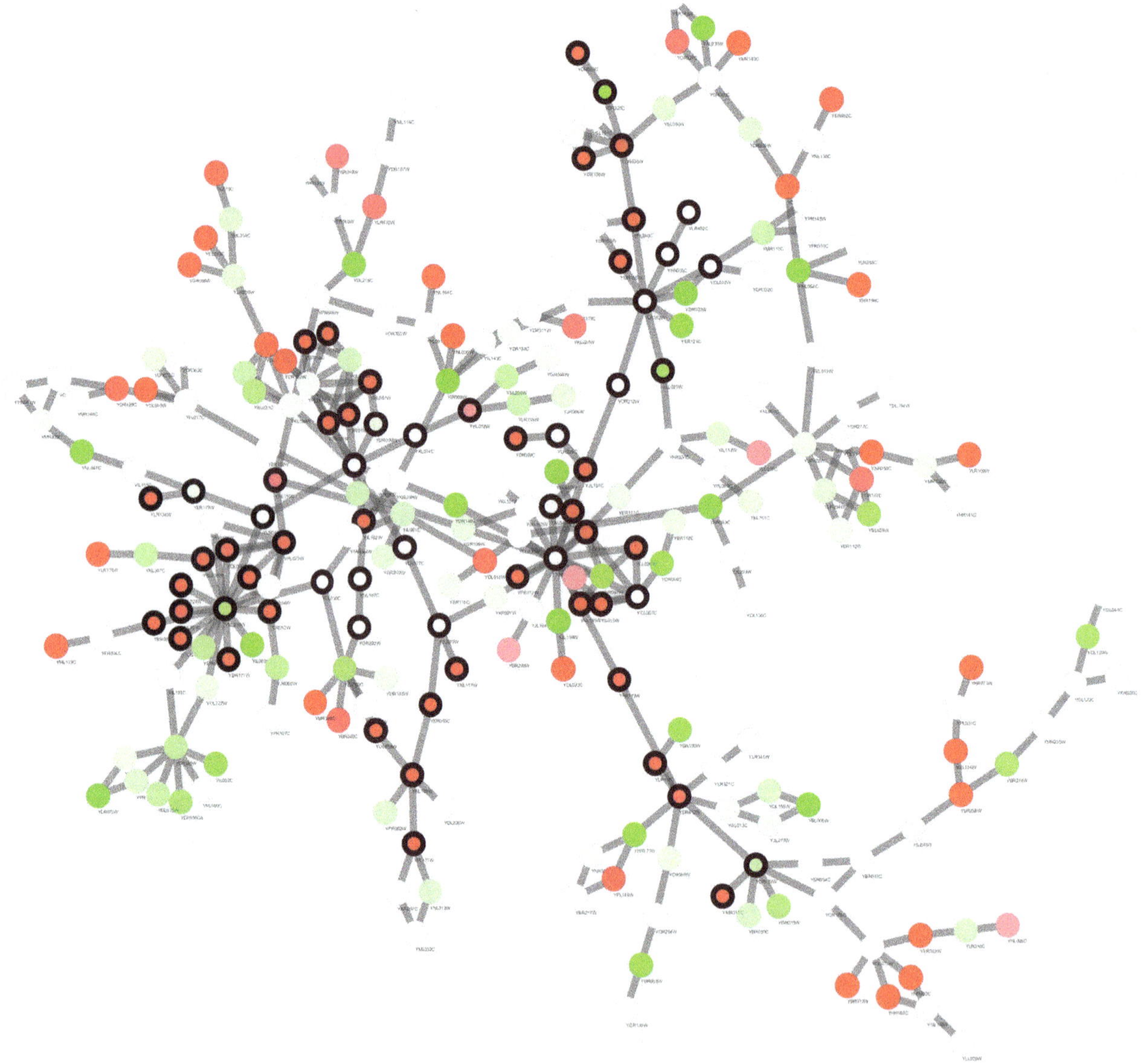

Fig. 3 Visualization of Module 1 with maximized active score S_A detected by the proposed algorithm in network 1. *Node color* and *border* are set the same as Fig. 2. Module contains the majority of *red* nodes that are connected densely, indicating high activity. Notice that compared to small separated modules identified by jActiveModule shown in Fig. 2, this module tends to connect small areas of *red* nods by including linkage nodes with *white* or *light green* color. Although these intermediate nodes shows only modest changes in expression, they serve as bridges for cross-talk between functional groups, or as transcription factors that regulate other genes

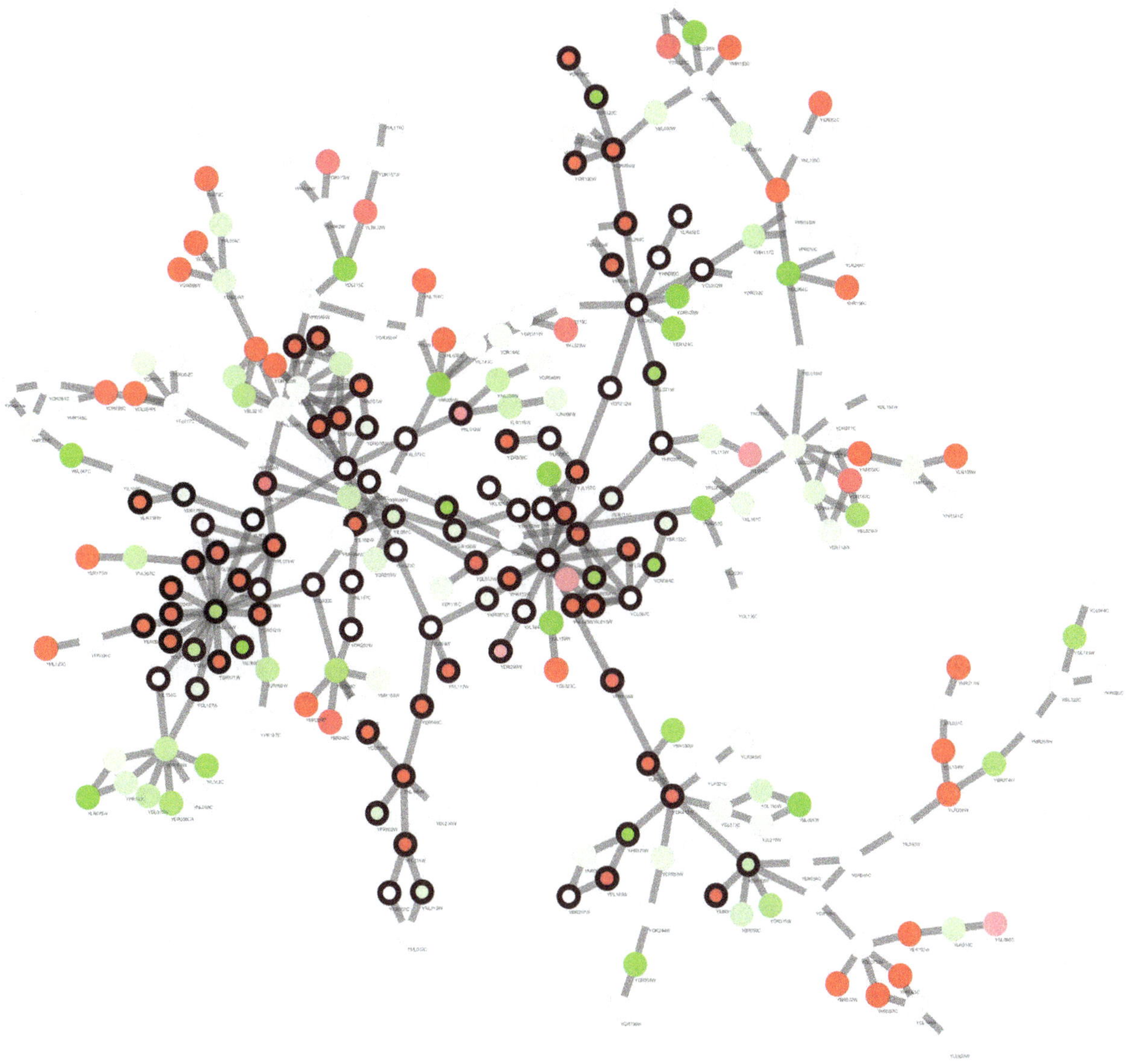

Fig. 4 Visualization of Module 2 which is the knee point of the Pareto front with optimal trade-off between S_A and R_A detected by the proposed algorithm in network 1. *Node color* and *border* are set the same as Fig. 2. Compared to Fig. 3, this module expands broader as R_A gets higher

a large connected subgraph. Even for Module 1 where the active module score is maximised, because of the integration of the prior knowledge, highly active areas are more likely to be connected together by intermediate nodes that might not be significantly differential expressed, but serve as a bridge for cross-talk between neighboring functional areas.

By visualisation of those Pareto solutions (figures not shown), we found that as the solution on Pareto front moves from maximum active score to maximum pathway coverage score, such intermediate nodes appear with higher frequency. We also found that, as R_A gets higher, detected module expands from a small core area with high activity to a broad area with more varied functional groups while still keeping overall activity. This result indicates that by using prior knowledge, we are able to reveal underlying mechanisms that link different activities in the network.

While all the three modules are significantly enriched in the GO term "galactose catabolic process via UDP-galactose" (corresponding p-value 5.15×10^{-03}, 1.63×10^{-02} and 4.48×10^{-02}, respectively), annotations for Module 1 (the extreme point with maximum activity score S_A) are more densely related with galactose metabolic process. On the other hand, for Module 3 with maximum KEGG pathway coverage score R_A, core annotations remain the same while additional annotations concerning essential biological processes increases. However, it is worth noting that, all the additional annotations can be reasonably related to the cellular response to disturbance in galactose utilization pathway.

The most interesting module is Module 2, which represents the optimal trade-off between prior knowledge and novel information from data. It is worth noting from Tables 3 and 2 that, even it is a knee point solution, Module 2 has a slightly worse S_A but much higher R_A than all the modules identified by JActiveModule. We can also observe from Table 3 that, module 2 has a range of slightly broader annotations concerning metabolic process of galactose, pyruvate and gluconeogenesis, which are highly relevant to galactose untilization pathways [37].

Analysis of network 2

Parameters of BUM model a and λ to fit p-value distribution are estimated as shown in Fig. 5. Threshold τ is calculated at given FDR level. See Table 1 details of parameters.

The proposed algorithm is applied to network 2 with threshold $R_{ratio} = 0.8$ for KEGG pathway coverage score, resulting in a set of 12 Pareto solutions. Solutions on the Pareto front are chosen for gene ontology analysis on biological process. Surprisingly, all identified modules shows a high consistency in the annotation on drug reaction, which exactly reflects the cellular response for yeast under the diclofenac treatment. Three genes (YDR406W, YOR153W and YOR153W, all act as ATP-binding transporter, for detailed functional explanation, see caption in Fig. 6) that play an important role in the cellular reaction and resistance to drug treatment are discovered in all the 12 modules, indicating the accuracy and robustness of searching algorithm.

Similar to the analysis methods for results in network 1, 3 representative modules on Pareto front with different trade-off between active score S_A and pathway coverage score R_A are select for gene ontology annotation (see Table 4) and visualization (Fig. 6). From Table 4 we can see that as pathway score R_A increases, size of module increases and the annotation includes a larger range of biological processes. As drug reaction is considerably complicated response that involves a series of up or down regulation in related function groups such as protein kinase pathway, ribosome biogenesis, rRNA processing and zinc-responsive genes [32], the enriched annotation in modules with higher R_A provides a guidance of deciding which functional groups to look into as it combines both prior knowledge from existing interaction database and novel information from gene expression data specific for given experimental conditions.

Conclusion

An integrated multi-objective approach has been proposed for active module identification. The algorithm is motivated by the idea that incorporating prior information into data-driven method would provide new insights into sophisticated biological processes. We also designed an constraint based on algebraic connectivity to ensure the connectedness of the identified active modules.

We first applied our algorithm on a small molecular interaction network, which identified a set of Pareto solutions that represents different trade-off between prior knowledge and novel information from data. Gene Ontology analysis results show that it successfully identifies modules with relevant and reasonable biological

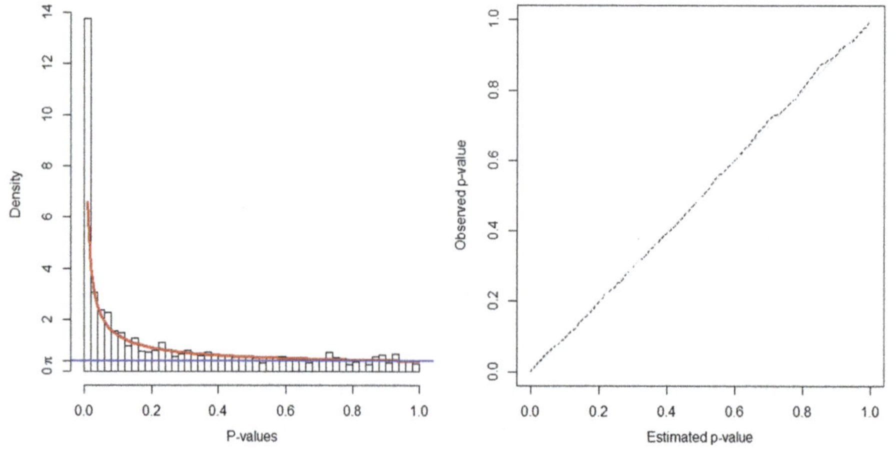

Fig. 5 BUM model estimation on network 2. Histogram of p-values with fitted BUM model and a Q-Q plot of estimated and empirical distribution of p-values for network 2. As the network size increases, estimation becomes more accurate

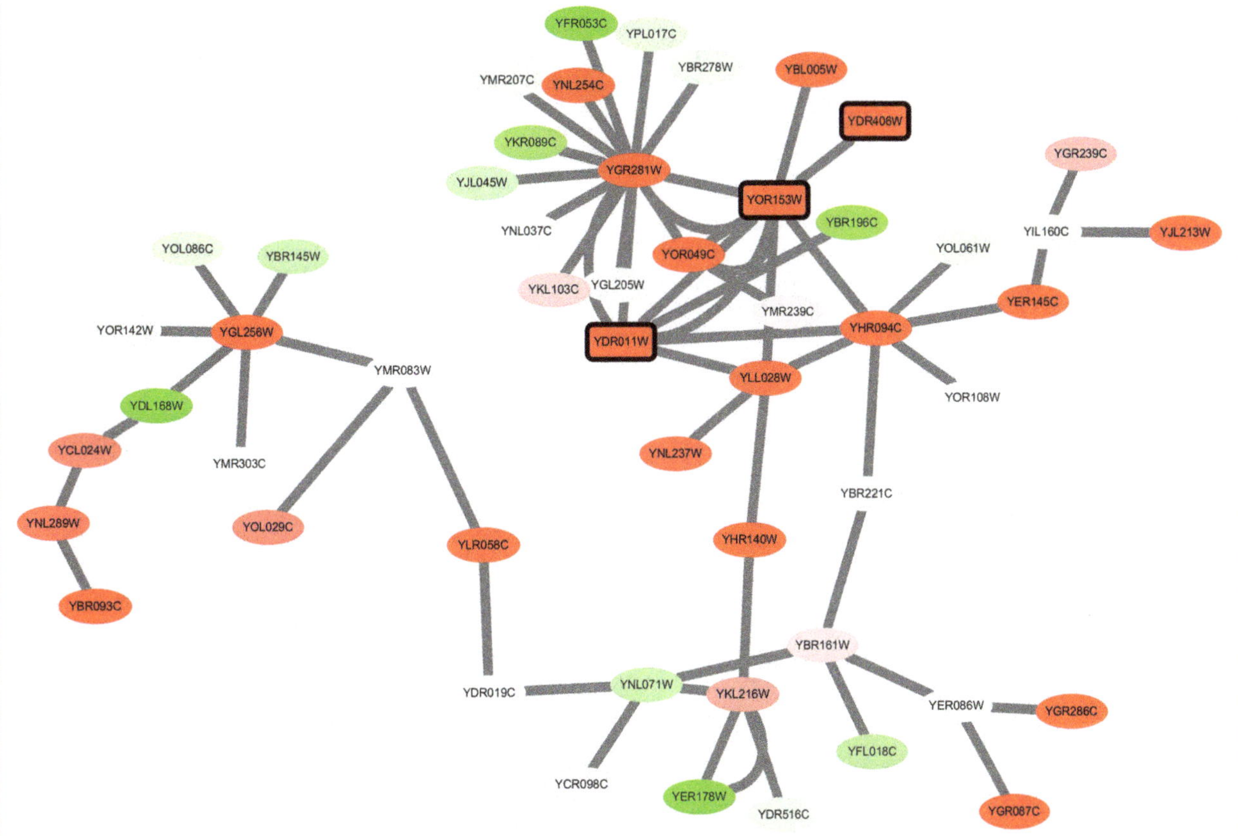

Fig. 6 Visualization of module 3 identified by the proposed algorithm in network 2. Each node represents for a gene. The setting for node color is the same with Fig. 2. The turning point between red and green is set to the value $\tau = 7.71 \times 10^{-6}$. Three *rectangle shaped nodes* with *black border* are genes involved in drug export and are highly consistent in all modules. YDR406W is an ATP-binding cassette multidrug transporter. YDR011W is a ATP-binding cassette transporter. YOR153W is also an ATP-binding cassette multidrug transporter. The three genes serve as an important role in yeast's resistance to diclofenac

Table 4 Gene ontology results of 3 modules on Pareto front detected by the proposed algorithm in network 2

Module	Size	S_A	R_A	Typical GO terms	*p*-value
1	34	91.01	0	Drug export	1.79×10^{-03}
				Cellular response to drug	4.71×10^{-02}
2	39	57.56	4	Drug export	2.84×10^{-03}
3	62	46.332	8	Drug export	1.21×10^{-02}
				Amino acid catabolic process to alcohol via Ehrlich pathway	8.65×10^{-09}
				Ethanol metabolic process	3.71×10^{-06}
				NADH oxidation	3.73×10^{-03}
				Glycolytic process	4.34×10^{-03}
				Fermentation	1.40×10^{-02}
				Macromolecule metabolic process	2.51×10^{-02}

interpretations. The algorithm was applied to the second network, The approach is then applied on a microarray dataset from diclofenac-treated yeast cells and identify modules to elucidate the molecular mechanisms of diclofenac toxicity and resistance. The algorithm identifies accurate and consistent modules with biological function densely related to given cellular response, proving that the integrated approach for network construction is feasible and that the proposed algorithm is able to identify biologically meaningful modules in large scale network.

Abbreviations
BUM: Beta-uniform mixture; FDR: False discovery rate; GEO: Gene expression omnibus; GO: Gene ontology; KEGG: Kyoto encyclopedia of genes and genomes

Acknowledgements
Not applicable.

Declarations
This article has been published as part of *BMC Systems Biology* Volume 11 Supplement 2, 2017. Selected articles from the 15th Asia Pacific Bioinformatics Conference (APBC 2017): systems biology. The full contents of the supplement are available online https://bmcsystbiol.biomedcentral.com/articles/supplements/volume-11-supplement-2.

Funding
This paper was supported by European Union Seventh Framework Programme (FP7 / 2007-2013; grant agreement number NMP4-LA-2013-310451). The publication costs for this article was also funded by European Union Seventh Framework Programme (FP7 / 2007-2013; grant agreement number NMP4-LA-2013-310451).

Authors' contributions
WC participated in the design of the study, performed all programming work and carried out the analysis. JL participated in the design of the study. SH participated in the design of the study and guided the project. All authors read and approved the final manuscript.

Competing interests
The authors declare that they have no competing interests.

Author details
[1]School of Computer Science, University of Birmingham, Edgbaston, B15 2TT Birmingham, UK. [2]Key Laboratory of Intelligent Perception and Image Understanding of Ministry of Education, Xidian University, Xi'an, 710071 Shaanxi, People's Republic of China.

References
1. Allison DB, Cui X, Page GP, Sabripour M. Microarray data analysis: from disarray to consolidation and consensus. Nat Rev Genet. 2006;7(1):55–65.
2. Barabasi AL, Oltvai ZN. Network biology: understanding the cell's functional organization. Nat Rev Genet. 2004;5(2):101–13.
3. Barabási AL, Gulbahce N, Loscalzo J. Network medicine: a network-based approach to human disease. Nat Rev Genet. 2011;12(1):56–68.
4. Gross AM, Ideker T. Molecular networks in context. Nat Biotechnol. 2015;33(7):720–1.
5. Liu Y, Tennant DA, Zhu Z, Heath JK, Yao X, He S. Dime: a scalable disease module identification algorithm with application to glioma progression. PloS ONE. 2014;9(2):86693.
6. Guo Z, Li Y, Gong X, Yao C, Ma W, Wang D, Li Y, Zhu J, Zhang M, Yang D, et al. Edge-based scoring and searching method for identifying condition-responsive protein–protein interaction sub-network. Bioinformatics. 2007;23(16):2121–128.
7. Wang YC, Chen BS. Integrated cellular network of transcription regulations and protein-protein interactions. BMC Syst Biol. 2010;4(1):1.
8. Muraro D, Simmons A. An integrative analysis of gene expression and molecular interaction data to identify dys-regulated sub-networks in inflammatory bowel disease. BMC Bioinforma. 2016;17(1):1.
9. Ma H, Schadt EE, Kaplan LM, Zhao H. Cosine: Condition-specific sub-network identification using a global optimization method. Bioinformatics. 2011;27(9):1290–8.
10. Shlomi T, Cabili MN, Herrgård MJ, Palsson BØ, Ruppin E. Network-based prediction of human tissue-specific metabolism. Nat Biotechnol. 2008;26(9):1003–10.
11. Barabási AL, Albert R. Emergence of scaling in random networks. Science. 1999;286(5439):509–12.
12. Newman ME. Modularity and community structure in networks. Proc Natl Acad Sci. 2006;103(23):8577–582.
13. Albert R, Jeong H, Barabási AL. Error and attack tolerance of complex networks. Nature. 2000;406(6794):378–82.
14. Hartwell L, Hopfield J, Leibler S, Murray A. From molecular to modular cell biology. Nature. 1999;402:c47–c52. This fundamental article defines the concept of modularity in cell biology. CAS ISI PubMed Article.
15. Jia G, Cai Z, Musolesi M, Wang Y, Tennant DA, Weber RJ, Heath JK, He S. Community detection in social and biological networks using differential evolution. Learn Intell Optim. 2012;71:71-85.
16. Huang Q, White T, Jia G, Musolesi M, Turan N, Tang K, He S, Heath JK, Yao X. Community detection using cooperative co-evolutionary differential evolution. In: International Conference on Parallel Problem Solving from Nature. Springer; 2012. p. 235–44.
17. Mitra K, Carvunis AR, Ramesh SK, Ideker T. Integrative approaches for finding modular structure in biological networks. Nat Rev Genet. 2013;14(10):719–32.
18. He S, Zhu Z, Jia G, Tennant D, Huang Q, Tang K, Heath J, Musolesi M, Yao X. Cooperative co-evolutionary module identification with application to cancer disease module discovery. IEEE Transactions on Evolutionary Computation. 2016;1–1.
19. Ideker T, Ozier O, Schwikowski B, Siegel AF. Discovering regulatory and signalling circuits in molecular interaction networks. Bioinformatics. 2002;18(suppl 1):233–40.
20. Hwang T, Park T. Identification of differentially expressed subnetworks based on multivariate anova. BMC Bioinforma. 2009;10(1):1.
21. Klammer M, Godl K, Tebbe A, Schaab C. Identifying differentially regulated subnetworks from phosphoproteomic data. BMC Bioinforma. 2010;11(1):1.
22. Dittrich MT, Klau GW, Rosenwald A, Dandekar T, Müller T. Identifying functional modules in protein–protein interaction networks: an integrated exact approach. Bioinformatics. 2008;24(13):223–31.
23. Pounds S, Morris SW. Estimating the occurrence of false positives and false negatives in microarray studies by approximating and partitioning the empirical distribution of *p*-values. Bioinformatics. 2003;19(10):1236–42.
24. Benjamini Y, Hochberg Y. Controlling the false discovery rate: a practical and powerful approach to multiple testing. J R Stat Soc Ser B Methodol. 1995;57:289–300.
25. Kanehisa M, Goto S. Kegg: kyoto encyclopedia of genes and genomes. Nucleic Acids Res. 2000;28(1):27–30.
26. KEGG REST-style entry for *Saccharomyces cerevisiae*. http://rest.kegg.jp/link/sce/pathway. Accessed Mar 2016.
27. Deb K, Pratap A, Agarwal S, Meyarivan T. A fast and elitist multiobjective genetic algorithm: Nsga-ii. IEEE Trans Evol Comput. 2002;6(2):182–97.
28. Pétrowski A. A clearing procedure as a niching method for genetic algorithms. In: Evolutionary Computation, 1996., Proceedings of IEEE International Conference On. IEEE; 1996. p. 798–803.
29. Shannon P, Markiel A, Ozier O, Baliga NS, Wang JT, Ramage D, Amin N, Schwikowski B, Ideker T. Cytoscape: a software environment for integrated models of biomolecular interaction networks. Genome Res. 2003;13(11):2498–504.
30. jActiveModules in Cytoscape App Store. http://apps.cytoscape.org/apps/jactivemodules. Accessed Oct 2015.
31. NCBI Gene Exprssion Omnibus - GSE29331. http://www.ncbi.nlm.nih.gov/geo/query/acc.cgi?acc=GSE29331. Accessed Apr 2016.
32. van Leeuwen JS, Vermeulen NP, Vos JC. Involvement of the pleiotropic drug resistance response, protein kinase c signaling, and altered zinc homeostasis in resistance of saccharomyces cerevisiae to diclofenac. Appl Environ Microbiol. 2011;77(17):5973–980.
33. GEO2R. http://www.ncbi.nlm.nih.gov/geo/geo2r/. Accessed May 2016.
34. BioGRID (The Biological General Repository for Interaction Datasets). http://thebiogrid.org/. Accessed May 2016.
35. Beisser D, Klau GW, Dandekar T, Müller T, Dittrich MT. Bionet: an r-package for the functional analysis of biological networks. Bioinformatics. 2010;26(8):1129–30.
36. Consortium GO, et al. Gene ontology consortium: going forward. Nucleic Acids Res. 2015;43(D1):1049–56.
37. Berg JM, Tymoczko JL, Stryer L. Biochemistry. New York: W H Freeman; 2002.

13

Stochastic modeling and simulation of reaction-diffusion system with Hill function dynamics

Minghan Chen†, Fei Li†, Shuo Wang and Young Cao*

Abstract

Background: Stochastic simulation of reaction-diffusion systems presents great challenges for spatiotemporal biological modeling and simulation. One widely used framework for stochastic simulation of reaction-diffusion systems is reaction diffusion master equation (RDME). Previous studies have discovered that for the RDME, when discretization size approaches zero, reaction time for bimolecular reactions in high dimensional domains tends to infinity.

Results: In this paper, we demonstrate that in the 1D domain, highly nonlinear reaction dynamics given by Hill function may also have dramatic change when discretization size is smaller than a critical value. Moreover, we discuss methods to avoid this problem: smoothing over space, fixed length smoothing over space and a hybrid method.

Conclusion: Our analysis reveals that the switch-like Hill dynamics reduces to a linear function of discretization size when the discretization size is small enough. The three proposed methods could correctly (under certain precision) simulate Hill function dynamics in the microscopic RDME system.

Keywords: Reaction diffusion master equation (RDME), Hill function, Stochastic simulation, Hybrid method

Background

Cell reproduction requires elaborate spatial and temporal coordination of crucial events, such as DNA replication, chromosome segregation, and cytokinesis. In cells, protein species are well organized and regulated throughout their life cycles. Theoretical biologists have been using classic chemical reaction rate laws with deterministic ordinary differential equations (ODEs) and partial differential equations (PDEs) to model molecular concentration dynamics in spatiotemporal biological processes. However, wet-lab experiments in single cell resolution demonstrate that biological data present considerable variations from cell to cell. The variations arise from the fact that cells are so small that there exist only one or two copies of genes, tens of mRNA molecules and hundreds or thousands of protein molecules [1–3]. At this scale, the traditional way of modeling molecule "concentration" is not applicable. Noise in molecule populations cannot be neglected, as noise may play a significant role in the overall dynamics inside a cell. Therefore, to accurately model the cell cycle mechanism, discrete and stochastic modeling and simulation should be applied.

A convenient strategy to build a stochastic biochemical model is to break a deterministic model into a list of chemical reactions and simulate them with Gillespie's stochastic simulation algorithm (SSA) [4, 5]. One of the major difficulties in this conversion strategy lies in the propensity calculation of reactions. Gillespie's SSA is well defined for mass action rate laws. However, in many biochemical models, in addition to mass action rate laws, other phenomenological reaction rate laws are often used. For example, the Michaelis-Menten equation [6] and Hill functions [7] are widely used in biological models to model the fast response to signals in regulatory control systems. Although theoretically these phenomenological

*Correspondence: ycao@cs.vt.edu
†Equal contributors
Department of Computer Science, Virginia Tech, Blacksburg, VA 24061, USA

rate laws may be generated from elementary reactions with mass action rate laws, in practice the detailed mechanisms behind these phenomenological rate laws are not well known and may not be very important. Stochastic modeling and simulation with these phenomenological rate laws are sometimes inevitable.

In recent years, stochastic modeling and simulation for spatiotemporal biological systems, particularly reaction-diffusion systems, have captured more and more attention. Several algorithms and tools [8–11] to model and simulate reaction-diffusion systems have been proposed. These methods can be categorized into two theoretical frameworks: the spatially and temporally continuous Smoluchowski modeling framework [12] and the compartment-based modeling framework, formulated as the spatially discretized reaction-diffusion master equation (RDME) [13, 14]. The Smoluchowski framework [12, 15, 16] stores the exact position of each molecule and is mathematically fundamental, whereas the RDME is coarse-grained and better suited for large scale simulations [17]. In RDME, the spatial domain is discretized into small compartments. Within each compartment, molecules are considered "well-stirred". Under the RDME scheme, diffusion is modeled as continuous time random walk on mesh compartments, while reactions fire only among molecules in the same compartment. Stochastic dynamics of the chemical reactions in each compartment is governed by the chemical master equation (CME) [18, 19]. Yet, the CME is computationally impossible to solve for most practical problems. Stochastic simulation methods were then applied to generate realizations of system trajectories. It has been well established that the discretization compartment size for RDME should be smaller than the mean free path of the reactions for the compartment to be considered as well-stirred [20]. In addition, it has been proved that the RDME of bi-molecular reactions in 3D domain becomes incorrect and yields unphysical results when the discretization size approaches microscopic scale [21–23].

In this paper, we focus on the stochastic modeling of reaction-diffusion systems with reaction rate laws given by Hill functions. In the Results section, we present our numerical analysis on a toy model of reaction-diffusion system with Hill function dynamics. We will show that the RDME framework of the Hill function dynamics has serious simulation defects when the discretization size approach microscopic limit: When the discretization size is small enough, the typical switching pattern of Hill dynamics becomes linear to the input signal (and the discretization size). Later, we propose potential solutions for the discretization of the reaction-diffusion systems with Hill function rate laws. Finally, we conclude this paper with a discussion on RDME for general nonlinear functions and the hybrid method.

Caulobacter modeling

Caulobacter crescentus captures great interest in the study of asymmetric cell division. When a *Caulobacter* cell divides, it produces two functionally and morphologically distinct daughter cells. The asymmetric cell division of *Caulobacter crescentus* requires elaborate temporal and spatial regulations [24–27]. In literature [28–30], four essential "master regulators" of the *Caulobacter* cell cycle, DnaA, GcrA, CtrA and CcrM, have been identified. These master transcription regulators determine the dynamics of around 200 genes. They oscillate temporally to drive the dynamics of cell cycle. Among them, the molecular mechanisms governing CtrA functions have been well studied. The simulation we are concerned with in this paper is also related to this CtrA module. So we give a brief introduction to it.

In swarmer cells, a two-component phosphorelay system (with both CckA and ChpT) phosphorylates the CtrA. Then the chromosomal origin of replication (*Cori*) is bound by the phosphorylated CtrA (CtrAp) to inhibit the initiation of chromosome replication [31]. Later during the swarmer-to-stalked transition period, CtrAp gets dephosphorylated and degraded, allowing the initiation of chromosome replication again. Thus the CtrA has important impact on the chromosome replication in our model, and should be well regulated.

The regulation of CtrA is achieved by the histidine kinase CckA through the following pathway. An ATP-dependent protease, ClpXP, degrades CtrA [32, 33] and is localized to the stalk pole by CpdR. As the nascent stalked cell progresses through the cell cycle, CpdR is phosphorylated by CckA/ChpT, losing it polar localization, and consequently losing its ability to recruit ClpXP protease for CtrA degradation. In addition, CtrA is reactivated through CckA/ChpT phosphorylation [34]. Moreover, the regulatory network of the histidine kinases CckA is influenced by a non-canonical histidine kinase, DivL [35]. DivL promotes CckA kinase, which then phosphorylates and activates CtrA in the swarmer cell. During the swarmer-to-stalked transition period, DivL activity is down-regulated, thereby inhibiting CckA kinase activity. As a result, dephosphorylation and degradation of CtrA trigger the initiation of chromosome replication.

In order to study the regulatory network in *Caulobacter crescentus*, Subramanian et al. [26, 27] developed a deterministic model with six major regulatory proteins. The deterministic model provides robust switching between swarmer and stalked states. Figure 1 (left) demonstrates the total population change during the *Caulobacter crescentus* cell cycle with this deterministic model. In the swarmer stage (from $t = 0$ to 30 min), the CtrA is phosphorylated at a high population level, which inhibits the initiation of chromosome replication. During the swarmer-to-stalked transition period (from $t = 30$ to

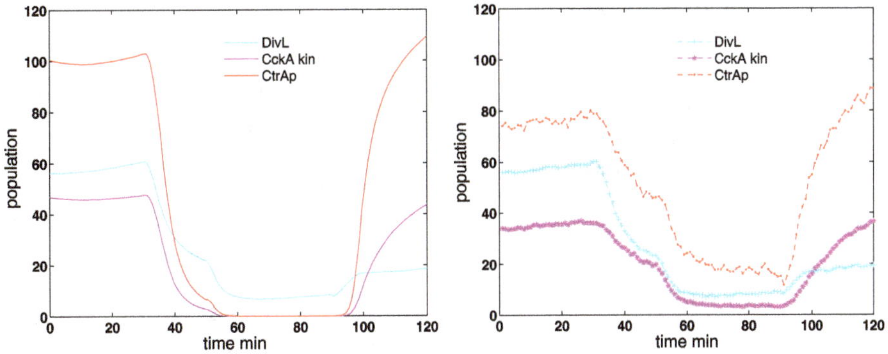

Fig. 1 The population oscillation of CtrAp during *Caulobacter crescentus* cell cycle. Left figure shows the simulation result of deterministic model and the right figure shows the stochastic simulation result. In the swamer stage ($t = 0 \sim 30$min), the CtrA is phosphorylated and at high population level state, which inhibits the initiation of chromosome replication. During swarmer-to-stalked transition ($t = 30 \sim 50$min), the CtrAp population quickly switch to low state, allowing the consequent initiation of chromosome replication in the stalked stage

50 min), the CtrAp population quickly drops to a low level, allowing the consequent initiation of chromosome replication in the stalked stage.

In stochastic simulation of the spatiotemporal model of this regulatory network, the phosphorylated CtrA (CtrAp) population switch from a high level in swarmer stage to a low level in stalked stage is not as sharp as expected, shown in Fig. 1 (right). On the other hand, the DivL population level from the stochastic simulation seems similar to that from the deterministic simulation. A simple analysis suggests that the Hill function dynamics, which models the up regulation of CckA kinase activity by DivL, might be the culprit. Further investigation leads to the discovery of the Hill function limitation at small discretization sizes, as analyzed in the next section.

Methods

Reaction diffusion master equation

Before we plunge into Hill functions in reaction-diffusion systems, we will first briefly review mathematical modeling and simulation methods of spatially inhomogeneous stochastic systems.

The dynamics of a spatially inhomogeneous stochastic system has been considered as governed by the reaction-diffusion master equation (RDME), developed in an early work of Gardiner [13]. The RDME framework partitions the spatial domain into small compartments, such that molecules within each compartment can be considered well-stirred. Assume a biochemical system of N species $\{S_1, S_2, \ldots, S_N\}$ and M reactions within a spatial domain Ω, which is partitioned into K grids V_k, $k = 1, 2, \ldots, K$. For simplicity, we assume that the space Ω is one dimensional (1D). Each species population, as well as the reactions in the system will have a local copy for each compartment. The state of the reaction-diffusion system at any time t is represented by the vector state vector $X(t) = \{X_{1,1}(t), X_{1,2}(t), \ldots, X_{1,K}(t), \ldots, X_{n,k}(t), \ldots, X_{N,K}(t)\}$, where $X_{n,k}(t)$ denotes the molecule population of species S_n in the grid V_k at time t. Reactions in each compartment is governed by the Chemical Master Equation (CME), while diffusion is modeled as random walk across neighboring compartments. Each reaction channel R_j in any compartment k can be characterized by the *propensity function* $a_{j,k}$ and the *state change vector* $\nu_j \equiv (\nu_{1j}, \nu_{2j}, \ldots, \nu_{Nj})$. The dynamics of the diffusion of species S_i from compartment V_k to V_j is formulated by the *diffusion propensity function* $d_{i,k,j}$ and the *diffusion state change vector* $\mu_{k,j}$ similarly. $d_{i,k,j}(x)dt$ gives the probability that, given $X_{i,k}(t) = x$, one molecule of species S_i at grid V_k diffuses into grid V_j in the next infinitesimal time interval $[t, t + dt)$. If $j = k \pm 1$, then $d_{i,k,j}(x) = \frac{D}{h^2}x$, where D is the diffusion rate coefficient and h is the characteristic length, also called discretization size, of a grid; Otherwise $d_{i,k,j} = 0$. The state change vector $\mu_{k,j}$ is a vector of length K with -1 in the k-th position, 1 in the j-th position and 0 everywhere else.

With the reaction-diffusion propensity functions and state change vectors, the RDME completely depicts the dynamics of the system:

$$
\begin{aligned}
&\frac{\partial P(\mathbf{x}, t|\mathbf{x}_0, t_0)}{\partial t} \\
&= \sum_{k=1}^{K}\sum_{j=1}^{M} \left(a_{j,k}(\mathbf{x} - \nu_{j,k})P(\mathbf{x} - \nu_{j,k}, t|\mathbf{x}_0, t_0) - a_{j,k}(\mathbf{x})P(\mathbf{x}, t|\mathbf{x}_0, t_0)\right) \\
&+ \sum_{i=1}^{N}\sum_{k=1}^{K}\sum_{j=1}^{K} \left(-d_{i,k,j}(x_{ik})P(\mathbf{x}, t|\mathbf{x}_0, t_0)\right. \\
&\left.+ d_{i,k,j}(X_{ik} - \mu_{k,j})P(X_{11}, \ldots, X_{ik} - \mu_{k,j}, \ldots, X_{N,K}, t|\mathbf{x}_0, t_0)\right),
\end{aligned} \quad (1)
$$

where $P(\mathbf{x}, t|\mathbf{x}_0, t_0)$ denotes the probability that the system state $X(t) = \mathbf{x}$, given that $X(t_0) = \mathbf{x}_0$. The RDME is a set of ODEs that gives one equation for every possible state. It is both theoretically and computationally intractable to solve RDME for practical biochemical systems due to the huge number of possible combinations of states. Instead of solving RDME for the time evolution of the probabilities, we can construct numerical realizations of $\mathbf{X}(t)$. A popular method to construct the trajectories of a reaction-diffusion system is to simulate each diffusive jumping and chemical reaction event explicitly. With enough trajectory realizations, we can derive the distribution of each state vector at different times.

The RDME model have been used as an approximation of the Smoluchowski framework in the mesoscopic scale. Furthermore, researches have discovered that in the microscopic limit, bimolecular reactions may be eventually lost when the grid size becomes infinitely small in the three dimensional domain [21, 23]. The RDME framework requires that the two reactant molecules for a bimolecular reaction must be in the same compartment in order to fire a reaction. Intuitively, we may realize that with more discrete compartments, it is less likely for the two molecules to encounter each other at the same compartment in a high dimensional domain. In order to model the reaction-diffusion system with RDME in the microscopic limit, Radek and Chapman [22] derived a formula of mesh-dependent reaction propensity correction for bimolecular reactions when the discretization size h is larger than a critical size h_{crit}. This reaction propensity correction formula fails when the discretization size h is smaller than this critical value. Recently, Isaacson [36] proposed a convergent RDME framework (cRDME). In the cRDME framework the diffusion is modeled exactly as in the RDME, while the bimolecular reaction occurs with a nonzero propensity, as long as the distance of the two reactant molecules is less than the reaction radius as defined in the Smoluchowski framework.

In conclusion, the discretization size for the RDME framework should be small enough to avoid discretization error. Yet when the mesh size is less than a critical value, the RDME may become inaccurate for the loss of bimolecular reactions in high dimensional domains. In this paper we will demonstrate that discretization size in space also has great influence on Hill function dynamics in reaction-diffusion systems. The switch-like Hill dynamics breaks even in a 1D domain when the discretization size is small.

Hill function

The Hill function [7], as well as the Michaelis-Menten function [6] are widely used in enzyme kinetics modeling. In molecular biology, enzymes catalyze biochemical substrates into products, while remaining unchanged. The enzyme kinetics reactions are usually formulated as

$$E + S \underset{k_{-1}}{\overset{k_1}{\rightleftharpoons}} ES \overset{k_2}{\rightarrow} E + P \tag{2}$$

Leonor Michaelis and Maud Leonora Menten proposed the "quasi-steady state" assumption and formulated the reaction rate equation for the enzyme kinetics, which is mostly referred to as the "Michaelis-Menten" equation. With the conservation law and the quasi-steady state assumption, the Michaelis-Menten equation is given as

$$\frac{d[P]}{dt} = V_{max}\frac{[S]}{K_M+[S]}, \tag{3}$$

with $V_{max} = k_2[E]_0$ being the maximum reaction rate and $K_m = \frac{k_{-1}+k_2}{k_1}$ being the Michaelis constant.

Sometimes one substrate molecule can have several enzyme binding sites and multiple bindings (cooperative binding) with enzymes are required to activate the substrate.

$$S + nE \underset{k_{-1}}{\overset{k_1}{\rightleftharpoons}} SE_n \overset{k_2}{\rightarrow} nE + P \tag{4}$$

In real biological models, the binding of the n enzyme molecules to a substrate does not take place at once but in a succession of steps. Using the quasi-steady state assumption and conservation laws, the Hill function that formulates the reaction dynamics is given as

$$\frac{d[P]}{dt} = V_{max}\frac{[E]^n}{{K_m}^n+[E]^n}, \tag{5}$$

with V_{max} as the maximum reaction rate, K_m as the Michaelis constant, and n as the Hill coefficient. The Hill function is widely used to model "step-regulated" reaction as an activity switch.

Results

To simplify the analysis, a toy model of a reaction-diffusion system in one dimension is constructed. As demonstrated in Fig. 2, in the toy model an enzyme species E (typically a transcription factor) is constantly synthesized and degraded. The enzyme E further upregulates the DNA expression of a product P. The synthesis rate of P is formulated as a Hill function.

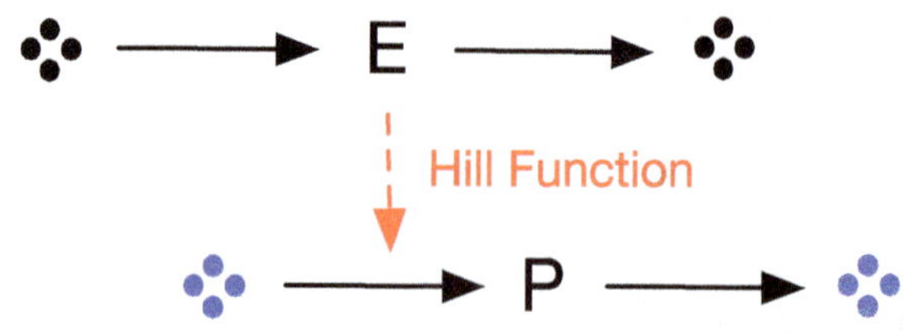

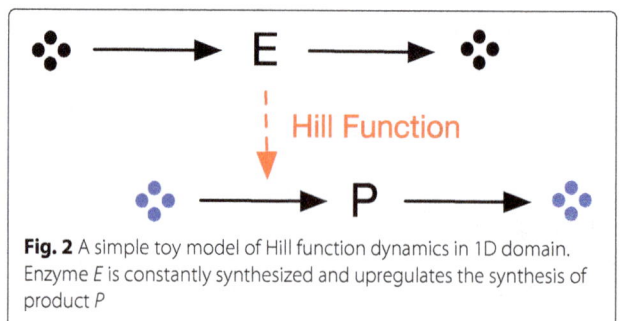

Fig. 2 A simple toy model of Hill function dynamics in 1D domain. Enzyme E is constantly synthesized and upregulates the synthesis of product P

Assume a spatial domain of size L is equally partitioned into K compartments with size $h = L/K$ for each. The list of reactions and reaction propensities in each compartment are given as

$$\begin{array}{llll} \emptyset \rightarrow E_i, & a_1 & = & k_s \cdot h; \\ E_i \rightarrow \emptyset, & a_2 & = & k_d \cdot E_i; \\ \emptyset \xrightarrow{E_i} P_i, & a_3 & = & k_{syn} \cdot h \frac{E_i^4}{(K_m \cdot h)^4 + E_i^4}; \\ P_i \rightarrow \emptyset, & a_4 & = & k_{deg} \cdot P_i; \\ E_i \rightarrow E_{i\pm1}, & a_5 & = & 2\frac{D_E}{h^2} E_i; \\ P_i \rightarrow P_{i\pm1}, & a_6 & = & 2\frac{D_P}{h^2} P_i; \end{array} \tag{6}$$

The parameters k_s, k_d are the synthesis, degradation rates, respectively, for enzyme species E, and similarly k_{syn}, k_{deg} are those for product P. K_m is the Michaelis constant in the Hill function.

In the one-dimensional domain, the enzyme E is constantly synthesized and degraded. At the equilibrium state, the distribution of the total population of E is given by the Poisson distribution,

$$P_E(n) = \frac{\alpha^n}{n!} e^{-\alpha}, \tag{7}$$

where $\alpha = \frac{k_s}{k_d} L$ denotes the mean value of the total number of enzyme E molecules in the domain.

For an individual compartment (bin), consider the probability $P_E^{(i)}(n)$ that an individual bin i contains n molecules of enzyme E. At the equilibrium state, enzyme E is homogeneously distributed in the system. The probability that each molecule of E stays in a certain bin i is given by $p = 1/K$. The probability that, of all the E molecules in the domain, none is in bin i is approximated by

$$\begin{aligned} P_E^{(i)}(0) &= P_E(0) + P_E(1)(1 - \tfrac{1}{K}) + P_E(2)(1 - \tfrac{1}{K})^2 \\ &\quad + \ldots + P_E(N)(1 - \tfrac{1}{K})^N + \ldots \\ &= \sum_{n=0}^{N} e^{-\alpha} \frac{\alpha^n}{n!} \left(1 - \tfrac{1}{K}\right)^n \\ &= e^{-\alpha/K}. \end{aligned} \tag{8}$$

The other probability terms are not important in the analysis.

With the distribution of the enzyme molecular population, the mean reaction propensity for the synthesis of protein P in the i-th bin is

$$\langle a_{syn}^i \rangle = k_{syn} h \sum_{n=0}^{\infty} \frac{n^4}{(K_m \cdot h)^4 + n^4} P_E^{(i)}(n). \tag{9}$$

Notice that when $n = 0$, the Hill function is zero, and when the discrete bin size h is small, the Hill function approaches one quickly if $n \geq 1$. For example, when $K_m \cdot h \leq 0.5$ the Hill function $\frac{n^4}{(K_m \cdot h)^4 + n^4} \geq 0.94$ for $n \geq 1$. Therefore, upper and lower bounds for the product P synthesis propensity, when $k_m \cdot h \leq 0.5$, are

$$0.94 k_{syn} \cdot h \sum_{n=1}^{\infty} P_E^{(i)}(n) \leq \langle a^{syn} \rangle \leq k_{syn} \cdot h \sum_{n=1}^{\infty} P_E^{(i)}(n). \tag{10}$$

Hence, when the discretization size h is small enough, the propensity for the product P synthesis reaction can be approximated as

$$\begin{aligned} \langle a_{syn}^{(i)} \rangle &\approx k_{syn} \cdot h \cdot \sum_{n=1}^{\infty} P_E^{(i)}(n) \\ &= k_{syn} \cdot h \cdot (1 - P_E^{(i)}(0)) \\ &= k_{syn} \cdot h \cdot (1 - e^{-\alpha/K}). \end{aligned} \tag{11}$$

When the discretization size h is small and K is large, the mean reaction propensity can be further approximated as

$$\langle a_{syn}^{(i)} \rangle \approx k_{syn} \cdot h \cdot \alpha / K. \tag{12}$$

Notice that α/K is the mean population of enzyme E in the i-th bin. The Hill function of the product P synthesis is now reduced to a *linear function* of the enzyme E population in the i-th bin.

Furthermore, from (12) the mean population of product P in the bin i is

$$\langle P^{(i)} \rangle = \frac{k_{syn} \cdot h}{k_{deg}} \frac{\alpha}{K}, \tag{13}$$

and the total product P population in all K bins is

$$\begin{aligned} \langle P \rangle &= \frac{k_{syn} \cdot L}{k_{deg}} \frac{k_s \cdot L}{k_d} \frac{1}{K} \\ &= \frac{k_{syn}}{k_{deg}} \cdot \alpha \cdot h. \end{aligned} \tag{14}$$

Equation 14 shows that the total population of product P is a linear function of α, the mean population of E and $h = L/K$, the discretization size. With finer discretization, less product P is produced. Figure 3 shows the histograms and the mean values of the product P population with different discretization sizes. The histograms show that with finer discretization, the population histograms shift further to the left.

The log-log plot (Fig. 3, right) shows that when the discretization size is small enough, the total product P population is a linear function of discretization size. The slope of the log-log plot is about 1.0 at small discretization size h, regardless of K_m.

Moreover, simulation results show that when the mean enzyme E population is less than the constant K_m in the Hill function ($K_m > \alpha$), the population of product P increases slightly before the Hill function dynamics breaks at small discretization sizes. Note that the Hill function dynamics show a concave shape with respect to enzyme E population when the enzyme E population is smaller than the Michaelis constant K_m. Therefore, it is reasonable that the product P population in this reaction-diffusion model

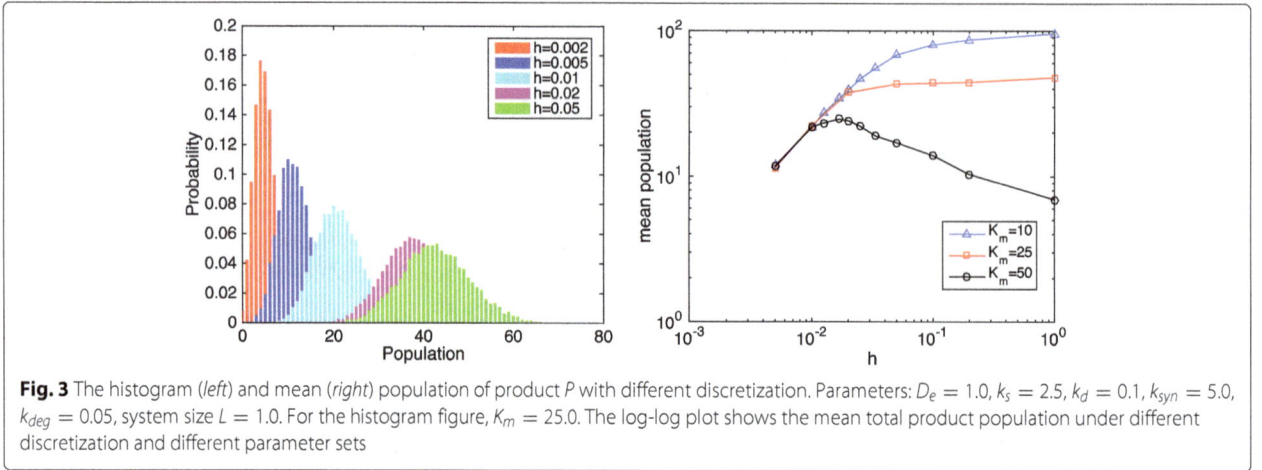

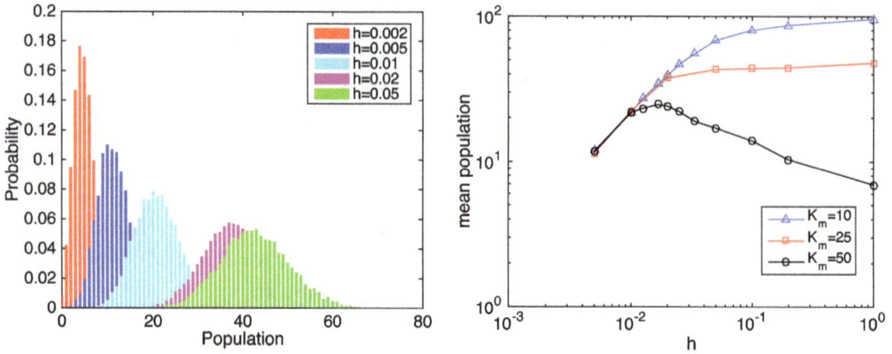

Fig. 3 The histogram (*left*) and mean (*right*) population of product *P* with different discretization. Parameters: $D_e = 1.0$, $k_s = 2.5$, $k_d = 0.1$, $k_{syn} = 5.0$, $k_{deg} = 0.05$, system size $L = 1.0$. For the histogram figure, $K_m = 25.0$. The log-log plot shows the mean total product population under different discretization and different parameter sets

increases slightly when the Michaelis constant K_m is larger than the mean enzyme E population α.

The numerical analysis above makes two approximations

$$\begin{cases} \frac{n^4}{(K_m \cdot h)^4 + n^4} \approx 1, \text{for } n \geq 1; \\ e^{-\alpha/K} \approx 1 - \alpha/K. \end{cases} \quad (15)$$

to get the linear relation. Assuming an error tolerance of 5%, the two approximations can be simplified to

$$\begin{cases} K_m \cdot h < 0.5, \\ \alpha/K < 1/3. \end{cases} \quad (16)$$

Hence, when the discretization bin number

$$K > \max\{2LK_m, \quad 3\alpha\}, \quad (17)$$

the Hill dynamics reduce to a linear function.

Equivalently, in order for the Hill function dynamics to work well, the discretization number K should be less than or equal to this threshold. However, the coarse discretization from a small K leads to spatial error. Two potential solutions to this discretization dilemma are proposed next.

Discussion

From the previous analysis, the Hill dynamics in RDME systems reduces to a linear function due to the lack of intermediate states — the discrete population in each individual bin yields an integer value (0 or 1) for the Hill function. Thus a natural solution to it is to generate intermediate states by a smoothing technique that averages the population over neighboring bins when calculating the reaction propensity.

To model a RDME system in high dimensions with fine discretization, previous studies [21] have suggested relaxing the same-compartment reaction assumption and allowing reactions within neighboring compartments. The next subsection shows that allowing reactions within neighboring compartments is equivalent to smoothing over neighboring compartments.

Smooth over neighboring bins

A natural technique that bridges the discrete and continuous models is to smooth the spatial population by taking the average of neighboring bins. Consider first smoothing the enzyme E population within the neighboring m bins (including the bin itself) when calculating the reaction propensity.

Following previous analysis, the reaction probability for the synthesis of product P in the i-th bin is

$$\begin{aligned} \langle \hat{a}_{syn}^{(i)} \rangle &= k_{syn} \cdot h \sum_{\hat{n}=0}^{\infty} \left(\frac{(n/m)^4}{(K_m \cdot h)^4 + (n/m)^4} P_E^{(i)}(n; m) \right) \\ &= k_{syn} \cdot h \sum_{n=0}^{\infty} \left(\frac{n^4}{(m \cdot K_m \cdot h)^4 + n^4} P_E^{(i)}(n; m) \right), \end{aligned} \quad (18)$$

where $P_E^{(i)}(n;m)$ denotes the probability that the m neighboring bins of the i-th bin have a total enzyme E population of n. The interpretation of this equation is that the synthesis reaction in the i-th bin is interacting with the m neighboring bins and the propensity is calculated based on the total enzyme E population of all the neighboring bins. By probability theory,

$$P_E^{(i)}(0; m) = e^{-\alpha m/K}. \quad (19)$$

As before, only the term $P_E^{(i)}(0;m)$ is important.

In Eq. (18), for any fixed integer $m \geq 0$, there exists an $h \geq 0$, such that $m \cdot K_m \cdot h < 0.5$ and the Hill function is still approximately one. With such a discretization size h, the product P synthesis propensity can be approximated as

$$\begin{aligned} \langle \hat{a}_{syn}^{(i)} \rangle &\approx k_{syn} \cdot h \sum_{n=1}^{\infty} P_E^{(i)}(n; m), \\ &= k_{syn} \cdot h (1 - P_E^{(i)}(0; m)), \\ &= k_{syn} \cdot h (1 - e^{-\alpha m/K}) \\ &\approx k_{syn} \cdot h \cdot \alpha \cdot m/K. \end{aligned} \quad (20)$$

Again, with a fixed smoothing bin number m, the synthesis reaction propensity becomes linear in the mean enzyme E population $\alpha m/K$ of the m bins, and the mean population of product P in the system is

$$\langle P\rangle = \frac{k_{syn}\cdot L}{k_{deg}}\frac{k_s\cdot L}{k_d}\frac{m}{K}, \tag{21}$$

which is linear in m/K and the mean total enzyme E population α. The linear function can be achieved with an h such that

$$\begin{cases} m\cdot K_m\cdot h < 0.5, \\ m\cdot \alpha/K < 0.33. \end{cases} \tag{22}$$

Figure 4 plots the mean population of product P in the toy model with the smoothing technique and $m = 5$. Numerical results show that smoothing over a fixed number m of compartments gives a good solution for a certain range of discretization sizes. However, there always exists a small enough critical discretization size h_{crit} such that the Hill function dynamics reduce to a linear function when the discretization size is smaller than this h_{crit}. Moreover, fixed length smoothing, in the scenarios where the Michaelis constant K_m is larger than the mean enzyme E population α, gives a result closer to that of the deterministic simulation when the discretization sizes are not too small.

Convergent hill function dynamics in reaction-diffusion systems

The previous subsection demonstrates that a sufficiently small discretization size h will still break the Hill dynamics even with the strategy of smoothing over a fixed number of bins, thus the number of bins needs to vary with the discretization size.

Inspired by the convergent-RDME framework [36], a remedy for the failure of Hill function dynamics in reaction-diffusion systems is to smooth the population over bins within a certain distance.

From the analysis, a small smoothing length would cause the failure of the Hill function dynamics and a large smoothing length would degrade the spatial accuracy of the model. Based on the criteria of failure for the Hill function dynamics with fixed m, Eq. (22), we can choose the smallest m that would not result in failure for the Hill function dynamics, i.e., m such that neither of the two assumptions in the previous analysis are valid. This choice is

$$m = \lceil \max\left\{\frac{0.5}{K_m\cdot h}, \frac{0.33\cdot L}{\alpha\cdot h}\right\}\rceil. \tag{23}$$

Following the terminology in the convergent-RDME framework [36], the "reaction radius ρ" of the Hill function dynamics is defined as $\rho = m\cdot h$, where m is given in (23).

Figure 5 shows numerical results for the toy model in the reaction-diffusion system with different discretization sizes and with the convergent smoothing technique (m and h related by (23)). It is clear that the convergent smoothing technique gives very good simulation results for all h values.

Applying the fixed length smoothing technique to the DivL-CckA Hill function model in the *Caulobacter crescentus* cell cycle results in a sharp CtrAp population change during swarmer-to-stalked transition. Figure 6 shows the CtrAp trajectories from the deterministic model and stochastic model simulation results. The fixed length smoothing technique yields more CtrAp in the swarmer stage and less CtrAp in the stalked stage, which yields a sharp CtrAp population change during the swarmer-to-stalked transition as expected.

Conclusions

Motivated by the misbehavior of DivL-CckA dynamics in the stochastic simulation of the *Caulobacter crescentus* cell cycle model, a study of the Hill function dynamics in reaction-diffusion systems reveals that when the discretization size is small enough, the switch-like behavior

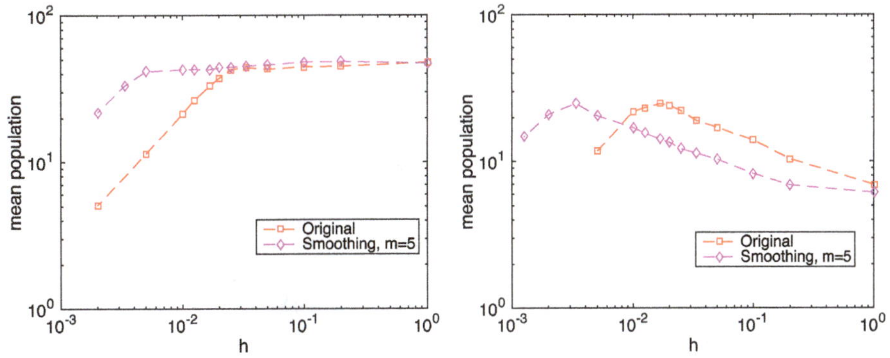

Fig. 4 The total population of product P with different discretization. Parameters: system size $L = 1.0$, $D_e = 1.0$, $k_s = 2.5$, $k_d = 0.1$, $k_{syn} = 5.0$, $k_{deg} = 0.05$. For the left figure $K_m = 25.0$, while for the right figure $K_m = 50.0$

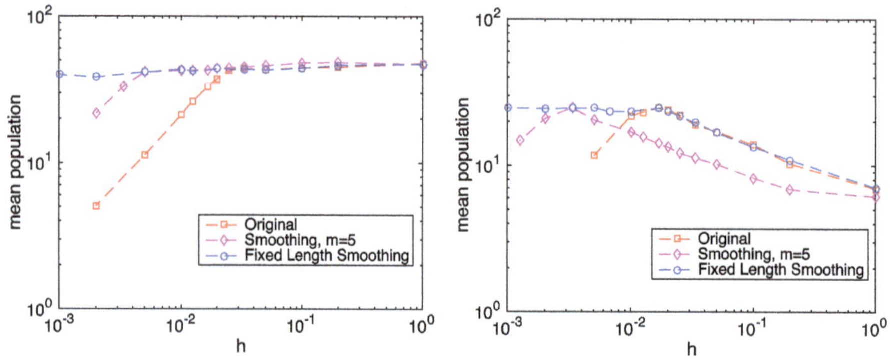

Fig. 5 The total population of product P with different discretization. Parameters: system size $L = 1.0$, $D_e = 1.0$, $k_s = 2.5$, $k_d = 0.1$, $k_{syn} = 5.0$, $k_{deg} = 0.05$. For the left figure $K_m = 25.0$, while for the right figure $K_m = 50.0$

of Hill function dynamics reduces to a linear function of input signal and discretization size. A proposed fixed length smoothing method, which allows chemical reactions to occur with reactant molecules within a distance of fixed length, the "reaction radius"of the Hill function dynamics, seems to give a very good remedy to this problem.

It is known that in high dimensions bimolecular reactions are lost with the RDME in the microscopic limit [21]. This work shows that one-dimensional Hill function dynamics in a RDME framework gives a similar challenge when the discretization size is small enough. The conjecture is that the problem lies in the RDME requirement that reactions only fire with the reactant molecules in the same discrete compartment.

Furthermore, this defect in RDME at the microscopic limit is believed to be a common scenario for all highly nonlinear reaction dynamics. Theoretical biologists have developed many highly nonlinear reaction dynamics that need special attention when converted to stochastic models.

Here we will extend our analysis and discuss a general situation in stochastic simulation of reaction diffusion systems. Suppose that we have a species X, whose population is represented by state variable x, and there is a particular reaction R:

$$\emptyset \xrightarrow{X} P, \tag{24}$$

in which X serves as an enzyme to produce P and the propensity function is represented by $f(x)$. For each X molecular, it can diffuse in a 1D domain with a small length L and with a diffusion coefficient D. Suppose the 1D domain is partitioned into K bins, thus the discretization size is $h = \frac{L}{K}$. The system can then be represented as a chain reaction

$$X_1 \underset{d}{\overset{d}{\rightleftharpoons}} X_2 \underset{d}{\overset{d}{\rightleftharpoons}} \cdots \underset{d}{\overset{d}{\rightleftharpoons}} X_K, \tag{25}$$

where $d = \frac{D}{h^2}$ is the jump rate corresponding to diffusion. The concerned reaction R could fire in any of the bins with propensity $f(x_i)$. Assume that L is small enough such that $\frac{D}{L^2}$ is very large and $d \gg \sum_{i=1}^{K} f(x_i)$ regardless of K. In that

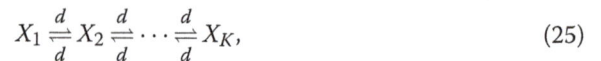

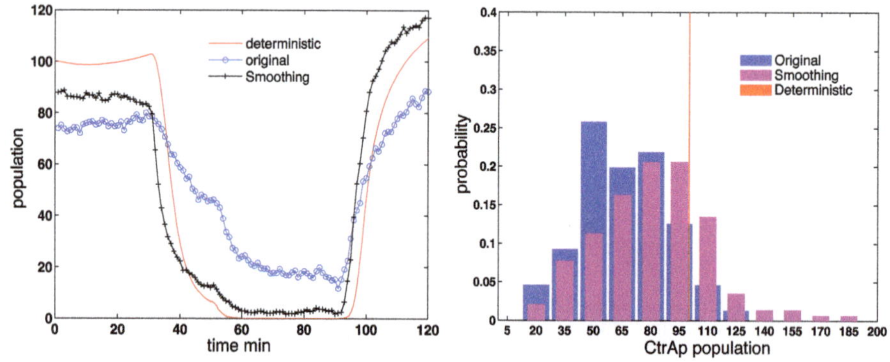

Fig. 6 The Comparison of CtrAp of deterministic model and the stochastic simulation results. *Left*: CtrAp population oscillation trajectory during *Caulobacter crescentus* cell cycle. *Right*: The histogram of CtrAp population in the swarmer cells ($t = 30min$). For model parameters, please refer to [27]

case, the chain reaction system (25) can be considered as a virtual fast system and the slow scale SSA [37] can be applied here. As a result, if the total population of X is n, in each bin, the mean value of x_i is given by

$$\langle x_i \rangle = \frac{n}{K}. \tag{26}$$

Then based on the theory of slow scale SSA, the propensity of the corresponding synthesis reaction (24) should be

$$\langle f(x_i) \rangle = \sum_{j=0}^{\infty} f(j) P(x_i = j), \tag{27}$$

where $P(\cdot)$ is the probability under the distribution when the virtual fast system (25) converges to stochastic partial equilibrium [37].

However, the propensity function converted directly from the deterministic model has a different form as $f(\langle x_i \rangle)$. Note that for a nonlinear function, such as the Hill function or the Michaelis Menten function,

$$\langle f(x_i) \rangle \neq f(\langle x_i \rangle). \tag{28}$$

(28) highlights the mismatch between the RDME framework and the deterministic model.

Hybrid method

In order to have a stochastic model that is consistent with its deterministic counterpart, the propensity function should take the form $f(\langle x_i \rangle)$. This motivates us to adopt the hybrid ODE/SSA method [38] and apply it to the reaction diffusion systems. This hybrid method was a simple idea. It was originally presented by Haseltine and Rawlings [38] and our implementation has some modification to make it fit better with the root finding function used in LSODAR [39]. Consider a system of N species (denoted by $\{S_1, \ldots, S_N\}$) and M reactions (denoted by $\{R_1, \ldots, R_M\}$). For each reaction R_j, there is a propensity function $a_j(x)$ and a state-change vector ν_j. We partition these M reactions into two subsets. The subset S_{slow} contains slow reactions, with index 1 to M_S, and is simulated by the SSA. The subset S_{fast} contains fast reactions, with index M_S+1 to M, and is formulated and solved by ODEs. The simulation of these two subsets is then combined as described below. Let τ be the jump interval of the next slow (stochastic) reaction, and μ be its reaction index. Set $t = 0$. The hybrid method simulate the system as follows:

1) Two uniform random numbers, r_1 and r_2 in $U(0,1)$, are generated.
2) Solve the ODE system for S_{fast} and find the root τ for the integral equation:

$$\int_t^{t+\tau} a_{tot}(\mathbf{x}, s) ds + \log(r_1) = 0, \tag{29}$$

where $a_{tot}(\mathbf{x}, t)$ is the sum of propensities of all reactions in S_{slow}. Because $\mathbf{x}$ varies with t in the ODE system, $a_{tot}(\mathbf{x}, t)$ is a function of t as well.
3) μ is selected as the smallest integer satisfying

$$\sum_{i=1}^{\mu} a_i(\mathbf{x}, t) > r_2 a_{tot}(\mathbf{x}, t). \tag{30}$$

4) Update $\mathbf{x} \leftarrow \mathbf{x} + \nu_\mu$.
5) Return to step 1) if stopping condition is not reached.

Note that our implementation is different from Haseltine and Rawling's original method in step 2. Suppose that the ODE system is given by

$$\mathbf{x}' = f(\mathbf{x}). \tag{31}$$

We add an integration variable z and the following equation to the ODE system.

$$z' = a_{tot}(\mathbf{x}), \quad z(t) = \log(r_1), \tag{32}$$

where we note that $\log(r_1)$ is negative and a_{tot} is always nonnegative. In the hybrid simulation, for each step we start from the current time t and numerically [39]

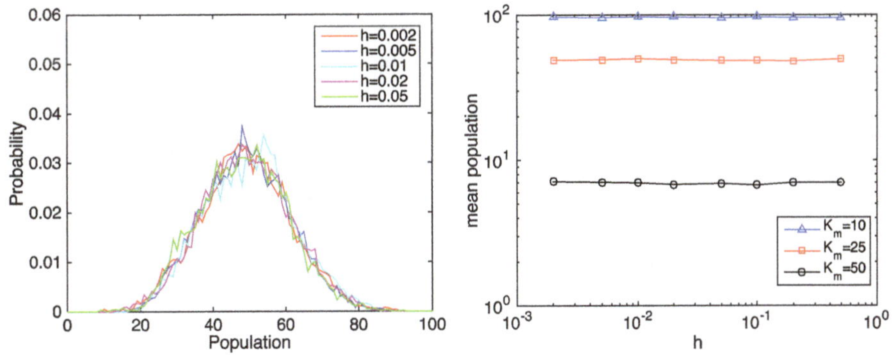

Fig. 7 The histogram (*left*) and mean (*right*) population of product P with different discretization, simulated by the hybrid method. Parameters: $D_e = 1.0$, $k_s = 2.5$, $k_d = 0.1$, $k_{syn} = 5.0$, $k_{deg} = 0.05$, system size $L = 1.0$. For the histogram figure, $K_m = 25.0$. The log-log plot shows the mean total product population under different discretization and different parameter sets

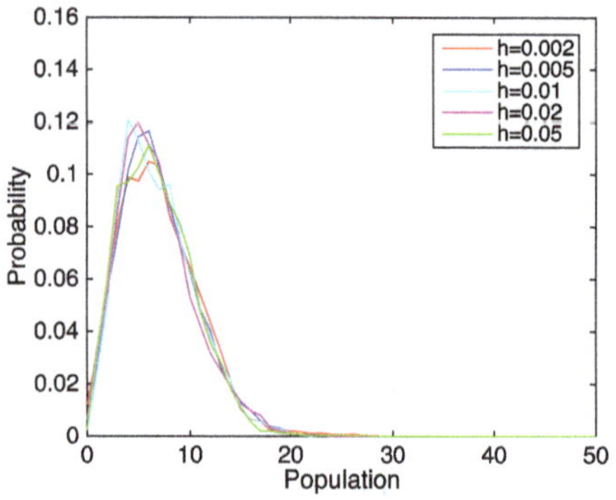

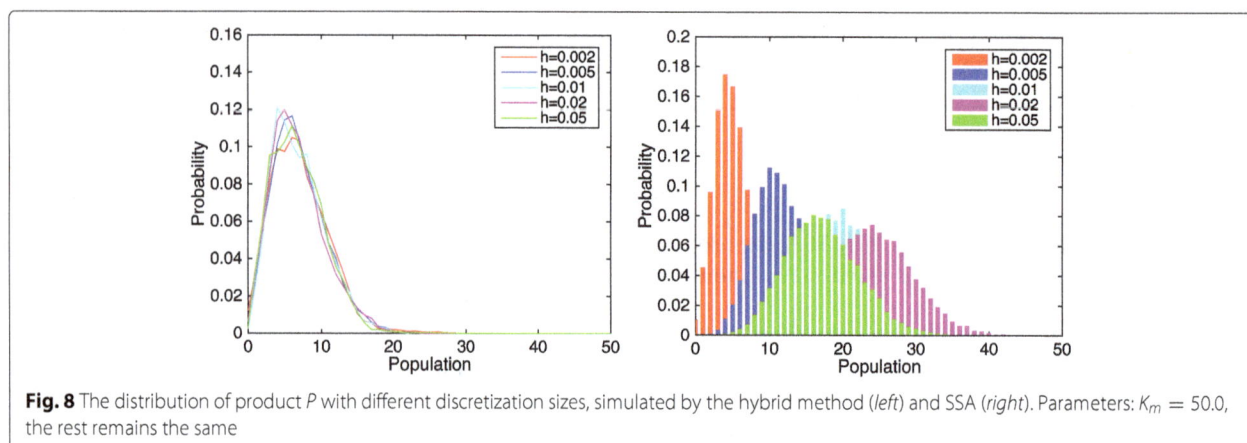

Fig. 8 The distribution of product P with different discretization sizes, simulated by the hybrid method (*left*) and SSA (*right*). Parameters: $K_m = 50.0$, the rest remains the same

integrate the original ODEs (31) and the extra integral Eq. (32). The integration stops when $z(t + \tau) = 0$. As a result, τ is the solution to (29). This procedure can be numerically simulated using standard ODE solvers combined with root-finding functions, such as the LSODAR [39]. Note that since z is an integration variable, one may choose to omit it from the error control mechanism [40]. Adding this extra variable will not greatly affect the efficiency.

We applied the hybrid method to the toy model (6). In our simulation, all diffusion events are partitioned into fast systems and solved by the ODE solver LSODAR, while chemical reactions are simulated by SSA under the hybrid framework described above. We test cases when $K_m = 10, 25, 50$ and Figs. 7 and 8 show the corresponding numerical results. It is obvious that in all three cases, the mean population remains horizontal even when the bin size decreased to the magnitude of 10^{-3}. In Fig. 8, the mean molecule of product P centers around seven under different discretization sizes, while results from SSA shift to the left as discretization size decreases.

Numerical results certainly suggest that the hybrid method has great potential in stochastic simulation of RD systems. We would like to note that great details still need to be studied, but that is not the focus for this paper.

Abbreviations

CME: Chemical master equation; ODE: Ordinary differential equation; PDE: Partial differential equation; RDME: Reaction diffusion master equation; SSA: Stochastic simulation algorithm

Funding

This work was partially supported by the National Science Foundation award DMS-1225160, CCF-0953590, CCF-1526666, and MCB-1613741. In particular, the publication of this paper is directly funded from CCF-1526666, and MCB-1613741.

Authors' contributions

FL initially realized the simulation error described in this paper. FL, MC and YC then designed the toy model and analyzed the numerical error caused by Hill function simulation. Later SW joined to help the implementation of hybrid simulation. FL and MC together drafted the manuscript, and YC gave critical revisions on the writing. All authors have read and approved the final manuscript.

Competing interests

The authors declare that they have no competing interests.

About this supplement

This article has been published as part of *BMC Systems Biology* Volume 11 Supplement 3, 2017: Selected original research articles from the Third International Workshop on Computational Network Biology: Modeling, Analysis, and Control (CNB-MAC 2016): systems biology. The full contents of the supplement are available online at http://bmcsystbiol.biomedcentral.com/articles/supplements/volume-11-supplement-3.

References

1. McAdams H, Arkin A. Stochastic mechanisms in gene expression. Proc Natl Acad Sci. 1997;94(3):814–9. http://www.pnas.org/content/94/3/814.
2. Fedoroff N, Fontana W. Small numbers of big molecules. Science. 2002;297(5584):1129–31. http://www.sciencemag.org/content/297/5584/1129.
3. Samoilov M, Plyasunov S, Arkin AP. Stochastic amplification and signaling in enzymatic futile cycles through noise-induced bistability with oscillations. Proc Natl Acad Sci U S A. 2005;102(7):2310–5. http://www.pnas.org/content/102/7/2310.
4. Gillespie DT. A general method for numerically simulating the stochastic time evolution of coupled chemical reactions. J Comput Phys. 1976;22(4):403–34.
5. Gillespie DT. Exact stochastic simulation of coupled chemical reactions. J Phys Chem. 1977;81(25):2340–61.
6. Michaelis L, Menten ML. Die Kinetik der Invertinwirkung. Biochem Z. 1913;49:333–69.
7. Hill AV. The possible effects of the aggregation of the molecules of haemoglobin on its dissociation curves. J Physiol. 1910;40(Suppl):iv–vii.
8. Hattne J, Fange D, Elf J. Stochastic reaction-diffusion simulation with MesoRD. Bioinformatics. 2005;21(12):2923–4.
9. Andrews SS, Bray D. Stochastic simulation of chemical reactions with spatial resolution and single molecule detail. Phys Biol. 2004;1:137–51.
10. van Zon JS, ten Wolde PR. Green's-function reaction dynamics: A particle-based approach for simulating biochemical networks in time and space. J Chem Phys. 2005;123(23):234910. http://scitation.aip.org/content/aip/journal/jcp/123/23/10.1063/1.2137716.

11. Novère NL, Shimizu TS. StochSim: modelling of stochastic biomolecular processes. Bioinformatics. 2001;17:575–6.
12. von Smoluchowski M. Zur kinetischen Theorie der Brownschen Molekularbewegung und der Suspensionen. Annalen der Physik. 1906;326(14):756–80. http://dx.doi.org/10.1002/andp.19063261405.
13. Gardiner CW, McNeil KJ, Walls DF, Matheson IS. Correlations in stochastic theories of chemical reactions. J Stat Phys. 1976;14:307–31.
14. Nicolis G, Prigogine I. Self-organization in nonequilibrium systems : from dissipative structures to order through fluctuations. New York: A Wiley-Interscience Publication; 1977. http://opac.inria.fr/record=b1078628.
15. doi M. Stochastic theory of diffusion-controlled reaction. J Phys A Math General. 1976;9(9):1479. http://stacks.iop.org/0305-4470/9/i=9/a=009.
16. Keizer J. Nonequilibrium statistical thermodynamics and the effect of diffusion on chemical reaction rates. J Phys Chem. 1982;86(26):5052–67. http://dx.doi.org/10.1021/j100223a004.
17. Fange D, Berg OG, Sjöberg P, Elf J. Stochastic reaction-diffusion kinetics in the microscopic limit. Proc Natl Acad Sci. 2010;107(46):19820–5. http://www.pnas.org/content/107/46/19820.
18. McQuarrie DA. Stochastic approach to chemical kinetics. J Appl Probab. 1967;4(3):413–78. http://www.jstor.org/stable/3212214.
19. Gillespie DT. A rigorous derivation of the chemical master equation. Physica A: Stat Mech Appl. 1992;188(1–3):404–25.
20. Baras F, Mansour MM. Reaction-diffusion master equation: a comparison with microscopic simulations. Phys Rev E. 1996;54:6139–48. http://link.aps.org/doi/10.1103/PhysRevE.54.6139.
21. Isaacson SA. The reaction-diffusion master equation as an asymptotic approximation of diffusion to a small target. SIAM J Appl Math. 2009;70(1): 77–111.
22. Erban R, Chapman SJ. Stochastic modelling of reaction-diffusion processes: algorithms for bimolecular reactions. Phys Biol. 2009;6(4): 046001. http://stacks.iop.org/1478-3975/6/i=4/a=046001.
23. Hellander S, Hellander A, Petzold L. Reaction-diffusion master equation in the microscopic limit. Phys Rev E. 2012;85:042901. http://link.aps.org/doi/10.1103/PhysRevE.85.042901.
24. Li S, Brazhnik P, Sobral B, Tyson JJ. A Quantitative Study of the Division Cycle of *Caulobacter crescentus*, Stalked Cells. PLoS Comput Biol. 2008;4(1):e9. http://dx.plos.org/10.1371%2Fjournal.pcbi.0040009.
25. Li S, Brazhnik P, Sobral B, Tyson JJ. Temporal controls of the asymmetric cell division cycle in *Caulobacter crescentus*. PLoS Comput Biol. 2009;5(8): 000463. http://dx.doi.org/10.1371%2Fjournal.pcbi.1000463.
26. Subramanian K, Paul MR, Tyson JJ. Potential role of a bistable histidine kinase switch in the asymmetric division cycle of *Caulobacter crescentus*. PLoS Comput Biol. 2013;9(9):003221. http://dx.doi.org/10.1371%2Fjournal.pcbi.1003221.
27. Subramanian K, Paul MR, Tyson JJ. Dynamical localization of DivL and PleC in the asymmetric Division cycle of *Caulobacter crescentus*: a theoretical investigation of alternative models. PLoS Comput Biol. 2015;11(7):004348.
28. Collier J, Murray SR, Shapiro L. DnaA couples DNA replication and the expression of two cell cycle master regulators. EMBO J. 2006;25(2):346–56.
29. Collier J, Shapiro L. Spatial complexity and control of a bacterial cell cycle. Curr Opin Biotechnol. 2007;18(4):333–40. http://www.sciencedirect.com/science/article/pii/S0958166907000894.
30. Holtzendorff J, Hung D, Brende P, Reisenauer A, Viollier PH, McAdams HH, et al. Oscillating global regulators control the genetic circuit driving a bacterial cell cycle. Science. 2004;304(5673):983–7. http://www.sciencemag.org/content/304/5673/983.
31. Quon KC, Yang B, Domian IJ, Shapiro L, Marczynski GT. Negative control of bacterial DNA replication by a cell cycle regulatory protein that binds at the chromosome origin. Proc Natl Acad Sci. 1998;95(1):120–5. http://www.pnas.org/content/95/1/120.
32. McGrath PT, Iniesta AA, Ryan KR, Shapiro L, McAdams HH. A dynamically localized protease complex and a polar specificity factor control a cell cycle master regulator. Cell. 2006;124(3):535–47. http://www.sciencedirect.com/science/article/pii/S0092867406000663.
33. Jenal U, Fuchs T. An essential protease involved in bacterial cell-cycle control. EMBO J. 1998;17(19):5658–69.
34. Iniesta AA, McGrath PT, Reisenauer A, McAdams HH, Shapiro L. A phospho-signaling pathway controls the localization and activity of a protease complex critical for bacterial cell cycle progression. Proc Natl Acad Sci. 2006;103(29):10935–40. http://www.pnas.org/content/103/29/10935.
35. Tsokos C, Perchuk B, Laub M. A dynamic complex of signaling proteins uses polar localization to regulate cell-fate asymmetry in *Caulobacter crescentus*. Developmental Cell. 2011;20(3):329–41.
36. Isaacson SA. A convergent reaction-diffusion master equation. J Chem Phys. 2013;139(5):054101. http://scitation.aip.org/content/aip/journal/jcp/139/5/10.1063/1.4816377.
37. Cao Y, Gillespie DT, Petzold LR. The slow-scale stochastic simulation algorithm. J Chem Phys. 2005;122(1):014116.
38. Haseltine EL, Rawlings JB. Approximate simulation of coupled fast and slow reactions for stochastic chemical kinetics. J Chem Phys. 2002;117(15):6959–69.
39. Hindmarsh AC. ODEPACK, a systematized collection of ODE solvers. IMACS Trans Sci Comput Amsterdam. 1983;1:55–64.
40. Petzold LR. A Description of DASSL: a differential/algebraic system solver, proceeding of the 1st IMACS World Congress. Montreal. 1982;1:65-68.

14

The use of genome-scale metabolic network reconstruction to predict fluxes and equilibrium composition of N-fixing versus C-fixing cells in a diazotrophic cyanobacterium, *Trichodesmium erythraeum*

Joseph J. Gardner and Nanette R. Boyle*

Abstract

Background: Computational, genome based predictions of organism phenotypes has enhanced the ability to investigate the biological phenomena that help organisms survive and respond to their environments. In this study, we have created the first genome-scale metabolic network reconstruction of the nitrogen fixing cyanobacterium *T. erythraeum* and used genome-scale modeling approaches to investigate carbon and nitrogen fluxes as well as growth and equilibrium population composition.

Results: We created a genome-scale reconstruction of *T. erythraeum* with 971 reactions, 986 metabolites, and 647 unique genes. We then used data from previous studies as well as our own laboratory data to establish a biomass equation and two distinct submodels that correspond to the two cell types formed by *T. erythraeum*. We then use flux balance analysis and flux variability analysis to generate predictions for how metabolism is distributed to account for the unique productivity of *T. erythraeum*. Finally, we used *in situ* data to constrain the model, infer time dependent population compositions and metabolite production using dynamic Flux Balance Analysis. We find that our model predicts equilibrium compositions similar to laboratory measurements, approximately 15.5% diazotrophs for our model versus 10-20% diazotrophs reported in literature. We also found that equilibrium was the most efficient mode of growth and that equilibrium was stoichiometrically mediated. Moreover, the model predicts that nitrogen leakage is an essential condition of optimality for *T. erythraeum*; cells leak approximately 29.4% total fixed nitrogen when growing at the optimal growth rate, which agrees with values observed *in situ*.

Conclusion: The genome-metabolic network reconstruction allows us to use constraints based modeling approaches to predict growth and optimal cellular composition in *T. erythraeum* colonies. Our predictions match both *in situ* and laboratory data, indicating that stoichiometry of metabolic reactions plays a large role in the differentiation and composition of different cell types. In order to realize the full potential of the model, advance modeling techniques which account for interactions between colonies, the environment and surrounding species need to be developed.

Keywords: Flux balance analysis, Constraints based modeling, Flux variability analysis, Nitrogen cycle, Diazotrophy, Marine cyanobacteria

* Correspondence: nboyle@mines.edu
Department of Chemical and Biological Engineering, Colorado School of Mines, Golden, CO 80401, USA

Background

Nitrogen serves a critical role in the metabolism of all organisms. As a key component in nucleic acids and proteins, it is required for healthy growth and it is often one of the most limiting nutrients for optimal yield. Human intervention via the Haber-Bosch process for the production of ammonia has greatly shifted the global nitrogen cycle, however many ecosystems still rely heavily on biological nitrogen fixation. One such ecosystem is in the open ocean, which is a nutrient-limited environment and organisms that thrive here have evolved to thrive in deplete conditions. *Trichodesmium* is a genus of filamentous diazotrophic (nitrogen fixing) cyanobacteria that not only flourishes in this environment but provides bio-available nitrogen for surrounding species. *Trichodesmium* is responsible for fixing roughly 100 $TgNy^{-1}$ of nitrogen annually (42% of global N fixation) [1] and has been reported to 'leak' 30-50% of the nitrogen it fixes [2]. The genus is ubiquitous in marine environments; it is found in environments as diverse as the Mediterranean Sea [3], the Pacific Ocean [4–6], and the Great Barrier Reef where it has implications not only as a source of nitrogen, but also as a center for eutrophication [7]. It dwells primarily near the surface [8] and can swell to occupy acres of the ocean or sea. Despite its prominence in the global nitrogen cycle, most research efforts have focused on *in situ* sampling and therefore little has been done to model and or predict the effect of different environmental factors on the growth and nitrogen fixation rates in *Trichodesmium.*

Trichodesmium is a colonial cyanobacteria which grows in multicellular filaments called trichomes, each containing about 130 cells [9]. *Trichodesmium* is a non-heterocystous cyanobacterium which means it does not employ specialized cells (heterocysts) for nitrogen fixation. Instead, nitrogen fixation and photosynthesis can occur within the same cell. Most non-hetrocystous cyanobacteria separate oxygen producing photosynthesis from nitrogenase by using temporal separation; they fix nitrogen at night when the cellular metabolism is in respiration mode (consuming carbohydrates stored during the day by photosynthesis). *Trichodesmium* is unique in its mechanism to fix nitrogen, it fixes nitrogen during the day while simultaneously fixing carbon via photosynthesis. Respiration rates in *Trichodesmium* are reported to be higher than other cyanobacteria, which ensures a micro- or anaerobic environment and thus minimizes the potential poisoning of nitrogenase by oxygen [10, 11]. Nitrogenase is only expressed in a subset (10-20%) of cells consecutively arranged in the middle of the trichome. These diazotrophic cells only express photosystem I because photosystem II produces oxygen [10, 12–15]. Current characterization of *Trichodesmium* is limited predominantly to population level observations due to its genetic intractability and difficulty to culture. While several laboratory studies investigating the complex genome [16–18], transcriptome [19, 20], and proteome [21] have been published, most relate to populations level or sparse *in situ* studies in diverse, non-ideal growth conditions. A handful of other recent studies report on the morphology/structure of the cells [8, 10, 22, 23] and how cells respond to iron, nickel, and other nutrient stresses [24–27]. Despite the availability of these studies, they are limited in scope and do not provide a complete picture of *Trichodesmium* on a cellular scale. The long doubling time (57-98 h), low growth density (~100mg/L) [24, 28–30], and lack of genetic tools have limited laboratory based research on *Trichodesmium*, especially compared to other diazotrophic cyanobacteria such as *Anabaena* and *Cyanothece.*

This work presents the first genome-scale reconstruction of a colony forming diazotrophic cyanobacterium, *T. erythraeum*, a leader in marine nitrogen fixation. It models biological optimization of metabolic exchange and biomass creation through Flux Balance Analysis (FBA) and Flux Variability Analysis (FVA) [31] to predict the different metabolic behaviors of the two cell types formed by *T. erythraeum*, nitrogen fixing cells and photoautotrophic cells, constrained by laboratory or published data/observations. The models described in this work illustrate how *T. erythraeum* divides labor between two cells stoichiometrically and applies the first step towards a multi-objective framework of these bilaterally operating cells. These results are extended to understand overall population compositions and metabolite production rates to visualize what role metabolite passage plays in formation of these complex colonies via dynamic Flux Balance Analysis (dFBA) [32] and a population co-optimization algorithm. This model lays the foundation for future colonial cyanobacteria characterization and integration with *in situ* and transcriptomic data for *T. erythraeum.*

Results

Biomass composition and growth rates

Cells were grown in YBC-II medium in ambient air (79% N_2) or supplemented with either 100 μ mol KNO_3 as a nitrogen source. Growth was monitored by measuring total chlorophyll content (see Fig. 1a). These data were then used to calculate the growth rate and doubling time assuming exponential growth (Table 1).

The biomass composition of *T. erythraeum* was measured in order to formulate an accurate biomass formation equation. The composition of cells grown on both ambient air and nitrate were measured (see Table 2 as described in the methods section. The "soluble pool" is a collection of soluble metabolites that are well conserved between organisms for survival and includes small

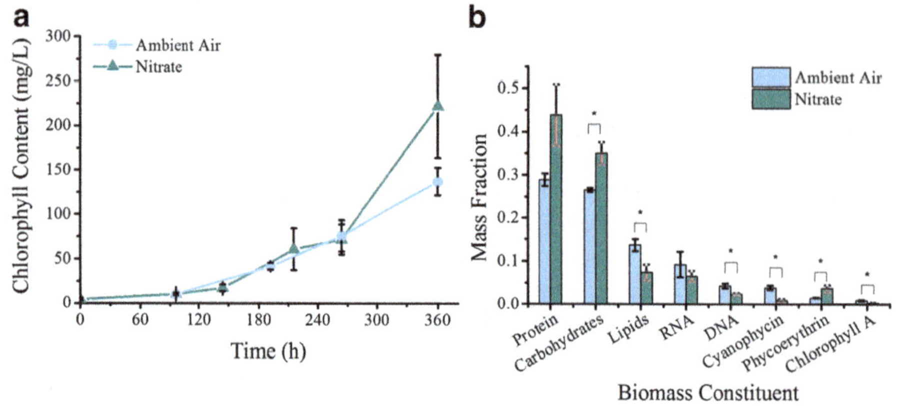

Fig. 1 Growth curves and biomass compositions of *T. erythraeum*. Cells were grown with either ambient air (circles/left and blue) or potassium nitrate (triangles/right and dark cyan) as the nitrogen source in YBCII medium and growth was monitored by measuring total chlorophyll a content. **a**) Growth curves of cells in different nitrogen sources and computational growth. Error bars represent standard deviation from 3 biological replicates. **b**) Biomass composition of *T. erythraeum*. The major elements of biomass were measured directly from cultures grown on diatomic nitrogen (ambient air) or potassium nitrate. Error bars represent standard error from 6 biological replicates

sugars, energy carrying molecules, and other small molecules. To accurately represent proteins and lipids, the amino acid and fatty acid composition of cells were measured as well (see Additional file 8: Tables S1 and S2). We attempted to grow cells on ammonium because we hypothesized that this would remove the necessity of diazotroph formation and enable all the cells in the trichome to be carbon-fixing. This would allow biomass measurement of photoautotrophs directly; however, growth on ammonium in the laboratory was not possible as seems to be consistent with some other cyanobacteria. Therefore we used the composition of cells grown on N_2 as the average biomass composition for all subsequent modeling efforts.

Network reconstruction and manual curation

Reconstructing a complete genome-scale metabolic network from a genomic sequence required several iterations. The first draft of the network was created using an automated genome-scale model algorithm, the SEED RAST [33] and contained 956 reactions. Automated metabolic network reconstructions rely on homology to well characterized and annotated model organisms such as *E. coli* or *S. cerevisiae*. Therefore, the unique metabolic pathways for photosynthesis and nitrogen fixation were not accounted for in the initial draft. The initial automated draft had several gaps and missing reactions, therefore several iterations of manual curation were necessary to fill in missing reactions and infer the presence

Table 1 Growth rates and doubling times of *T. erythraeum*

Nitrogen source	Growth rate (d^{-1})	Doubling time (h)
Ambient Air	$0.0108 \pm 5.14 \times 10^{-4}$	64.4 ± 5.10
KNO_3	$0.0120 \pm 5.65 \times 10^{-4}$	58.1 ± 2.86

Reported error is standard error $\left(\frac{\sigma}{\sqrt{n}}\right)$ where $n = 3$ biological replicates

of reactions that were not predicted based on homology to model organisms. First, we focused on closing the balance on reactions associated with biomass and cellular energy pathways (light harvesting, ATP cycling, and Redox reactions); this was done by referencing metabolic pathway databases such as BioCyc [34] and KEGG [35]. We then compared our draft network to genome scale reconstructions of other related organisms, including *Cyanothece* ATCC 51142 [36], *Synechocystis* sp. PCC 6803 [37], *Synechococcus* sp. PCC 7002 [38], *Arabidopsis thaliana* [39], *Phaeodactylum tricornutum* [40], and *Chlamydomonas reinhardtii* [41, 42]. Through comparison, we added more photosynthesis-specific metabolic reactions which were predicted to be present in the genome based on BLAST results. Finally, transport reactions included in the model were selected based on proteomic data or diffusion (CO_2, H_2O, N_2, etc.) [16]. Manual curation efforts built the model out to a maximum of 1035 reactions; closer inspection of the reactions revealed that several were predicted by the SEED algorithm but had no significant homology to the *T. erythraeum* genome and were non-essential, therefore they were removed. The current draft of the genome-scale metabolic model (Additional file 1: Table S1 and Additional file 2: Table S2) for *T. erythraeum* contains 971 reactions: 1 biomass formation equation (based on experimental data), 9 macromolecule synthesis and condensation reactions, 27 exchange reactions, 38 transport reactions (validated by proteomics data), and 907 metabolic reactions. These reactions involve 647 unique genes and 986 metabolites. Despite our manual curation efforts, the model still has 215 dead end reactions: 113 are involved in lipid, amino acid, or pyrimidine/purine metabolism and are bypassed by summary reactions (see Additional file 3 for a complete list).

Table 2 Biomass composition of *T. erythraeum*

Metabolite	Mass fraction (g/g DW)		Biomass coefficient (mmol/g DW)	
	N_2	KNO_3	N_2	KNO_3
Protein	0.289	0.438	2.12×10^{-4}	2.66×10^{-4}
Phycoerythrin[a]	1.54×10^{-2}	3.67×10^{-2}	2.64×10^{-2}	4.46×10^{-4}
Cyanophycin[a]	3.80×10^{-2}	9.33×10^{-2}	6.96×10^{-2}	4.31×10^{-2}
Carbohydrate	0.265	0.351	4.59×10^{-1}	5.33×10^{-1}
RNA	9.18×10^{-2}	6.51×10^{-2}	2.88×10^{-3}	2.46×10^{-3}
DNA	4.28×10^{-2}	2.40×10^{-2}	1.39×10^{-3}	1.09×10^{-3}
Lipids	0.1370	7.40×10^{-2}	3.89×10^{-3}	3.00×10^{-3}
Phycocyanin[b]	2.60×10^{-2}	3.67×10^{-2}	4.45×10^{-2}	5.37×10^{-2}
Chlorophyll[b]	8.91×10^{-3}	0.424×10^{-3}	9.99×10^{-3}	7.37×10^{-3}
Soluble Pool	2.86×10^{-2}	2.86×10^{-2}	3.79×10^{-2}	3.79×10^{-2}
Total	0.914	1.04	-	

The biomass equation is the molar concentration of the metabolite predicted by the computational molar mass and uses the values from the ambient air (N_2) grown cultures. The "Soluble Pool" is a collection of soluble metabolites that are more or less conserved between organisms for survival (including small sugars, energy carrying molecules, etc.) [a]Subset of protein measurement. [b]Subset of lipid measurement

Manual curation of the model led us to identify 9 genes which are present in the genome based on homology but not annotated, 5 genes encoding enzymes with related functions we assume to be promiscuous and 1 gene which is required for the production of biomass but was not present in the genome based on homology (see Table 3).

Flux balance analysis

A *T. erythraeum* trichome is made up of cells with two distinct metabolic modes: photoautotrophic and diazotrophic. Each cell type was modeled separately and thus required a different set of constraints to define the cell type. The specific constraints applied to each cell type based on literature and experimental data are provided in the methods section (Table 6). Growth associated ATP demand is assumed to be identical to *Cyanothece* sp. ATCC 51142 [36]: 544 mmol ATP (g DW h)$^{-1}$. Maintenance energy, represented by the reaction EN_ATP: $ATP + H_2O \rightarrow ADP + H^+ + Pi$ was adjusted until the predicted growth rate matched published experimental growth rates (0.0146 h^{-1}) [29]. Maintenance energy demands were found to be: 64.3 mmol ATP (g DW h)$^{-1}$ for photoautotrophs and 67.2 mmol ATP (g DW h)$^{-1}$ for diazotrophs. We hypothesize that this number is significantly higher than heterotrophic bacteria for two reasons: (i) maintaining a micro-aerobic or anaerobic environment in the cells for nitrogenase is energetically demanding and (ii) we do not constrain the photon absorption beyond the amount provided to the cells in the laboratory despite knowledge that cells aren't 100% efficient at light harvesting. The COBRA toolbox [43] was used to evaluate the biomass yields for each cell type separately subject to published growth rates, carbon dioxide uptake, and nitrogenase activity (see Table 4). An .m file to run the model in Matlab with appropriate constraints has been included in the supplemental files (Additional file 4).

Flux maps were generated in order to visualize carbon trafficking within the cell for both cell types (illustrated in Fig. 2 with data in Additional file 5: Table S5) subject to the constraints specified in Table 6. As expected, the model for a photoautotrophic cell exhibited high flux through the non-oxidative pentose phosphate pathway and gluconeogenesis. Moreover, the highest recorded flux is through ribulose-1,5-bisphosphate carboxylase/oxygenase (RuBisCO), the enzyme responsible for carbon fixation. The TCA and glyoxylate cycles are partially inactive; the majority of energy is produced through lower glycolysis or photosynthesis and the TCA Cycle's main utility is in precursor biosynthesis. The diazotroph, on the other hand, has higher flux in the pathways associated with respiration: the oxidative pentose phosphate pathway and TCA cycle in particular displayed substantial activity. The glyoxylate shunt also has high flux, indicating crucial differences in how metabolism is regulated in the different cell types.

Another utility of a FBA model is the ability to predict essential genes. We performed an *in silico* gene knockout analysis and identified 275 genes as essential in phototrophic cells and 253 in diazotrophic cells (see red-coded genes in Additional file 6: Genes Table S6). Essential genes are frequently linked to biomass relevant compound synthesis (like pigments and amino acids), carbon and/or nitrogen fixation, or glycolysis. Genes and reactions which decrease growth rate but were not lethal were frequently linked to central carbon processing. These gene knockout results are corroborated by

Table 3 Unannotated metabolic reactions in the *T. erythraeum* genome but included in the model based on homology to related organisms and/or to close gaps for biomass formation

Pathway	Proposed function	E.C. Number	Gene	Annotated function	Closest organism
Newly/Improved Annotated					
Amino Acid Metabolism	L-alanine: glyoxylate aminotransferase	2.6.1.44	Tery_3167	Serine: glyoxylate transaminase	*Leptolyngbya* sp. NIES 3755
	L-serine: pyruvate aminotransferase	2.6.1.44, 2.6.1.45, 2.6.1.51	Tery_3167	Serine: glyoxylate transaminase	*Leptolyngbya* sp. NIES 3755
	L-aspartase	4.3.1.1	Tery_1328	Fumarase	*Nitrosococcus oceani*
	L-arogenate: 2-oxoglutarate aminotransferase	2.6.1.79	Tery_0293	L-aspartate aminotransferase	*Pleurocapsa* sp. PCC 7327
	L-threonine ammonium-lyase	4.3.1.19	Tery_4742	Pyridoxal-5'-phosphate -dependent enzyme, beta subunit/ cysteine synthase A	*Zymomonas mobilis* subsp. NRRL B-12526
Isoprenoid Synthesis	Tocopherol phytyltransferase	2.5.1.117	Tery_3881	Homogentisate phytyltransferase	*Nostoc* sp. NIES-3756
Pigment Metabolism	Chlorophyllide-a: $NADP^+$ oxidoreductase	1.1.5.31.3.1.75	18445Tery_3563	NmrA-like	*Amborella trichopoda*
Secondary Carbon Metabolism	Citramalate synthase	2.3.1.182	Tery_2253	2-Isopropylmalate synthase	*Dehalococcoides mccartyi VS*
Sulfur Metabolism	O-succinylhomoserine thiol lyase	2.5.1.48	Tery_0352	8-Amino-7-oxononanoate synthase	*Nostoc* sp. PCC 7524
Tricarboxylic Acid Cycle	Isocitrate lyase	4.1.3.1	Tery_4268	2,3-Dimethylmalate lyase/ methylisocitrate lyase	*Candidatus Nitrososphaera gargensis*
Assumed Promiscuous					
Amino Acid Metabolism	4-Hydroxyglutamate transaminase	2.6.1.23	Tery_0293	L-aspartate aminotransferase	*T. erythraeum*
Cofactor and Energy Carrier Metabolism	Dihydroneopterin P_i dephosphorylase	3.6.1.1	Tery_1519	Inorganic diphosphatase	
	Dihydroneopterin PPP_i dephosphorylase	3.6.1.1	Tery_1519	Inorganic diphosphatase	
Lipid Metabolism/Secondary Carbon Metabolism	Glycoaldehyde dehydrogenase	1.2.1.21	Tery_2599	Aldehyde dehydrogenase	
Nucleotide Metabolism	3'-5'-Nucleotide phosphodiesterase: cAMP	3.1.4.17	Many	3'-5'-Nucleotide phosphodiesterase: NMP	
Missing (but Essential) Gene					
Cofactor and Energy Carrier Metabolism	(R)-Pantoate:$NADP^+$ 2-oxidoreductase	1.1.1.169	None	None	None

reaction analysis, where reactions are removed from the model instead of genes; this analysis found 370 reactions essential in a photoautotroph and 363 in a diazotroph (see red coded reactions in Additional file 6: Reactions Table S6). Most reactions overlapped, but 6 carbon fixation reactions and 3 gluconeogenesis reactions were unique to essentiality in photoautotrophs while ammonium output and 2 nitrogen fixation reactions were unique to diazotrophs. Unfortunately, *T. erythraeum* has not been reported to be genetically tractable and

Table 4 Predicted yields and selected fluxes for T. erythraeum

Cell type	Carbon Uptake (moles C/ g DW)	NH_4^+ Flux (mole NH_4^+/ mole C)	Biomass yield (g DW/mole C)	Biomass yield (g DW/mole N)
Diazotroph	0.0572	0.204	17.5	55.3
Photoautotroph	0.0643	-0.0996	15.6	156

Biomass and exchange differences between the two cell types are a result of different sources of energy. The carbon mass percent (45.8%) is identical for both cell types because the same biomass formation equation was used (based on growth on N_2)

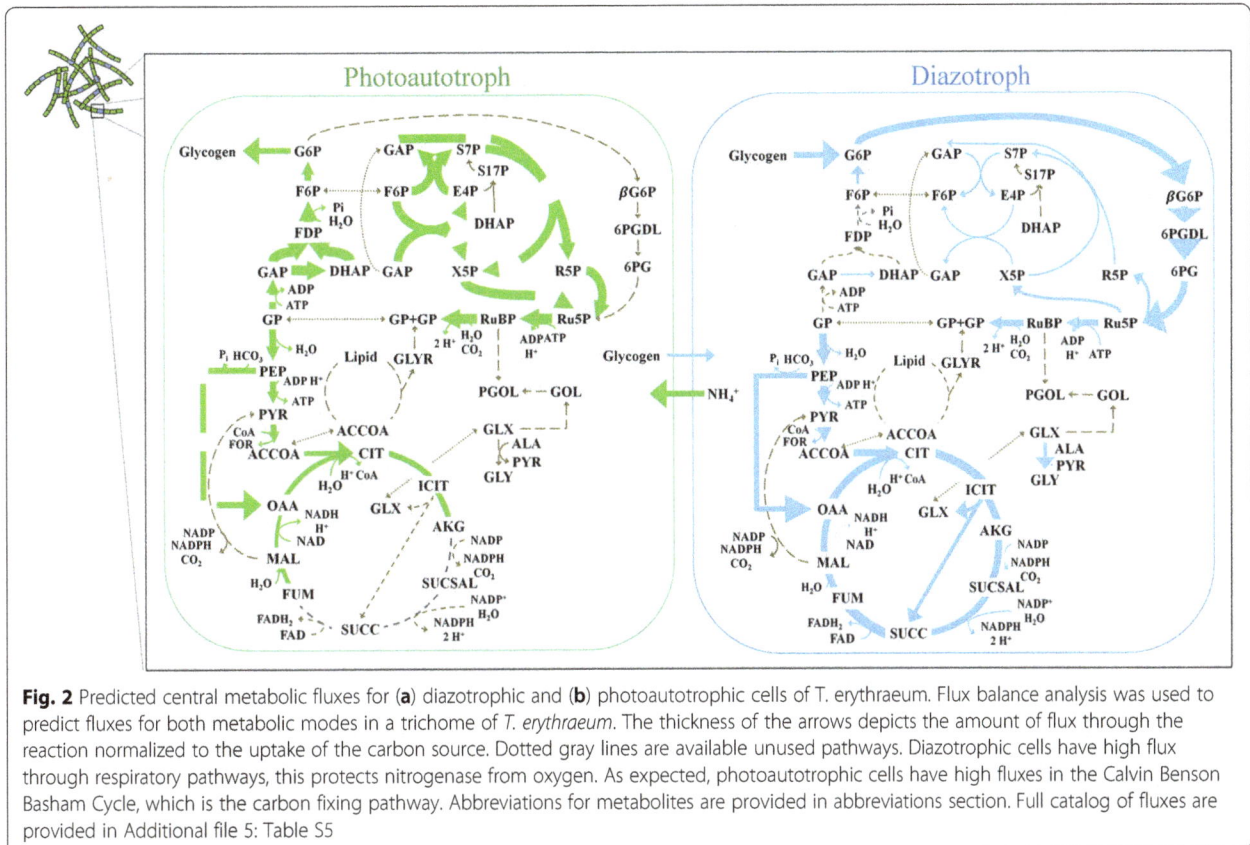

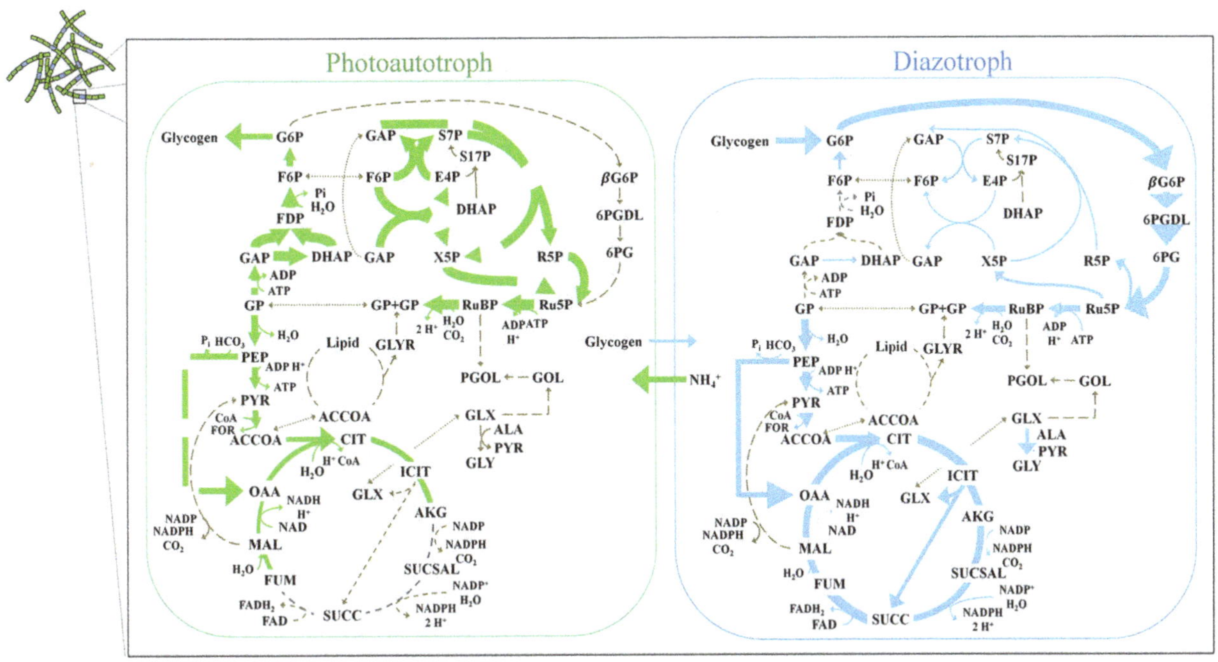

Fig. 2 Predicted central metabolic fluxes for (**a**) diazotrophic and (**b**) photoautotrophic cells of T. erythraeum. Flux balance analysis was used to predict fluxes for both metabolic modes in a trichome of *T. erythraeum*. The thickness of the arrows depicts the amount of flux through the reaction normalized to the uptake of the carbon source. Dotted gray lines are available unused pathways. Diazotrophic cells have high flux through respiratory pathways, this protects nitrogenase from oxygen. As expected, photoautotrophic cells have high fluxes in the Calvin Benson Basham Cycle, which is the carbon fixing pathway. Abbreviations for metabolites are provided in abbreviations section. Full catalog of fluxes are provided in Additional file 5: Table S5

therefore it is not possible to experimentally validate these results.

Flux variability analysis (FVA)

Flux balance analysis uses optimization techniques to predict fluxes in the cell, but most (if not all) genome-scale models are underdetermined so there exists the possibility of multiple flux distributions that lead to the same solution (in our case biomass/growth rate). To evaluate the flexibility in our model, we performed flux variability analysis (FVA) to estimate how much variability a particular reaction can tolerate and still give the same solution. Enzymes, selected from important pathways and by comparing different metabolic distributions in the diazotrophic and photoautotrophic FBA, are shown in Fig. 3 and a table of bounds provided as Additional file 7: Table S7. Overall, photoautotrophic cells display tighter metabolic regulation and have required flux through every major pathway except for nitrogen fixation. In comparison, the diazotrophic model displays high variability through these central pathways with the exception of nitrogenase function. For the photoautotrophic model, two genes exhibit fully nonzero flux: phosphoglycerate kinase (PGK) and ribulose-1,5-bisphosphate carboxylase/oxygenase (RuBisCO). Otherwise, two genes, D-fructose-1,6-bisphosphate D-glyceraldehyde 3-phosphate lyase (FPAL) in upper glycolysis, and sedoheptulose-7-phosphate: D-glyceraldehyde-3-phosphate (TAGSFE) in the pentose phosphate pathway exhibit wide ranges with TAGSFE possibly functioning in both directions while the same optimum is achieved. The complete tricarboxylic acid (TCA) cycle is invariably nonfunctional for energy metabolism in optimization of the photoautotrophs (ACCOASYN, MDH, ICLY, and ICIT) and only functions for precursor biosynthesis. Diazotrophs ultimately demonstrate more variability due to their energy source as more redundancy exists in carbon oxidation than in photosynthesis: all reactions can accommodate zero flux except nitrogen fixation. FPAL and TAGSFE both show reversibility (same as photoautotroph with respect to TAGSFE). Although there are similar patterns in the different cell types, energy metabolism is significantly different.

Dynamic flux balance analysis

Dynamic Flux Balance Analysis (dFBA) was conducted to predict how well the model performed when compared to laboratory experiments using the same data to constraint the model as reported earlier (see Table 6). No additional constraints on diatomic nitrogen or carbon dioxide uptake were applied except for the demand from the other cell type We used the dFBA model to test a number of different growth conditions and hypotheses. First, to predict the

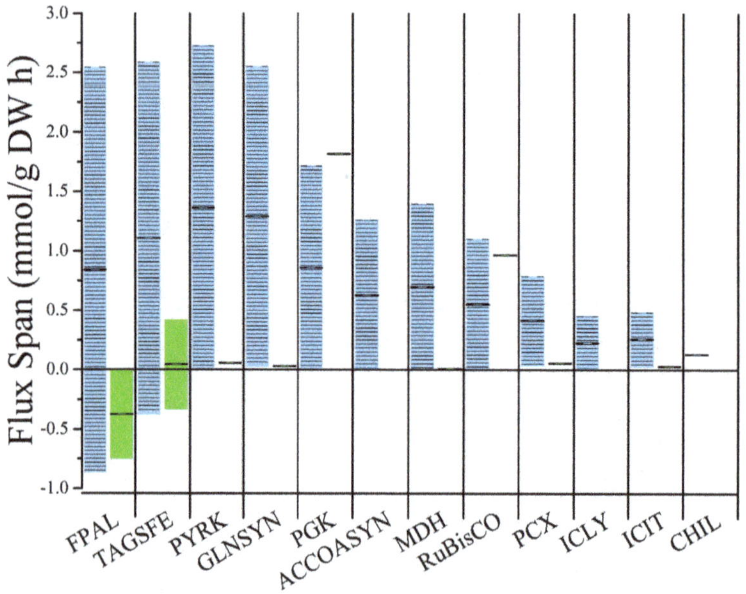

Fig. 3 Allowable variation in flux for central metabolic reactions. Bars visualize the flux variability through important pathways. . Blue/lined bars are diazotrophic fluxes and green/solid bars are photoautotrophic fluxes. High variability implies adaptable responses while low variability implies a narrow essentiality for biomass and energy generation. Greater variability was displayed between cell types in similar pathways including glycolysis (FPAL, PYRK, PGK), the Calvin Cycle (TAGSFE, RuBisCO) and nitrogen processing (GLNSYN) while smaller variability was through the TCA cycle (ACCOASYN, MDH, ICIT) and the glyoxylate shunt (ICLY). Non-zero fluxes were found for the photoautotroph in PGK and RuBisCO while reversibility was found in both cell types for FPAL and TAGSFE. Greater variabilities through carbon processing were due to redundancies in cellular processing, while less variability or non-zero variability was found in non-redundant pathways like carbon and nitrogen fixation. Abbreviations are found under "Enzymes" in the Abbreviations section

equilibrium trichome composition, the model was run with an inoculum of equal parts photoautotrophic and diazotrophic cells. Once equilibrium was determined, initial concentrations of each cell type were calculated from experimental data and the simulation was re-run to validate that the most efficient growth and metabolite production occurred at this equilibrium. The resulting colony fractions to which the cells invariably converged were found to be 0.15:0.85 diazotrophs:photoautotrophs, similar to experimental evidence [10]. To understand how the initial inoculum effects the lag phase of the cells, we ran simulations for a number of different inoculums. Given that the algorithm requires a sequential progression of metabolism (with the photoautotroph calculating its metabolism and then the diazotroph reacting), a "grace period" of 3 time steps was built in where the cells could borrow substrates from an arbitrary cache in their environment so that they could grow initially. However, if a population is unable to produce metabolites or biomass after those three consecutive time steps, the cells were assumed to be unviable and were terminated. How the growth rate evolves over time to reach the experimental value (0.0146 h^{-1}) is shown in Fig. 4. The expected growth rate is obtained rapidly for near-equilibrium starting populations (15:85, 20:80) and more slowly for populations far

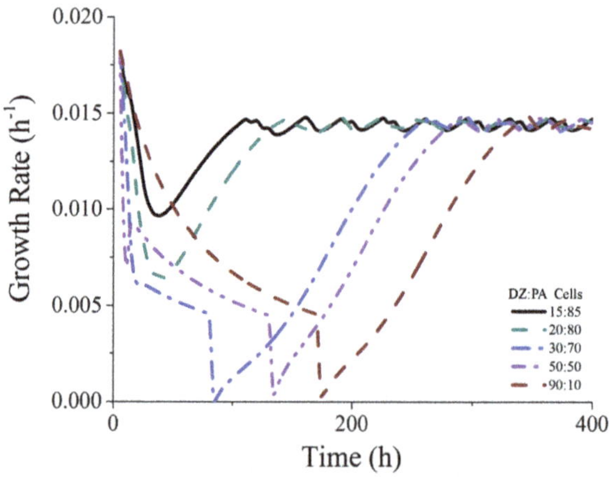

Fig. 4 Growth rate evolution at different initial compositions. The growth rate between time points, taken by measuring biomass generated by dFBA at $t_i = t$ and $t_f = t + 5$, was plotted versus time. Each line represents a different initial composition as listed on the graph (15:85 corresponds to 0.1 diazotroph:0.9 photoautotroph, etc.). All other conditions were held equal, and carbon uptake and nitrogen fixation were adapted from physiological data over 400 h intervals. The gray line represents the initial composition being set to the equilibrium composition. Growth was unachievable at initial compositions of 0:100 and 100:0

from equilibrium (90:10). Populations of 0:100 and 100:0 were nonviable. More detailed figures of how the model predicts composition changes when far from equilibrium is shown in Fig. 5 with initial equality of cell types. Cells appear to have three distinct growth phases, the first where they are adjusting their ratios to move closer toward equilibrium (lag phase) while not able to leak metabolites. When the cells are near equilibrium (at ~0.22 diazotroph: 0.78 photoautotroph), they are able to grow exponentially, but are unable to leak ammonium consistent with their exponential growth (Fig. 5c). When they finally reach equilibrium, growth and ammonium leakage is exponential, indicating most efficient growth. In the dFBA model, photoautotrophic cells are unable to grow without diazotrophic cells and vice versa; however, if the models are optimized assuming nutrient unlimited environments (unrestricted availability of glycogen or ammonium), the photoautotroph can grow at a rate of 0.0182 h^{-1} and the diazotroph can grow at a maximum rate of 0.0188 h^{-1}. From these results, the equilibrium found in nature and this study represents the best-case growth scenario for *T. erythraeum*; this growth appears to be stoichiometrically and metabolically motivated instead of regulated through other means.

This experiment generated several important parameters to compare against literature. Intercellular metabolite production, cell fraction, and percent released of total fixed nitrogen were all predicted using dFBA. The model was run for 1000 iterations with different inoculum compositions to investigate the effect of initial compositions of population development. The generated values for an ideal population (beginning with equilibrium concentrations of each cell type) are summarized in Table 5. The model over predicts nitrogenase flux (0.490 mol N_2 (g DW h)$^{-1}$ for the model versus 0.132 mol N_2 (g DW h)$^{-1}$ for published measurements but is similar to literature values for carbon dioxide exchange (0.922 mol CO_2 (g DW h)$^{-1}$ for the model versus 0.927 mol CO_2 (g DW h)$^{-1}$ for laboratory results) [29]. The model resulted in a total biomass of 13.8 mg/L when simulated at equilibrium starting composition, compared to the 10–40 mg/L found in our laboratory experiments. Variability in growth occurred as illustrated in Fig. 6: growth rates peaked at 0.0146 h^{-1} for growth rate as expected due to

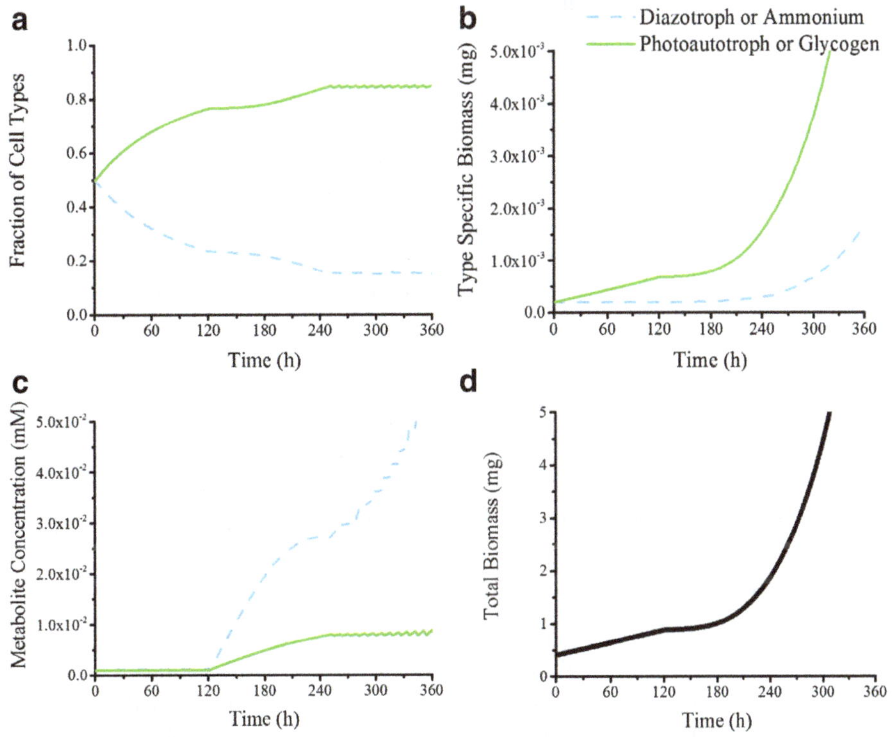

Fig. 5 Computational population rates for *T. erythraeum*. Dotted blue lines indicate diazotroph or fixed nitrogen, green lines indicate photoautotroph or glycogen. **a** Fraction of population for each cell type. Three phases of growth are present: linear redistribution of cells to create enough photoautotrophs, steady preferential allocation to photoautotrophs to drive biomass generation, and achievement of equilibrium. Equilibrium is 0.1544 diazotroph and 0.8456 photoautotroph. **b** Growth rates of each cell type. Biomass is modeled using a batch reactor model with growth rate determined by FBA using the genome-scale reconstruction and time steps of 1 h. **c** Medium concentration for metabolites. These indicate the metabolite accumulation in the medium as determined by the amount of metabolite produced by each cell type less the metabolite consumed plus the amount of metabolite already existing in solution. **d** Total growth rate of population. This is the total growth rate and is plotted with experimental growth rates in Fig. 1. It is calculated by adding the two biomasses in *B* together

Table 5 Productivities of *T. erythraeum* according to literature, laboratory experiments, and the dFBA model

Source	Population fraction (Diazotroph: Photoautotroph)	Growth rate (h^{-1})	Doubling Time (h)	% Nitrogen released	Nitrogenase flux (mmol N (g DW h)$^{-1}$)	CO_2 Uptake (mmol CO_2 (g DW h)$^{-1}$)
Literature	0.1:0.9 – 0.2:0.8 [10]	0.0146 [29]	47.5 [29]	7.7 [62] – 52 [22]	0.132 [29]	0.927 [29]
Boyle Lab	N/A	$0.0108 \pm 8.53 \times 10^{-4}$	64.4 ± 5.10	N/A	N/A	N/A
dFBA Model	0.1544: 0.8456	0.0146	47.5	39.7%	0.490	0.922

N_E corresponds to exported nitrogen in the form of ammonium

constraints (47.5 h for doubling time) at initial cell fractions of 0.15 diazotrophs: 0.85 photoautotrophs). Initial diazotroph concentrations at zero resulted in no growth while photoautotroph dominated inoculums had the minimum nonzero values of 2.80 10^{-4} h^{-1}. Fixed nitrogen release rate distributions and growth rate distributions through all iterations are illustrated in Fig. 7a and b respectively.

Finally, dFBA was conducted to investigate the effect of different nitrogen sources on the growth rate of *T. erythraeum*. All other constraints on the mode were held constant and the resulting growth curves are given in Fig. 8. As expected, growth rates increase with increasing levels of nitrogen reduction and carbon in the nitrogen source. Unfortunately, growth on nitrogen sources other than N_2 or NO_3 have proven to be problematic and there is no evidence of *T. erythraeum*'s ability to utilize other nitrogen sources.

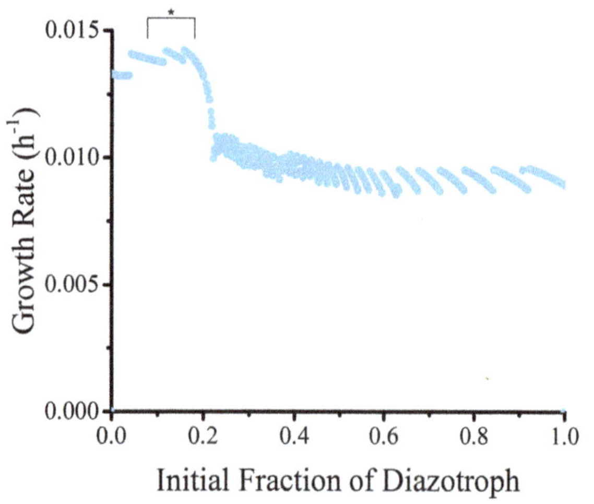

Fig. 6 Growth rate at different initial compositions. The average growth rate dependent on different relative values of diazotroph to photoautotrophic cells. 1000 iterations were generated with random values of initial biomass for each cell type. The initial fraction was calculated using the ratio of these two randomly generated numbers and dFBA was run for 400 h to simulate population development over that time period. Where growth rate was zero over three or more time steps (1 h each) or the cells were unable to manufacture their own nutrients, cell death was assumed and occurred at zero for initial fraction of diazotroph. Optimal growth was found at 0.1544 and suboptimal non-zero growth was found with diazotroph dominated initial populations. The bracket and asterisk refers to the literature predicted equilibrium concentration of cells

Discussion

In this work, we present a genome-scale metabolic network reconstruction of *T. erythraeum* which has been experimentally validated and used to predict growth under a variety of conditions.

Experimental data

An important aspect of any metabolic model is to collect experimental data to aid in the development of the model (biomass equation) and for validation (metabolic production and cellular equilibrium). Major biomass constituents were measured experimentally with direction from literature to define the scope of the biomass constituents (such as inclusion of biliproteins and cyanophycin [8, 23]). It should be noted that *T. erythraeum* differs from other nitrogen fixing organisms and cyanobacteria. Species like *Cyanothece* and *Anabaena* either temporally regulate nitrogen fixation using circadian rhythms or through spatial segregation by forming special cells called heterocysts. *T. erythraeum*, while it does exhibit some circadian regulation [44], can fix nitrogen at all times, day or night. It does form two cell types, but the only evidence of different structures is indicated by accumulation of starches [23]. Cyanophycin granules exist, but are distributed throughout the cells, indicating that they are nitrogen storage compounds rather than structural devices. After assessing the literature based behavior of *T. erythraeum* and its differences from other bacteria, biomass was evaluated when grown on nitrogen or nitrate. The existence of two separate cell types (diazotrophic and photoautotrophic) within a trichome implies a locally nitrogen limited and nitrogen replete environment. We sought to mimic these effects. As expected, the nitrogen provided to the cell culture has a significant impact on biomass composition, and cells partition their carbon in different ways depending on availability of reduced nitrogen. Since the reduction of diatomic nitrogen to ammonia requires a large amount of energy (16 ATP), it is likely that the cells activate their nitrogen sparing mechanisms in order to limit the amount of nitrogen needed for growth. This is evident in the ratio between carbohydrates and lipids

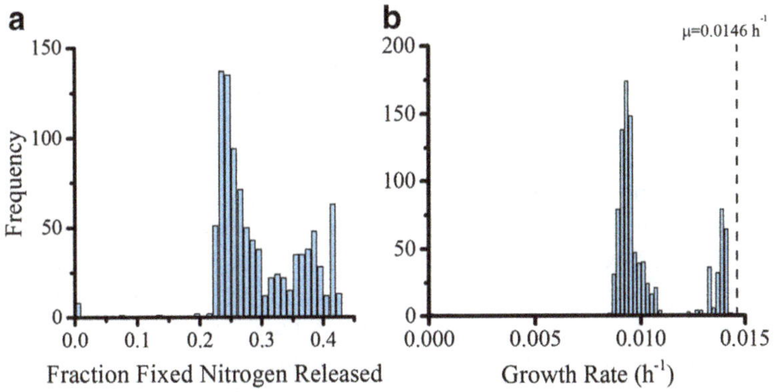

Fig. 7 Nitrogen and biomass production based on variable initial cell concentrations. 1000 iterations were run with randomly generated initial biomasses of each cell type and run for 400 h to simulate laboratory measured behavior. **a** Fraction of fixed nitrogen released calculated by the average amount of fixed nitrogen accumulated in the medium of a Batch Reactor model divided by the average fixation rate over the time period. Values are clustered between 25% and 45% except for non-growth cases where no nitrogen was released because death was assumed. **b** Growth Rate over all iterations. Non-growth cases resulted in zero, but otherwise a bimodal distribution existed. This can be coupled with 0.1544:0.8456 diazotroph: photoautotroph ratio determined by simulation.photoautotroph ratio determined by simulation. The dotted black line refers to laboratory measured growth at 0.0146 h^{-1}

(Fig. 1a), which is significantly lower – only 54% – in cells grown on diatomic nitrogen compared to those grown on nitrate. This is similar to reports of algae which accumulate lipids in response to nitrogen starvation [45–50]. Here, we assume that growth on diatomic nitrogen causes the cells to act like nitrogen is limiting. Cells grown on nitrate have an increased protein content, again highlighting that the cell is no longer acting as if nitrogen is limiting. This increase in protein is also commensurate with an increase in the major photosynthetic pigment-protein complex phycoerythrin. Cultures grown on diatomic nitrogen convert their fixed nitrogen to higher quantities of cyanophycin, a molecule used for nitrogen storage. Meanwhile, this study also sought to characterize the laboratory growth rate of *T. erythraeum* grown on different sources of nitrogen. Unfortunately, *T. erythraeum* showed a reticence to grow on ammonium in the laboratory and has no evidence of a putative transporters for amino acids, commensurate with heuristic knowledge of other cyanobacteria, which limited this experimentation. Even so, growth rates are similar to previous laboratory measurements (0.0108 h^{-1} for the Boyle Lab and 0.0146 h^{-1} from [29]) but are dissimilar to *in situ* studies with values measured as low as 1.46×10^{-4} h^{-1} in the North Atlantic [22]. This is not surprising, it is well known that the open ocean is extremely nutrient deplete and thus a challenging growth environment. Also, our measurements of growth on different nitrogen sources do not indicate a

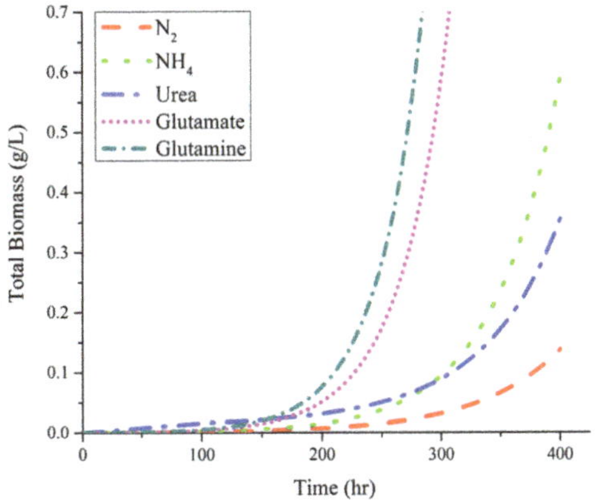

Compound	μ (h^{-1})	Carbon Atoms	Nitrogen Atoms
N_2	0.0145	0	2
NH_4^+	0.0182	0	1
Urea	0.0151	1	2
Glutamate	0.0243	5	1
Glutamine	0.0262	5	2

Fig. 8 Predicted *in silico* growth rates on different nitrogen and nitrogen/carbon sources. Each line represents an excess of a compound. The slowest rate is given by ambient nitrogen which is the same set of values used for the computational curve. The accompanying table relays growth rates and nitrogens or carbons in the source

statistically significant disparity in growth rate based on nitrogen source despite significant changes in cellular composition. As such, the data implies that nitrogen availability is not the major limiting factor in cellular growth. This is supported by literature which demonstrates that *T. erythraeum* is much more efficient in higher CO_2 concentrations [51]. This means that the major effect of nitrogen source is on cellular composition and implies that carbon fixation is the growth limiting function of *T. erythraeum*.

Metabolic network reconstruction

Equipped with a high-level biological understanding of the organism, a genome scale reconstruction was built using the annotated genome, enzyme databases, models of related organisms, and the laboratory biomass measurements. The first draft of the genome-scale metabolic network for *T. erythraeum* was created using the SEED RAST algorithm [33] followed by manual curation. Manual curation was conducted by adding all possible reactions predicted by databases, similar organisms, and gaps based on data in primary literature and publicly available databases [52]. This was pruned to the current model (986 metabolites, 647 unique genes and 971 reactions) by vetting each reaction through BLAST and essentiality. These numbers are comparable to metabolic network reconstructions of similar organisms (Additional file 8: Table S3). Reactions relating to central metabolism, including carbon fixation, glycolysis, the TCA cycle, photosynthesis and the pentose phosphate pathway, are conserved between *T. erythraeum* and the other cyanobacteria listed in Additional file 8: Table S3. Biomass equations were based on the *Cyanothece* ATCC 51142 model [36] for lipid and protein formation. The reactions were mass balanced to predict molar masses, and the biomass measurements were interpreted using these computational predictions. Finally, closure was obtained through proteomic assessment of transport reactions in conjunction with the metabolic network and biomass equations. Overall, the manually curated model of *T. erythraeum* is a strong summary of the majority of metabolic processes and relates well to previous models.

One of the more challenging aspects of building a metabolic network of a non-model organism is the lack of (or completeness of) genome annotation. However, in performing a genome-scale reconstruction, shortcomings in annotation can be identified. In order to achieve closure, 16 reactions were added that are not linked directly to a gene in the *T. erythraeum* genome by assuming enzyme promiscuity, hypothetical protein function, or a lack of evidence for its existence. To validate or refute the presence of these genes in the genome, we did extensive BLAST analysis with related organisms with more complete annotations (other cyanobacteria, *E. coli* and *A. thaliana*). We identified 5 genes encoding enzymes which had evidence of the desired function but were annotated differently (Table 3); we assumed these enzymes to be promiscuous and perform the necessary functions to fill gaps in the network. These assumptions were possible because the genes are often found to have redundant or generic function (the promiscuous genes include phosphatases and transaminases). We identified an additional 10 genes which were previously unannotated but are predicted to be present in the *T. erythraeum* based on homology to related organisms (Table 3). Most of these genes encode enzymes associated with the glyoxylate cycle, aminotransferases, and amino acid metabolism. Again, some of these enzymes have traditionally exhibited generic behavior (like transaminases), but other enzymes were unannotated due to their single-function synthetic purposes (like amino acid synthesis). Interestingly, enzymes associated with the glyoxylate cycle show the most similarity with ammonium oxidizing bacteria. The aminotransferases and amino acid metabolism enzymes, which generate some of the carbon molecules that participate in the glyoxylate cycle, show most similarity to other colony forming cyanobacteria like *Leptolyngbya* and *Nostoc*. This indicates that nitrogen metabolism may have impacts on peripheral carbon metabolism that merits future investigation. Finally, one enzyme which was necessary for the model to produce biomass, (R)-pantoate: $NADP^+$ 2-oxidoreductase (E.C. 1.1.1.169), was not found to be present in the genome. This implies that *T. erythraeum* may have evolved an alternative pathway which has not yet been identified, or that the synthetic route for (R)-pantoate has significant structural differences. Main pathways included in the model along with the several of the newly annotated enzymes are presented in Fig. 9. Confidence levels in the model reflect the strength of prediction of the enzymes, with ones referring to the enzymes discussed above.

Predicting central metabolic fluxes

With a vetted series of metabolic interactions, the model was used to study whole-cell behavior for both diazotrophs and photoautotrophs separately using flux balance analysis (FBA). The two cell types share the same genome but have distinct metabolisms based on differences in regulation due to cellular differentiation: diazotrophic cells provide biologically available nitrogen for the community and photoautotrophic cells provide a reduced carbon source, in the form of glycogen, to the diazotrophs. To model metabolic fluxes for each type of cell, we developed two sets of strict constraints (see Table 6) to reflect the laboratory constraints measured by literature [29]. We fit the predicted growth

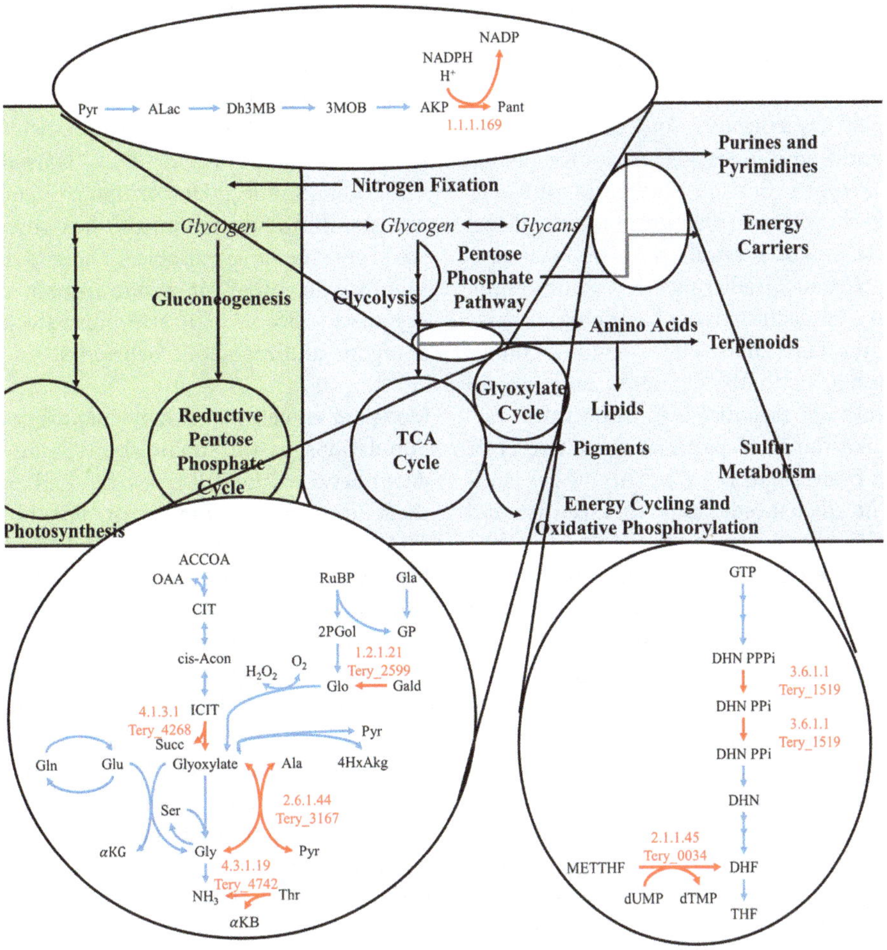

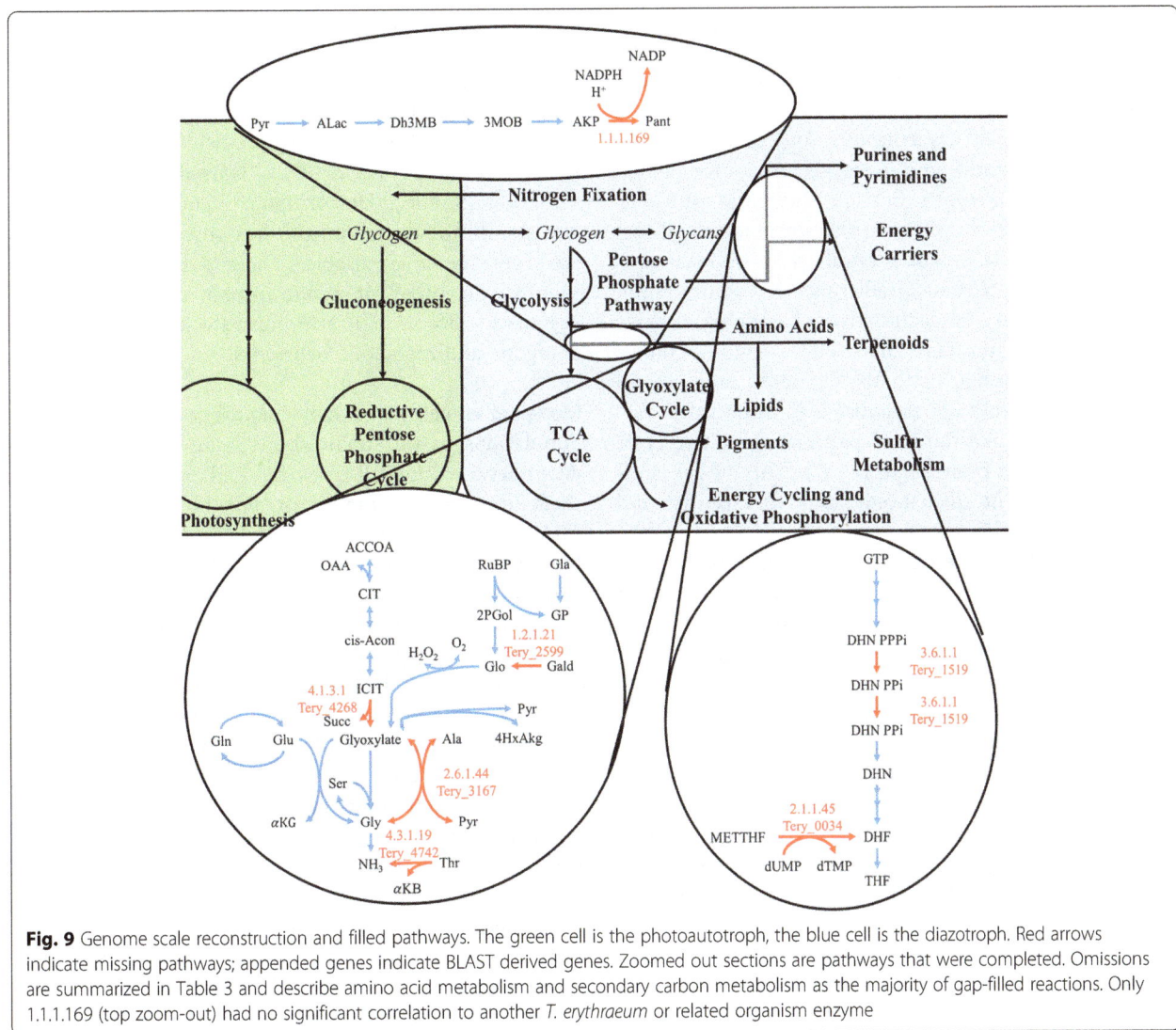

Fig. 9 Genome scale reconstruction and filled pathways. The green cell is the photoautotroph, the blue cell is the diazotroph. Red arrows indicate missing pathways; appended genes indicate BLAST derived genes. Zoomed out sections are pathways that were completed. Omissions are summarized in Table 3 and describe amino acid metabolism and secondary carbon metabolism as the majority of gap-filled reactions. Only 1.1.1.169 (top zoom-out) had no significant correlation to another *T. erythraeum* or related organism enzyme

Table 6 Constraints for each cell type in model simulations for FBA and FVA

Parameter	Diazotroph	Photoautotroph
Carbon Uptake (mmol C (g DW)$^{-1}$ h^{-1})	0.927 (glycogen) [29]	0.927 (CO_2) [29]
Nitrogen Uptake (mmol N (g DW)$^{-1}$ h^{-1})	Unlimited (N_2)	Unlimited (NH_4^+)
Nitrogenase Flux (mmol (g DW)$^{-1}$ h^{-1})	0.132	0
Maintenance Energy (mmol ATP (g DW)$^{-1}$ h^{-1})	64.3	67.2
hv^{PSI}	80	$80 - hv^{PSII}$
hv^{PSII}	0	$80 - hv^{PSI}$
Growth Rate (h^{-1})	0.0146 [29]	0.0146 [29]
Metabolite Output	NH_4^+	Glycogen
Objective Function	Maximize biomass	Maximize biomass

rate from the model by changing maintenance energy and used metabolite production (of the metabolites exchanged between cells) as the objective equation to ensure satisfaction of both objectives. As expected, flux distributions were specific to cell type. Carbon fixation is central to photoautotrophic function and is evident by high flux through the Calving Benson Bassham cycle. For diazotrophs, the absence of Photosystem II requires a functional TCA cycle for production of cellular energy. Interestingly, the oxidative pentose phosphate pathway was preferred over glycolysis, likely due to minimizing carbon oxidation and/or redox balancing. This same phenomenon explains the activation of the glyoxylate shunt which has evolved to conserve carbon [53–56]. The flux maps envision crucial, but expected, differences in metabolism between cell types that are similar to previous findings.

Gene and reaction essentiality were also performed using the same constraints and agree well with flux balance analysis (see Additional file 6: Table S6) and gene essentiality in related organisms. Essential genes include carbon fixation, biomass synthesis, and central metabolism (gluconeogenesis and lower glycolysis) for photoautotrophs and nitrogen fixation, biomass synthesis, and carbon oxidative cycles for diazotrophic cells. Most notably, ammonium export is required for biomass production in diazotrophs, predicting that ammonium leakage is necessary for redox/energy balancing in these cells. These findings correlated with previous studies which find that photosynthetic pigments and certain amino acid pathways are essential [42] while some carbon metabolism like the later parts of the TCA cycle are dispensable in cyanobacteria [57]. The model is effective, then, in *in silico* predictions of essential and conserved metabolic motifs.

The flux maps we present here represent only a single possible solution due to the underdetermined system used for optimization. To assess the bounds and possible adaptations of the model, FVA was conducted and visualized for important pathways (Fig. 3). The photoautotrophic cells showed much tighter bounds for central metabolic processes than their diazotrophic counterparts, likely because their metabolites (glycogen) require roughly 64 ATP/mol as opposed to ammonium which requires 8 ATP/mol (16/mol N_2) plus additional maintenance energy. Consistent with gene essentiality and intuition, Ribulose-1,5-bisphosphate carboxylase/oxygenase (RuBisCO) requires a non-zero flux to maintain optimal behavior. Phosphoglycerate kinase (PGK) – which is in lower glycolysis/gluconeogenesis – is a less intuitive non-zero reaction, but the reaction represents an important step in both optimal carbon oxidation and reduction. For both cell types, D-fructose-1,6-bisphosphate D-glyceraldehyde 3-phosphate lyase (FPAL) and sedohep tulose-7-phosphate: D-glyceraldehyde-3-phosphate (TAG SFE), which correspond to reactions in upper glycolysis and the non-oxidative pentose phosphate pathway, show high variability; TAGSFE can even function reversibly while maintaining optimality. Otherwise, the photoautotrophic cells have zero or near-zero flux for all TCA_Cycle enzymes (acetyl-coA synthase: ACCOASYN, malate dehydrogenase: MDH, isocitrate lyase: ICLY, and bifunctional aconitate hydratase 2/2-methylisocitrate dehydratase: ICIT) because of a lack of electron donors available for carbon oxidation. Glutamine synthase (GLNSYN) displays a small flux for ammonium metabolism and incorporation while PEP carboxylase (PCX) can accommodate flux using its role as an alternative to RuBisCO for single-carbon incorporation. In the diazotrophic cells, pyruvate kinase (PYRK) and glutamine synthase (GLNSYN) exhibit high flux capacities due to their substrates being precursors for many essential metabolites in the cell and the energetic latitude given by the reduced carbon supply, but are irreversible. TCA cycle has intermediate flux capacity because the reactions in it represent best-case processing for carbon oxidation, while the glyoxylate shunt (isocitrate lyase, or ICLY) is used for carbon conservation [53–56]. Diazotrophic metabolism only has one significant deviation: the flux through the nitrogenase enzyme is invariable. These data are similar to findings for other photoautotrophic and diazotrophic organisms [36, 58, 59] and illustrate similar conserved energetic and metabolic behaviors.

Modeling equilibrium colony composition

Simulations to this point showed the metabolic differences between the cell types, but had not accommodated their function as a community. For this reason, dynamic FBA (dFBA) was used to expand the study to simulate how gene-predicted metabolism causes higher order community behaviors. Previous studies have used dFBA for similar applications, including diauxic growth in *E. coli* and the competition between *Rhodoferax* and *Geobacter* [32, 60, 61], but this study presents the idea of dynamic allocation of cells based on metabolite production to simulate need-based differentiation. Since all cells are defined by the same genome, differentiate based on the needs of the community and are invested in the viability of surrounding cells, there is a third level of interaction within these communities. To model this, diazotrophs and photoautotrophs were treated like separate, symbiotic, interdependent populations that could achieve perfect diffusion of substrate. Then, based on the deficiency of one metabolite, cells would be allocated from the deficiency generating cell (the consumer) in proportion to the shortage to correct it. If there was only metabolite surplus, all excess cells from the diazotroph would be allocated towards photoautotrophic production which was implied to drive biomass growth. This was run with both this and the opposite assumption (that all excess would be allocated to the diazotroph) with identical results, meaning that this assumption was an artifact of the algorithm rather than a pivotal administration. Importantly, this algorithm requires the photoautotrophs to act by generating biomass and enough glycogen for the prior generation plus a predictive amount based on growth rate before the diazotroph reacts and creates ammonium. Therefore, the model can inappropriately fail since metabolites are not being simultaneously synthesized. The model was "seeded" with initial ammonium and glycogen substrates that served to prevent this from affecting the model with a "grace period" of three time steps to allow the model to correct the initial "loan" from the environment. Once the algorithm was developed, it was used to interrogate population efficiency by determining equilibrium compositions of cells, predicting metabolite excretion fluxes, and to predict the effect of changing environments on the cell (especially for conditions that are not possible experimentally). The laboratory imposed constraints used for FBA were converted to relaxed, optimal constraints to reflect the lower growth rates measured in our lab. When optimized independently, assuming replete nutrients for each cell type, both cells grow much faster than in limited, co-dependent conditions (0.0182 h^{-1} for photoautotrophs and 0.0188 h^{-1} for diazotrophs versus 0.0146 h^{-1} for the population). This is logical: when nutrients are limited overall growth is limited. Another observation is that photoautotrophic growth is slower even ideal conditions, reinforcing the hypothesis that biomass generation is limited more by carbon than nitrogen.

To determine equilibrium behavior by the community and how well the model could predict the behaviors we observed, we conducted dFBA using this algorithm over 360 h with several different starting conditions. First, a 50:50 starter culture was used to model both how a colony might achieve equilibrium through cellular differentiation and the resulting equilibrium. In order to achieve efficiency, three phases govern cell response to metabolite deficiencies as illustrated by Fig. 5. First, photoautotrophs are generated to correct for carbon deficiency over the first 120 h (steeper slopes in Fig. 5a and linear slopes in Fig. 5a and d) and without excretion of metabolites (Fig. 5c). Meanwhile, the diazotrophs are tasked with producing enough fixed nitrogen to support the community, but all biomass generation is diverted towards photoautotrophs because the "seed" carbon or nitrogen is rapidly depleted. In the second phase from hours 120 to 250, where the cells are near but not at equilibrium, the cells redistribute more slowly as the community is able to support itself without reaching a deficiency while producing some metabolites (Fig. 5c). Finally, after about 250 h, equilibrium is reached where more photoautotrophs will deplete the environment of fixed nitrogen and fewer will reduce biomass generation. This is also the stage where metabolite production is most rapid, and ammonium and glycogen are exchanged between cells and the environment exponentially Fig. 5. These results are encouraging because they justify our observation that the model converges to an equilibrium regardless of initial composition, and the equilibrium falls within experimental and *in situ* levels.

Convergence toward equilibrium was investigated further by incrementally changing initial biomass ratios and plotting growth rate versus time. Rapid convergence by an inoculum of 15:85 diazotroph:photoautotroph is expected and promotes this as ideal behavior by cellular populations (Fig. 4). The convergence by other inoculums reinforces the idea of an equilibrium composition and shows the inefficiency of other starting populations. Finally, since the model assumes no diffusional limitation and similarly structured cells, this convergence to an equilibrium demonstrates that metabolic interactions play a large role in community regulation of *T. erythraeum*. We repeated this analysis 1000 times to see how the composition of the inoculum affected equilibrium composition of the cells (Fig. 6). In agreement with the single simulation results from above, starting with compositions that were not close to equilibrium resulted in suboptimal growth (for excess diazotrophs) or no growth (for excess photoautotrophs). "Death" (defined as a growth rate of zero for three time steps or more), was strictly contained within the initial phase of growth when the inoculum was 0:100 or 100:0. This is because either nitrogen generation or glycogen generation is too low to support the population before efficient biomass production is achieved and the cells do not have enough exogenous resources to correct this. The equilibrium composition and nitrogen release fraction are both close to observed data: a starter culture closer to the cell composition measured in the ocean (10–20% diazotrophic cells and 80–90% photoautotrophic cells [10, 12–15]) grows nearest the constrained growth rate, while compositions further from that equilibrium result in depressed rates. The full distribution of possible growth rates can be seen in Fig. 7b and illustrates a bimodal distribution. The separation between the two peaks can be explained by the extent of domination by either cell type. When diazotrophs dominate, suboptimal growth occurs; when photoautotrophs dominate, the cells operate optimally or die due to deficient nitrogen production. Figure 7a depicts the range of nitrogen release rates dependent on initial inoculum and display a fairly narrow range. In non-death situations, it is consistent within 25% and 45% of total fixed nitrogen, well within the literature recorded values of 7.7% [62] –52% [22]. Across all growth rates, nitrogen leakage is 29.4% of total fixed nitrogen, but for optimal growth rates ($\frac{\mu}{\mu_{max}} \geq 0.9$), nitrogen leakage is 37.5%. Again, nitrogen leakage appears essential to efficiency in biomass generation as evidenced both by observation and simulation. The repetition of the simulation show that an equilibrium composition similar to observation was invariable and is mediated by metabolism while nitrogen is a required side-effect of this optimal growth.

The model was also used to simulate growth conditions that have not been possible to perform in the laboratory. To date, *T. erythraeum* has not been reported to grow on more reduced nitrogen sources; however, we are able to predict how growth on these other N sources using our model (Fig. 8). With increasingly more reduced nitrogen sources and higher carbon content ($N_2 > NH_4^+ > Urea > Glutamate > Glutamine$), the growth rate increases from 0.0146 h^{-1} to 0.0262 h^{-1} with the exception of urea, which has the second lowest growth rate (0.0151 h^{-1}) due to its requirement of ATP consuming active transport and a byproduct of moderately useful CO_2. In the Boyle laboratory, *T. erythraeum* only grows on N_2 or nitrate and this general trend is also seen in other laboratory data. *T. erythraeum*, therefore, lives in a very narrow optimum between two important objectives: carbon and nitrogen fixation. The availability of carbon (or lack thereof) particularly limits growth. In reality, light also has an important role in growth rate because a few centimeters below the surface of the ocean light becomes severely limited. Diffusion, predation, colony shape, and other influential factors can also limit the cell's ability to grow optimally. Even so, biomass generation predictions correlate well between with laboratory data and predicted equilibrium compositions of cells are close to observed values. In order to more effectively leverage this model for colony modeling more sophisticated methods should be used, however, the model has already enabled genome discovery and investigation into the metabolic regulation of this unique and significant cyanobacterium.

Conclusions

A genome-scale metabolic network reconstruction was performed for the filamentous diazotrophic cyanobacterium, *T. erythraeum*. This organism has a prominent role in the global nitrogen cycle; it is responsible for 42% of the annual biological nitrogen fixation and secretes between 7.7% [62] and 52% [22] of the nitrogen it fixes into the ocean, providing an important source of bioavailable nitrogen to other organisms. This model was then subjected to constraints based modeling, such as FBA, FVA and dFBA to investigate the effect of changing environmental conditions and initial cell composition on equilibrium cell compositions and to predict inter- and intra-cellular fluxes. Our simulations indicate that cells exhibit traditional metabolic diversions to conserve carbon in nitrogen-fixing cells (diazotrophs) and to conserve energy in carbon fixing cells (photoautotrophs). From simulations with unconstrained carbon/nitrogen uptake, it appears that photoautotrophs have a lower optimal growth rate than diazotrophs; however in nature, these conditions would never exist. Diazotrophs rely on photoautotrophs for their required carbon and cannot grow in their absence. The model predicts an optimal growth rate for a trichome made up of 15.4% diaztroph, this agrees well with published reports of 10–20% [10, 12–15]. Thus, the model is capable of accurately predicting the required cellular compositions for optimal growth, implying that the composition of cells within a trichome is largely determined by stoichiometry. The success of this and other correlations between our model and laboratory and *in situ* measurements infers that the hypothesis that metabolites are the main influence for cell differentiation for *T. erythraeum* is correct.

The genome-scale metabolic network reconstruction and subsequent model simulations lay the foundation for further interrogation of the metabolism of *T. erythraeum*. This globally significant diazotroph plays an integral role in the ocean, providing a much needed nitrogen source in a very deplete environment. The modeling techniques we have employed above perform well in terms of predicting the growth of a trichome in an ideal environment; but to fully capture the role of *T. erythraeum* in the ocean, more advanced modeling techniques must be developed. In its native environment, *T. erythraeum* interacts with many other species and therefore multiscale models which can capture not only the interaction of the cells with their environment but interactions between cells (identical and other species) must be developed. These types of multi-scale models may also prove useful in modeling and predicting the effect of rising temperatures and carbon dioxide levels in the atmosphere.

Methods

Cell cultivation

Trichodesmium erythraeum IMS101 cells were acquired from the Bigelow Laboratory for Ocean Sciences (East Boothbay, ME, USA). They were grown in an New Brunswick (Hamburg, Germany) Innova 44R incubator at 24 °C with 80 $\mu E/m^2/s$ with 12h light/12h dark cycles. Cells were grown in artificial seawater YBC-II medium [44] at pH 8.15-8.20. Where specified, KNO_3 was added to final concentration of 100 µM. All chemicals were obtained from Sigma-Aldrich (St. Louis, MO). Growth rate was monitored by measuring chlorophyll absorbance [63] from 50 mL of culture every two days.

Biomass composition

Total biomass mass was determined by dry weight analysis, cells were filtered with a Whatman 0.22 μm cellulose-nitrate filter and dried in an oven overnight.

Protein quantitation

Total protein was quantified using the Pierce BCA Protein Assay Kit (Waltham, MS, USA). Cyanophycin and phycoerythrin are both protein-based compounds, so to avoid double counting, their mass fractions were subtracted from the total protein content. Proteins were hydrolyzed, derivatized and analyzed on an Agilent 5973 Mass Detector with an Agilent 6890N Network GC System on an HP-5 column to determine the amino acid composition following the method by Antoniewicz et al. [64].

Carbohydrate quantitation

Carbohydrates were measured colorimetrically using the anthrone method [65] against glycogen as a standard.

Cyanophycin quantitation

Cyanophycin is a primary nitrogen storage molecule for diazotrophs and is polymerized arginine and aspartamine in a 1:1 ratio [66]. Cyanophycin was extracted by disrupting 740 μL of 250 mL cells concentrated to 2 mL via filtration and rinsing with TE buffer with 2.70 mg/mL lysozyme overnight at 37 °C, centrifuging at 16,100 x G for 5 min, and resuspending the pellet in 1 mL of 0.01 M HCl (in which cyanophycin is soluble) for 2 h. The extraction was repeated on the pellet, the supernatant fractions were combined, and cyanophycin was quantified colorimetrically using the Sakaguchi reaction [67].

Phycoerythrin quantitation

Phycoerythrin is a protein-pigment complex that is abundant in *T. erythraeum* and absorbs light for photosynthesis. It is soluble in neutral and slightly basic environments and the supernatant resulting from disruption of 740 μL of cells concentrated from 250 mL to 2 mL via 2.70 mg/mL lysozyme was collected and measured spectrophotometrically at 455 nm, 565 nm, and 592 nm, and quantitated using the equation $[R-PE] = 0.12 \times [(A_{565} - A_{592}) - 0.20 \times (A_{455} - A_{592})]$ [68].

Lipid quantitation

Lipids were extracted using the Bligh and Dyer method [69]. Chlorophyll a and phycocyanin are both lipids, so their masses were subtracted from the total lipid to avoid double counting.

Total lipids were assumed to be made up of the four major subclasses: SQDGs, MGDGs, DGDGs, and PGPs [70]. Fatty acid composition was determined via GC/MS to determine the relative carbon length that constituted lipids in total. The carbon length ratios (C_{18}:C_{16}:C_{14}, etc.) was assumed to be the same distribution for every lipid subclass. Otherwise, lipid pathways were consistent with the methods described in the *Cyanothece ATCC51422* model [36] and the relative amounts of lipid subclasses (SQDG:MGDG:DGDG:PGP) were assumed to be consistent with *Cyanothece.*

Fatty acid methyl esters (FAMEs) were analyzed using GC/MS methods previously described for *Synechococcus sp. PCC7002* [71].

Chlorophyll A quantitation

The only chlorophyll pigment *T. erythraeum* contains is chlorophyll a, which was extracted using an 80% acetone/20% methanol solvent according to Harris [63]. First, 250 mL cells were filtered using a 0.45 μm glass fiber filter and were then washed with 2 mL solvent to yield a 2 mL green solution. The resulting liquid was centrifuged at 4 °C at 13,100 x G for 10 min and the 1 mL of the supernatant was diluted to 2 mL and assayed spectrophotometrically through measurement at 646.6 nm, 663.6 nm, and 750 nm, and using the equation: $[ChlA] = [0.01776(A_{646.6} - A_{750}) + 0.00734(A_{663.6} - A_{750})]$.

Phycocyanin quantitation

2.70 mg.mL lysozyme was used to disrupt 740 μL cells concentrated from 250 mL to 2 mL via filtration overnight and phycocyanin was measured in the resulting medium according to established techniques [72, 73].

DNA and RNA quantitation

DNA and RNA were extracted using MoBio UltraClean Microbial Isolation Kits (Carlsbad, CA, USA) with 50 mL cells concentrated to 2 mL, of which 740 μL was disrupted using 2.70 mg/mL lysozyme overnight at 37 °C/proteinase K for 2 h at 55 °C instead of bead disruption. The concentration of DNA and RNA were assessed spectrophotometrically using 1 cm path-length extinction coefficients [74].

Genome scale metabolic network reconstruction

The genome scale network reconstruction was based on the publically available genome sequence [75] and an automated annotation program [33]. This first draft was manually curated using primary literature [76, 77], enzyme/metabolic databases [34, 35, 78, 79], and comparison to similar organisms [36, 38, 59, 80]. Reactions were considered reversible unless indicated by a database, an annotation in a similar organism, or significant thermodynamic infeasibility to be otherwise. For conserved or metabolically necessary reactions that were not contained in a database, NCBI's BLAST with an acceptable e-score of 1e-6 [81] algorithm was used and compared to related genera like *Leptolyngbya*, *Anabaena*, *Synechococcus*, and *Cyanothece* to identify unannotated genes. Where these organisms failed to predict genome similarity, proteins displaying the function as the unannotated, or insufficiently annotated, reaction were BLASTed against all available organisms. This resulted in similarities to the organisms in Table 3, and included *Nitrosococcus oceani*, *Pleurocapsa* sp. PCC 7327, *Zymomonas mobilis* subsp. NRRL B-12526, *Amborella trichopoda*, *Dehalococcoides mccartyi VS*, *Nostoc* sp. PCC 7524, and *Candidatus Nitrosophaera gargensis.* Photosynthesis was modeled similarly to other cyanobacterial genome scale reconstructions [36, 38, 59, 80] by converting the complex protein interactions to a series of redox reactions with photons as an initial substrate and subsequent reduction reactions being passed to energy carriers and ATP synthase. Gene-protein-reaction associations were assumed using sequence data in conjunction with existing models and protein structures to predict conditionality of their assignments (e.g., if a reaction required a protein complex, the GPRs were assigned an *and* operator; redundant genes were assigned an *or* operator). These were done with consultation of KEGG [35], BioCyc [34], BRENDA [78], and CyanoBase [79]. Finally, transport reactions included in the model were validated based on proteomic data or diffusion (CO_2, H_2O, N_2, etc.) [16]. Manual curation efforts built the model out to 1035 reactions; closer inspection of the reactions revealed that several were predicted by the SEED algorithm but had no significant homology to the *T. erythraeum* genome and were non-essential, therefore they were removed. Irreversibility was assumed where databases indicated significant energetic unfavorability (such as in redox reactions) or where other models demonstrated consistent unidirectionality. Reactions were finally elementally and charge balanced using built-in COBRA functionality in conjunction with the PubChem Database [82]. Small molecules with known properties like pyruvate were used to build out into summary biomass reactions and predict summary macromolecule and biomass constitution. Confidence intervals were added to the model to reflect the source of the reported gene linking: fours were assigned to the transport reactions because of their proteomic data, twos were assigned to all reactions with significant sequence similarity to known protein reactions, and ones were assigned to all gap-filling reactions or reactions with insignificant similarity. The model was assessed using built-in COBRA functionality to determine Type III Reactions by closing all exchange reactions and evaluating the flux variability of the internal reactions. This resulted in no variability, meaning that futile cycles were eliminated [83].

Model simulations

Defining constraints

The genome scale metabolic network was used to predict fluxes using FBA [84]. The constraints applied to the model were determined from experimental data or laboratory or *in situ* data found in literature [22, 29]. Two different sets of constraints were used in order to model the different cell types: diazotrophic and photoautotrophic. Diazotrophic involved Photosystem II deactivation [85], nitrogenase activation, and carbon/nitrogen sources available for uptake. Photoautotrophic cells uptake CO_2 and ammonium while exporting glycogen, diatomic oxygen, and diatomic nitrogen. Available nutrients, trace metals, and ions which are allowed to be used for growth were restricted according to the YBC-II medium formulation. Table 6 details the differences in constraints for each modeled cell type. An upper limit for photon uptake was set to 80 μEinsteins because this represents the total light provided to the cells in the laboratory. Then, biomass related ATP was set to values similar to *Cyanothece* sp. ATCC 51142, another photosynthetic diazotroph [36], at 544 mmol ATP (g DW h)$^{-1}$ and maintenance related ATP was set so that the growth rate matched literature derived physiological data [29]. The model was constrained by the growth rate provided by literature using dFBA to define biomass maintenance energy constraints [29] and then concurrently constraining nitrogenase and carbon dioxide uptake rates from the same study. The maintenance flux for photoautotrophs was determined to be 64.3 mmol (g DW h)$^{-1}$ and 67.2 mmol (g DW h)$^{-1}$ for diazotrophs by matching the growth rate in the dFBA model to the growth rate measured in literature. This is significantly higher than other reported values, but is justified by the lack of substantial light restrictions. It is assumed that light uptake restrictions and maintenance ATP are both contained within this constraint. All other nitrogen and carbon sources except for those specified in Table 6 were set to zero.

Problem formulation for FVA and FBA

FBA and FVA were conducted using COBRA toolbox functions [43]. The optimization problem was constructed around the steady-state assumption with objectives of biomass generation subject to different cell type constraints specified by Table 6. Table 7 shows the abbreviations and notations for equations.

The cell-specific uptake and export rules as well as the seawater constraints generated a suite of constraints of the form (where DZ means diazotroph, PA means photoautotroph, and M means macroscopic):

The cell-specific uptake and export rules as well as the seawater constraints generated a suite of constraints of the form:

$$-1000 \leq v_s^i \leq 0 \; \forall s \in \mathcal{S}, \; \forall i \quad (1)$$

Diazotroph Cell Bounds:

$$\begin{aligned} -1000 \leq v_n^{DZ} &\leq 0 \; \forall n \in \{N_2, Glycogen, , O_2\} \\ 0 \leq v_\mu^{DZ} &\leq 1000 \; \forall \mu \in \{NH_4^+, CO_2\} \\ v_{PSII}^{DZ} &= 0 \\ v_{chIL}^{DZ} &= 0.206 \end{aligned} \quad (2)$$

Phtotoautotroph Cell Bounds:

Table 7 Variables and notations used in simulations

Variables	Identifiers	Sets	Indices
$\boldsymbol{\alpha}$: allocation coefficient	': uncorrected value	$\mathcal{C}$: consumers	γ: consumer of glycogen/consumer cell in allocation
C: concentration	0: initial	$\mathcal{P}$: producers	c: producer of glycogen
f: fraction of cell type	γ: consumed metabolite for photoautotroph	$\mathcal{S}$: seawater nutrients	i: species
$\hat{\mu}$: maximum growth rate	c: produced metabolite for photoautotroph		μ: consumer of ammonium
$\boldsymbol{\mu}$: growth rate	*chlL*: nitrogenase		n: producer of ammonium
$\boldsymbol{v}$: flux	*cons*: consumer		π: producer cell in allocation
N: number of steps	δ: differentiation pool		
$\boldsymbol{\sigma}$: portion control for allocation	DZ: diazotroph		
S: stoichiometric matrix	f: final		
t: time	m: metabolite		
X: biomass	n: step		
$\boldsymbol{Y_{N\rightarrow Env}}$: fraction of nitrogen released to the environment	PA: photoautotroph		
	prod: producer		
	PSI/II: Photosystem I/II		
	T: total		

$$
\begin{aligned}
-1000 \le \boldsymbol{v}^{PA}_{NH_4^+} \le 0 \\
0 \le \boldsymbol{v}^{\mathbf{PA}}_{c} \le 1000\ \forall c \in \{N_2, Glycogen\} \\
0 \le \boldsymbol{v}^{PA}_{PSII} \le 1000 \\
\boldsymbol{v}^{PA}_{CO_2} = -0.927 \\
\boldsymbol{v}^{PA}_{chlL} = 0
\end{aligned} \tag{3}
$$

It should be noted that seawater ingredients are assumed to be replete in this model.

Steady state was assumed to make the model solvable:

$$
\boldsymbol{S} \cdot \boldsymbol{v} = 0 \tag{4}
$$

Essentially, this means that mass is conserved and all inputs exit either through the biomass equation or through export. The second major assumption was that the only outputs would be biomass, glycogen, or ammonium; energy or mass balancing mechanisms, such as through organic acid spilling, were assumed to be negligible because of efficient, exponential growth conditions.

Gene deletions, reaction deletions, and Dead End reaction analysis were run using these constraints. These problem formulations were solved using the COBRA toolbox [43] in conjunction with the Gurobi (Houston, TX, USA) solver in MatLab (Natick, MS, USA).

Dynamic FBA (dFBA)

dFBA was executed with the aid of the Dynamic Multispecies Metabolic Modeling (DyMMM) framework as a template [61]. Both this model and the DyMMM model built on traditional dFBA which estimates solutions to the time-dependent equations governing population growth and substrate utilization. This is done using FBA methods to determine interactive fluxes (exchange, biomass, and objective reaction fluxes) from a genome scale. Metabolites were assumed to perfectly diffuse since the relevant study was assumed to center around metabolism and differentiation of cells in response to shortage. The growth rate is also sufficiently slow as to reduce the effects of transport versus diffusion phenomena in ideal, exponential conditions. To model the differentiation between cell types, dFBA allowed a portion of a period's growth rate to be proportionally allocated to the underproducing cell. If there was no metabolite deficiency in a period, excess growth of all cells was allocated to photoautotrophic cells. Otherwise, the carbon dioxide uptake, nitrogen uptake, and metabolite export rates were constrained only by the requirements of the population (therefore eliminating the strict carbon dioxide and nitrogenase constraints). Only the biomass maintenance energy was constrained to match laboratory growth rates with initial ideal (equilibrium) concentrations of cells at biomass similar to initial biomass measured in laboratory studies. These maintenance fluxes became 64.3 mmol ATP (g DW)$^{-1}$ h^{-1} for the photoautotroph and 67.2 mmol ATP (g DW)$^{-1}$ h^{-1} for the diazotroph.

$$\max_{C^{i}_{Glycogen},t,\mu^{i},v^{PA}_{CO_2}},v_{m^{DZ}},f_{DZ}\left(\mu^{PA}\ and\ v^{DZ}_{NH_4^+}\right) \quad (5)$$

$$s.t.\left[C^{PA}_{Glycogen}\right]_t \geq \left[\left|C^{DZ}_{Glycogen}\right|\right]_{t-1}$$

$$v^{DZ}_{Glycogen} \leq \frac{1}{6}\frac{v^{PA}_{CO_2}}{f_{DZ}}$$

For the purpose of this program, the final constraint is estimated (knowing that μ and Δt are non-negative) by:

$$\left[C^{PA}_{Gly}\right]_t = \left(\mu^{DZ}\Delta t + 1\right)\left[\left|C^{DZ}_{Glycogen}\right|\right]_{t-1}\left[\frac{X^{DZ}_f}{X^{PA}_f}\right]_{t-1} \quad (6)$$

The dFBA model was built assuming Batch Reactor behavior which has the design equation:

$$\frac{dX}{dt} = \mu X \quad (7)$$

Assuming constant growth rate:

$$X'_f = X_0 e^{\mu\Delta t} \quad (8)$$

$$\Delta X' = X'_f - X_0 \quad (9)$$

Cells were allocated to a pool of differentiated cells from the preliminary $\Delta X'$, meaning that a cell can only lose as much biomass as it gains in a period. The amount of cells was determined according to the substrate shortage in a period.

$$C^{T}_{cons,m} = \sum_{\gamma}\left(\Delta X^{\gamma}\cdot v^{\gamma}_{m}\cdot X'^{\gamma}_{f}\cdot\Delta t\right)\ \forall\gamma\in C_m$$

$$C^{T}_{prod,m} = \sum_{\pi}\left(\Delta X^{\pi}\cdot v^{\pi}_{m}\cdot X'^{\pi}_{f}\cdot\Delta t\right)\forall\pi\in P_m$$

$$\sigma^{\gamma} = \frac{C^{\gamma}_{cons,m}}{C^{T}_{cons,m}} = \frac{X'^{\gamma}_{f}\cdot\Delta t\cdot\left|v^{\gamma}_{m}\right|}{\sum_{\gamma}\left(\Delta X^{\gamma}\cdot v^{\gamma}_{m}\cdot X'^{\gamma}_{f}\cdot\Delta t\right)}\forall\gamma\in C_m$$

$$\sigma\pi = \frac{C^{\pi}_{cons,m}}{C^{T}_{cons,m}} = \frac{X'^{\pi}_{f}\cdot\Delta t\cdot\left|v^{\pi}_{m}\right|}{\sum_{\gamma}\left(\Delta X^{\pi}\cdot v^{\pi}_{m}\cdot X'^{\pi}_{f}\cdot\Delta t\right)}\forall\pi\in P_m$$

$$\alpha = \frac{S^{T}_{prod,m} + S^{T}_{cons,m}}{S^{T}_{cons,m}} \quad (10)$$

The differentiated cell pool, X_δ, was determined by applying the allocation coefficient to the productive cells according to their relative production coefficient:

$$X_\delta = \sum_{\Upsilon}\left(\alpha\cdot\sigma^{\gamma}\cdot\Delta X'\gamma\right)\ \forall\gamma\in C_m \quad (11)$$

$$X^{\gamma}_{f} = \Delta X'^{\gamma} + \sigma^{\gamma}\cdot X_\delta + X^{\gamma}_{0}\ \forall\gamma\in C_m \quad (12)$$

And cells are "dealt" according to those same coefficients for consumers:

$$X^{\pi}_{f} = \Delta X^{\pi} + \sigma^{\pi}\cdot X_\delta + X^{\pi}_{0}\ \forall\pi\in P_m \quad (13)$$

Adjusted values are calculated for final concentrations and, during the next iteration, the new initial values are the final values from the previous iteration. This ends at a defined maximum time (400 h) and cell fractions are calculated as a function of biomass of cell type i divided by total biomass. Convergence was assessed using randomized initial compositions, assuming the same total initial biomass amount, and comparing the incremental growth rate, or biomass generation at each time step, as it changed over the time period. Incremental growth rate (μ_t), total growth rate (μ_T), and fixed nitrogen yield ($Y_{N\rightarrow Env}$) were calculated by:

$$\mu t = \frac{\ln\left(\frac{\sum_i X^{i}_{t}}{\sum_i X^{i}_{t-1}}\right)}{1}$$

$$\mu T = \frac{\ln\left(\frac{\sum_i X^{i}_{f}}{\sum_i X^{i}_{0}}\right)}{\Delta t} \quad (14)$$

$$Y_{N\rightarrow Env} = \frac{\Delta C_{NH_4^+}}{C_{NH_4^+,prod}} \quad (15)$$

Since multiple steady states exist for population composition, growth rate, and substrate release rate, randomized initial cell concentrations (both in terms of composition and initial biomass) were generated using MatLab over 1000 iterations. Also, FBA was run individually on each cell type once more to determine the growth rate of each cell type in replete (independent, unlimited) conditions.

The hypothetical viability of *T. erythraeum* on different nitrogen sources was assessed by "opening" reaction boundaries for exchange reactions corresponding to the nutrient and assuming the nutrient was in excess. Then, the dFBA simulation was run assuming equilibrium starting composition. The growth profiles and growth rates were recorded in the same way as previous simulations. Where proteomics data did not infer transport of a particular compound, a simple, diffusion style transporter was temporarily added to the model.

Additional files

Additional file 1: Genome scale metabolic reconstruction for *T. erythraeum*. This file contains the full SBML formatted genome scale reconstruction of *Trichodesmium erythraeum* (iTery101) described by this study without constraints.

Additional file 2: List of reactions and metabolites included in the model. This file contains the more readable format of the Genome scale reconstruction with a catalog of reactions in one tab and of metabolites in the other, supplemented by relevant information for both sets.

Additional file 3: This file contains the list of dead end reactions generated by the study, not including exchange reactions. This means all reactions which include a metabolite that cannot be otherwise resolved (through reaction to another compound) is included. Dead ends are ignorant to photoautotroph or diazotroph because of the similar genetic background.

Additional file 4: MakeTERY function to establish constraints and create separate models. This file contains all relevant constraints for the two different cell types. By default, it also generates the optimal solutions subject to literature constraints as described in the FBA and FVA sections of the paper.

Additional file 5: Flux reactions for ideal, exponential growth with fixed biomass/growth rate and objective metabolite production. This file contains the entire simulations data used to generate Fig. 2 and based on the flux balance analysis described by the study. It includes the entire catalog of reactions, the corresponding flux for both diazotroph and photoautotroph, and the reaction formula.

Additional file 6: List of effects of gene and reaction deletion identified by *in silico* knockout analysis. This file contains the list of effects from each gene and reaction deletion in separate tabs. Gene/reaction deletions that result in lethality are colored red, those that reduce function are colored yellow, and those that cause no change are white. The single green reaction (deletion of biomass maintenance, EN_ATP) improves gene function. Reactions have associated genes listed alongside and genes have associated reactions listed alongside. There are reports for both diazotrophs and photoautotrophs. Data is reported as ratio (growth rate of knockout divided by "wild type" growth rate), raw knockout growth rate, and change in growth rate calculated using the following equation: $\Delta\mu\{\} = = \mu WT - \mu mutant \mu WT$

Additional file 7: List of lower and upper bounds of flux identified by flux variability analysis. Flux Variability for ideal, exponential growth with fixed biomass/growth rate and objective metabolite production with the same bounds as Flux Balance Analysis. Included are lower and upper bounds for acceptable fluxes along with the total range for each reaction. Reactions are organized by the "Euclidean Range", the square root of the ranges squared, which give an image for the most consistently variable and non-variable reactions between cell types.

Additional file 8: Table S1. Average protein composition of *T. erythraeum*. Proteins were hydrolyzed and amino acid concentrations were measured using gas chromatography/mass spectrometry. The molar fraction from ambient air (N_2) was used for the protein and lipid assembly equations. Starred quantities were derived from a previous study [1] because our method was not able to detect them. **Table S2**. Average lipid composition of *T. erythraeum*. Extracted lipids then analyzed as fatty acid methyl esters on a gas chromatograph/ mass spectrometer. Relative amounts of subclasses of lipids were assumed from previous literature [2]. **Table S3**: Comparison to related genome-scale metabolic network reconstructions. Summary figures for related genome scale reconstructions for relevant photosynthetic organisms.

Abbreviations

General
BLAST: Basic local alignment search tool; COBRA: Constraint based reconstruction and analysis; dFBA: Dynamic flux balance analysis; DW: Dry weight; DyMMM: Dynamic multispecies metabolic model; FAME: Fatty acid methyl ester; FBA: Flux balance analysis; FVA: Flux variability analysis; GC/MS: Gas chromatography/mass spectrometry; PSI: Photosystem I; PSII: Photosystem II; RAST: Rapid annotation using subsystems technology; TCA: Tricarboxylic acid

Enzymes
TAGSFE: sedoheptulose-7-phosphate: D-glyceraldehyde-3-phosphate (E.C. 2.2.1.2); PCX: PEP carboxylase (E.C. 4.1.1.31); MDH: Malate dehydrogenase (1.1.1.37); PYRK: Pyruvate kinase (E.C. 2.7.1.40); ICLY: Isocitrate lyase (E.C. 4.1.3.1); ACCOASYN: Acetyl-CoA synthetase (E.C. 6.2.1.1); PDH: Pyruvate dehydrogenase (E.C. 1.2.4.1); RuBisCO: D-ribulose-5-phosphate 1-phosphotransferase (E.C. 2.7.1.19); *chlL*: nitrogenase (E.C. 1.18.6.1); GLNSYN: L-glutamate: ammonium ligase (E.C. 6.3.1.2); PGK: 3-phospho-D-glycerate 1-phosphotransferase (E.C. 2.7.2.3); ATP: ATP synthase (E.C. 3.6.3.14); FPAL: D-fructose-1,6-bisphosphate D-glyceraldehyde 3-phosphate lyase (E.C. 4.1.2.13); ICIT: bifunctional aconitate hydratase 2/2-methylisocitrate dehydratase (E.C. 4.2.1.3, E.C. 4.2.1.4)

Lipid subclasses
DGDG: Digalactosyldiacylglycerol; MGDG: Monogalactosyldiacylglycerol; PG: Phosphatidylglycerol; SQDG: Sulfoquinovosyldiacylglycerol

Metabolites
6PG: 6-phospho-D-gluconate; 6PGDL: 6-phosph-D-glucono-1,5-lactone; AcCoA: Acetyl-CoA; AKG/αKG: α-ketoglutarate/2-oxoglutarate; ALA: L-alanine; cAMP: cyclic-AMP; CIT: Citrate; CoA: Coenzyme-A; DHAP: Dihydroxyacetone phosphate; E4P: Erythrose-4-phosphate; F6P: Fructose-6-phosphate; FDP: Fructose 1,6-diphosphate; FOR: Formate; FUM: Fumarate; G6P: Glucose-6-phosphate; GAP: Glyceraldehyde 3-phosphate; GLX: Glyoxylate; GLY: Glycine; GLYR: Glycerate; GOL: Glycerol; GP: 3-phosphoglycerate; ICIT: Isocitrate; MAL: Malate; NMP: Nucleotide monophosphate; OAA: Oxaloacetate; PEP: Phospho*enol*pyruvate; PGOL: Phosphoglycolate; PYR: Pyruvate; R5P: Ribose-5-phosphate; Ru5P: Ribulose-5-phosphate; RuBP: Ribulose 1,5-bisphosphate; S17P: Sedoheptulose 1,7-bisphosphate; S7P: Sedoheptulose 7-phosphate; SUCC: Succinate; SUCSAL: Succinic semialdehyde; X5P: Xylulose 5-phosphate; βG6P: β-glucose-6-phosphate

Acknowledgements
We would like to thank Dr. Bri-Mathias Hodge for guidance with computational aspects of the project and Ben Miller for initial experimentation while protocols were first being applied as well as the Bigelow Laboratory for Ocean Sciences, East Boothbay, ME, for helping with *T. erythraeum* culturing.

Funding
We are grateful to the Coors Foundation and Colorado School of Mines for funding this research. Neither funding body had a role in the design of the study or the collection, analysis or interpretation of results.

Authors' contributions
NRB conceived the study. JJG reconstructed the metabolic network from genome sequences and wrote/modified code for modeling. JJG also performed all the experiments required for measuring biomass composition. NRB and JJG both participated in troubleshooting, analyzing data, writing and revising the manuscript. All authors read and approved the final manuscript.

Competing interests
The authors declare that they have no competing interests.

References

1. Berman-Frank I, Lundgren P, Falkowski P. Nitrogen fixation and photosynthetic oxygen evolution in cyanobacteria. Res Microbiol. 2003;154: 157–64.
2. Glibert PM, Bronk DA. Release of dissolved organic nitrogen by marine diazotrophic cyanobacteria, *Trichodesmium* spp. Appl Environ Microbiol. 1994;60:3996–4000.
3. Spatharis S, Skliris N, Meziti A, Kormas KA, Weyhenmeyer G. First record of a *Trichodesmium erythraeum* bloom in the Mediterranean Sea. Can J Fish Aquat Sci. 2012;69:1444–55.
4. Letelier RM, Karl DM. Role of *Trichodesmium* spp. in the productivity of the subtropical North Pacific Ocean. Mar Ecol Prog Ser. 1996;133:263–73.
5. Bowman TE, Lancaster L. A bloom of the planktonic blue-green alga, *Trichodesmium erythraeum*, in the Tonga Islands. Limnol Oceanogr. 1965;10: 291–3.
6. Montoya JP, Holl CM, Zehr JP, Hansen A, Villareal TA, Capone DG. High rates of N_2 fixation by unicellular diazotrophs in the oligotrophic Pacific Ocean. Nature. 2004;430:1027–32.
7. Bell PR, Uwins PJ, Elmetri I, Phillips JA, Fu F-X, Yago AJ. Laboratory culture studies of *Trichodesmium* isolated from the great Barrier Reef Lagoon, Australia. Hydrobiologia. 2005;532:9–21.
8. van Baalen C, Brown Jr RM. The ultrastructure of the marine blue green alga, *Trichodesmium erythraeum*, with special reference to the cell wall, gas vacuoles, and cylindrical bodies. Arch Mikrobiol. 1969;69:79–91.
9. Ohki K, Taniuchi Y. Detection of nitrogenase in individual cells of a natural population of *Trichodesmium* using immunocytochemical methods for fluorescent cells. J Oceanogr. 2009;65:427–32.
10. Berman-Frank I, Lundgren P, Chen Y-B, Küpper H, Kolber Z, Bergman B, Falkowski P. Segregation of Nitrogen Fixation and Oxygenic Photosynthesis in the Marine Cyanobacterium *Trichodesmium*. Science. 2001;294:1534–7.
11. Capone DG, Zehr JP, Paerl HW, Bergman B, Carpenter EJ. *Trichodesmium*, a Globally Significant Marine Cyanobacterium. Science. 1997;276:1221–9.
12. Paerl HW. Spatial Segregation of CO_2 Fixation in *Trichodesmium* spp.: Linkage to N_2 fixation potential. J Phycol. 1994;30:790–9.
13. Lin S, Henze S, Lundgren P, Bergman B, Carpenter EJ. Whole-cell immunolocalization of nitrogenase in marine diazotrophic cyanobacteria, *Trichodesmium* spp. Appl Environ Microbiol. 1998;64:3052–8.
14. Janson S, Carpenter EJ, Bergman B. Compartmentalisation of nitrogenase in a non-heterocystous cyanobacterium: *Trichodesmium contortum*. FEMS Microbiol Lett. 1994;118:9–14.
15. Fredriksson C, Bergman B. Nitrogenase quantity varies diurnally in a subset of cells within colonies of the non-heterocystous cyanobacteria *Trichodesmium* spp. Microbiology. 1995;141:2471–8.
16. Walworth N, Pfreundt U, Nelson WC, Mincer T, Heidelberg JF, Fu F, Waterbury JB, del Rio TG, Goodwin L, Kyrpides NC. *Trichodesmium* genome maintains abundant, widespread noncoding DNA *in situ*, despite oligotrophic lifestyle. Proc Natl Acad Sci. 2015;112:4251–6.
17. Berman-Frank I, Bidle KD, Haramaty L, Falkowski PG. The demise of the marine cyanobacterium, *Trichodesmium* spp., via an autocatalyzed cell death pathway. Limnology Oceanography. 2004;49:997–1005.
18. Sudek S, Haygood MG, Youssef DT, Schmidt EW. Structure of trichamide, a cyclic peptide from the bloom-forming cyanobacterium *Trichodesmium erythraeum*, predicted from the genome sequence. Appl Environ Microbiol. 2006;72:4382–7.
19. Pfreundt U, Kopf M, Belkin N, Berman-Frank I, Hess WR: The primary transcriptome of the marine diazotroph *Trichodesmium erythraeum* IMS101. *Scientific reports* 2014, 4.
20. El-Shehawy R, Lugomela C, Ernst A, Bergman B. Diurnal expression of *hetR* and diazocyte development in the filamentous non-heterocystous cyanobacterium *Trichodesmium erythraeum*. Microbiology. 2003;149:1139–46.
21. Sandh G, Ran L, Xu L, Sundqvist G, Bulone V, Bergman B. Comparative proteomic profiles of the marine cyanobacterium *Trichodesmium erythraeum* IMS101 under different nitrogen regimes. Proteomics. 2011;11:406–19.
22. Mulholland MR, Capone DG. The nitrogen physiology of the marine N_2-fixing cyanobacteria *Trichodesmium* spp. Trends Plant Sci. 2000;5:148–53.
23. Sandh G, El-Shehawy R, Díez B, Bergman B. Temporal separation of cell division and diazotrophy in the marine diazotrophic cyanobacterium *Trichodesmium erythraeum* IMS101. FEMS Microbiol Lett. 2009;295:281–8.
24. Garcia NS, Fu F, Sedwick PN, Hutchins DA. Iron deficiency increases growth and nitrogen-fixation rates of phosphorus-deficient marine cyanobacteria. ISME J. 2015;9:238–45.
25. Dyhrman S, Chappell P, Haley S, Moffett J, Orchard E, Waterbury J, Webb E. Phosphonate utilization by the globally important marine diazotroph *Trichodesmium*. Nature. 2006;439:68–71.
26. Ho T-Y. Nickel limitation of nitrogen fixation in *Trichodesmium*. Limnol Oceanogr. 2013;58:112–20.
27. Sohm JA, Mahaffey C, Capone DG. Assessment of relative phosphorus limitation of *Trichodesmium* spp. in the North Pacific, North Atlantic, and the north coast of Australia. Limnol Oceanogr. 2008;53:2495.
28. Rodriguez IB, Ho T-Y. Diel nitrogen fixation pattern of *Trichodesmium*: the interactive control of light and Ni. Sci Rep. 2014;4.
29. Hutchins D, Fu F-X, Zhang Y, Warner M, Feng Y, Portune K, Bernhardt P, Mulholland M. CO_2 control of *Trichodesmium* N_2 fixation, photosynthesis, growth rates, and elemental ratios: Implications for past, present, and future ocean biogeochemistry. Limnol Oceanogr. 2007;52:1293–304.
30. Mulholland MR, Capone DG. Stoichiometry of nitrogen and carbon utilization in cultured populations of *Trichodesmium* IMS101: Implications for growth. Limnol Oceanogr. 2001;46:436–43.
31. Mahadevan R, Schilling C. The effects of alternate optimal solutions in constraint-based genome-scale metabolic models. Metab Eng. 2003;5:264–76.
32. Mahadevan R, Edwards JS, Doyle FJ. Dynamic flux balance analysis of diauxic growth in *Escherichia coli*. Biophys J. 2002;83:1331–40.
33. Overbeek R, Olson R, Pusch GD, Olsen GJ, Davis JJ, Disz T, Edwards RA, Gerdes S, Parrello B, Shukla M. The SEED and the rapid annotation of microbial genomes using subsystems technology (RAST). Nucleic Acids Res. 2014;42:D206–14.
34. Caspi R, Foerster H, Fulcher CA, Kaipa P, Krummenacker M, Latendresse M, Paley S, Rhee SY, Shearer AG, Tissier C. The MetaCyc database of metabolic pathways and enzymes and the BioCyc collection of pathway/genome databases. Nucleic Acids Res. 2008;36:D623–31.
35. Kanehisa M, Sato Y, Kawashima M, Furumichi M, Tanabe M. KEGG as a reference resource for gene and protein annotation. Nucleic Acids Res. 2016;44:D457–62.
36. Vu TT, Stolyar SM, Pinchuk GE, Hill EA, Kucek LA, Brown RN, Lipton MS, Osterman A. Fredrickson JK, Konopka AE: Genome-scale modeling of light-driven reductant partitioning and carbon fluxes in diazotrophic unicellular cyanobacterium *Cyanothece* sp. ATCC 51142. PLoS Comput Biol. 2012;8: e1002460.
37. Knoop H, Gründel M, Zilliges Y, Lehmann R, Hoffmann S, Lockau W, Steuer R. Flux Balance Analysis of Cyanobacterial Metabolism: The Metabolic Network of *Synechocystis* sp. PCC 6803. PLoS Comput Biol. 2013;9:e1003081.
38. Vu TT, Hill EA, Kucek LA, Konopka AE, Beliaev AS, Reed JL. Computational evaluation of *Synechococcus* sp. PCC 7002 metabolism for chemical production. Biotechnol J. 2013;8:619–30.
39. de Oliveira Dal'Molin CG, Quek L-E, Palfreyman RW, Brumbley SM, Nielsen LK. AraGEM, a genome-scale reconstruction of the primary metabolic network in *Arabidopsis*. Plant Physiol. 2010;152:579–89.
40. Levering J, Broddrick J, Dupont CL, Peers G, Beeri K, Mayers J, Gallina AA, Allen AE, Palsson BO, Zengler K. Genome-scale model reveals metabolic basis of biomass partitioning in a model diatom. PLoS ONE. 2016;11:e0155038.
41. Boyle N, Morgan J. Flux balance analysis of primary metabolism in *Chlamydomonas reinhardtii*. BMC Syst Biol. 2009;3:4.
42. Chang RL, Ghamsari L, Manichaikul A, Hom EF, Balaji S, Fu W, Shen Y, Hao T, Palsson BØ, Salehi-Ashtiani K. Metabolic network reconstruction of *Chlamydomonas* offers insight into light-driven algal metabolism. Mol Syst Biol. 2011;7:518.
43. Schellenberger J, Que R, Fleming RM, Thiele I, Orth JD, Feist AM, Zielinski DC, Bordbar A, Lewis NE, Rahmanian S. Quantitative prediction of cellular metabolism with constraint-based models: the COBRA Toolbox v2. 0. Nat Protoc. 2011;6:1290–307.
44. Chen Y-B, Zehr JP, Mellon M. Growth and Nitrogen Fixation of the Diazotrophic Filamentous Nonheterocytous Cyanobacterium *Trichodesmium* sp. IMS 101 in Defined Media: Evidence for a Circadian Rhythm. J Phycology. 1996;32:916–23.
45. Yamaberi K, Takagi M, Yoshida T. Nitrogen depletion for intracellular triglyceride accumulation to enhance liqefaction yield of marine microalgal cells into a fuel oil. J Mar Biotechnol. 1997;6:44–8.
46. Li Y, Horsman M, Wang B, Wu N, Lan CQ. Effects of nitrogen sources on cell growth and lipid accumulation of green alga *Neochloris oleoabindans*. Biotechnol Products Process Eng. 2008;81:629–36.
47. Li Y, Han D, Hu G, Dauvillee D, Sommerfeld M, Ball S, Hu Q. Chlamydomonas starchless mutant defective in ADP-glucose pyrophosphorylase hyper-accumulates triacylglycerol. Metab Eng. 2010;12:387–91.

48. Boyle NR, Page MD, Liu B, Blaby IK, Casero D, Kropat J, Cokus SJ, Hong-Hermesdorf A, Shaw J, Karpowicz SJ. Three acyltransferases and nitrogen-responsive regulator are implicated in nitrogen starvation-induced triacylglycerol accumulation in Chlamydomonas. J Biol Chem. 2012;287: 15811–25.
49. Blaby IK, Glaesener AG, Mettler T, Fitz-Gibbon ST, Gallaher SD, Liu B, Boyle NR, Kropat J, Stitt M, Johnson S. Systems-level analysis of nitrogen starvation–induced modifications of carbon metabolism in a Chlamydomonas reinhardtii starchless mutant. Plant Cell. 2013;25:4305–23.
50. Miller R, Wu G, Deshpande RR, Vieler A, Gärtner K, Li X, Moellering ER, Zäuner S, Cornish AJ, Liu B, et al. Changes in Transcript Abundance in Chlamydomonas reinhardtii following Nitrogen Deprivation Predict Diversion of Metabolism. Plant Physiol. 2010;154:1737–52.
51. Levitan O, Rosenberg G, Setlik I, Setlikova E, Grigel J, Klepetar J, Prasil O, Berman-Frank I. Elevated CO_2 enhances nitrogen fixation and growth in the marine cyanobacterium *Trichodesmium*. Glob Chang Biol. 2007;13:531–8.
52. Förster J, Famili I, Fu P, Palsson BØ, Nielsen J. Genome-scale reconstruction of the *Saccharomyces cerevisiae* metabolic network. Genome Res. 2003;13: 244–53.
53. Kornberg H. The role and control of the glyoxylate cycle in *Escherichia coli*. Biochem J. 1966;99:1.
54. Flavell RB, Woodward DO. Metabolic role, regulation of synthesis, cellular localization, and genetic control of the glyoxylate cycle enzymes in *Neurospora crassa*. J Bacteriol. 1971;105:200–10.
55. Schwender J, Seemann M, Lichtenthaler HK, Rohmer M. Biosynthesis of isoprenoids (carotenoids, sterols, prenyl side-chains of chlorophylls and plastoquinone) via a novel pyruvate/glyceraldehyde 3-phosphate non-mevalonate pathway in the green alga *Scenedesmus obliquus*. Biochem J. 1996;316:73–80.
56. Eastmond PJ, Germain V, Lange PR, Bryce JH, Smith SM, Graham IA. Postgerminative growth and lipid catabolism in oilseeds lacking the glyoxylate cycle. Proc Natl Acad Sci. 2000;97:5669–74.
57. Rubin BE, Wetmore KM, Price MN, Diamond S, Shultzaberger RK, Lowe LC, Curtin G, Arkin AP, Deutschbauer A, Golden SS. The essential gene set of a photosynthetic organism. Proc Natl Acad Sci. 2015;112:E6634–43.
58. Boyle NR, Morgan JA. Flux balance analysis of primary metabolism in Chlamydomonas reinhardtii. BMC Syst Biol. 2009;3:1.
59. Montagud A, Navarro E, de Córdoba PF, Urchueguía JF, Patil KR. Reconstruction and analysis of genome-scale metabolic model of a photosynthetic bacterium. BMC Syst Biol. 2010;4:1.
60. Zomorrodi AR, Maranas CD. OptCom: a multi-level optimization framework for the metabolic modeling and analysis of microbial communities. PLoS Comput Biol. 2012;8:e1002363.
61. Zhuang K, Ma E, Lovley DR, Mahadevan R. The design of long-term effective uranium bioremediation strategy using a community metabolic model. Biotechnol Bioeng. 2012;109:2475–83.
62. Mulholland MR, Bernhardt PW, Heil CA, Bronk DA, O'Neil JM. Nitrogen fixation and release of fixed nitrogen by *Trichodesmium* spp. in the Gulf of Mexico. Limnol Oceanogr. 2006;51:1762–76.
63. Harris EH, Stern DB, Witman G. The Chlamydomonas sourcebook. Amsterdam: Elsevier; 2009.
64. Antoniewicz MR, Kelleher JK, Stephanopoulos G. Accurate assessment of amino acid mass isotopomer distributions for metabolic flux analysis. Anal Chem. 2007;79:7554–9.
65. Yemm E, Willis A. The estimation of carbohydrates in plant extracts by anthrone. Biochem J. 1954;57:508.
66. Simon RD. Cyanophycin granules from the blue-green alga *Anabaena cylindrica*: a reserve material consisting of copolymers of aspartic acid and arginine. Proc Natl Acad Sci. 1971;68:265–7.
67. Messineo L. Modification of the Sakaguchi reaction: spectrophotometric determination of arginine in proteins without previous hydrolysis. Arch Biochem Biophys. 1966;117:534–40.
68. Dumay J, Morançais M, Nguyen HPT, Fleurence J. Extraction and purification of R-phycoerythrin from marine red algae. Nat Prod Mar Algae Methods Protoc. 2015;1308:109–17.
69. Bligh EG, Dyer WJ. A rapid method of total lipid extraction and purification. Can J Biochem Physiol. 1959;37:911–7.
70. Harwood JL. Membrane lipids in algae. In Lipids in photosynthesis: structure, function and genetics. Cardiff: Springer; 1998:53–64.
71. Work VH, Radakovits R, Jinkerson RE, Meuser JE, Elliott LG, Vinyard DJ, Laurens LM, Dismukes GC, Posewitz MC. Increased lipid accumulation in the *Chlamydomonas reinhardtii* sta7-10 starchless isoamylase mutant and increased carbohydrate synthesis in complemented strains. Eukaryot Cell. 2010;9:1251–61.
72. Boussiba S, Richmond AE. C-phycocyanin as a storage protein in the blue-green alga *Spirulina platensis*. Arch Microbiol. 1980;125:143–7.
73. Subramaniam A, Carpenter EJ, Karentz D, Falkowski PG. Bio-optical properties of the marine diazotrophic cyanobacteria *Trichodesmium* spp. I. Absorption and photosynthetic action spectra. Limnol Oceanogr. 1999;44:608–17.
74. Barbas CF, Burton DR, Scott JK, Silverman GJ. Quantitation of DNA and RNA. Cold Spring Harb Protoc. 2007;2007:pdb. ip47.
75. Nordberg H, Cantor M, Dusheyko S, Hua S, Poliakov A, Shabalov I, Smirnova T, Grigoriev IV, Dubchak I. The genome portal of the Department of Energy Joint Genome Institute: 2014 updates. Nucleic Acids Res. 2014;42:D26–31.
76. Zhang S, Bryant DA. Biochemical validation of the glyoxylate cycle in the cyanobacterium *Chlorogloeopsis fritschii* strain PCC 9212. J Biol Chem. 2015; 290:14019–30.
77. Zhang S, Bryant DA. The tricarboxylic acid cycle in cyanobacteria. Science. 2011;334:1551–3.
78. Scheer M, Grote A, Chang A, Schomburg I, Munaretto C, Rother M, Söhngen C, Stelzer M, Thiele J, Schomburg D. BRENDA, the enzyme information system in 2011. Nucleic Acids Res. 2010:doi: doi: 10.1093/nar/gkq1089.
79. Fujisawa T, Okamoto S, Katayama T, Nakao M, Yoshimura H, Kajiya-Kanegae H, Yamamoto S, Yano C, Yanaka Y, Maita H. CyanoBase and RhizoBase: databases of manually curated annotations for cyanobacterial and rhizobial genomes. Nucleic Acids Res. 2014;42:D666–70.
80. Triana J, Montagud A, Siurana M, Fuente D, Urchueguía A, Gamermann D, Torres J, Tena J, de Córdoba PF, Urchueguía JF. Generation and evaluation of a genome-scale metabolic network model of *Synechococcus elongatus* PCC7942. Metabolites. 2014;4:680–98.
81. Altschul SF, Gish W, Miller W, Myers EW, Lipman DJ. Basic local alignment search tool. J Mol Biol. 1990;215:403–10.
82. Kim S, Thiessen PA, Bolton EE, Chen J, Fu G, Gindulyte A, Han L, He J, He S, Shoemaker BA. PubChem substance and compound databases. *Nucleic acids research* 2015:doi: 10.1093/nar/gkv951
83. Thiele I, Palsson BØ. A protocol for generating a high-quality genome-scale metabolic reconstruction. Nat Protoc. 2010;5:93–121.
84. Orth JD, Thiele I, Palsson BO. What is flux balance analysis? Nat Biotech. 2010;28:245–8.
85. Küpper H, Ferimazova N, Šetlík I, Berman-Frank I. Traffic lights in *Trichodesmium*. Regulation of photosynthesis for nitrogen fixation studied by chlorophyll fluorescence kinetic microscopy. Plant Physiol. 2004;135: 2120–33.

15

Predicting network modules of cell cycle regulators using relative protein abundance statistics

Cihan Oguz[1*], Layne T. Watson[2,3,4], William T. Baumann[5] and John J. Tyson[1]

Abstract

Background: Parameter estimation in systems biology is typically done by enforcing experimental observations through an objective function as the parameter space of a model is explored by numerical simulations. Past studies have shown that one usually finds a set of "feasible" parameter vectors that fit the available experimental data equally well, and that these alternative vectors can make different predictions under novel experimental conditions. In this study, we characterize the feasible region of a complex model of the budding yeast cell cycle under a large set of discrete experimental constraints in order to test whether the statistical features of relative protein abundance predictions are influenced by the topology of the cell cycle regulatory network.

Results: Using differential evolution, we generate an ensemble of feasible parameter vectors that reproduce the phenotypes (viable or inviable) of wild-type yeast cells and 110 mutant strains. We use this ensemble to predict the phenotypes of 129 mutant strains for which experimental data is not available. We identify 86 novel mutants that are predicted to be viable and then rank the cell cycle proteins in terms of their contributions to cumulative variability of relative protein abundance predictions. Proteins involved in "regulation of cell size" and "regulation of G1/S transition" contribute most to predictive variability, whereas proteins involved in "positive regulation of transcription involved in exit from mitosis," "mitotic spindle assembly checkpoint" and "negative regulation of cyclin-dependent protein kinase by cyclin degradation" contribute the least. These results suggest that the statistics of these predictions may be generating patterns specific to individual network modules (START, S/G2/M, and EXIT). To test this hypothesis, we develop random forest models for predicting the network modules of cell cycle regulators using relative abundance statistics as model inputs. Predictive performance is assessed by the areas under receiver operating characteristics curves (AUC). Our models generate an AUC range of 0.83-0.87 as opposed to randomized models with AUC values around 0.50.

Conclusions: By using differential evolution and random forest modeling, we show that the model prediction statistics generate distinct network module-specific patterns within the cell cycle network.

Keywords: Parameter optimization, Differential evolution, Ensemble modeling, Machine learning, Random forests, Budding yeast, Cell cycle, Systems biology

Background

In systems biology research, mathematical models of sufficient predictive power allow researchers to interrogate biological systems under a wide variety of experimental conditions that may be difficult to achieve in the laboratory. Such in-silico experiments may lead to discoveries that affect life in important ways, for example, in understanding the molecular basis of certain diseases and in designing drugs for their treatment [1, 2]. What makes a model reliably predictive? Before using a model for predictive purposes, it is essential to show that the model is capable of reproducing major known experimental trends. In other words, incorporation of experimental data into a model by parameter optimization is a critical first step. Due to limitations in direct experimental measurements of kinetic parameters, a common approach is to estimate

*Correspondence: cihanoguzvt@gmail.com
[1]Department of Biological Sciences, Virginia Tech, Blacksburg VA, 24061 USA
Full list of author information is available at the end of the article

all unknown model parameters by minimizing the difference between model simulations and experimental data [3]. This approach often generates a set of parameter vectors with equivalent (or comparable) performance. Such parametric uncertainty can be used to advantage by extracting information about critical and dispensable parts of a model using global sensitivity analysis or identifying the most informative future experiments. This information can be used to constrain the model's parameters [4] or to refine the model's structure [5].

Creating an ensemble of parameter vectors with similar (or identical) performance (with respect to a known set of experimental observations) is especially useful when one would like to predict the potential outcome(s) of novel experimental designs. We refer the reader to [6] for a comprehensive survey of experimental design studies (with an emphasis on objective function formulations) from several fields including systems biology. More recent work in the area of experimental design within systems biology includes a study that compares the performances of several alternative methods with and without a predetermined network topology [7] and a novel framework for model selection implemented for both stochastic and deterministic models [8].

In the literature, "ensemble modeling" is a common term used to describe studies of multiple models [5, 9, 10] or a single model with multiple parameter vectors [11]. Here, we focus on the latter case with a complex model of the budding yeast cell cycle (more than 100 model parameters). Of special interest to us is parameter space exploration with a discontinuous objective function that is the sum of many discrete constraints. Recent work in ensemble modeling includes using simulated annealing with a multi objective function to extract robust and fragile model features [12], implementation of Metropolis Monte Carlo and multi ellipsoidal sampling [11], exploration of parameter space by adaptive sparse grids with control objectives [13], and identifying model fragilities with random walks [14]. More recently, ensembles of parameter vectors were generated to understand parameter adaptations underlying phenotypic transitions [15] with an application in pharmacological intervention [16]. In [17], Rumschinski introduced a set-based framework for detecting incorrect model hypotheses and refining parameter estimates with the help of infeasibility certificates and a bisection algorithm that identifies parts of parameter spaces consistent with incomplete and noisy experimental data. This approach was illustrated using two simple models with four species and 3–5 parameters. More recently, Rodriguez-Fernandez et al. implemented a mixed-integer nonlinear programming (MINLP) formulation to simultaneously perform model selection and parameter estimation using in silico generated data of homeostasis in *E. coli* [18]. For this biological system, the authors identified the best model among 1700 nested models in a computationally efficient manner rather than fully analysing each candidate model separately. Starting with 21 model parameters, the resulting solution showed that parameters were precisely estimated, while identifiability issues and scalability to models of larger complexity were mentioned as limitations of this model identification approach [18].

A common element in these ensemble modeling studies is the use of time-series data for optimizing parameters and for exploring the parameter space for alternative "feasible" vectors that provide acceptable fits to the data. Here, we use an ensemble modeling methodology for complex models when the constraining data are not quantitative time-series of model variables (which are often unavailable in experimental studies of cell physiology) but discrete qualitative observations (in our case, the observed phenotypes of many different yeast strains carrying mutations of cell cycle genes). In addition, the model we consider is much more complex, with many more adjustable parameters and much more experimental data, than the models studied in the work cited above.

Ensemble modeling with qualitative constraints has recently been explored by Pargett et al. [19], who combined "optimal scaling" and gradient-based multi-objective optimization for incorporating a heterogeneous set of experimental constraints into ODE models of stem cell regulation in *Drosophila*. Starting from a core model with 10 states and 18 unknown parameters, the authors generated several additional models by considering alternative connections between components of the regulatory network. Following the parameter optimization step, experimental design was implemented (based on ranking the predictive variances of measurements) in order to decrease the uncertainty of model parameter values and model structure. Each candidate model was represented by ensembles of optimal parameter vectors and Pareto optimality was used for comparing model performance and for identifying informative experiments.

In [20], temporal logics (typically used with discrete models) was implemented to express the dynamical features of a continuous (ODE-based) model of an enzymatic reaction network involved in cancer. Furthermore, global robustness and sensitivity analysis was used for identifying the boundaries between distinct regions of the model's parameter space (producing different states such as stable steady states and oscillations) and for generating several novel biological insights regarding system's dynamics [20].

For a recent review on the use of qualitative data for estimating the parameters of continuous models, we refer the reader to [21]. This review covers the application of alternative data normalization techniques depending on the nature of the experimental data at hand (qualitative vs. quantitative), formulation of multi-objective optimization using heterogeneous experimental data sets, and Pareto optimality based analysis of tradeoffs between such multiple objectives.

The proposed approach in this paper extends our recent work on parameter optimization of a complex model of the budding yeast cell cycle [22]. Starting from an ensemble of optimally performing parameter vectors, we propose several ways to explore the parameter space for more such vectors. In this search, our aim is to find parameter vectors with diverse predictions (i.e., an extended range of predictions for the phenotypes of novel genetic strains). We demonstrate that differential evolution (DE) [23], which is a metaheuristic method, can effectively find feasible parameter vectors with extended predictive ranges provided an additional feasibility criterion (in addition to the criterion of optimal model performance) is enforced so that the search does not get stuck in a small region of parameter space. We show how DE can be forced to widen the range of predictions during the search for optimal parameter vectors.

The application of DE in similar contexts include [24] in which DE is hybridized with Kalman Filter for improving the parameter estimation accuracy compared to pure DE and genetic algorithm (GA) based approaches. In [24], simple models of glycolysis and the cell cycle, with artificially generated noisy time series data, are used to demonstrate the improved performance of the hybrid approach. More recently, the 18 parameters of an ODE-based dynamic model of endocytosis are optimized with several metaheuristic methods including DE under different observability settings (complete vs. incomplete observability of system variables), multiple levels of measurement noise, and with real and artificially generated time series data [25]. In this study, DE turned out to be the best performer in terms of estimation accuracy and convergence speed while practical parameter identifiability problems suggested the need for additional experimental data to further constrain the model's parameters. Recent studies on the use of metaheuristic methods in a wide range of science and enginnering applications are surveyed in [26] with more than 200 references (including the applications of several DE variants). An earlier review paper focuses on the application of metaheuristic methods to systems biology problems [27] including experimental design [28–30] and parameter identifiability [31–33].

Our modified-DE approach generates an ensemble of feasible parameter vectors (i.e., vectors that satisfy a maximum number of discrete experimental constraints) with a broad "range of predictions" (i.e., vectors that extend the number of different phenotypic patterns predicted for a predefined set of mutant yeast strains). We then use this ensemble to test whether relative protein abundance predictions are influenced by the topology of the cell cycle regulatory network by ranking cell cycle regulators in our model with respect to their cumulative variability scores. The results suggest that the statistics of these predictions may be generating patterns specific to individual network modules. To test this hypothesis, we develop random forest models for predicting the network modules of cell cycle regulators using relative protein abundance statistics as model inputs. Our overall approach that ties the statistical features of model predictions to the modules of the cell cycle network, starting from optimizing the settings of DE for exploring the feasible region of the model in the parameter space is summarized in Fig. 1.

Methods

Problem formulation

The cell cycle is the ordered sequence of events that govern cell growth, replication of the cell's genome, and division into two daughter cells that are capable of repeating this cycle in successive generations [34, 35]. The four phases of the cell cycle are DNA synthesis (S phase) and mitosis (M phase) separated by two gaps (G1 and G2). G1, S, G2 and M phases progress sequentially in a repeated manner, which is crucial to maintaining a constant number of chromosomes per cell after each cycle of DNA replication and cell division. Furthermore, the duration of a single cell cycle (i.e., from birth to division) has to be balanced (on average) with the time needed for doubling the amounts of all other cellular components. If this condition is not met (i.e., the mass doubling time is substantially different from the cell cycle time), then average cell size becomes progressively smaller or larger leading to cell death. In addition, a number of "checkpoints" prevent G1-S-G2-M progression in cases such as DNA damage or improper alignment of replicated chromosomes on the mitotic spindle. All of these features of cell cycle progression are controlled by the periodic activation of cyclin-dependent kinases (CDKs) [34]. Since the fundamental molecular mechanisms governing the activation of CDKs are similar among all eukaryotes, an improved understanding of cell cycle controls has potential benefits far beyond the intrinsic challenge of unraveling this complex molecular control system.

To this end, we have proposed a variety of deterministic, stochastic and hybrid models of the CDK control mechanism in budding yeast cells and other eukaryotes [36–41]. Using the model in [40, 41], comprised of 26 ODEs and 126 kinetic parameters, we previously proposed a method for optimizing the parameter values under 119 qualitative experimental constraints [22]. (The parameters and variables of this model are listed in Additional file 1: Tables S1 and S2, respectively.) This model includes three classes of variables (or regulatory proteins).

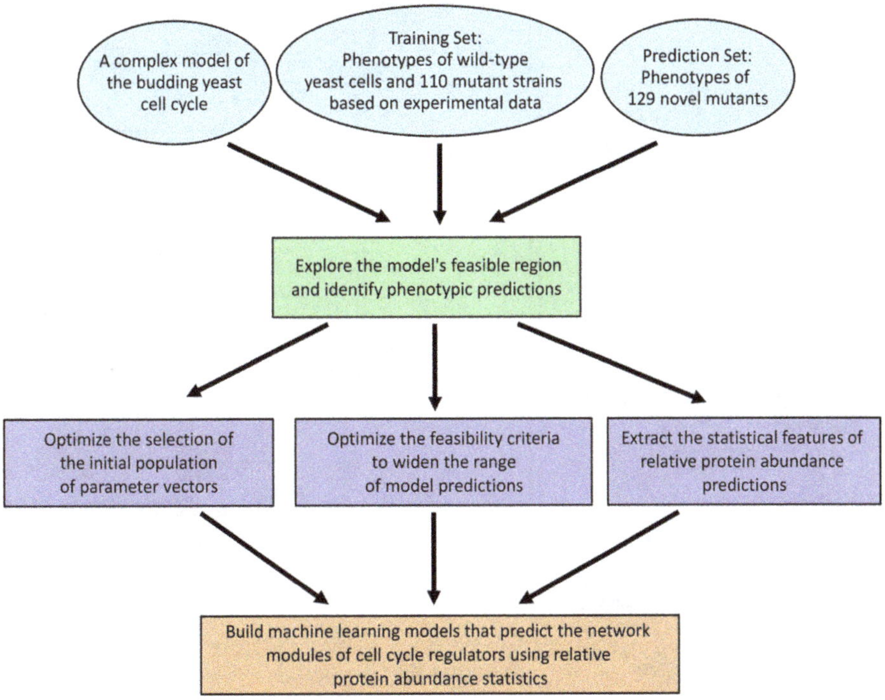

Fig. 1 Schematic of the implemented modeling strategy for predicting the network modules of cell cycle regulators using relative protein abundance statistics

Class-1 variables are modeled by mass action kinetics of transcription factor activity and proteolytic degradation, whereas Class-2 variables (fractions of proteins in their active forms) are modeled by sigmoidal functions representing the phosphorylation and dephosphorylation reactions. On the other hand, Class-3 variables (or protein complexes) are modeled by maximum or minimum functions based on the quasi steady state assumption due to the fast time scales associated with these complex formation processes. The regulatory network represented by this model is composed of three distinct modules of proteins (START, S/G2/M, and EXIT) as shown in Fig. 2. The cell cycle events that take place in each module are summarized below.

- START module: START (or G1/S transition) is an event in G1 phase when a new round of DNA synthesis and mitosis are committed by a cell. The most critical step of the START transition is the translocation of Whi5, a stoichiometric inhibitor of SBF and MBF (transcription factors of Cln2 and Clb5 synthesis modeled as a single variable named SBF), from nucleus to cytoplasm. In early G1, SBF is not active since it is inhibited by Whi5. As the cells grow, Cln3 and Bck2 concentrations rise enough to phosphorylate Whi5 (inhibitor of SBF), and as a result SBF becomes active, promoting Cln2 and Cln5 synthesis. Increasing concentrations of Cln2, Cln3, and Clb5 support progression of bud emergence.
- S/G2/M module: Increasing Cln2 concentration following the START transition leads to phosphorylation and degradation of CKI. As a result of this, Clb5 is released. The active form of Clb5 promotes DNA synthesis, further inhibiting CKI through phosphorylation. Cln2 and Clb5 inhibit Cdh1 (responsible from Clb2 degradation) and Clb2 concentration increases resulting in the activation of Mcm1 (transcription factor of Clb2), and further Clb2 accumulation. By phosphorylating and inactivating SBF, Clb2 also halts the synthesis of Cln2 and Clb5 and the cells get ready for mitotic exit. Activation of APC by Clb2 and the cooperation of APC with Cdc20 are some of the key steps required for metaphase-anaphase transition and mitotic cyclin degradation. For Clb2 and Clb5 to be degraded, APC has to be phosphorylated and spindle assembly checkpoint needs to be released. Both of these processes are driven by Clb2.
- EXIT module: Activation of Cdc14 is the most critical event in the EXIT module since it is essential for exit from mitosis and return to G1 state. Cdc14 dephosphorylates several proteins previously phosphorylated by CDKs in S/G2/M, thereby leading to the activation of Cdh1 and CKI, as well as the

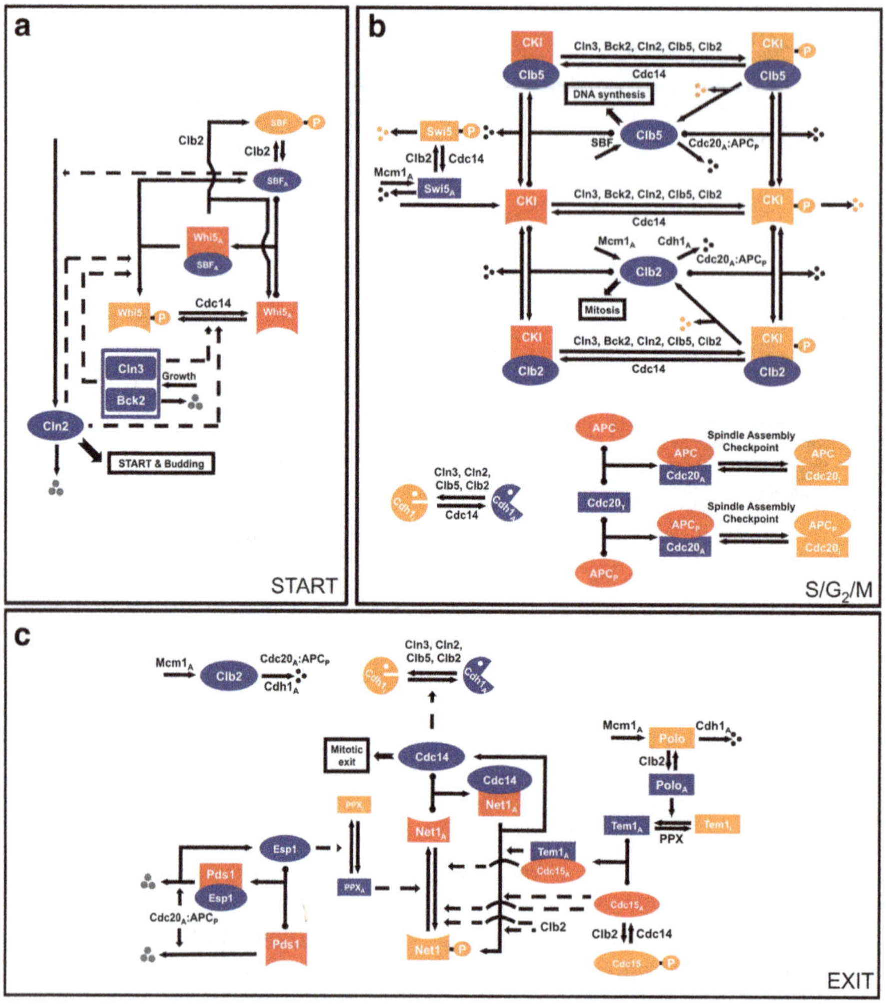

Fig. 2 Wiring diagram of the budding yeast cell cycle network (from [40]). The network consists of three modules, namely START (in **a**), S/G2/M (in **b**) and EXIT (in **c**). *Red* and *blue* icons represent components that are in their active forms and orange icons represent components that are inactive. *Solid lines* represent chemical reactions (synthesis and degradation, phosphorylation and dephosphorylation, association and dissociation), whereas *dashed lines* represent activating or inhibitory influences of components on the chemical reactions. For simplicity, some interactions are not shown in the figures

repression of Clb2 and Clb5. Two pathways, namely FEAR (Cdc fourteen early anaphase release) and MEN (mitotic exit network), are involved in the activation of Cdc14. The release of Esp1 from Pds1 (through Cdc20 activity) in the FEAR pathway leads to chromatid separation and phosphorylation of Net1. As a result, Cdc14 is released from Net1:Cdc14 complex and free Cdc14 drives exit from mitosis. In order for budding yeast cells to return to G1 state by the robust phosphorylation of Net1, the FEAR pathway is supported by the MEN pathway through the activation of Cdc15 and Tem1 that form a complex (MEN). This results in the full release of Cdc14, activation of Cdh1, complete degradation of Clb2, as well as the stabilization of CKI and a fully restored G1 phase.

In [22], starting from an initial parameter vector that captured 72 of the 119 experimental phenotypes in the Training Set, we improved the number of captured phenotypes to 111. In the process, the optimization algorithm produced more than 3000 parameter vectors that captured the same 111 phenotypes of the Training Set. We call this collection an ensemble of "feasible" parameter vectors. (The ranges of model parameter values in this ensemble are given in Additional file 1: Tables S3 and S4.) In this paper, our goal is to extend the ensemble of feasible parameter vectors identified by [22] to maximize the range of model predictions for a specific group of novel mutant strains (the Prediction Set). These mutants were not included in the Training Set because their phenotypes have not yet been characterized experimentally.

Mutant strains in the Prediction Set originate from the elimination of certain phosphorylation and dephosphorylation reactions that were predicted to be critical [22] only in certain gene deletion backgrounds (not in wild type background) as shown in Table 1. We first set these rates to zero one-by-one to create nine single-mutant strains. The background is wild type (WT) for these strains. In the second step, we create double-mutant strains by setting these nine rates to zero in pairs, which results in 36 additional strains. Finally, we generate triple mutants by following the same strategy (84 more strains) resulting in a total of 129 novel strains in the Prediction Set (Additional file 1: Table S5). The initial conditions (species concentrations) for simulating these strains come from the cell state right after the last division in the WT simulations. For all simulations (listed in Additional file 1: Tables S5 and S6), Euler's method with 0.05 min step size is used to integrate the model equations. The total simulation time per mutant (or WT) is 2000 min.

The range of model predictions

With m as the total number of feasible parameter vectors and n as the vector dimension (total number of parameters in the model), a collection of parameter vectors that capture the 111 phenotypes (Additional file 1: Table S7) out of the 119 total phenotypes in the Training Set defines an $m \times n$ feasible ensemble matrix.

$$\mathbf{X} = \begin{bmatrix} x_1^{(1)} & x_2^{(1)} & \cdots & x_n^{(1)} \\ x_1^{(2)} & x_2^{(2)} & \cdots & x_n^{(2)} \\ \vdots & \vdots & \ddots & \vdots \\ x_1^{(m)} & x_2^{(m)} & \cdots & x_n^{(m)} \end{bmatrix} \quad (1)$$

Here, $x_j^{(i)}$ is the value of the jth parameter in the ith parameter vector of $\mathbf{X}$, which also generates an $m \times l$ prediction matrix.

$$\mathbf{P} = \begin{bmatrix} p_1^{(1)} & p_2^{(1)} & \cdots & p_l^{(1)} \\ p_1^{(2)} & p_2^{(2)} & \cdots & p_l^{(2)} \\ \vdots & \vdots & \ddots & \vdots \\ p_1^{(m)} & p_2^{(m)} & \cdots & p_l^{(m)} \end{bmatrix}, \quad (2)$$

where $p_j^{(i)} \in \{0, 1, 2\}$ characterizes the phenotype for the jth novel genetic strain for the ith parameter vector and l is the total number of novel strains. Phenotype values are set according to the following rules. If, during the simulation of a novel strain, cell size exceeds 25 (arbitrary units) at any time, then the strain's phenotype is inviable ($p_j^{(i)} = 2$). On the other hand, if cell size at last division is within 5% of the cell sizes at the two previous divisions, then the phenotype is viable ($p_j^{(i)} = 1$). Finally, if the model generates cycles of multiple periodicity and cell size at division oscillates between values that differ by more than 5%, then the phenotype is "multiply periodic" ($p_j^{(i)} = 0$). The number $S(\mathbf{P})$ of unique rows in $\mathbf{P}$ is defined as the range of the prediction vectors in $\mathbf{P}$. As we explore different schemes for computing prediction matrices, we compute S values for the ensembles created by these schemes. For each ensemble generation scheme, the sampling efficiency (e_S) is computed as S/n_{tot}, where n_{tot} is the total number of samples taken from the parameter space. This measure allows us to compare different ensemble generation schemes based on the ranges of phenotypic predictions they produce.

Table 1 Phosphorylation and dephosphorylation reactions that induce synthetic lethality upon their elimination

Eliminated reaction (rate constant)	Single mutation strains that are viable before and inviable after setting the rate constant to zero
Whi5 phosphorylation by Bck2 (kp_{i5k2})	*cln3Δ*, Multicopy *BCK2*, *cdh1Δ*, *sic1Δ*, *swi5Δ*, *CLB5-dbΔ*, *net1-ts*, *GAL-CLB2*, *APC-A*
CKI phosphorylation by Cln2 ($e_{ki,n2}$)	*bck2Δ*, *GAL-SIC1*, *net1-ts*, *APC-A*
CKI phosphorylation by Clb2 ($e_{ki,b2}$)	*GAL-CLN3*, *cdh1Δ*, *GAL-CLB5*, *CLB1 clb2Δ*
CKI dephosphorylation by Cdc14 ($kdp_{ki,14}$)	*bck2Δ*, *cdh1Δ*, *GAL-CLB2*, *APC-A*
Whi5 phosphorylation by Cln3 (kp_{i5n3})	*bck2Δ*, *cdh1Δ*, *APC-A*
SBF phosphorylation by Clb2 (kp_{bfb2})	*cdh1Δ*, *CLB5-dbΔ*, *APC-A*
Whi5 phosphorylation by Cln2 (kp_{i5n2})	*bck2Δ*, *APC-A*
Whi5 dephosphorylation by Cdc14 (kdp_{i514})	*APC-A*
Net1 dephosphorylation by PPX ($kdp_{net,px}$)	Multicopy *CDC15*

Upon setting a phosphorylation or dephosphorylation rate constant to zero as specified in the left column, viability is lost in several single mutation strains (specified in the right column). These rate constants are eliminated to create the single, double, and triple mutants (a total of 129 novel mutant strains in the Prediction Set). The phenotypes and the relative abundances of species in these mutant simulations are the predictions of the model

Based on our previous study [22], which demonstrated that DE is an effective tool for exploring the parameter space of our high dimensional model given a discrete multi objective function (i.e, the number of phenotypes in the Training Set captured by the model), we continue using DE, this time for identifying the range of model predictions. While searching for an implementation of DE to meet this objective with efficient sampling, we encounter technical limitations with the standard implementation of DE that is typically used for parameter optimization, and we surmount these limitations by (i) improving the selection of the ensemble that serves as the starting point of DE, and (ii) adding new constraints to DE that force the method to search for feasible parameter vectors expanding the range of model predictions.

Differential Evolution

Let E^D denote real D-dimensional Euclidean space, and let $x = (x_1, \ldots, x_D) \in E^D$ be a vector of parameter values. The vector x includes both the 126 kinetic constants in the model and the 26 ODE initial conditions ($D = 152$). For each vector $x \in E^D$ proposed by the optimization algorithm, we calculate the phenotype $p_j^{(i)} \in \{0, 1, 2\}$ (for the jth strain for the ith parameter vector) for each of the 119 yeast strains in the Training Set. The objective function $O(x)$ is an integer-valued function that counts the number of phenotypes in the Training Set that are correctly captured by the model, given the parameter values in the vector x.

In DE, parameter vectors are propagated from generation to generation by processes of mutation, crossover, and selection. Each generation (indexed by $t = 0, 1, \ldots$) consists of N parameter vectors $x^{(t,i)}$. Hence, the real number $x_j^{(t,i)}$ is the value of the jth parameter in the ith parent in the tth generation. Let $u^{(t,i)}$ be the trial parameter vector born from the ith parent in the tth generation, whose components are constructed in two steps called "mutation" and "crossover". Then, given the parent parameter vector $x^{(t,i)}$ and trial parameter vector $u^{(t,i)}$, a decision is made as to which one is propagated to generation $t + 1$.

The steps of DE are described below.

1. Mutation. First, for each i, $1 \leq i \leq N$, we create a "mutant" vector

$$v^{(t,i)} = x^{(t,i)} + F \cdot d^{(t,i)} = x^{(t,i)} + F \cdot \left(x^{(t,i')} - x^{(t,i'')}\right) \tag{3}$$

by perturbing a parental parameter vector $x^{(t,i)}$, where the perturbation vector $d^{(t,i)}$ is the difference between the parameter vectors of two distinct additional parents i' and i'' chosen at random from the tth generation of parents, and $0 < F < 1$ ($F = 0.1$ in this study).

2. Crossover. For each i ($1 \leq i \leq N$) and j ($1 \leq j \leq D$), and uniform [0, 1] random variables $U_{i,j}$, define the offspring by

$$u_j^{(t,i)} = \begin{cases} v_j^{(t,i)} & , 0 \leq U_{i,j} \leq C, \\ x_j^{(t,i)} & , \text{otherwise.} \end{cases} \tag{4}$$

We choose the "crossover probability" $C = 0.5$ so that neither parental values nor mutant values are given an advantage during the crossover step.

3. Selection. The next generation parent $x^{(t+1,i)}$ is either the parent $x^{(t,i)}$ or the trial vector $u^{(t,i)}$. As DE explores the parameter space under different settings in this study, depending on the settings of the particular DE run, we impose three distinct feasibility criteria for selection, which are described below.

- Feasibility Criterion 1 (FC_1): Trial vector $u^{(t,i)}$ satisfies FC_1 if the model it defines captures the 111 phenotypes listed in Additional file 1: Table S7 out of the 119 phenotypes in the Training Set. FC_1 is always enforced by DE for creating Ensembles 1 through 16 in Table 2. For each ensemble generation scheme, the efficiency of sampling in terms of identifying parameter vectors that satisfy FC_1 (e_{FC_1}) is computed as n_{FC_1}/n_{tot}, where n_{FC_1} is the number of parameter vectors that satisfy FC_1 and n_{tot} is the total number of samples taken from the parameter space.
- Feasibility Criterion 2 (FC_2): FC_2 requires that trial vector $u^{(t,i)}$ can only replace parent vector $x^{(t,i)}$ if $u^{(t,i)}$ leads to an expansion in the feasible region's estimated volume. For this, we compute the estimated volumes of two Ensembles $\mathbf{X_1}$ and $\mathbf{X_2}$. The first ensemble $\mathbf{X_1}$ consists of all the parent vectors of the current tth generation of DE (all satisfying FC_1) including $x^{(t,i)}$. This ensemble excludes $u^{(t,i)}$ since it is not a parent vector. The second ensemble $\mathbf{X_2}$ includes $u^{(t,i)}$ in addition to all the parent vectors excluding $x^{(t,i)}$. FC_2 dictates that the trial vector $u^{(t,i)}$ can only replace $x^{(t,i)}$ if the estimated volume of the second ensemble is greater than the estimated volume of the first one ($V(\mathbf{X_2}) > V(\mathbf{X_1})$). (We describe our approach for estimating the volume spanned by an ensemble of parameter vectors in Section 1 of the Additional file 2: Supplementary Text.) With ensemble creation Schemes 4 to 7 in Table 2, DE enforces FC_2 together with FC_1 so that a trial vector replaces the corresponding parent if and only if the trial vector that reproduces the 111 target phenotypes of the

Table 2 Ensembles of feasible vectors generated with different schemes

Ensemble #	Scheme #	Ensemble size	Selection of the initial DE population from Ensemble 1	Feasibility criteria used in selection step of DE	# generations per DE run	# DE runs	S (Range of predictions)
1		3146	-	Parameter vectors satisfy FC_1	Ensemble generated in optimization [22]	-	30
2	1	243	-	Parameter vectors satisfy FC_1	Ensemble extracted from 50,000 LHS samples	-	51
3	2	7143	Randomly selected parameter vectors	FC_1	400	1	6
4	3	1893	$V_{max}(10)$	FC_1	400	1	41
5	4	1594	$V_{max}(10)$	FC_1 and FC_2	1600	1	64
6	4	1326	$V_{max}(10)$	FC_1 and FC_2	1600	1	69
7	5	3405	$V_{max}(123)$	FC_1 and FC_2	1600	1	94
8	5	3753	$V_{max}(123)$	FC_1 and FC_2	1600	1	80
9	6	2207	S_{max} & $V_{max}(123)$	FC_1 and FC_2	1600	1	117
10	6	1842	S_{max} & $V_{max}(123)$	FC_1 and FC_2	1600	1	95
11	7	3704	S_{max} & $V_{max}(123)$	FC_1, FC_2, and FC_3	1600	1	112
12	7	3481	S_{max} & $V_{max}(123)$	FC_1, FC_2, and FC_3	1600	1	133
13	8	4280	S_{max} & $V_{max}(123)$	FC_1 and FC_3	1600	1	313
14	8	4550	S_{max} & $V_{max}(123)$	FC_1 and FC_3	1600	1	367
15	7	15520	S_{max} & $V_{max}(123)$	FC_1, FC_2, and FC_3	2200	4	293
16	8	15050	S_{max} & $V_{max}(123)$	FC_1 and FC_3	2200	4	671

Parameter ranges used for LHS are from Ensemble 1. Parameter vectors in all ensembles capture the phenotypes listed in Additional file 1: Table S7, while missing the phenotypes in Additional file 1: Table S8. S: The range of the phenotypic prediction vectors generated per ensemble (unique rows of the prediction matrix **P**). $V_{max}(10)$: Biased selection is used to expand the estimated volume spanned by the initial population with respect to the axes of the ten most critical parameters (Table 3). $V_{max}(123)$: Biased selection is used to expand the estimated volume spanned by the initial population with respect to the axes of 123 kinetic parameters. S_{max}: Biased selection is used to enhance the initial population's range of phenotypic predictions. The prediction ranges for all ensembles can be reproduced using Additional file 4 (simulation code), and Additional files 5, 6, 7, 8 and 9 (Ensembles 1 through 16)

Training Set, and leads to an expansion in the feasible region's estimated volume.

- Feasibility Criterion 3 (FC_3): FC_3 requires that trial vector $u^{(t,i)}$ can only replace parent vector $x^{(t,i)}$ if $u^{(t,i)}$ yields a prediction vector for the 129 mutant strains of the Prediction Set that has not been derived from any parent vector up through the tth generation of DE. In other words, if a trial vector $u^{(t,i)}$ satisfies FC_1, $u^{(t,i)}$ replaces its parent $x^{(t,i)}$ if and only if the prediction vector $\hat{\mathbf{p}}$ generated by $u^{(t,i)}$ is not among the rows of the prediction matrix generated by all the parent vectors up through the point of generation of $u^{(t,i)}$. For creating Ensembles 11, 12, and 15 in Table 2, DE enforced all three criteria so that a trial vector replaces the corresponding parent if and only if the trial vector defines a model that captures the 111 target phenotypes of the Training Set, leads to an expansion in the feasible region's estimated volume, and produces a new phenotypic prediction vector for the 129 novel mutants in the Prediction Set. Ensembles 13, 14, and 16 are created by enforcing only first and the third criteria.

Results and discussion

Exploring the parameter space with Latin hypercube sampling

Our starting ensemble in this study is derived from the 3415 feasible parameter vectors identified in [22]. The size of this ensemble is reduced by 8% since only 3146 of these vectors are FC_1-feasible when truncated to 32-bit IEEE single precision. (We are eliminating parameter vectors that are very sensitive with respect to FC_1.) We call this collection of vectors "Ensemble 1". (Throughout this paper, parameter vectors are considered feasible only if their truncated 32-bit values are also feasible.) Applying Ensemble 1 to the Prediction Set, we generate 30 unique prediction vectors.

We explore this initial feasible region by Latin hypercube sampling (LHS). The bounds of the hypercube are formed by the minimum and maximum values of each parameter from Ensemble 1. 50,000 samples are generated as described in Section 2 of the Additional file 2: Supplementary Text. Out of these sample vectors, only 243 (0.5% of the total) are FC_1-feasible. These feasible vectors form Ensemble 2, which produces 51 unique prediction vectors; a 70% improvement (51/30) in the total range of predictions (previously defined as the number of unique prediction vectors).

Exploring the parameter space with DE

The results of LHS point out the possibility of finding feasible parameter vectors with a wider range of model predictions compared to those of Ensemble 1. We next investigate the possibility of using DE to identify alternative feasible ensembles with wider prediction ranges.

First, we created an initial random selection of 19 parameter vectors from Ensemble 1. (The population size of 19 is dictated by computational limitations imposed by the complexity of the model and the size of the Training Set [22]). Starting from this initial population of parameter vectors, DE explores the parameter space with mutation, crossover, and selection operations (described in Methods). (Rather than maximizing the total number of captured phenotypes by the model as we did previously [22], we only look for parameter vectors that capture the the 111 phenotypes listed in Additional file 1: Table S7 while missing the remaining eight phenotypes (Additional file 1: Table S8). Such vectors are feasible according to FC_1 as described earlier). In 400 generations, DE generates 7143 vectors (Ensemble 3 in Table 2) whose truncated 32-bit values satisfy FC_1. Despite its large size, Ensemble 3 yields only six unique prediction vectors for the 129 strains in the Prediction Set.

Why did DE perform so poorly compared to LHS even though, in our previous study, it was superior to random sampling in optimizing model performance (capturing phenotypes in the Training Set)? The answer comes from a comparison of the volumes the parameter space that are spanned by Ensembles 2 and 3. Ensemble 3 has an estimated volume that is 83 orders of magnitude smaller than that of Ensemble 2. In other words, DE zooms into a much smaller region of parameter space than LHS.

Following this observation, we conjectured that selecting the volume covered by the initial population of DE in a systematic way, rather than a random way, might improve the performance of the search. Therefore, we next choose an initial DE population such that the estimated volume spanned by the population vectors is maximized with respect to the axes of the ten most critical model parameters listed in Table 3. The details of the procedure for picking such a population are described in the Additional file 2: Supplementary Text (Section 3). A DE run for 400 generations, starting with this new initial population, finds 1893 feasible vectors (Ensemble 4), which account for 41 unique prediction vectors. This six-fold improvement compared to Ensemble 3 (6 vs. 41) shows that the outcome is highly dependent on the selection of the initial population, and supports the proposed scheme for maximizing the volume of the initial population of parameter vectors. We also note that Ensemble 4, although four-fold smaller than Ensemble 3 in terms of the total number of feasible parameter vectors, generates a much wider range of predictions.

Nonetheless, the range of the predictions generated by Ensemble 4 is less than the range generated by Ensemble 2 (LHS). Why is this the case? The answer lies in

Table 3 The ten most critical model parameters

Parameter name
Total amount of Cdc14
SPN synthesis rate
Total amount of Esp1
Total amount of Net1
Degradation rate of Cdc20
PPX inactivation by Esp1
Efficiency of Cdc14-Net1 complex (RENT) formation
Time scale for protein activation
Net1 phosphorylation by Clb2
Total amount of Mcm1

Based on the sensitivity analysis in [22], the listed model parameters had the largest effects on the objective function (number of phenotypes captured by the model) upon perturbations. Criticality decreases from top to bottom

the evolution of the volume spanned by the trial vectors generated during DE. Figure 3 (black line for Ensemble 4) shows that as DE progresses, the estimated volume spanned by the most recent feasible vectors, which serve as the parent vectors producing trial vectors in DE, continually shrinks as the generations pass. Details of the computation of this dynamic estimated volume are in Section 4 of the Additional file 2: Supplementary Text. One way to prevent this shrinkage is to increase the value of F in Eq. 3. However, increasing the value of F from 0.1 to 1 leads to a 37–64 fold drop in the sampling efficiency e_{FC_1} with Schemes 2 and 3 (both schemes described in Table 2).

Therefore, to prevent this drop in dynamic volume, we introduce a new constraint (FC_2) as described in the Methods section. To enforce FC_2, the estimated volumes of two distinct ensembles are computed every time a new trial vector that satisfies FC_1 is found. The first ensemble includes all parameter vectors satisfying FC_1 until that point of DE, except the newest trial vector generated. Hence, this ensemble includes the trial vector's competitor: the parent vector. The second ensemble is generated by including the trial vector instead of the parent vector, with the remaining members being identical to those of the first ensemble. If the estimated volume of the second ensemble is greater than that of the first one, the trial vector replaces the parent vector in the next generation of DE in the search for feasible vectors. Otherwise, the parent vector is not replaced, but the trial vector is recorded since it satisfies FC_1 and its predictions for the phenotypes of the Prediction Set are evaluated after DE is complete. Succinctly, FC_2 allows a trial vector to replace a parent only if it leads to an expansion of the feasible region. As shown in Fig. 3 (green and red lines), this new feasibility criterion prevents the volume of the feasible region from shrinking as the generations pass (two independent DE runs). Additional file 3: Figure S1 (blue line) and Additional file 1: Table S9 show that without this volume maximization strategy, the ranges of nearly all parameters are diminished after 400 generations. On the other hand, with estimated volume maximization, the majority of the

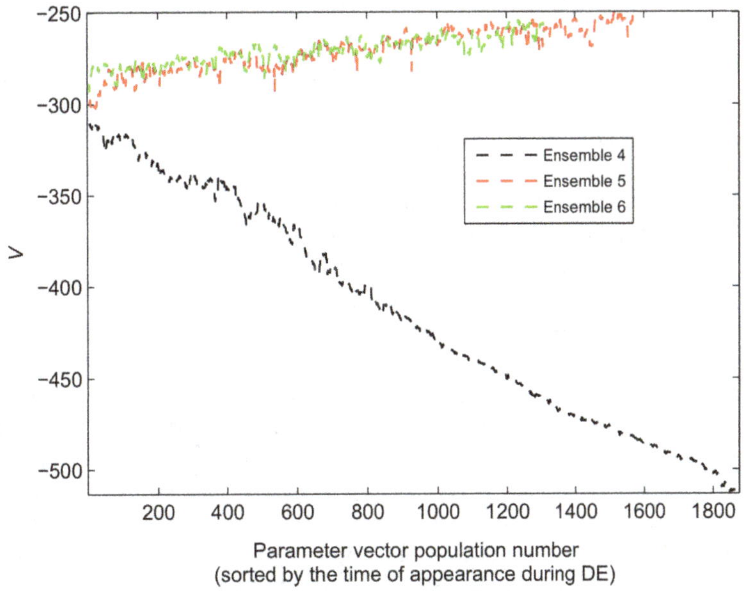

Fig. 3 Dynamic evolution of the estimated volume V spanned by the parameter vectors generated during different DE realizations. Details regarding the computation of the estimated volume are provided in Section 4 of the Additional file 2: Supplementary Text. Ensembles 5 and 6 are generated by the Scheme 4 that uses FC_1 and FC_2, whereas Ensemble 4 is generated by Scheme 3 which only uses FC_1 as its feasibility criteria

parameters have about a 10% variation after 400 generations (green line in Additional file 3: Figure S1). These parameter ranges are calculated by dividing the maximum parameter values by the minimum parameter values among all parent vectors at the 400th generation. Due to this improvement in the parameter ranges, we allow DE to explore for an additional 1200 generations. Additional file 3: Figure S1 (red line) shows that about 10% range for most parameters is still preserved among the parent vectors in the 1600th generation. We perform two realizations of DE with this approach for 1600 generations, thereby creating two more ensembles (Ensembles 5 and 6 in Table 2). These ensembles generate 64 and 69 unique phenotypic prediction vectors, respectively.

A further improvement comes from selecting the initial DE population to maximize the volume spanned by the vectors with respect to all 123 kinetic parameters rather than just the 10 most critical parameters. (Note that the kinetic parameters ks_{n2}, f, and MDT have fixed values in Additional file 1: Table S3.) Two independent DE realizations for 1600 generations produce an average of 87 unique phenotypic prediction vectors (Ensembles 7 and 8 in Table 2) further increasing the range of predictions compared to those of Ensembles 5 and 6. This is also a significant improvement over LHS (51 unique prediction vectors) even though DE required about 30,000 samples (1600 generations $\times$ 19 vectors) to identify $\sim$ 70% wider (87/50) range of predictions compared to 50,000 LH samples. Having gotten DE to a point where it is more efficient than random sampling in terms of exploring the feasible region, we next seek ways to improve the performance of DE even further.

Increasing the phenotypic diversity of the initial population of DE

As we previously stated, 3146 feasible parameter vectors from the initial DE optimization run on the Training Set [22] (Ensemble 1) generate 30 unique phenotypic prediction vectors for the Prediction Set. Interestingly, 97% of these vectors generate only five of the total 30 prediction vectors as shown in Additional file 3: Figure S2. Due to this, the initial population of parameter vectors used in the last two DE runs (Scheme 5 in Table 2) produces a total of four unique prediction vectors all of which are in this set of five dominant prediction vectors. In other words, the diversity of the initial population in terms of phenotypic predictions is very low, only 13% (4/30) of the diversity is utilized. Therefore, to increase this diversity, we select the initial population of feasible parameter vectors such that each one generates a different prediction vector (a total of 19) for the 129 strains in the Prediction Set. While using this initial selection scheme, we also maximize the estimated volume spanned by the selected vectors (Scheme 6 in Table 2). The details of this diversification procedure are in the Additional file 2: Supplementary Text (Section 5). This strategy further expands the range of predictions, with two independent runs (each for 1600 generations) increasing the average number of unique prediction vectors from 87 to 106 (Table 2). Thus, improved predictive diversity among the parent parameter vectors in the initial population results in feasible vectors (generated during DE) that are predictively more diverse.

Enforcing an increased range of predictions during DE

In order to explore the phenotypic prediction space of the model further, we enforce a third criterion during DE. With this new criterion, a parent parameter vector is only replaced by a trial vector if the trial vector generates a new prediction vector, one not heretofore generated by any feasible parameter vector during this DE run. (For the descriptions of parent and trial parameter vectors, refer to Methods section.) In other words, with this modification, the trial parameter vector has to satisfy three constraints to replace the parent vector. It should reproduce 111 phenotypes in Additional file 1: Table S7 (FC_1), increase the estimated volume of the feasible region upon replacing the parent vector (FC_2), and generate a new prediction vector (FC_3). Two independent realizations with this new scheme (for 1600 generations) increase the average number of unique predictions from 106 to 122.5 (average of Ensembles 11 and 12 in Table 2). We note that since the occurrence of a trial vector that satisfies the first two criteria is not very frequent (less than 10% among the samples generated by DE), simulating the 129 mutant strains of the Prediction Set on-the-fly (during DE) adds negligible computational time compared to the time required to run DE for 1600 generations with the 119 phenotypes in the Training Set.

Since our major goal in this study is to devise a method that discovers as many unique phenotypic prediction vectors as possible, we next drop the second feasibility criterion FC_2 (maximization of the feasible region's estimated volume during DE) but keep the first and third criteria (FC_1 and FC_3). As shown in Additional file 3: Figure S3 (green line) and Additional file 1: Table S10, even though FC_2 is dropped, DE is still able to keep some parametric variability among the feasible vectors after 1600 generations. This variability is due to the presence of FC_3 that indirectly forces diversity in parameter values by guiding the search towards new prediction vectors. More importantly, dropping FC_2 results in an average of 340 unique phenotypic prediction vectors (Ensembles 13 and 14 in Table 2), almost a 200% increase (122.5 to 340) in the range of predictions. Hence, not enforcing the second feasibility criterion allows us to exploit DE's search capability for expanding the range of predictions. e_S value of Scheme 8, computed as the number of unique prediction vectors found per sample taken in the parameter space, is equal

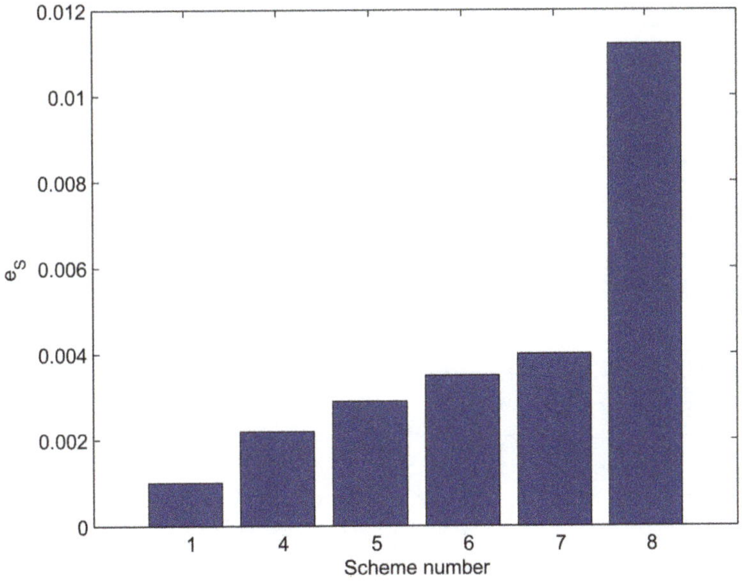

Fig. 4 Comparison of e_S with different schemes. e_S is the efficiency of sampling computed as the ratio between the range of phenotypic predictions (S) and the total number of samples taken from the parameter space (n_{tot}). LHS is used in Scheme 1, whereas DE is used in the remaining schemes. The detailed settings used for ensemble generation with each scheme are given in Table 2. For schemes 4–8, we average two e_S values computed from two independently generated ensembles (per scheme)

to 0.011 (340 unique prediction vectors found in 1600 generations × 19 parameter vectors per generation). The same efficiency value is 0.001 for random LHS (51 unique prediction vectors found in 50000 randomly generated parameter vectors), a 10-fold difference in favor of our DE based approach. Figure 4 provides a snapshot of the performances (e_S values) of different schemes. We also note that random selection of the initial population decreases the e_S value of Scheme 8 by 81%, whereas selecting an initial population with expanded volume (with respect to the axes of 123 kinetic parameters), but without enhanced predictive diversity, causes a 64% drop in Scheme 8's e_S value (results based on two DE runs for 1600 generations in both cases). These results show that the selection of the initial population of DE is critical for the efficient exploration of the prediction space of the model.

At this point, we have two top performing DE based schemes (Schemes 7 and 8 in Table 2) for exploring the prediction space of the model. Top performing Scheme 8 is illustrated in Fig. 5. Next, we will compare the performances of the two top performing schemes in a more thorough way, using the aggregates of ensembles from several DE runs with a higher number of generations per DE run. Then, from these ensembles, we will extract future experiments for which the model produces wide (or narrow) prediction ranges. Our goal will be to differentiate between the strong predictions of the model (e.g., novel phenotypes that are viable regardless of the parameter vector location in the feasible region) and the model predictions with some variability within the feasible region of the model's parameter space.

Comparison of the two most efficient ensemble generation schemes

In order to compare the performances of Schemes 7 and 8 more thoroughly, we perform four DE runs with each scheme (2200 generations per run). As shown in Table 2, Scheme 8 produces 671 unique prediction vectors (from 15050 feasible parameter vectors in Ensemble 16), whereas the number of unique prediction vectors is 293 for Scheme 7 (from 15520 feasible parameter vectors in Ensemble 15), reiterating our previously stated conclusion that Scheme 8 is more efficient in exploring the phenotypic prediction space. Lower performance of Scheme 7 suggests that maximizing the feasible estimated volume during DE (through FC_2) may have no benefit.

However, in this section, we will show that Scheme 7 outperforms Scheme 8 in terms of a "robustness" measure based on parametric perturbations to be defined. After each of these perturbations, we simulate the model to check if the outcome of the simulation (mutant phenotype) is the same as the phenotype before the perturbation. For this robustness analysis, we limit our focus to the ten most critical model parameters (Table 3) and the ten most fragile phenotypes (Table 4), which were previously identified by the sensitivity analysis in [14].

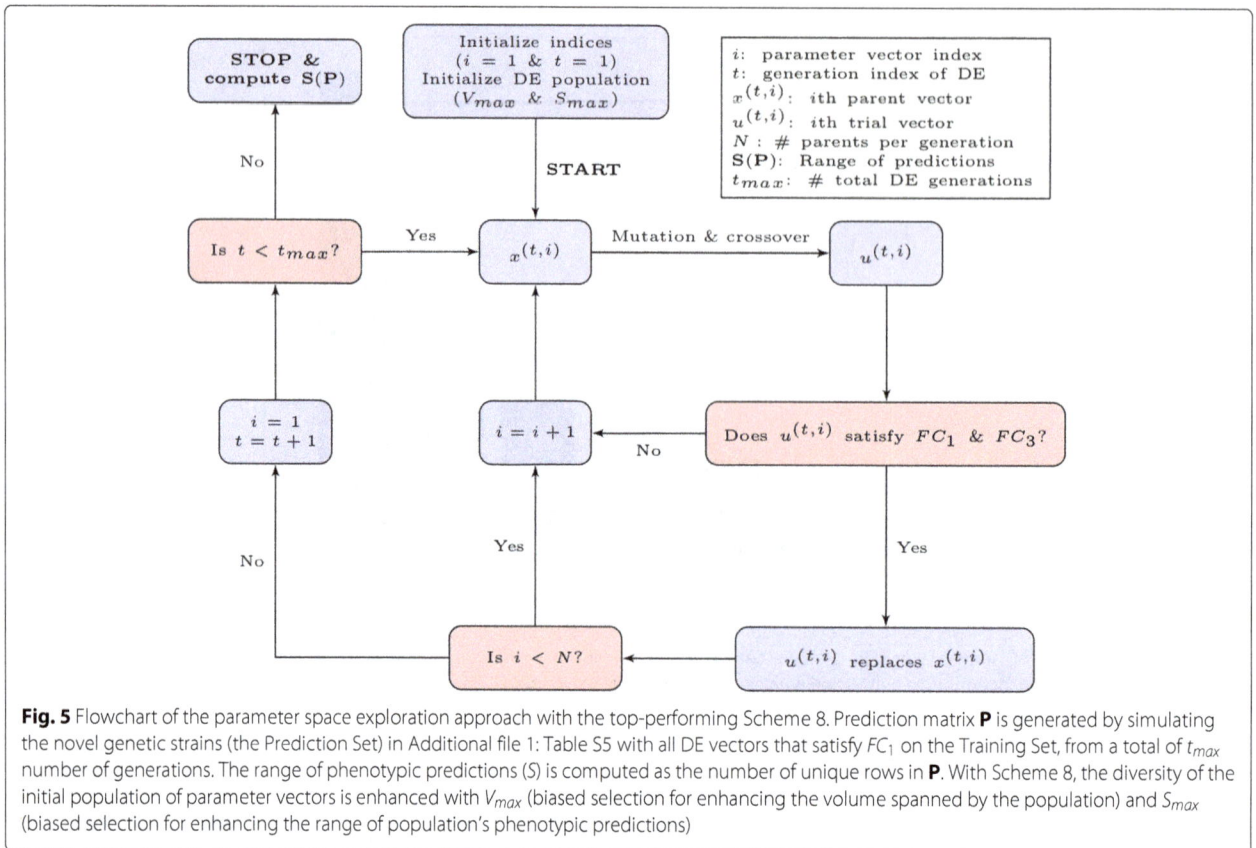

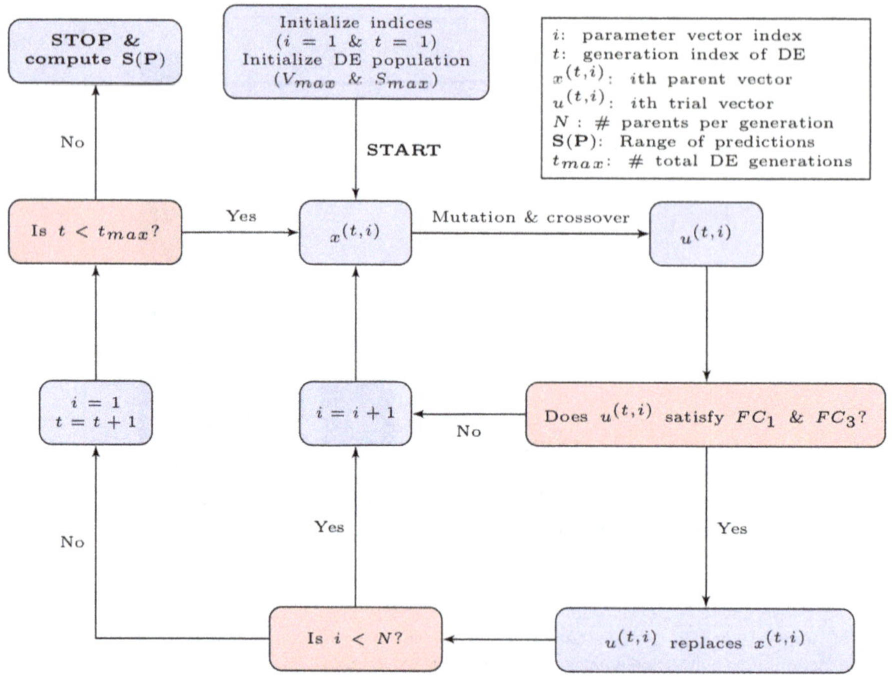

Fig. 5 Flowchart of the parameter space exploration approach with the top-performing Scheme 8. Prediction matrix **P** is generated by simulating the novel genetic strains (the Prediction Set) in Additional file 1: Table S5 with all DE vectors that satisfy FC_1 on the Training Set, from a total of t_{max} number of generations. The range of phenotypic predictions (S) is computed as the number of unique rows in **P**. With Scheme 8, the diversity of the initial population of parameter vectors is enhanced with V_{max} (biased selection for enhancing the volume spanned by the population) and S_{max} (biased selection for enhancing the range of population's phenotypic predictions)

Each critical parameter is perturbed ±20%, ±40%, ±60%, ±80% (eight perturbation levels) from its nominal value and also set to zero (the ninth perturbation level). Each of these individual perturbations defines a new parameter vector. With each new vector, each of the ten fragile phenotypes is simulated (initial conditions come from the WT simulation as described before). In each simulation, the phenotype derived from the feasible vector before the perturbation is either maintained or lost. We have 900 simulations (9 perturbation levels × 10 perturbed parameters × 10 simulated phenotypes) that are used to quantify the robustness of each parameter vector. The robustness score of the ith parameter vector is defined by

$$\hat{R}_i = \sum_{j=1}^{10}\sum_{k=1}^{10}\sum_{l=1}^{9} R_{i,j,k,l}, \tag{5}$$

where j is number of the critical parameter that is perturbed, k is number of the fragile mutant that is simulated, l is the number of the perturbation level, and $R_{i,j,k,l}$ is 0 (1) if the fragile phenotype from Table 4 is lost (maintained) after the parametric perturbation. The highest robustness score within an ensemble of m parameter vectors is

$$\breve{R} = \max_{1\leq i\leq m} \hat{R}_i \,. \tag{6}$$

$\hat{R}_i = 900$ is the highest possible robustness score for a feasible parameter vector that satisfies FC_1 prior to perturbations. One way to compare different ensembles in terms of robustness is to compare the distributions of $\hat{R}_i$.

Table 4 The ten most fragile phenotypes

Phenotype #	Phenotype name
61	*CLB2-db*Δ multicopy *SIC1* (Viable)
18	*cln1*Δ *cln2*Δ *cdh1*Δ (Viable)
63	*CLB2-db*Δ *clb5*Δ *clb6*Δ in galactose (Viable)
105	*cdc15*Δ *net1-ts cdh1*Δ (Viable)
56	*GAL-CLB2 cdh1*Δ (Inviable)
20	*cln1*Δ *cln2*Δ *cdh1*Δ *GAL-CLN2* (Viable)
59	*CLB2-db*Δ in galactose (Inviable)
77	*APC-A* (Viable)
78	*APC-A sic1*Δ (Viable)
73	*CLB5-db*Δ *pds1*Δ (Viable)

Based on the sensitivity analysis in [22], these are the phenotypes that are most often lost (i.e., incorrectly simulated) when perturbations are applied to individual model parameters in feasible parameter vectors. Fragility decreases from top to bottom

In addition, each parameter vector i and fragile phenotype-critical parameter pair (k, j) define the robustness score

$$\tilde{R}_{i,j,k} = \sum_{l=1}^{9} R_{i,j,k,l}, \tag{7}$$

which helps us differentiate between different ensemble generation schemes listed in Table 2 in terms of the robustness linked to particular phenotype-parameter pairings. The maximum robustness of such a pair in an ensemble of parameter vectors is defined by

$$\overline{R}_{j,k} = \max_{1 \leq i \leq m} \tilde{R}_{i,j,k}. \tag{8}$$

As shown in Fig. 6a, feasible parameter vectors in Ensemble 15 (produced with Scheme 7) generate a bimodal distribution of robustness $\hat{R}$, computed for each feasible vector. Ensemble 15's first mode with low robustness overlaps with Ensembles 1 and 16 (both ensembles have a unimodal distribution of $\hat{R}$). On the other hand, Ensemble 15's second mode with higher robustness has no overlap with the two other ensembles' distributions. Hence, maximizing the estimated volume of the feasible region through the mutation and crossover operations of DE leads to the discovery of feasible points in the parameter space with superior robustness. The maximum robustness value $\check{R}$ among Ensemble 15 is 672, but only 512 among Ensemble 1, a 31% improvement with Ensemble 15 (generated by Scheme 7). On the other hand, $\check{R}$ is 514 among Ensemble 16 (generated by Scheme 8), approximately equal to the $\check{R}$ value among Ensemble 1. In addition, as depicted in Fig. 6b, c, and d, Scheme 7 improves the maximum robustness among 70 critical parameter-fragile phenotype pairs, whereas the number of such pairs is only 21 for Scheme 8. Here, the maximum robustness (per ensemble) is quantified by a $\overline{R}$ value per parameter-phenotype pair (Eq. 8).

From these results, we conclude that by forcing the DE search to expand the range of predictions, one can explore the prediction space effectively as demonstrated by Scheme 8's superior predictive diversity over its alternatives (Table 2). However, by forcing DE to maximize the feasible region's estimated volume, it is possible to improve the robustness of the model in reproducing

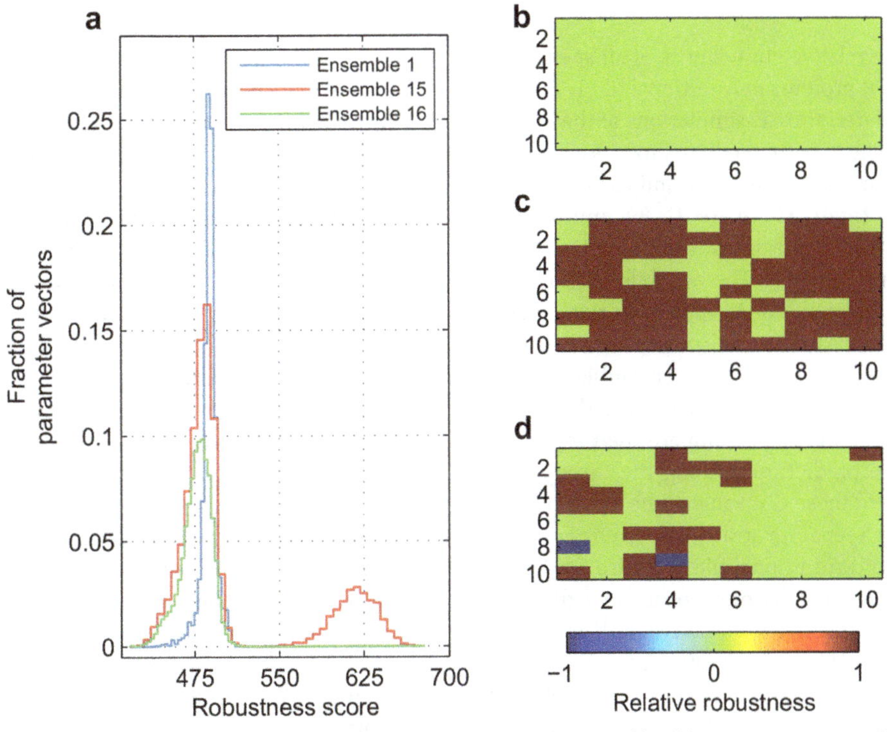

Fig. 6 Robustness-based comparison of different ensembles. **a** Robustness score distributions of feasible parameter vectors in Ensembles 1, 15, and 16. Each parameter vector's robustness is computed by perturbing the ten most critical model parameters with nine distinct perturbation levels to simulate the ten most fragile phenotypes (Table 4). The total number of perturbations that do not lead to phenotype losses in these 900 simulations is recorded as the robustness score $\hat{R}$ per feasible parameter vector. **b-d** Comparison of the maximum robustness $\overline{R}$ per phenotype-parameter pair among Ensembles 1 in **b**, 15 in **c**, and 16 in **d**. The relative robustness in Ensemble 15 or 16 is -1 (1) if the particular robustness value is lower (higher) than Ensemble 1

experimentally verified phenotypes, but at the expense of predictive diversity. Therefore, one should select the appropriate scheme for parameter space exploration, depending on one's preference between higher robustness (Scheme 7) or diversity of model predictions (Scheme 8). Higher robustness against parametric perturbations may be enforced in some cases. For instance, one may need to modify the values of parameters in feasible vectors in order to capture additional experimental constraints while still capturing the original data [42], and this would favor the selection of Scheme 7 over Scheme 8.

Relative protein abundance predictions

Up to this point, we have only considered the phenotypic prediction range for the 129 mutant strains in the Prediction Set. Next, we consider predictions of relative protein abundances. In simulations, the time average concentration of a protein represents the model's prediction for that protein's abundance in an asynchronous population of budding yeast cells. For theoretical and experimental reasons, it is better to focus on relative protein abundances, i.e., the ratio of the abundance of one protein with respect to another. Relative abundances of proteins are typically measured by Western Blotting [43] or mass spectrometry [44]. Relative abundance measurements have been useful in estimating the parameters of systems biology models in the past [45, 46].

We compute the relative abundances of all species (cell size and 25 different proteins in Additional file 1: Table S2) over 2000 min in deterministic simulations of the 86 novel mutants that are consistently predicted to be viable by the parameter vectors in Ensembles 1, 15, and 16 (about 33000 feasible vectors in total). There are 91, 89, and 86 viable mutants (among the 129 strains in the Prediction Set) in these ensembles, respectively. The variability of each relative abundance prediction is quantified by its coefficient of variation (CV=standard deviation/mean) across the feasible parameter vectors within each ensemble. In order to show the effectiveness of characterizing the feasible region beyond Ensemble 1, we compare the ranges of all relative abundance predictions among the three ensembles (after collecting these CV values (one value per relative abundance) in a separate array for each ensemble). As shown in Fig. 7a and Additional file 3: Figure S4, Ensembles 15 and 16 generated by our parameter exploration schemes 7 and 8, respectively, exhibit significantly wider CV distributions than Ensemble 1 once again demonstrating the capacity of our DE-based approach to explore the parameter space. According to Additional file 1: Table S11, both the mean and standard deviation values of CV distributions from Ensembles 15 and 16 are consistently greater than double those from Ensemble 1. Figure 7b-d show an example, where the ranges of model predictions made by Ensembles 15 and 16 (Fig. 7c and d) for two relative abundances are significantly wider and much less sparse in the prediction space compared to Ensemble 1 (Fig. 7b).

As we ranked the 86 novel viable mutants in terms of decreasing value of a prediction variability statistic generated by Ensemble 16, namely the sum of relative abundance CV's predicted for each mutant strain, we observed that the ten highest ranked strains with most variability (Table 5) are composed of three double mutants and seven triple mutants (no single mutant), suggesting that the increased number of mutations in a genetic strain provide model predictions with wider relative abundance ranges. Figure 8a and Additional file 1: Table S12 confirm this trend. Here, we see that a higher prediction variability statistic is associated with double and triple mutants compared to single mutants. Histograms of the CV distributions for the WT strain and the five highest ranked novel mutant strains (Fig. 8b-g) indicate that these mutants generate predictions with significantly higher variabilities compared to the WT strain. Interestingly, these five double and triple mutants have common mutations (Table 5). For instance, even though Mutant 57 has an additional mutation compared to Mutant 11, the distributions of the CV values of the predicted relative abundances are almost identical. Hence, this additional mutation does not increase the prediction variability in the relative abundance measurements. A similar trend is observed with Mutant 21 (double mutant) and Mutants 90 and 85 (triple mutants), once again indicating a common mutation pair that is responsible for the wide prediction ranges. Mutant 21 is created from two single mutations (Mutants 2 and 6 in Additional file 1: Table S5). As depicted in Fig. 8h and Additional file 1: Table S13, these two individual mutations synergize upon creating Mutant 21 and generate overall ranges of predictions (each CV value corresponds to the range of one prediction) wider than either of the single mutations alone. These analyses highlight the usefulness of our approach to designing genetic strains that generate informative model predictions. For instance, the pair of relative abundance predictions shown in Fig. 7d have CV values that are higher than 0.40 among Ensemble 16. In contrast, the two relative abundance predictions shown in Additional file 3: Figure S5 have CV values that are less than 0.01 among the same ensemble. Hence, the presented parameter space exploration approach enables us to differentiate between informative genetic strains with high prediction variability (Fig. 7b, c, and d) and the genetic strains that generate model predictions with low variabilities (Additional file 3: Figure S5).

Similar approaches have been used in two previous model driven experimental design studies [47, 48]. In [47], Dong et al. presented an experimental design process called "Computing Life" and illustrated it for the biological clock of *Neurospora crassa.* At each experimental design cycle, the authors chose the Maximally Informative Next

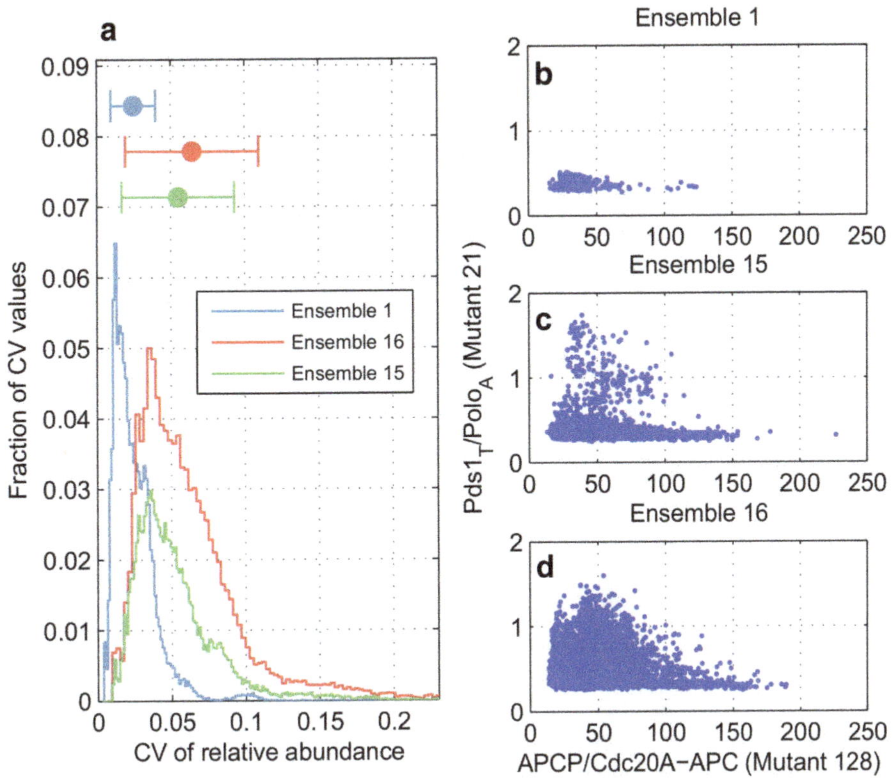

Fig. 7 Relative abundance predictions from different ensembles. **a** Distributions of CV values of the relative abundance predictions generated by three different ensembles of parameter vectors. Mean ± standard deviation for each distribution (listed in Additional file 1: Table S11) is depicted by a single horizontal bar. The extreme values of these distributions are shown in more detail in Additional file 3: Figure S4. **b-d** The displayed relative abundance predictions (with high variability) are generated by Ensembles 1 (in b), 15 (in c), and 16 (in d). CV values of these predictions are of 0.18/0.51/0.53 (x-axis) and 0.076/0.27/0.41 (y-axis) among Ensemble 1/15/16

Table 5 The ten novel phenotypes with highest predictive variance

Mutant #	Mutation 1	Mutation 2	Mutation 3
90	$e_{ki,n2} = 0$	$kp_{bfb2} = 0$	$kdp_{i514} = 0$
21	$e_{ki,n2} = 0$	$kp_{bfb2} = 0$	-
57	$kp_{i5k2} = 0$	$e_{ki,b2} = 0$	$kdp_{i514} = 0$
11	$kp_{i5k2} = 0$	$e_{ki,b2} = 0$	-
85	$e_{ki,n2} = 0$	$kp_{i5n3} = 0$	$kp_{bfb2} = 0$
58	$kp_{i5k2} = 0$	$e_{ki,b2} = 0$	$kdp_{net,px} = 0$
81	$e_{ki,n2} = 0$	$kdp_{ki,14} = 0$	$kp_{bfb2} = 0$
128	$kp_{bfb2} = 0$	$kdp_{i514} = 0$	$kdp_{net,px} = 0$
42	$kp_{bfb2} = 0$	$kdp_{net,px} = 0$	-
106	$e_{ki,b2} = 0$	$kp_{bfb2} = 0$	$kdp_{net,px} = 0$

Based on the relative abundance predictions generated by the parameter vectors in Ensemble 16, these mutants mutants have the highest variability values (i.e., Mutant 90 has the largest sum of CV values from the relative abundance predictions among the 129 novel mutants). Variability decreases from top to bottom

Experiment (MINE) from a large set of potential network models and microarray experiments using a criterion that enforced maximal independence between observables. This analysis identified several genes (from a total of 11,000 genes) under the direct control of a key clock oscillator and also discovered a link between this clock and ribosome biogenesis. In [48], Donahue et al. implemented a sparse grid approximation using polynomials to explore their objective function (based on time series data) in order to discriminate simultaneously between uncertainties in model structure and in parameter values (without an initially determined feasible region). One disadvantage of the sparse grid search is the required smoothness of the objective function, whereas typically rugged objective function landscapes [49] are observed for large and nonlinear network models. This is especially the case in our study where many discrete experimental constraints determine the feasibility of model parameter vectors. For detailed theoretical discussions regarding the use of prediction variability statistics in model-based experimental design, we refer the reader to two excellent reviews [4, 6].

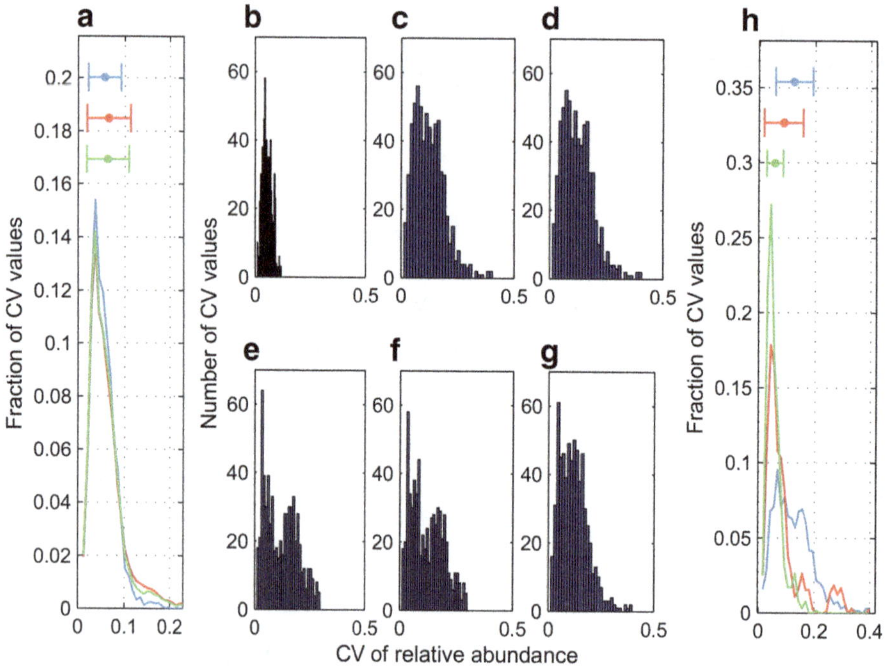

Fig. 8 CV values of the relative abundance predictions from different mutants. **a** Smoothened distributions of CV values of the relative abundance predictions generated by single mutants (*blue curve*), double mutants (*green curve*), and triple mutants (*red curve*) among the 129 novel mutants. Mean ± standard deviation for each distribution (listed in Additional file 1: Table S12) is depicted by a single horizontal bar. These predictions are generated by the parameter vectors in Ensemble 16. **b-g** CV values of the relative abundance predictions for the WT strain (in **b**) and the five most informative (highest ranked based on a prediction variability statistic) novel mutants (Mutant 90 in **c**, Mutant 21 in **d**, Mutant 57 in **e**, Mutant 11 in **f**, and Mutant 85 in **g**. These predictions are generated by the parameter vectors in Ensemble 16. **h** Distributions of CV values of the relative abundance predictions generated by Mutant 21 (a double mutant shown in *blue curve*), which is a combination of two single mutants: Mutant 2 (*green curve*) and Mutant 6 (*red curve*). Mean ± standard deviation for each distribution (listed in Additional file 1: Table S13) is depicted by a single horizontal bar. These predictions are generated by the parameter vectors in Ensemble 16

As shown in Fig. 9, starting from Ensemble 16, one can also refine the feasible ranges of parameters upon incorporating a relative abundance measurement into the model. Here, we see that low values of ki_{10}, $CDC14_T$, and ks_{spn}, and medium values of γ and γ_{ki} produce high values (greater than 120) of the relative abundance APCP/Cdc20A-APC in Mutant 128 (relative abundance measurement with the highest variability based on its CV value). Hence, it is possible to use new data from an experiment that is predicted to be informative and eliminate some of the parameter vectors in the feasible ensemble. In other words, future measurements selectively taken based on model predictions by targeting highly variable relative protein abundances within particular mutants could be useful for reducing parametric uncertainty. However, we did not perform experimental design in our study since it is outside our scope.

Ranking cell cycle proteins and biological processes in terms of prediction variability

In order to study the potential relationships between the variabilities of relative abundance predictions linked to individual cell cycle proteins and the topology of the cell cycle network, we first identified the total variability associated with each of the 26 proteins. To this end, for each protein, we computed the sum of the CV values for each protein abundance ratio with that particular protein in its numerator. We refer to this sum as the "variability score" of the protein. (We have verified that our ranking of proteins based on their variability scores does not depend on whether we use the protein in the numerator or denominator in the summation process (data not shown)).

Next, we ranked the cell cycle proteins with respect to their total variability scores (Table 6). Five of the seven species in the EXIT module were in the group of "low variability" proteins (the bottom half of the group). In contrast, four of the five proteins within the START module were within the "high variability" category (the top half). The less variable nature of the EXIT module aligns with our previous study, which identified the EXIT module as the most fragile network module [22], as well as an experimental study which showed the cell cycle was least tolerant to overexpression of *CDC14* (a major regulator in the EXIT module) among 31 cell cycle genes

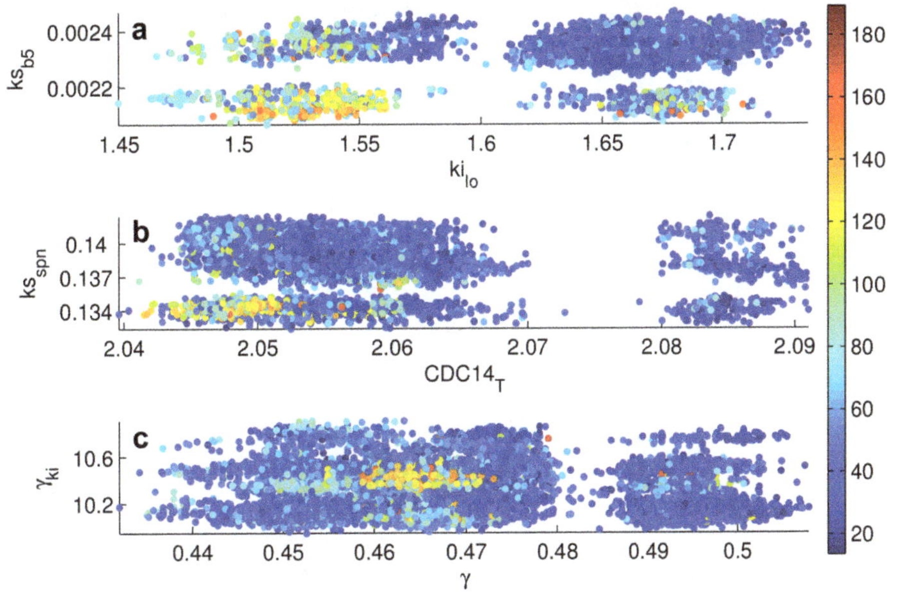

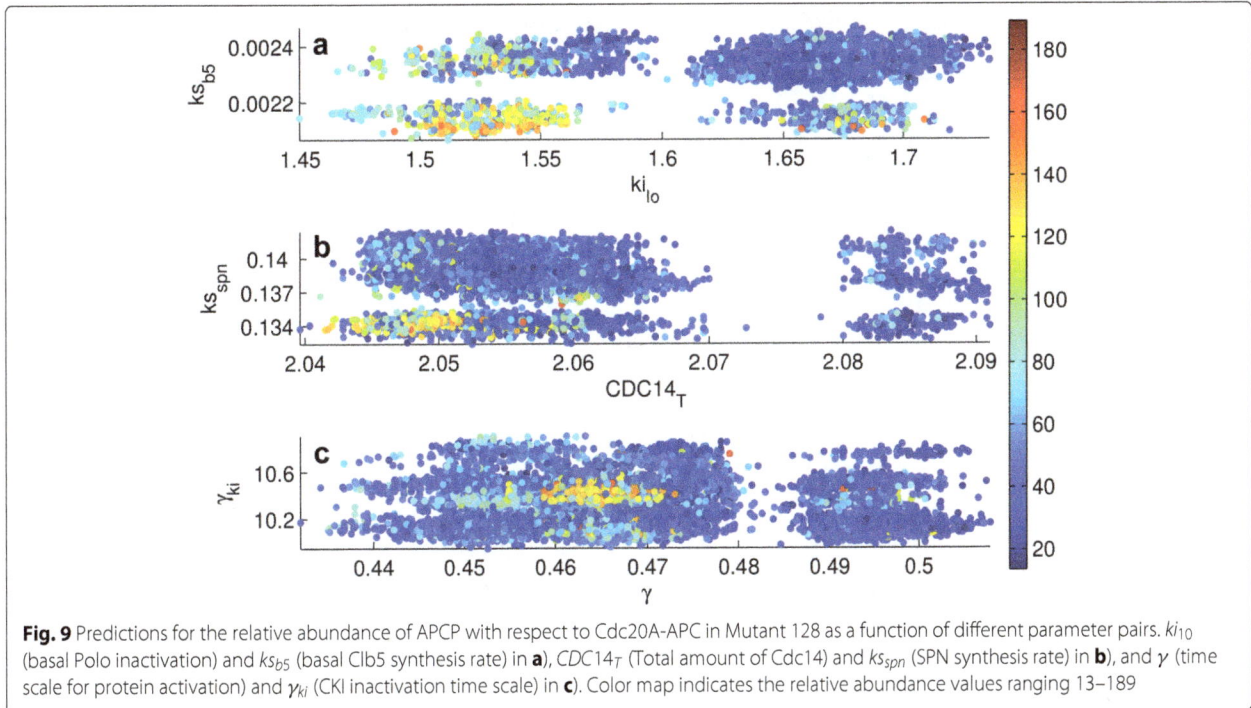

Fig. 9 Predictions for the relative abundance of APCP with respect to Cdc20A-APC in Mutant 128 as a function of different parameter pairs. ki_{lo} (basal Polo inactivation) and ks_{b5} (basal Clb5 synthesis rate) in **a**), $CDC14_T$ (Total amount of Cdc14) and ks_{spn} (SPN synthesis rate) in **b**), and γ (time scale for protein activation) and γ_{ki} (CKI inactivation time scale) in **c**). Color map indicates the relative abundance values ranging 13–189

studied by Moriya et al. [50]. Similarly, two other proteins in the EXIT module, namely *NET1* and *PDS1* were among the more fragile genes (ranked 8th and 10th among 31 genes in terms of cell cycle's tolerance limit to their overexpression) in [50] in agreement with the "low variability" status of these proteins in our model (Table 6).

Ten regulators in the S/G2/M module, on the other hand, were evenly distributed among both categories. Two regulators in this module, namely Cdc20A-APC and Cdc20A-APCP had strikingly different predictive variability scores. Cdc20A-APC complex has the highest score of 231.4), whereas Cdc20A-APCP complex was ranked 20th with a score of 115.58. These two complexes are responsible from the degradation of Clb5, Clb2, and Pds1 through ubiquitin-mediated proteolysis [51]. Interestingly, Cdc20A-APCP is 9.3, 3.8, and 6.5–fold more potent than Cdc20A-APC (based on the average parameter values in Ensemble 16) in terms of degrading Clb5, Clb2, and Pds1, respectively. Hence, a potent (or critical) regulator turned out to have less predictive variability compared to a weaker regulator in our model once again pointing to a potential relationship between the cell cycle network and the variability scores of individual model variables (more critical variables have less predictive variance). After ranking cell cycle proteins by denominator-based formation of relative abundance-network module pairs (i.e., each relative abundance is matched to the module of the protein in its denominator), we computed the Pearson correlation coefficient between the vectors formed from the order of the two rankings (numerator-based vs. denominator-based) as 0.99. Hence, the ranking of cell cycle proteins was independent of the way network modules were assigned to relative abundance values.

Next, we compiled all of the gene ontology based biological processes [52] associated with the cell cycle proteins and ranked them using the variability score associated with each regulator (Table 7). In cases where a biological process was associated with more than one protein, we computed the mean and standard deviation of the variability scores associated with each process. According to Table 7, the biological processes with the largest predictive variability values (85th-100th percentile range), were identified as the regulation of cell size and regulation (both negative and positive) of the G1/S transition. These processes are known to be closely tied to each other [53]. Budding yeast cells that are exceptionally small at birth than others spend more time in G1 before entering S phase due to an experimentally verified size threshold requirement [54]. Later studies showed that this size control mechanism acts for the most part in daughter cells (in our simulations, the "daughter cell" is the smaller cell at each asymmetric division) through multiple daughter-specific transcription factors [55] and also showed that this mechanism is "imperfect" [56] since cell size at birth is not perfectly correlated with the length of G1 phase. It is also thought that size fluctuations can not be compensated in a single cycle due to the imperfect

Table 6 Cell cycle regulators ordered in terms of their variability scores (decreasing from top to bottom)

Rank	Regulator	Variability score	Percentile	Category	Network module	Class label
1	Cdc20A-APC	231.4	100	High variability	S/G2/M	2
2	Cln3	208.3	94	High variability	START	1
3	BUD	195.71	91	High variability	-	-
4	APCP	191.2	87	High variability	S/G2/M	2
5	WHI5dep	160.63	82	High variability	START	1
6	SBFdep	156.26	79	High variability	START	1
7	$Polo_A$	154.48	75	High variability	EXIT	3
8	Cln2	152.8	71	High variability	START	1
9	Tem1	147.02	67	High variability	EXIT	3
10	ORI	146.3	64	High variability	-	-
11	$Clb5_T$	139.93	60	High variability	S/G2/M	2
12	CKI_T	129.58	56	High variability	S/G2/M	2
13	$Cdh1_A$	124.2	52	High variability	S/G2/M	2
14	SPN	122.95	48	Low variability	-	-
15	$Polo_T$	122.16	44	Low variability	EXIT	3
16	Bck2	120.28	40	Low variability	START	1
17	$Pds1_T$	119.01	37	Low variability	EXIT	3
18	$Clb2_T$	118.21	33	Low variability	S/G2/M	2
19	PPX	117.53	29	Low variability	EXIT	3
20	Cdc20A-APCP	115.58	25	Low variability	S/G2/M	2
21	V (Mass)	114.18	21	Low variability	-	-
22	CKI_P	111.99	17	Low variability	S/G2/M	2
23	Cdc15	109.39	13	Low variability	EXIT	3
24	Net1dep	109.19	10	Low variability	EXIT	3
25	$CDC20_T$	109.03	6	Low variability	S/G2/M	2
26	$Swi5_T$	107.48	1	Low variability	S/G2/M	2

nature of size control [53] and we hypothesize that this factor plays into the high values of model prediction variability associated with the relative abundances of proteins that regulate size control and the G1/S transition. Aligned with this trend, Di Talia et al. [56], observed that cell size at birth is significantly variable with CV values around 0.2 for both daughters and mothers. Hence, our identification of "cell size" and "regulation of G1/S transition" as the biological processes associated with the highest values of predictive variability is consistent with previous experimental literature.

Based on Table 7, the biological processes associated with the smallest predictive variability values (1st-12th percentile range) were identified as the positive regulation of transcription involved in exit from mitosis (and also its simpler form "regulation of exit from mitosis"), mitotic spindle assembly checkpoint, and negative regulation of cyclin-dependent protein kinase by cyclin degradation. These processes are associated with Swi5 (the transcription factor for CKI), Net1 (stoichiometric inhibitor of Cdc14), Cdc15 (responsible for Net1 phosphorylation), and Cdc20 (required for Clb5, Clb2, and Pds1 degradation) that all play critical roles for mitotic exit which is the cell cycle network module with least predictive variability as we previously stated.

The findings that we summarize in this section, when taken together, suggest that the statistics generated from the model predictions are influenced by the topology of the cell cycle network and that these statistics may also be generating distinct patterns that are specific to individual network modules. In order to test this hypothesis, we next implemented the "random forest" classification method and developed statistical models to predict the network modules in which individual cell cycle regulators operate (i.e., biological functions of these regulators) using model prediction statistics.

Table 7 Biological processes ordered in terms of predictive variability which decreases from top to bottom

Rank	Biological process	Variability score	Percentile
1	Regulation of cell size	164.38±42.18	100
2	Negative regulation of transcription involved in G1/S transition of mitotic cell cycle	160.63±0.00	95
3	Regulation of transcription involved in G1/S transition of mitotic cell cycle	157.89±71.29	91
4	Positive regulation of transcription involved in G1/S transition of mitotic cell cycle	156.26±0.00	85
5	Positive regulation of transcription from RNA polymerase II promoter	156.26±0.00	85
6	Regulation of cyclin-dependent protein kinase activity	154.81±38.41	81
7	Mitotic spindle orientation checkpoint	147.02±0.00	71
8	Exit from mitosis	147.02±0.00	71
9	Establishment of mitotic spindle localization	147.02±0.00	71
10	Regulation of mitotic spindle assembly	139.93±0.00	64
11	Positive regulation of DNA replication	139.93±0.00	64
12	G1/S transition of mitotic cell cycle	130.11±13.89	60
13	Positive regulation of spindle pole body separation	129.07±15.36	54
14	G2/M transition of mitotic cell cycle	129.07±15.36	54
15	Positive regulation of protein ubiquitination	124.20±0.00	47
16	Negative regulation of spindle pole body separation	124.20±0.00	47
17	Regulation of cell cycle	120.28±0.00	40
18	Positive regulation of gene expression	120.28±0.00	40
19	Mitotic sister chromatid segregation	119.01±0.00	36
20	Regulation of mitotic spindle elongation	118.21±0.00	30
21	Negative regulation of protein dephosphorylation	118.21±0.00	30
22	Positive regulation of mitotic metaphase/anaphase transition	116.62±10.73	23
23	Activation of APC-Cdc20 complex activity	116.62±10.73	23
24	Protein phosphorylation	109.39±0.00	16
25	Mitotic cytokinesis	109.39±0.00	16
26	Regulation of exit from mitosis	109.29±0.14	12
27	Negative regulation of cyclin-dependent protein kinase by cyclin degradation	109.03±0.00	6
28	Mitotic spindle assembly checkpoint	109.03±0.00	6
29	Positive regulation of transcription involved in exit from mitosis	107.48±0.00	1

Predicting biological functions (or network modules) of cell cycle regulators using relative abundance statistics

In order to predict the biological functions (or network modules) of cell cycle regulators using relative abundance statistics, we implemented the random forest classification method using the Statistics and Machine Learning Toolbox^{TM} of Matlab® [57]. For each relative abundance (a total of 47850 relative abundances with finite CV values), four features were used for predicting the network modules of individual cell cycle proteins, namely the mean, standard deviation, and CV values of the particular relative abundance and the ID-number of the viable novel mutant (of the 129 strains in the Predictive Set plus the wild type strain) that is simulated to generate the relative abundance prediction. The true class of each relative abundance was identified as the network module to which the protein in the numerator belonged. (We later tested if the predictive accuracy significantly changed when the denominator was taken as the reference point for identifying the true class labels and found out that our predictive ability was not dependent on this choice.) Predictive accuracy is computed by generating receiver operating characteristic (ROC) curves (true positive rate vs. the false positive rate obtained using several classifier output thresholds) and quantifying the areas under these curves (AUC) for each network module as the positive class vs. the negative class generated by combining the remaining two modules (i.e., START module vs. S/G2/M and EXIT modules, S/G2/M module vs. START and EXIT modules, and EXIT module vs. START and S/G2/M modules). We performed 100 runs (per set of features or model inputs) and reported the average AUC and its p-value based on a Z-test with respect to a random model with two classes (i.e., AUC=0.5) [58], an approach commonly taken for computing the statistical significance of AUC in ROC based predictive modeling studies. When the p-value computed from the AUC is less than 0.05, the predictive performance measured by the AUC value is deemed statistically significant. We also generated randomized models by permuting the class labels (or network modules) attached to each relative abundance in 100 independent realizations. The p-values associated with the predictive performances of these randomized models were expected to be higher than 0.05 in order to verify the statistical significance achieved by the non-randomized models trained and tested by the true network modules associated with all the relative abundances.

Per decision tree, approximately 64% of the samples are retained to be used for model training, whereas the remaining samples are used for model testing. These test samples are referred to as "out-of-bag" (OOB) samples, whereas the training samples are expanded by bootstrapping [59] (or sampling with replacement) up to the sample size of the original data [60] prior to model training. Classification of the test samples are based on the complete ensemble of trees (a total of 100 trees) with a voting scheme. For example, a test sample (i.e., the protein in the numerator of a relative abundance) is predicted to be in the "START" module if the number of trees that predict this outcome is higher than the ones that predict the protein's network module as "S/G2/M" or "EXIT".

As shown in Table 8, random forest models developed using model prediction statistics were highly predictive of network modules (START, S/G2/M and EXIT) in which the cell cycle regulators operate with an average AUC of 0.83–0.87 (with less than 0.01% variability and p-values of zero). Furthermore, the randomized models generated by permuting the network modules attached to relative abundances had no predictive value indicated by AUC values around 0.5 (and p-values around 0.5), typical of a coin-flipping process with two possible system states (e.g., START module vs. S/G2/M or EXIT). Hence, the predictive performances of models trained with the correct (or non-random) network module-relative abundance matching were statistically significant.

Recent studies have indicated that abundances of proteins are regulated in a biological function-dependent manner [61–63]. For example, in general, production and degradation rates of regulatory proteins are trained by evolution to quickly respond to certain stimuli, whereas proteins produced by housekeeping genes and structural proteins that are critical for the integrity of an organism are relatively more stable [61]. Furthermore, it is now also clear that protein abundance signatures are shaped not only by transcriptional and post-transcriptional regulation [64] but also by translation and post-translational regulation, which play prominent roles in determining both dynamic and steady-state behaviours of protein abundances [61, 62, 65]. The cell cycle model used in our study takes into account all of these individual modes of regulation and successfully predicts the network modules of individual cell cycle regulators (related to their biological functions) from model prediction statistics. This outcome demonstrates the critical importance of developing comprehensive and accurate models of important biological processes (such as cell cycle control) for correctly predicting various dynamic and steady-state behaviours shaped by a complex interplay between several modes of regulation. Generating correct predictions despite such complexity holds the key to elucidating critical components and their interactions in complex biological networks in a context-dependent manner.

Conclusions

Previously [22], we demonstrated a practical approach for fitting a complex dynamical model of the budding yeast cell cycle [40, 41] to a large set of qualitative experimental observations (viability/inviability of mutant strains of yeast). Taking a further step in this work, we characterize the feasible region of this model in order to test whether the statistical features of relative protein abundance predictions are influenced by the topology of the cell cycle regulatory network.

Using differential evolution (DE), we generate an ensemble of feasible parameter vectors that reproduce the phenotypes (viable or inviable) of wild-type yeast cells and 110 mutant strains (we call these 111 strains the Training Set). We use this ensemble to predict the phenotypes of 129 mutants (the Prediction Set) for which experimental data is not available. We identify 86 novel mutants that are predicted to be viable and then rank the cell cycle proteins in terms of their contributions to cumulative variability of relative protein abundance predictions. Of the three modules in the cell cycle control system (START, S/G2/M, and EXIT), the EXIT module (the most fragile module identified in [22]) has the least predictive variability, whereas the START module has the highest predictive variability. When we compile all of the gene ontology based biological processes associated with the cell cycle proteins

Table 8 Predictive performances of the random forest models developed using relative abundance statistics along with the *p*-values corresponding to mean AUC values in 100 independent realizations (STD corresponds to standard deviation)

Positive class	AUC (Mean±STD)	*p*-value	AUC (Mean±STD) with randomized modules	*p*-value with randomized modules
START	0.8667±0.0004	<1.0E-15	0.4996±0.0046	0.55
S/G2/M	0.8326±0.0005	<1.0E-15	0.5003±0.0038	0.46
EXIT	0.8366±0.0005	<1.0E-15	0.5008±0.0038	0.40

Here, for each relative abundance, the network module of the cell cycle regulator in the "numerator" is used as the true class label of the relative abundance for model training and testing. The results were practically identical (less than 0.01 change in AUC values) when the regulator in the "denominator" was used as the true class label

in the model, we identify that the proteins involved in "regulation of cell size" and "regulation of G1/S transition" contribute most to predictive variability, whereas proteins involved in "positive regulation of transcription involved in exit from mitosis", "mitotic spindle assembly checkpoint", and "negative regulation of cyclin-dependent protein kinase by cyclin degradation" contribute the least. These results suggest that the statistics of these predictions may be generating patterns specific to individual network modules (START, S/G2/M, and EXIT). To test this hypothesis, we develop random forest models for predicting the network modules of cell cycle regulators using relative abundance statistics as model inputs. Predictive performance is assessed by the areas under receiver operating characteristics curves (AUC). Our models generate an AUC range of 0.83-0.87 as opposed to randomized models with AUC values around 0.50. By using differential evolution and random forest modeling, we show that the model prediction statistics generate distinct network module-specific patterns within the cell cycle network.

Additional files

Additional file 1: Supplementary Tables. This pdf file includes 13 tables referred to in the main text.

Additional file 2: Supplementary Text. This pdf file includes detailed descriptions of certain aspects of our study including the computation of the estimated volume spanned by an ensemble of parameter vectors (Section 1), using LHS for generating an ensemble of parameter vectors (Section 2), selection of the initial DE population that spans a large volume (Section 3), computation of the estimated volume spanned by the most recent subensemble of parameter vectors (Section 4), selection of the initial DE population that spans a large volume and has a large prediction range (Section 5), alternative parameter space exploration methods (Section 6), the impact of precision on the number of identified feasible parameter vectors (Section 7), the impact of additional normalization on the contributions of individual parameters to the feasible region's volume (Section 8), the impact of additional normalization on the contributions of individual parameters to the robustness score (Section 9), the impact of the viability criteria on the model prediction range (Section 10), the choice of ODE solver (Section 11), a potential biological application of the parameter space exploration approach (Section 12), and discussion regarding the most critical model parameters (Section 13), and the most fragile phenotypes (Section 14).

Additional file 3: Supplementary Figures. This pdf file includes five additional figures.

Additional file 4: Simulation code. This ZIP file includes a Matlab script (getpredictionrange.m), a C subroutine (for solving the ODEs), and additional files that simulate the model with different ensembles (by the execution of getpredictionrange.m) to compute the prediction range and protein abundances for all mutant strains in the Prediction Set for a given ensemble (ensembles included as additional ZIP files and getpredictionrange.m currently set to load and use Ensemble 1 for the computations).

Additional file 5: Ensemble 1. This ZIP file includes all of the parameter vectors in Ensemble 1.

Additional file 6: Ensembles 2 through 9. This ZIP file includes all of the parameter vectors in Ensembles 2 through 9.

Additional file 7: Ensembles 10 through 14. This ZIP file includes all of the parameter vectors in Ensembles 10 through 14.

Additional file 8: Ensemble 15. This ZIP file includes all of the parameter vectors in Ensemble 15.

Additional file 9: Ensemble 16. This ZIP file includes all of the parameter vectors in Ensemble 16.

Abbreviations

AUC: Area under the curve; CV: Coefficient of variation; DE: Differential evolution; LHS: Latin hypercube sampling; ODE: Ordinary differential equation; WT: Wild type

Acknowledgements

Cihan Oguz performed this work when he was employed at Virginia Tech. The content is solely the responsibility of the authors and does not necessarily represent the official views of the National Institutes of Health, the Department of Health and Human Services, or the United States government.

Funding

Research reported in this publication was supported by the National Institute of General Medical Sciences of the National Institutes of Health under award number R01 GM078989-07 to JJT and WTB. We are grateful to the Advanced Research Computing Lab at Virginia Tech for computing resources. The funders had no role in study design, data collection and analysis, decision to publish, or preparation of the manuscript. The content is solely the responsibility of the authors and does not necessarily represent the official views of the National Institutes of Health, the Department of Health and Human Services, or the United States government.

Authors' contributions

Conceived the study: CO. Developed the methodology, performed the simulations and analysis: CO. Wrote the paper: CO, WTB, LTW, JJT. All authors read and approved the final manuscript.

Competing interests
The authors declare that they have no competing interests.

Author details
[1]Department of Biological Sciences, Virginia Tech, Blacksburg VA, 24061 USA. [2]Department of Computer Science, Virginia Tech, 24061 Blacksburg VA, USA. [3]Department of Mathematics, Virginia Tech, 24061 Blacksburg VA, USA. [4]Department of Aerospace and Ocean Engineering, Virginia Tech, 24061 Blacksburg VA, USA. [5]Department of Electrical and Computer Engineering, Virginia Tech, 24061 Blacksburg VA, USA .

References

1. Butcher EC, Berg EL, Kunkel EJ. Systems biology in drug discovery. Nat Biotechnol. 2004;22(10):1253–9.
2. Nelander S, Wang W, Nilsson B, She QB, Pratilas C, Rosen N, Gennemark P, Sander C. Models from experiments: combinatorial drug perturbations of cancer cells. Mol Syst Biol. 2008;4(216):1–11.
3. Gutenkunst RN, Waterfall JJ, Casey FP, Brown KS, Myers CR, Sethna JP. Universally sloppy parameter sensitivities in systems biology models. PLoS Comput Biol. 2007;3(10):189.
4. Kreutz C, Timmer J. Systems biology: experimental design. FEBS J. 2009;276(4):923–42.
5. Kuepfer L, Peter M, Sauer U, Stelling J. Ensemble modeling for analysis of cell signaling dynamics. Nat Biotechnol. 2007;25(9):1001–6.
6. Franceschini G, Macchietto S. Model-based design of experiments for parameter precision: State of the art. Chem Eng Sci. 2008;63(19):4846–72.
7. Meyer P, Cokelaer T, Chandran D, Kim KH, Loh PR, Tucker G, Lipson M, Berger B, Kreutz C, Raue A, et al. Network topology and parameter estimation: from experimental design methods to gene regulatory network kinetics using a community based approach. BMC Syst Biol. 2014;8(1):13.
8. Silk D, Kirk PD, Barnes CP, Toni T, Stumpf MP. Model selection in systems biology depends on experimental design. PLoS Comput Biol. 2014;10(6): 1003650.
9. Schaber J, Baltanas R, Bush A, Klipp E, Colman-Lerner A. Modelling reveals novel roles of two parallel signalling pathways and homeostatic feedbacks in yeast. Mol Syst Biol. 2012;8(622):1–17.
10. Tran LM, Rizk ML, Liao JC. Ensemble modeling of metabolic networks. Biophys J. 2008;95(12):5606–17.
11. Jia G, Stephanopoulos G, Gunawan R. Ensemble kinetic modeling of metabolic networks from dynamic metabolic profiles. Metabolites. 2012;2(4):891–912.
12. Song SO, Chakrabarti A, Varner JD. Ensembles of signal transduction models using Pareto optimal ensemble techniques (POETs). Biotechnol J. 2010;5(7):768–80.
13. Noble SL, Buzzard GT, Rundell AE. Feasible parameter space characterization with adaptive sparse grids for nonlinear systems biology models. In: American Control Conference (ACC), 2011. New York: IEEE; 2011. p. 2909–14.
14. Dayarian A, Chaves M, Sontag ED, Sengupta AM. Shape, size, and robustness: feasible regions in the parameter space of biochemical networks. PLoS Comput Biol. 2009;5(1):1000256.
15. Tiemann C, Vanlier J, Hilbers P, van Riel N. Parameter adaptations during phenotype transitions in progressive diseases. BMC Syst Biol. 2011;5(1): 174.
16. Tiemann CA, Vanlier J, Oosterveer MH, Groen AK, Hilbers PA, van Riel NA. Parameter trajectory analysis to identify treatment effects of pharmacological interventions. PLoS Comput Biol. 2013;9(8):1003166.
17. Rumschinski P, Borchers S, Bosio S, Weismantel R, Findeisen R. Set-base dynamical parameter estimation and model invalidation for biochemical reaction networks. BMC Syst Biol. 2010;4(1):69.
18. Rodriguez-Fernandez M, Rehberg M, Kremling A, Banga JR. Simultaneous model discrimination and parameter estimation in dynamic models of cellular systems. BMC Syst Biol. 2013;7(1):76.
19. Pargett M, Rundell AE, Buzzard GT, Umulis DM. Model-based analysis for qualitative data: an application in drosophila germline stem cell regulation. PLoS Comput Biol. 2014;10(3):1003498.
20. Donzé A, Fanchon E, Gattepaille LM, Maler O, Tracqui P. Robustness analysis and behavior discrimination in enzymatic reaction networks. PloS ONE. 2011;6(9):24246.
21. Pargett M, Umulis DM. Quantitative model analysis with diverse biological data: applications in developmental pattern formation. Methods. 2013;62(1):56–67.
22. Oguz C, Laomettachit T, Chen KC, Watson LT, Baumann WT, Tyson JJ. Optimization and model reduction in the high dimensional parameter space of a budding yeast cell cycle model. BMC Syst Biol. 2013;7(1):53.
23. Price KV, Storn RM, Lampinen JA. Differential Evolution: A Practical Approach to Global Optimization. Natural Computing Series. Berlin: Springer; 2005.
24. Chong CK, Mohamad MS, Deris S, Shamsir MS, Choon YW, Chai LE. Improved differential evolution algorithm for parameter estimation to improve the production of biochemical pathway. Intl J Interactive Multimedia Artif Intell. 2012;1(5):22–9.
25. Tashkova K, Korošec P, Šilc J, Todorovski L, Džeroski S. Parameter estimation with bio-inspired meta-heuristic optimization: modeling the dynamics of endocytosis. BMC Syst Biol. 2011;5(1):159.
26. Mahdavi S, Shiri ME, Rahnamayan S. Metaheuristics in large-scale global continues optimization: A survey. Inf Sci. 2015;295:407–28.
27. Sun J, Garibaldi JM, Hodgman C. Parameter estimation using metaheuristics in systems biology: a comprehensive review. Comput Biol Bioinformatics IEEE/ACM Trans. 2012;9(1):185–202.
28. Banga JR, Versyck KJ, Van Impe JF. Computation of optimal identification experiments for nonlinear dynamic process models: a stochastic global optimization approach. Ind Eng Chem Res. 2002;41(10):2425–30.
29. Rodriguez-Fernandez M, Mendes P, Banga JR. A hybrid approach for efficient and robust parameter estimation in biochemical pathways. Biosystems. 2006;83(2):248–65.
30. Balsa-Canto E, Alonso AA, Banga JR. Computational procedures for optimal experimental design in biological systems. IET Syst Biol. 2008;2(4): 163–72.
31. Ashyraliyev M, Jaeger J, Blom JG. Parameter estimation and determinability analysis applied to drosophila gap gene circuits. BMC Syst Biol. 2008;2(1):83.
32. Audoly S, Bellu G, D'Angio L, Saccomani MP, Cobelli C. Global identifiability of nonlinear models of biological systems. Biomed Eng IEEE Trans. 2001;48(1):55–65.
33. Zak DE, Gonye GE, Schwaber JS, Doyle FJ. Importance of input perturbations and stochastic gene expression in the reverse engineering of genetic regulatory networks: insights from an identifiability analysis of an in silico network. Genome Res. 2003;13(11):2396–405.
34. Morgan DO. The Cell Cycle: Principles of Control. London: New Science Press; 2007.
35. Mitchison JM. The Biology of the Cell Cycle. London: Cambridge University Press; 1971.
36. Chen KC, Csikasz-Nagy A, Gyorffy B, Val J, Novak B, Tyson JJ. Kinetic analysis of a molecular model of the budding yeast cell cycle. Mol Biol Cell. 2000;11(1):369–91.
37. Chen KC, Calzone L, Csikasz-Nagy A, Cross FR, Novak B, Tyson JJ. Integrative analysis of cell cycle control in budding yeast,. Mol Biol Cell. 2004;15(8):3841–62. doi:10.1091/mbc.E03-11-0794.
38. Singhania R, Sramkoski RM, Jacobberger JW, Tyson JJ. A hybrid model of mammalian cell cycle regulation. PLoS Comput Biol. 2011;7(2): 1001077.
39. Kraikivski P, Chen KC, Laomettachit T, Murali T, Tyson JJ. From start to finish: computational analysis of cell cycle control in budding yeast. npj Syst Biol Appl. 2015;1:15016.
40. Laomettachit T. Mathematical modeling approaches for dynamical analysis of protein regulatory networks with applications to the budding yeast cell cycle and the circadian rhythm in cyanobacteria. PhD thesis, Virginia Institute of Technology. 2011. http://scholar.lib.vt.edu/theses/available/etd-11072011-021528/.

41. Laomettachit T, Chen KC, Baumann WT, Tyson JJ. A model of yeast cell-cycle regulation based on a standard component modeling strategy for protein regulatory networks. PloS ONE. 2016;11(5):0153738.
42. Donahue MM, Buzzard GT, Rundell AE. Robust parameter identification with adaptive sparse grid-based optimization for nonlinear systems biology models. In: American Control Conference, 2009. ACC'09. New York: IEEE; 2009. p. 5055–060.
43. Taylor SC, Berkelman T, Yadav G, Hammond M. A defined methodology for reliable quantification of western blot data. Mol Biotechnol. 2013;55(3):217–26.
44. Oda Y, Huang K, Cross FR, Cowburn D, Chait BT. Accurate quantitation of protein expression and site-specific phosphorylation. Proc Natl Acad Sci. 1999;96(12):6591–6.
45. Bucher J, Riedmaier S, Schnabel A, Marcus K, Vacun G, Weiss T, Thasler W, Nüssler A, Zanger U, Reuss M. A systems biology approach to dynamic modeling and inter-subject variability of statin pharmacokinetics in human hepatocytes. BMC Syst Biol. 2011;5(1):66.
46. Shankaran H, Zhang Y, Tan Y, Resat H. Model-based analysis of HER activation in cells co-expressing EGFR, HER2 and HER3. PLoS Comput Biol. 2013;9(8):1003201.
47. Dong W, Tang X, Yu Y, Nilsen R, Kim R, Griffith J, Arnold J, Schüttler HB. Systems biology of the clock in neurospora crassa. PloS ONE. 2008;3(8): 3105.
48. Donahue M, Buzzard G, Rundell A. Experiment design through dynamical characterisation of non-linear systems biology models utilising sparse grids. IET Syst Biol. 2010;4(4):249–62.
49. Lucia A, DiMaggio PA, Depa P. Funneling algorithms for multiscale optimization on rugged terrains. Ind Eng Chem Res. 2004;43(14):3770–81.
50. Moriya H, Shimizu-Yoshida Y, Kitano H. In Vivo Robustness Analysis of Cell Division Cycle Genes in *Saccharomyces cerevisiae*. PLOS Genet. 2010;6(4):. doi:10.1371/journal.pgen.002011.
51. Shirayama M, Tóth A, Gálová M, Nasmyth K. Apccdc20 promotes exit from mitosis by destroying the anaphase inhibitor pds1 and cyclin clb5. Nature. 1999;402(6758):203–7.
52. Dwight SS, Harris MA, Dolinski K, Ball CA, Binkley G, Christie KR, Fisk DG, Issel-Tarver L, Schroeder M, Sherlock G, et al. Saccharomyces genome database (sgd) provides secondary gene annotation using the gene ontology (go). Nucleic Acids Res. 2002;30(1):69–72.
53. Turner JJ, Ewald JC, Skotheim JM. Cell size control in yeast. Curr Biol. 2012;22(9):350–9.
54. Johnston G, Pringle J, Hartwell LH. Coordination of growth with cell division in the yeast saccharomyces cerevisiae. Experimental Cell Res. 1977;105(1):79–98.
55. Di Talia S, Wang H, Skotheim JM, Rosebrock AP, Futcher B, Cross FR. Daughter-specific transcription factors regulate cell size control in budding yeast. PLoS Biol. 2009;7(10):1000221.
56. Di Talia S, Skotheim JM, Bean JM, Siggia ED, Cross FR. The effects of molecular noise and size control on variability in the budding yeast cell cycle. Nature. 2007;448(7156):947–51.
57. MATLAB. Version 8.1 (R2013a). Natick: The MathWorks Inc.; 2013.
58. Hanley JA, McNeil BJ. The meaning and use of the area under a receiver operating characteristic (ROC) curve. Radiology. 1982;143(1):29–36.
59. Efron B. Bootstrap methods: another look at the jackknife. Annals Stat. 1979;7(1):1–26.
60. Dasgupta A, Sun YV, König IR, Bailey-Wilson JE, Malley JD. Brief review of regression-based and machine learning methods in genetic epidemiology: the genetic analysis workshop 17 experience. Genet Epidemiol. 2011;35(S1):5–11.
61. Vogel C, Marcotte EM. Insights into the regulation of protein abundance from proteomic and transcriptomic analyses. Nat Rev Genet. 2012;13(4): 227–32.
62. Schwanhäusser B, Busse D, Li N, Dittmar G, Schuchhardt J, Wolf J, Chen W, Selbach M. Global quantification of mammalian gene expression control. Nature. 2011;473(7347):337–42.
63. Vogel C, de Sousa Abreu R, Ko D, Le SY, Shapiro BA, Burns SC, Sandhu D, Boutz DR, Marcotte EM, Penalva LO. Sequence signatures and mrna concentration can explain two-thirds of protein abundance variation in a human cell line. Mol Syst Biol. 2010;6(1):400.
64. Plotkin JB. Transcriptional regulation is only half the story. Mol Syst Biol. 2010;6(1):406.
65. Maier T, Schmidt A, Güell M, Kühner S, Gavin AC, Aebersold R, Serrano L. Quantification of mrna and protein and integration with protein turnover in a bacterium. Mol Syst Biol. 2011;7(1):511.

Permissions

All chapters in this book were first published in SB, by BioMed Central; hereby published with permission under the Creative Commons Attribution License or equivalent. Every chapter published in this book has been scrutinized by our experts. Their significance has been extensively debated. The topics covered herein carry significant findings which will fuel the growth of the discipline. They may even be implemented as practical applications or may be referred to as a beginning point for another development.

The contributors of this book come from diverse backgrounds, making this book a truly international effort. This book will bring forth new frontiers with its revolutionizing research information and detailed analysis of the nascent developments around the world.

We would like to thank all the contributing authors for lending their expertise to make the book truly unique. They have played a crucial role in the development of this book. Without their invaluable contributions this book wouldn't have been possible. They have made vital efforts to compile up to date information on the varied aspects of this subject to make this book a valuable addition to the collection of many professionals and students.

This book was conceptualized with the vision of imparting up-to-date information and advanced data in this field. To ensure the same, a matchless editorial board was set up. Every individual on the board went through rigorous rounds of assessment to prove their worth. After which they invested a large part of their time researching and compiling the most relevant data for our readers.

The editorial board has been involved in producing this book since its inception. They have spent rigorous hours researching and exploring the diverse topics which have resulted in the successful publishing of this book. They have passed on their knowledge of decades through this book. To expedite this challenging task, the publisher supported the team at every step. A small team of assistant editors was also appointed to further simplify the editing procedure and attain best results for the readers.

Apart from the editorial board, the designing team has also invested a significant amount of their time in understanding the subject and creating the most relevant covers. They scrutinized every image to scout for the most suitable representation of the subject and create an appropriate cover for the book.

The publishing team has been an ardent support to the editorial, designing and production team. Their endless efforts to recruit the best for this project, has resulted in the accomplishment of this book. They are a veteran in the field of academics and their pool of knowledge is as vast as their experience in printing. Their expertise and guidance has proved useful at every step. Their uncompromising quality standards have made this book an exceptional effort. Their encouragement from time to time has been an inspiration for everyone.

The publisher and the editorial board hope that this book will prove to be a valuable piece of knowledge for researchers, students, practitioners and scholars across the globe.

List of Contributors

Marko Djordjevic
Institute of Physiology and Biochemistry, Faculty of Biology, University of Belgrade, Studentski trg 16, 11000 Belgrade, Serbia

Andjela Rodic
Institute of Physiology and Biochemistry, Faculty of Biology, University of Belgrade, Studentski trg 16, 11000 Belgrade, Serbia
Multidisciplinary PhD program in Biophysics, University of Belgrade, Belgrade, Serbia

Bojana Blagojevic and Magdalena Djordjevic
Institute of Physics Belgrade, University of Belgrade, Belgrade, Serbia

Evgeny Zdobnov
Department of Genetic Medicine and Development, University of Geneva and Swiss Institute of Bioinformatics, Geneva, Switzerland

Diala Abd-Rabbo and Stephen W. Michnick
Département de Biochimie et Médecine Moléculaire, Université de Montréal, C.P. 6128, Succursale centre-ville, Montréal, Québec H3C 3J7, Canada
Centre Robert-Cedergren, Bio-Informatique et Génomique, Université de Montréal, C.P. 6128, Succursale centre-ville, Montréal, Québec H3C 3J7, Canada

Brian K. Chu, Margaret J. Tse and Royce R. Sato
Department of Chemical Engineering and Materials Science, University of California Irvine, Irvine, CA, USA

Elizabeth L. Read
Department of Chemical Engineering and Materials Science, University of California Irvine, Irvine, CA, USA
Department of Molecular Biology and Biochemistry, University of California Irvine, Irvine, CA, USA

Hyundoo Jeong and Byung-Jun Yoon
Department of Electrical and Computer Engineering, Texas A&M University, College Station, TX, USA

Francisco Saitua, Paulina Torres, José Ricardo Pérez-Correa and Eduardo Agosin
Department of Chemical and Bioprocess Engineering, School of Engineering, Pontificia Universidad Católica de Chile, Avenida Vicuña Mackenna 4860, Santiago, Chile

Michael Vilkhovoy, Mason Minot and Jeffrey D. Varner
Department of Chemical and Biomolecular Engineering, Cornell University, 14853 Ithaca, NY, USA

David M. Bassen and Jonathan T. Butcher
Department of Biomedical Engineering, Cornell University, 14853 Ithaca, NY, USA

Saskia Trescher, Jannes Münchmeyer and Ulf Leser
Knowledge Management in Bioinformatics, Computer Science Department, Humboldt-Universität zu Berlin, Unter den Linden 6, 10099 Berlin, Germany

Luis U. Aguilera and Ursula Kummer
Department of Modeling of Biological Processes, COS Heidelberg / Bioquant, Heidelberg University, Im Neuenheimer Feld 267, 69120 Heidelberg, Germany

Christoph Zimmer
BIOMS (Center for Modeling and Simulation in the Biosciences), Heidelberg University, Im Neuenheimer Feld 267, 69120 Heidelberg, Germany

Piero Dalle Pezze and Nicolas Le Novère
The Babraham Institute, Babraham Campus, Cambridge CB22 3AT, UK

Thomas J. Snowden
Department of Mathematics and Statistics, University of Reading, RG6 6AX, Reading, UK
Certara QSP, University of Kent Innovation Centre, CT2 7FG Canterbury, UK

Marcus J. Tindall
Department of Mathematics and Statistics, University of Reading, RG6 6AX, Reading, UK
The Institute for Cardiovascular and Metabolic Research (ICMR), University of Reading, RG6 6AX Reading, UK

Piet H. van der Graaf
Leiden Academic Centre for Drug Research, Universiteit Leiden, NL-2333 CC Leiden, Netherlands
Certara QSP, University of Kent Innovation Centre, CT2 7FG Canterbury, UK

Leonid V. Omelyanchuk and Alina F. Munzarova
Institute of Molecular and Cellular Biology, Novosibirsk, Russia
Novosibirsk State University, Novosibirsk, Russia

Weiqi Chen and Shan He
School of Computer Science, University of Birmingham, Edgbaston, B15 2TT Birmingham, UK

Jing Liu
Key Laboratory of Intelligent Perception and Image Understanding of Ministry of Education, Xidian University, Xi'an, 710071 Shaanxi, People's Republic of China

Minghan Chen, Fei Li, Shuo Wang and Young Cao
Department of Computer Science, Virginia Tech, Blacksburg, VA 24061, USA

Joseph J. Gardner and Nanette R. Boyle
Department of Chemical and Biological Engineering, Colorado School of Mines, Golden, CO 80401, USA

Cihan Oguz and John J. Tyson
Department of Biological Sciences, Virginia Tech, Blacksburg VA, 24061 USA

Layne T. Watson
Department of Computer Science, Virginia Tech, 24061 Blacksburg VA, USA
Department of Mathematics, Virginia Tech, 24061 Blacksburg VA, USA
Department of Aerospace and Ocean Engineering, Virginia Tech, 24061 Blacksburg VA, USA

William T. Baumann
Department of Electrical and Computer Engineering, Virginia Tech, 24061 Blacksburg VA, USA

Index

www.ingramcontent.com/pod-product-compliance
Lightning Source LLC
Chambersburg PA
CBHW061244190326
41458CB00011B/3570
9781641160643